AF388908

Anticancer Properties of Natural and Derivative Products

Anticancer Properties of Natural and Derivative Products

Editors

José Antonio Lupiáñez
Amalia Pérez-Jiménez
Eva E. Rufino-Palomares

MDPI • Basel • Beijing • Wuhan • Barcelona • Belgrade • Manchester • Tokyo • Cluj • Tianjin

Editors
José Antonio Lupiáñez
Department of Biochemistry and
Molecular Biology,
Facultad de Ciencias,
Universidad de Granada
Spain

Amalia Pérez-Jiménez
Department of Zoology,
Faculty of Sciences,
University of Granada
Spain

Eva E. Rufino-Palomares
Department of Biochemistry and
Molecular Biology I, Faculty of
Sciences, University of Granada
Spain

Editorial Office
MDPI
St. Alban-Anlage 66
4052 Basel, Switzerland

This is a reprint of articles from the Special Issue published online in the open access journal *Molecules* (ISSN 1420-3049) (available at: https://www.mdpi.com/journal/molecules/special_issues/anticancer_np).

For citation purposes, cite each article independently as indicated on the article page online and as indicated below:

LastName, A.A.; LastName, B.B.; LastName, C.C. Article Title. *Journal Name* **Year**, *Volume Number*, Page Range.

ISBN 978-3-0365-2588-4 (Hbk)
ISBN 978-3-0365-2589-1 (PDF)

Contents

About the Editors

José Antonio Lupiáñez was born in Almería (Spain). He studied Biology at the University of Granada between 1967 and 1975, obtaining a BSc degree in 1972, an MSc degree in 1973 and a PhD degree in 1975. In 1973, he spent a semester as a predoctoral fellow at Cardiff University. In 1978, he obtained a postdoctoral Fulbright Fellowship in the Pharmacology department of the Indiana University School of Medicine under the supervision of Professor S.R. Wagle. Upon his return to Spain, he obtained a position as Associate Professor and created the research group "Drugs, Environmental Toxics and Cellular Metabolism". In 1990, he received the position of Full Professor. He has directed a score of doctoral theses and is the author of over 200 publications. He is a member of a number of scientific societies, including the Spanish Society of Biochemistry and Molecular Biology, European Society of Biochemistry, New York Academy of Sciences and American Chemical Society. In October 2018, he was appointed as Permanent Professor Emeritus.

Amalia Pérez-Jiménez obtained her degree in Environmental Sciences in 2001 and her PhD in 2008. Currently, she has published 46 articles in high-impact journals indexed in JCR with a research line focused on the study of nutrition, physiology, and biochemistry of fish, in addition to the use of natural compounds to improve animal and human health. Furthermore, she has also published three book chapters in relevant publishers. In addition, she has participated in numerous national and international conferences, and she has directed three research projects, having participated in more than twenty international and national projects. She is responsible for master's projects in different official masters and final degree projects in Environmental Sciences, Biochemistry and Biology degrees. Finally, she has been the head of the research group RNM156-Fish Nutrition and Feeding since 2018. A complete list of her publications can be found at her ORCID profile.

Eva E. Rufino-Palomares graduated in Biology from the University of Granada (UGR). In 2005, she obtained a pre-doctoral scholarship in public competition with which she obtained her doctorate in 2009. Later, after a research contract (2009–2011), she obtained a position as Assistant Professor in the Department of Biochemistry and Molecular Biology I of the UGR (2011), where she is currently senior lecturer. She has a total of 35 scientific contributions and has been on the research team of twelve long-term research projects, and as principal for three. She has more than fourteen years' teaching experience, having directed three doctoral, twenty final master projects, and fourteen final degree projects, and has been responsible for eleven research initiation scholarships for undergraduate and postgraduate students. She has also directed two teaching innovation projects and she is Secretary of the Academic Commission for the degree of biochemistry.

Preface to "Anticancer Properties of Natural and Derivative Products"

Many natural products consist of bioactive compounds synthesized by terrestrial and marine plants, microorganisms and animals, whose main objective is to prevent them from attacks by predators and/or pathogens. Traditionally, since ancient times, different cultures have used these compounds for the prevention and treatment of various human diseases. During the last few years especially, it has been reported that most of these phytochemicals possess a variety of interesting and significant biological properties, such as analgesic, antiallodynic, antidiabetic, antioxidant, antiparasitic, antimicrobial, antiviral, antiatherogenic, anti-inflammatory, antitumor, antiproliferative and normal growth stimulants, as well as significant cardioprotective and neuroprotective activity. This thematic book aims to collect and disseminate some of the most significant and recent contributions concerning the use of the natural compounds called phytochemicals, as well as some of their chemical derivatives for the prevention and treatment of cancer and other accompanying diseases.

On the other hand, in recent years, the synthesis of numerous chemical derivatives of these natural compounds has also intensified with the aim of enhancing their bioactive capacities. Among all these bioactivities, special attention has been paid to its antitumor capacity through the potential modulation of cancer initiation and growth, cell differentiation, apoptosis and autophagy, angiogenesis, and metastatic dissemination. In addition, a considerable number of studies have linked their anticancer effects to their anti-inflammatory and antioxidant activities. Reports on the biological activity of natural extracts will only be considered if accompanied by adequate chemical characterization.

The editors of this book wish to dedicate it in memoriam to our beloved teacher, colleague and friend, Professor Eduardo García Peregrín (1942–2021), teacher of more than forty promotions of biologists and pharmacists, an authentic example and reference to the Biochemistry of our University. In addition to his numerous scientific achievements, including being the first researcher to introduce the use of radioactive isotopes at our university in the 1960s, Eduardo was a professional who was highly valued by his colleagues, and a great mentor for many generations of students.

A good number of us had the opportunity not long ago to meet Eduardo and commemorate his retirement from the University of Granada. Many good stories were told, reminding us of his penchant for starting his talks with a wide smile, his love of music, being a staunch defender of the Mediterranean diet, a lover of good food and good wine, his qualities as a family man and his ability to find solutions to difficult personal problems. Although he had previously made forays into the world of scientific ethics, it was from 2011, once retired, that he directed all his intellectual efforts to the study of ethical problems in science, contributing to numerous press articles and scientific writings and attending a large number of meetings and thematic meetings. His ethical-theological vision of science was always accurate and well received by the scientific world.

All of us who had the pleasure of his time and company will greatly miss Eduardo. He leaves behind a fantastic legacy and an excellent example of how to live life, something that we must try to emulate and pass on to future generations.

The academic editors of this work thank the MDPI editorial group for providing the scientific space allowing us to share with future readers the feelings generated by the compilation of the research that make up this work. We would also like to congratulate all the participating authors, the reviewers, who have helped science to advance, and all the scientific and administrative staff of the *Molecules* Editorial Office for their extraordinary and excellent work. Finally, our special wish is that you all enjoy your reading.

José Antonio Lupiáñez, Amalia Pérez-Jiménez, Eva E. Rufino-Palomares
Editors

Editorial

Are Ancestral Medical Practices the Future Solution to Today's Medical Problems?

José A. Lupiáñez [1,*], Eva E. Rufino-Palomares [1,*] and Amalia Pérez-Jiménez [2,*]

[1] Department of Biochemistry and Molecular Biology I, Faculty of Sciences, University of Granada, Avenida Fuentenueva, 118071 Granada, Spain

[2] Department of Zoology, Physiology and Ecology, Faculty of Sciences, University of Granada, Avenida Fuentenueva, 118071 Granada, Spain

* Correspondence: jlcara@ugr.es (J.A.L.); evaevae@ugr.es (E.E.R.-P.); calaya@ugr.es (A.P.-J.); Tel.: +34-958-243089 (J.A.L.); +34-958-243252 (E.E.R.-P.); +34-958-249843 (A.P.-J.); Fax: +34-958-249945 (J.A.L. & E.E.R.-P. & A.P.-J.)

Citation: Lupiáñez, J.A.; Rufino-Palomares, E.E.; Pérez-Jiménez, A. Are Ancestral Medical Practices the Future Solution to Today's Medical Problems? *Molecules* **2021**, *26*, 4701. https://doi.org/10.3390/molecules26154701

Received: 29 July 2021
Accepted: 2 August 2021
Published: 3 August 2021

Publisher's Note: MDPI stays neutral with regard to jurisdictional claims in published maps and institutional affiliations.

Our cells and organs are threatened and, in most cases, constantly subjected to the aggression of numerous situations, both endogenous, characterized by unfavorable genetics, and exogenous, by deficient or inadequate nutrition, and even by a hostile environment; in most cases, they ultimately cause a cascade of degenerative and cardiovascular diseases, cancer, and infections, as well as those related to the metabolic syndrome, all of which eventually generate irreversible damage to the organism and, consequently, a significant deterioration in its survival. In many of these cases, exogenous treatment with essential biocomponents present in numerous species, mainly plants, can reverse this deterioration and, therefore, increase survival.

The year 2020 marked a millennium since the publication of the two books that laid the foundations of modern and current medicine—the volumes entitled, The Book of Healing and The Canon of Medicine, the latter commonly known as the "Avicenna Codex." Both books were published between 1014 and 1020 by the Persian Muslim physician, polymath, philosopher, scientist, and astronomer Abū 'Alī al-Husayn ibn' Abd Allāh ibn Sīnā, better known as Ibn Sina and westernized as Avicenna, as well as the "Prince of the Sages," the greatest of physicians, the Master par excellence, or the Third Master (after Aristotle and Al-Farabi). Although it is evident that Avicenna put into practice the first principles of surgery, his greatest contribution was the use of the main natural chemical components, derived from the plant and animal (kingdoms?) world, to achieve the cure of the most important diseases of the time as described in the second book of the Avicenna Codex, entitled De Medicinis Simplicibus: Pharmacologicae de Herbis Medicinalibus, which deals with the pharmacology of medicinal herbs and is intended for the study and use of natural medicines [1]. However, we must recognize that numerous types of plasters formed from plant extracts have been mainly used by humankind practically from the Metal Age (6000 BC–3300 BC) until almost the end of the Modern Age (1492 AD–1789 AD) and have been especially important in the development of the different traditional medicines of different cultures, both for Western and, especially, for Oriental medicines.

At present, and especially since the last 20 years, there is a real explosion in the search for natural chemical compounds, generally known as "phytochemicals," from the plant (kingdom?) world, both terrestrial and aquatic, capable of presenting important and abundant biological properties. The so-called phytochemicals are chemical compounds produced by many botanical species and which play an important role in the growth of the plant species themselves and also as participants in the defense against competitors, pathogens, or predators. The word "phytochemical" has been generally used to describe a number of botanical compounds that are being investigated for their effects on health, and many of them are being used both as drugs, in different traditional medicines, and also as poisons [2].

Among the enormous variety of phytochemicals currently characterized, we can find flavonoids, such as anthocyanins, flavones, flavanones, and flavanols (catechins, epicatechins, etc.), phytosterols, terpenoids, lignans, and stilbenes, the last two being considered an excellent source of polyphenols such as resveratrol [3]. All of them have been recognized with a large number of bioactive effects as nutraceuticals, essential nutrients, and even allelopathic, thus influencing the growth, survival, or reproduction of other organisms [4]. Among all these phytochemicals, several should be highlighted, namely, terpenoids and polyphenols, of which consequential research related to their bioactive capacities both in vitro and in vivo is being carried out [5]. Many of these compounds have properties such as anticancer, using different cancer lines, both solid and liquid [6–11], antiangiogenic [7,12–15], antioxidant [16–18], anti-inflammatory [19,20], cardio- and neuroprotective [21,22], antidiabetogenic [23,24], antifungal [25], antimicrobial and antiviral [26,27], antiparasitic [28], growth inducing [29–31], and enzyme inhibitory or activating [32,33], as well as modulators in the production of reducing equivalents whose role is essential to explain most of the processes of metabolic biosynthesis [34], of cellular and organic growth, nutrition and differentiation processes [35–38] as well as, of cellular detoxification processes [39] and oxygen-free radical scavenging [18,40].

Currently, special emphasis is put on the search for a great variety of chemical derivatives of all these phytochemicals, finding in many cases a significant increase in the different bioactive capacities with respect to those of the original compound [5]. Numerous chemical groups are being used in the synthesis of chemical derivatives of the most important triterpenoids in order to select derivatives that present a significant increase in efficiency in their bioactivity. Among them, acyl, aminoacyl, and dipeptidyl groups [26,27,41,42], pegylated and pegylated diamine derivatives [43–46], and even coumarin conjugates [47] stand out. In all cases, the anticancer, anti-inflammatory, antioxidant, or antiviral effects originally present in the molecules from which they are obtained are significantly increased.

This Special Issue entitled Anticancer properties of natural products and derivatives, (https://www.mdpi.com/journal/molecules/special_issues/anticancer_np, accessed on 30 November 2020), has tried to expose part of the works currently carried out in the field of natural products and their bioactivities, mainly anticancer, to the scientific world. It consists of a total of 15 publications, of which 11 are original articles and 4 are bibliographic reviews. All of them cover very different topics, both in terms of the type of active components or nutraceuticals and the organic source that provides them.

Propolis is a resinous mixture collected opportunistically by honeybees from various plant sources, such as tree buds, sap exudates, or other sources, which is then processed within the hive for use as a sealant for small holes to prevent infection. Although it has been used as a traditional and folk medicine for several millennia, numerous biological properties of propolis have recently been described, including cytotoxic, antiviral, antimicrobial, and antioxidant activities that promote the scavenging of oxygen free radicals (ROS) [48]. In this regard, Wezgowiec et al. [49] have studied the chemical composition and anticancer and anti-inflammatory activities, in vitro, of ethanolic and ethanol–hexane extracts of propolis from different Polish regions on tongue cancer cells (SCC-25). High concentrations of polyphenols and flavonoids seem to be responsible for these biological activities and for the differences between the activities of propolis from different locations. Administration of these extracts produced, among other effects, a significant reduction in mitochondrial and proliferative activity, together with a clear modification of oxygen-free radical scavenging activity. All these effects indicate a selective anticancer and anti-inflammatory potential, although, as the authors indicate, further study of the molecular mechanisms that explain this is necessary to obtain promising health benefits.

Adenosma bracteosum (Bonati) is a plant belonging to the group of tracheophytes, mainly present in the Southeast Asian region and whose extracts have been used in traditional Vietnamese medicine to cure liver diseases. The main active groups present are polyphenols, terpenoids, and flavonoids. Recently, ethanolic extracts have been shown to have significant antidiabetogenic activity [50]. In this context, Nguyen et al. [51] have analyzed the anticancer capacity of the different fractions derived from the ethanolic extract of this tracheophyte, using two cancer cell lines, one for lung carcinoma (NCI-H460) and the other for liver carcinoma (HepG2). Of all the fractions, the chloroform-derived one seems to be the most active being the active principles that provide this bioactivity are flavonoids, xantomicrol, and its oxygenated derivative, together with the triterpene, ursolic acid. Its most significant activities focus on modulating free radical levels and mitochondrial membrane potential. All these activities, together with potent cytotoxic activity, seem to be responsible for an increase in the levels and therefore the activity of caspase-3, which is ultimately responsible for the increase in cell apoptosis, suggesting that this plant offers a good opportunity to develop new anticancer drugs.

On many occasions, continuous treatment with radioactivity and chemotherapy to different types of cancer, in certain patients, can cause adverse side effects that, on too many occasions, generate great resistance to specific drugs in these tumors; therefore, it is necessary to find more effective and, mainly, less invasive pharmacological treatments. *Moringa oleifera* is a tree native to northern India for which important nutritional and pharmacological functions have been described, such as anti-inflammatory, antihypertensive, diuretic, hepatoprotective, hypocholesterolemic, antispasmodic, antiulcer, and antibacterial [52]. In this context, Luetragoon et al. [53] have investigated the anticancer effects of an active principle, 3-hydroxy-β-ionone, a sesquiterpenoid present in extracts of *Moringa oleifera* leaves. These authors have demonstrated the in vitro anticancer capacity of this compound in epidermoid carcinoma of the head and neck (SCC-15), by detecting cell cycle arrest in the G_2/M phase and a significant increase in cell apoptosis, thanks to an increase in caspase-3 levels, together with a decrease in the anti-apoptotic protein Bcl-2 and profound inhibition of cell migration after 6 h of treatment.

Among the active compounds present in the fruit and leaf of the olive tree (*Olea europaea* L.), the pentacyclic triterpenes stand out, along with, mainly due to their content, maslinic acid, of which a large number of beneficial health effects have been demonstrated on many occasions [5,8–10,16,29–32]. One of the most controversial functions of maslinic acid is its antioxidant capacity; thus, Mokhtari et al. [18] investigated this property in murine cutaneous melanoma cells (B16F10). In addition to the known selective cytotoxic effects of maslinic acid on cancer cells, it was also demonstrated that after provoking an oxidative stress situation by the addition of hydrogen peroxide (H_2O_2), the triterpene isolated from the olive tree protected tumor cells from concomitant oxidative damage by decreasing ROS levels and modifying the activities of different antioxidant enzymes such as superoxide dismutase (SOD), glutathione S-transferase (GST), and glutathione peroxidase (GPX), thus demonstrating a high antioxidant capacity of this triterpene and therefore its beneficial effects on health.

Elaeagnus angustifolia, commonly called paradise tree or Bohemian olive, is a shrub native to the Middle East whose floral extracts have been traditionally used to treat many diseases in that area. In their article, Jabeen et al. [54] reveal that flower extracts from the paradise tree were able to inhibit in vitro cell proliferation and modify cell cycle progression in two breast cancer lines (SKBR3 and ZR75-1) positive for the human epidermal growth factor receptor 2 (HER2) protein; furthermore, these extracts inhibited epithelial-mesenchymal transition (EMT), an important event for cancer invasion and metastasis, by increasing the levels of E-cadherin and β-catenin and inhibiting those of vimentin and fascin, as main marker molecules of cell invasiveness. Furthermore, these authors demonstrate the chemopreventive effects of these extracts by blocking HER2 activities and inactivating the JNK/1/2/3 signaling pathway.

As we have stated in this article, natural products play an important role in the development of new nutraceuticals that help to prevent and cure diseases, and in this sense, the triterpenes present in the olive tree (*Olea europaea* L.) are very relevant. An example of these compounds is uvaol, and a practically innovative study of this active principle was carried out by Bonel-Pérez et al. [15], in which they analyzed its anticancer and antimetastatic effects in vitro in a hepatocarcinoma cell line (HepG2). These authors show the main molecular responsible for these activities, focusing their study on the levels of the heat shock protein HSP60, on the levels of ROS, as well as those of the antiapoptotic Bcl2 and proapoptotic Bax proteins, together with a cellular arrest in the G_0/G_1 phase and downregulation of the AKT/PI3K and MAPK signaling pathway. These results constitute an interesting challenge in the treatment of this type of cancer.

Another important source of active ingredients, which are increasingly used in conventional oncology, are the so-called medicinal mushrooms, and although they act by interfering with tumor cells, disrupting both the development and progression of the disease, the mechanisms of action that cause this have yet to be elucidated. Roda et al. [55], using triple negative 4T1 mice administered in vivo with a mixture of these medicinal mushrooms, observed a significant reduction in both tumor density and the number of metastatic bodies, showing at the same time a decrease in inflammation and oxidative stress, both in the primary tumor and in metastases. These extraordinary effects molecularly implicate the p53/Bax versus Bcl2/PARP1/PCNA axis, forcing triple-negative breast cancer cells into apoptosis.

Pogostemon cablin, commonly known as patchouli, is a plant from which essential oil is extracted from the leaves, rich in sesquiterpenes, which has been used as an antiseptic since ancient times in Asia; in addition, its use as a pharmaceutical product prevents or cures various side effects such as fever, headache, pain, and inflammation. Other studies have revealed different bioactivities, such as antimicrobial and antiviral, anti-inflammatory, anti-aging, and antitumor. Therefore, Huang et al. [56] have investigated the in vitro and in vivo response of human hepatoma cells (HepG2 and MAHLAVU) to treatment with the essential oils of this plant. The in vitro antiproliferative effects are explained by the existence of high cytotoxicity, a cell arrest in G_0/G_1, together with an increase of apoptosis, both extrinsic and intrinsic, a decrease of the mitochondrial membrane potential, and a modulation of the Akt/mTOR pathway. In in vivo studies, they used BALB/c nude mice as a xenograft model, demonstrating that the administration of these essential oils suppressed tumor growth, owing to a significant reduction of the VEGF/VEGFR axis and an induction of apoptosis in tumor cells, prolonging the life of the mice.

Lactucoprim is a sesquiterpene lactone, a component of lactucarium, a milky liquid extracted from the wild lettuce, *Lactuca virosa*. It is also used as a sedative and analgesic. Its antiproliferative activity on U87Mg glioblastoma cells in vitro has been analyzed by Rotondo et al. [57]. In addition to a potent cytotoxic effect and cell cycle arrest in G_2/M, lactucoprim administration showed a significant reduction in cell growth and migration, as well as activation of autophagy. All these findings, together with an increase in apoptosis, owing to a decrease in procaspase-6 levels, an increase in PARP, and its positive participation in oxidative stress, allow us to affirm that lactucoprim can be considered as a promising adjuvant therapy in the treatment of this disease.

In the wild jungles of Costa Rica grow a large number of fungi that contain a large number of molecules with antitumor capacity. One of them, *Macrocybe titans*, contains a triacylglyceride called macrocybin, (which?) whose structure includes oleic acid, palmitic acid, and a more complex fatty acid with two double bonds. Using this active principle, Vilariño et al. [58] studied its anticancer activity in a xenograft with A549 tumors, achieving a significant reduction in tumor growth and a positive regulation of caveolin-1 expression, which explains the disassembly of the actin cytoskeleton in tumor cells.

The following article presents an example of how many of the chemical derivatives of a natural compound could increase both the ability to move specifically to their target and their biological activity. Grymel et al. [59] have synthesized, through the copper-catalyzed 1,3-dipolar azide-alkyne (CuAAC) cycloaddition reaction, new betulin derivatives with monosaccharides via a linker containing a heteroaromatic 1,2,3-triazole ring. These authors tested the in vitro cytotoxicity of all these derivatives using two cancer cell lines, one for human breast carcinoma (MCF-7) and the other for colorectal carcinoma (HCT-116). The main finding of this work is that the idea of adding sugar units to the betulin structure allows an optimized specific transfer of the glycoconjugate to the tumor cells.

Although, for now, we cannot consider phytochemicals and their metabolites as essential nutrients in humans, the fact is that more and more research, and a good example is this Special Issue, strongly links the fact that their intake leads to greater prevention of many diseases, including cancer. In this sense, in their article, Ferraz da Costa et al. [60] review both the molecular mechanisms of grape and red wine bioactive compounds and their metabolites in breast cancer, as well as chemoprevention and its treatment. It is very interesting to note the approach taken, relating the structure of the different compounds, flavonoids, monomeric catechins, proanthocyanidins, anthocyanins, anthocyanidins, and non-flavonoid phenolic compounds, such as resveratrol, to their metabolism and especially to their bioavailability. The review also includes an excellent discussion of in vitro, in vivo, and clinical trials on chemoprevention and therapy with these molecules.

Traditionally, chemotherapy and radiotherapy have been used in the treatment of cancer; however, it is necessary to discover new treatments that, on the one hand, are less aggressive for the organism and, on the other hand, more specific in order to recognize and differentiate tumor cells from those that are not. Possibly one of these novel treatments to eliminate different types of cancer is photodynamic therapy (PDT). The main requirement of this type of therapy is the use of photosensitizers (PS) and photoactivation using a specific wavelength of light in the presence of molecular oxygen. The combined action of these two elements is capable of exerting a cascade of molecular actions that end up modulating processes such as apoptosis, necrosis, and autophagy in tumor cells. Photoactive substances derived from medicinal plants have been shown to be safe in comparison with synthetic compounds, and although more and more natural compounds are being discovered that exhibit photosensitizing potential, it is necessary to continue along this path to find new, more active molecules that cover a broader spectrum. In this regard, Muniyandi et al. [61], in their review, put special emphasis on the importance of common photoactive groups (furanocoumarins, polyacetylenes, thiophenes, curcumin, alkaloids, and anthraquinones), their phototoxic effects, their anticancer activity, and their use as a potent PS to achieve an effective PDT result in the treatment of various types of cancers. Another review related to the use of photosensitizing compounds is presented by Verebová et al. [62]. These authors provide a comprehensive summary of the physical and chemical properties of photosensitizers of the hypericin type and their model composed of emodin, quinizarin, and danthron, and show us important antiviral, antifungal, antineoplastic and antitumor effects. They conclude their work by stating that these compounds can be used as potential agents in photodynamic therapy, especially in cancer therapy.

Following the common pattern of synthesizing chemical derivatives of natural molecules in the search for more biologically effective compounds, Professor Csuk's group [63] has screened and reviewed several triterpenoid derivatives of rhodamine. This compound belongs to a group of fluorescents, xanthene-based, fluorescein-derived, heterocyclic organic molecules that have traditionally been used as dyes and amplifying substances in dye lasers. In their study, these authors reveal the degree of cytotoxicity of these derivatives, all of which exhibit a low nanomolar range, combined with good tumor/non-tumor selectivity. The studies indicate that the homopiperazinyl spacer is more effective than the piperazinyl spacer, which allows them to state that the use of a homopiperazinyl spacer can be considered a promising candidate in biological studies.

As mentioned in this article, more and more new natural compounds with greater efficacy and biological selectivity are being sought, and all of us who form part of the medical–scientific community hope that some can be incorporated into the list of effective drugs that serve to save lives, especially at this time, when a pandemic such as COVID-19 is seriously affecting humanity, although much more severely in those countries with crucial problems in obtaining vaccines, but which, possibly, may have greater possibilities of obtaining these types of drugs. For all the contributions, we are deeply grateful for the effort and collaboration of all the authors who have made it possible, with their articles, to make this Special Issue a reality.

Author Contributions: Writing—original draft preparation, J.A.L.; Writing—review and editing. J.A.L., E.E.R.-P., and A.P.-J. All authors have agreed to the last version. All authors have read and agreed to the published version of the manuscript.

Funding: This work did not receive external funding.

Institutional Review Board Statement: Not applicable.

Informed Consent Statement: Not applicable.

Data Availability Statement: Not applicable.

Acknowledgments: The authors thank Alejandro L. Lupiáñez for the critical revision of the English version of the text.

Conflicts of Interest: The authors declare that they have no conflict of interest.

References

1. Al-Husayn, I.A.I.S. De Pharmacologicae de Herbis Medicinalibus. In *Avicenna Canon Medicinae, (Manuscript ms 2197)*, 2nd ed.; University Library of Bologna: Bologna, Italy, 1988; Volume 1020, pp. 156–416.
2. Molyneux, R.J.; Lee, S.T.; Gardner, D.R.; Panter, K.E.; James, L.F. Phytochemicals: The good, the bad and the ugly? *Phytochemistry* **2007**, *68*, 2973–2985. [CrossRef]
3. Dinkova-Kostova, A. Phytochemicals as Protectors against Ultraviolet Radiation: Versatility of Effects and Mechanisms. *Planta Medica* **2008**, *74*, 1548–1559. [CrossRef] [PubMed]
4. Gill, A.S.; Prasad, J.V.N.S. Allelopathic interactions in agroforestry systems. *Allelopath. Ecol. Agric. For.* **2000**, *13*, 195–207. [CrossRef]
5. Rufino-Palomares, E.E.; Pérez-Jiménez, A.; Reyes-Zurita, F.J.; García-Salguero, L.; Mokhtari, K.; Herrera-Merchán, A.; Medina, P.P.; Peragón, J.; Lupiáñez, J.A. Anti-cancer and Anti-angiogenic Properties of Various Natural Pentacyclic Tri-terpenoids and Some of their Chemical Derivatives. *Curr. Org. Chem.* **2015**, *19*, 919–947. [CrossRef]
6. Reyes-Zurita, F.J.; Centelles, J.J.; Lupiáñez, J.A.; Cascante, M. (2α,3β)-2,3-Dihydroxyolean-12-en-28-oic acid, a new natural triterpene from *Olea europea*, induces caspase dependent apoptosis selectively in colon adenocarcinoma cells. *FEBS Lett.* **2006**, *580*, 6302–6310. [CrossRef]
7. Laszczyk, M.N. Pentacyclic Triterpenes of the Lupane, Oleanane and Ursane Group as Tools in Cancer Therapy. *Planta Medica* **2009**, *75*, 1549–1560. [CrossRef]
8. Rufino-Palomares, E.E.; Reyes-Zurita, F.J.; García-Salguero, L.; Mokhtari, K.; Medina, P.P.; Lupiáñez, J.A.; Peragón, J. Maslinic acid, a triterpenic anti-tumoural agent, interferes with cytoskeleton protein expression in HT29 human colon-cancer cells. *J. Proteom.* **2013**, *83*, 15–25. [CrossRef]
9. Sánchez-Tena, S.; Reyes-Zurita, F.J.; Diaz-Moralli, S.; Vinardell, M.P.; Reed, M.; Garcia-Garcia, F.; Dopazo, J.; Lupiáñez, J.A.; Günther, U.; Cascante, M. Maslinic Acid-Enriched Diet Decreases Intestinal Tumorigenesis in ApcMin/+ Mice through Transcriptomic and Metabolomic Reprogramming. *PLoS ONE* **2013**, *8*, e59392. [CrossRef]
10. Reyes-Zurita, F.J.; Rufino-Palomares, E.E.; Medina, P.P.; García-Salguero, L.; Peragón, J.; Cascante, M.; Lupiáñez, J.A. Antitumour activity on extrinsic apoptotic targets of the triterpenoid maslinic acid in p53-deficient Caco-2 adenocarcinoma cells. *Biochimie* **2013**, *95*, 2157–2167. [CrossRef]
11. Reyes-Zurita, F.J.; Rufino-Palomares, E.E.; García-Salguero, L.; Peragón, J.; Medina, P.P.; Parra, A.; Cascante, M.; Lupiáñez, J.A. Maslinic Acid, a Natural Triterpene, Induces a Death Receptor-Mediated Apoptotic Mechanism in Caco-2 p53-Deficient Colon Adenocarcinoma Cells. *PLoS ONE* **2016**, *11*, e0146178. [CrossRef]
12. Chintharlapalli, S.; Papineni, S.; Ramaiah, S.K.; Safe, S. Betulinic Acid Inhibits Prostate Cancer Growth through Inhibition of Specificity Protein Transcription Factors. *Cancer Res.* **2007**, *67*, 2816–2823. [CrossRef] [PubMed]
13. Lin, C.-C.; Huang, C.-Y.; Mong, M.-C.; Chan, C.-Y.; Yin, M.-C. Antiangiogenic Potential of Three Triterpenic Acids in Human Liver Cancer Cells. *J. Agric. Food Chem.* **2011**, *59*, 755–762. [CrossRef]
14. Shanmugam, M.K.; Dai, X.; Kumar, A.P.; Tan, B.K.; Sethi, G.; Bishayee, A. Oleanolic acid and its synthetic derivatives for the prevention and therapy of cancer: Preclinical and clinical evidence. *Cancer Lett.* **2014**, *346*, 206–216. [CrossRef] [PubMed]
15. Bonel-Pérez, G.C.; Pérez-Jiménez, A.; Gris-Cárdenas, I.; Parra-Pérez, A.; Lupiáñez, J.; Reyes-Zurita, F.J.; Siles, E.; Csuk, R.; Peragón, J.; Rufino-Palomares, E.E. Antiproliferative and Pro-Apoptotic Effect of Uvaol in Human Hepatocarcinoma HepG2 Cells by Affecting G0/G1 Cell Cycle Arrest, ROS Production and AKT/PI3K Signaling Pathway. *Molecules* **2020**, *25*, 4254. [CrossRef] [PubMed]
16. Montilla, M.P.; Agil, A.; Navarro, M.C.; Jiménez, M.I.; García-Granados, A.; Parra, A.; Cabo, M.M. Antioxidant Activity of Maslinic Acid, a Triterpene Derivative Obtained from *Olea europaea*. *Planta Medica* **2003**, *69*, 472–474. [CrossRef]
17. Mokhtari, K.; Rufino-Palomares, E.E.; Pérez-Jiménez, A.; Reyes-Zurita, F.J.; Figuera, C.; García-Salguero, L.; Medina, P.P.; Peragón, J.; Lupiáñez, J.A. Maslinic Acid, a Triterpene from Olive, Affects the Antioxidant and Mitochondrial Status of B16F10 Melanoma Cells Grown under Stressful Conditions. *Evid. Based Complement. Altern. Med.* **2015**, *2015*, 1–11. [CrossRef]

18. Mokhtari, K.; Pérez-Jiménez, A.; García-Salguero, L.; Lupiáñez, J.A.; Rufino-Palomares, E.E. Unveiling the Differential Antioxidant Activity of Maslinic Acid in Murine Melanoma Cells and in Rat Embryonic Healthy Cells Following Treatment with Hydrogen Peroxide. *Molecules* **2020**, *25*, 4020. [CrossRef] [PubMed]
19. Marquez-Martin, A.; De La Puerta, R.; Fernandez-Arche, A.; Ruiz-Gutierrez, V.; Yaqoob, P. Modulation of cytokine secretion by pentacyclic triterpenes from olive pomace oil in human mononuclear cells. *Cytokine* **2006**, *36*, 211–217. [CrossRef] [PubMed]
20. Zhang, X.; Cao, J.; Zhong, L. Hydroxytyrosol inhibits pro-inflammatory cytokines, iNOS, and COX-2 expression in human monocytic cells. *Naunyn Schmiedeberg's Arch. Pharmacol.* **2009**, *379*, 581–586. [CrossRef]
21. Qian, Y.; Guan, T.; Tang, X.; Huang, L.; Huang, M.; Li, Y.; Sun, H. Maslinic acid, a natural triterpenoid compound from Olea europaea, protects cortical neurons against oxygen–glucose deprivation-induced injury. *Eur. J. Pharmacol.* **2011**, *670*, 148–153. [CrossRef]
22. Shaik, A.H.; Rasool, S.; Kareem, M.A.; Krushna, G.S.; Akhtar, P.M.; Devi, K.L. Maslinic Acid Protects Against Isoproterenol-Induced Cardiotoxicity in Albino Wistar Rats. *J. Med. Food* **2012**, *15*, 741–746. [CrossRef]
23. Khathi, A.; Serumula, M.R.; Myburg, R.B.; Van Heerden, F.; Musabayane, C.T. Effects of Syzygium aromaticum-Derived Triterpenes on Postprandial Blood Glucose in Streptozotocin-Induced Diabetic Rats Following Carbohydrate Challenge. *PLoS ONE* **2013**, *8*, e81632. [CrossRef] [PubMed]
24. Pérez-Jiménez, A.; Rufino-Palomares, E.E.; Fernández-Gallego, N.; Ortuño-Costela, M.C.; Reyes-Zurita, F.J.; Peragón, J.; García-Salguero, L.; Mokhtari, K.; Medina, P.P.; Lupiáñez, J.A. Target molecules in 3T3-L1 adipocytes differentiation are regulated by maslinic acid, a natural triterpene from *Olea europaea*. *Phytomedicine* **2016**, *23*, 1301–1311. [CrossRef] [PubMed]
25. Sultana, N.; Ata, A. Oleanolic acid and related derivatives as medicinally important compounds. *J. Enzym. Inhib. Med. Chem.* **2008**, *23*, 739–756. [CrossRef]
26. Parra, A.; Martin-Fonseca, S.; Rivas, F.; Reyes-Zurita, F.J.; Medina-O'Donnell, M.; Martinez, A.; García-Granados, A.; Lupiáñez, J.A.; Albericio, F. Semi-synthesis of acylated triterpenes from olive-oil industry wastes for the development of anticancer and anti-HIV agents. *Eur. J. Med. Chem.* **2014**, *74*, 278–301. [CrossRef]
27. Medina-O'Donnell, M.; Rivas, F.; Reyes-Zurita, F.J.; Cano-Muñoz, M.; Martinez, A.; Lupiáñez, J.A.; Parra, A. Oleanolic Acid Derivatives as Potential Inhibitors of HIV-1 Protease. *J. Nat. Prod.* **2019**, *82*, 2886–2896. [CrossRef] [PubMed]
28. Moneriz, C.; Mestres, J.; Bautista, J.M.; Diez, A.; Puyet, A. Multi-targeted activity of maslinic acid as an antimalarial natural compound. *FEBS J.* **2011**, *278*, 2951–2961. [CrossRef] [PubMed]
29. Fernández-Navarro, M.; Peragón, J.; Amores, V.; De La Higuera, M.; Lupiáñez, J.A. Maslinic acid added to the diet increases growth and protein-turnover rates in the white muscle of rainbow trout (*Oncorhynchus mykiss*). *Comp. Biochem. Physiol. Part. C Toxicol. Pharmacol.* **2008**, *147*, 158–167. [CrossRef]
30. Rufino-Palomares, E.E.; Reyes-Zurita, F.J.; Fuentes-Almagro, C.; De La Higuera, M.; Lupiáñez, J.A.; Peragón, J. Proteomics in the liver of gilthead sea bream (*Sparus aurata*) to elucidate the cellular response induced by the intake of maslinic acid. *Proteomics* **2011**, *11*, 3312–3325. [CrossRef]
31. Rufino-Palomares, E.; Reyes-Zurita, F.; García-Salguero, L.; Peragón, J.; De La Higuera, M.; Lupiáñez, J.A. Maslinic acid, a natural triterpene, and ration size increased growth and protein turnover of white muscle in gilthead sea bream (*Sparus aurata*). *Aquac. Nutr.* **2012**, *18*, 568–580. [CrossRef]
32. Reyes-Zurita, F.J.; Rufino-Palomares, E.E.; Lupiáñez, J.A.; Cascante, M. Maslinic acid, a natural triterpene from *Olea europaea* L., induces apoptosis in HT29 human colon-cancer cells via the mitochondrial apoptotic pathway. *Cancer Lett.* **2009**, *273*, 44–54. [CrossRef]
33. Reyes-Zurita, F.J.; Pachón-Peña, G.; Lizárraga, D.; Rufino-Palomares, E.E.; Cascante, M.; Lupiáñez, J.A. The natural triterpene maslinic acid induces apoptosis in HT29 colon cancer cells by a JNK-p53-dependent mechanism. *BMC Cancer* **2011**, *11*, 154. [CrossRef]
34. Barroso, J.B.; García-Salguero, L.; Peragón, J.; De la Higuera, M.; Lupiáñez, J.A. The influence of dietary protein on the kinetics of NADPH production systems in various tissues of rainbow trout (*Oncorhynchus mykiss*). *Aquaculture* **1994**, *124*, 47–59. [CrossRef]
35. Lupiáñez, J.A.; Adroher, F.J.; Vargas, A.M.; Osuna, A. Differential behaviour of glucose 6-phosphate dehydrogenase in two morphological forms of Trypanosoma cruzi. *Int. J. Biochem.* **1987**, *19*, 1085–1089. [CrossRef]
36. Adroher, F.J.; Osuna, A.; Lupiáñez, J.A. Differential energetic metabolism during Trypanosoma cruzi differentiation. II. Hexokinase, phosphofructokinase and pyruvate kinase. *Mol. Cell. Biochem.* **1990**, *94*, 71–82. [CrossRef]
37. Sánchez-Muros, M.J.; García-Rejón, L.; Lupiáñez, J.A.; De la Higuera, M. Long-term nutritional effects on the primary liver and kidney metabolism in rainbow trout (*Oncorhynchus mykiss*). II. Adaptive response of glucose 6-phosphate dehydrogenase activity to high-carbohydrate/low-protein and high-fat/non-carbohydrate diets. *Aquac. Nutr.* **1996**, *2*, 193–200. [CrossRef]
38. Barroso, J.B.; Peragón, J.; García-Salguero, L.; De la Higuera, M.; Lupiáñez, J.A. Carbohydrate deprivation reduces NADPH-production in fish liver but not in adipose tissue. *Int. J. Biochem. Cell Biol.* **2001**, *33*, 785–796. [CrossRef]
39. Sheehan, D.; Meade, G.; Foley, V.M.; Dowd, C.A. Structure, function and evolution of glutathione transferases: Implications for classification of non-mammalian members of an ancient enzyme superfamily. *Biochem. J.* **2001**, *360*, 1–16. [CrossRef] [PubMed]
40. Corpas, F.J.; Barroso, J.B.; Sandalio, L.M.; Palma, J.M.; Lupiáñez, J.A.; del Río, L.A. Peroxisomal NADP-Dependent Isocitrate Dehydrogenase. Characterization and Activity Regulation during Natural Senescence. *Plant. Physiol.* **1999**, *121*, 921–928. [CrossRef]

41. Parra, A.; Martin-Fonseca, S.; Rivas, F.; Reyes-Zurita, F.J.; Medina-O'Donnell, M.; Rufino-Palomares, E.E.; Martínez, A.; Garcia-Granados, A.; Lupiáñez, J.A.; Albericio, F. Solid-Phase Library Synthesis of Bi-Functional Derivatives of Oleanolic and Maslinic Acids and Their Cytotoxicity on Three Cancer Cell Lines. *ACS Comb. Sci.* **2014**, *16*, 428–447. [CrossRef]

42. Reyes-Zurita, F.J.; Medina-O'Donnell, M.; Ferrer-Martin, R.M.; Rufino-Palomares, E.E.; Martin-Fonseca, S.; Rivas, F.; Martínez, A.; García-Granados, A.; Pérez-Jiménez, A.; García-Salguero, L.; et al. The oleanolic acid derivative, 3-O-succinyl-28-O-benzyl oleanolate, induces apoptosis in B16–F10 melanoma cells via the mitochondrial apoptotic pathway. *RSC Adv.* **2016**, *6*, 93590–93601. [CrossRef]

43. Medina-O'Donnell, M.; Rivas, F.; Reyes-Zurita, F.J.; Martinez, A.; Martin-Fonseca, S.; Garcia-Granados, A.; Ferrer-Martín, R.M.; Lupiáñez, J.A.; Parra, A. Semi-synthesis and antiproliferative evaluation of PEGylated pentacyclic triterpenes. *Eur. J. Med. Chem.* **2016**, *118*, 64–78. [CrossRef] [PubMed]

44. Medina-O'Donnell, M.; Rivas, F.; Reyes-Zurita, F.J.; Martinez, A.; Galisteo-González, F.; Lupiáñez, J.A.; Parra, A. Synthesis and in vitro antiproliferative evaluation of PEGylated triterpene acids. *Fitoterapia* **2017**, *120*, 25–40. [CrossRef] [PubMed]

45. Medina-O'Donnell, M.; Rivas, F.; Reyes-Zurita, F.J.; Martinez, A.; Lupiáñez, J.A.; Parra, A. Diamine and PEGylated-diamine conjugates of triterpenic acids as potential anticancer agents. *Eur. J. Med. Chem.* **2018**, *148*, 325–336. [CrossRef]

46. Jannus, F.; Medina-O'Donnell, M.; Rivas, F.; Díaz-Ruiz, L.; Rufino-Palomares, E.E.; Lupiáñez, J.A.; Parra, A.; Reyes-Zurita, F.J. A Diamine-PEGylated Oleanolic Acid Derivative Induced Efficient Apoptosis through a Death Receptor and Mitochondrial Apoptotic Pathway in HepG2 Human Hepatoma Cells. *Biomolecules* **2020**, *10*, 1375. [CrossRef] [PubMed]

47. Vega-Granados, K.; Medina-O'Donnell, M.; Rivas, F.; Reyes-Zurita, F.J.; Martinez, A.; Álvarez de Cienfuegos, L.; Lupiáñez, J.A.; Parra, A. Synthesis and Biological Activity of Triterpene–Coumarin Conjugates. *J. Nat. Prod.* **2021**, *84*, 1587–1597. [CrossRef]

48. Toreti, V.C.; Sato, H.H.; Pastore, G.M.; Park, Y.K. Recent Progress of Propolis for Its Biological and Chemical Compositions and Its Botanical Origin. *Evid. Based Complement. Altern. Med.* **2013**, *2013*, 1–13. [CrossRef]

49. Wezgowiec, J.; Wieczynska, A.; Wieckiewicz, W.; Kulbacka, J.; Saczko, J.; Pachura, N.; Wieckiewicz, M.; Gancarz, R.; Wilk, K.A. Polish Propolis—Chemical Composition and Biological Effects in Tongue Cancer Cells and Macrophages. *Molecules* **2020**, *25*, 2426. [CrossRef]

50. Nguyen, N.H.; Pham, Q.T.; Luong, T.N.H.; Le, H.K.; Vo, V.G. Potential Antidiabetic Activity of Extracts and Isolated Compound from Adenosma bracteosum (Bonati). *Biomolecules* **2020**, *10*, 201. [CrossRef]

51. Nguyen, N.H.; Ta, Q.T.H.; Pham, Q.T.; Luong, T.N.H.; Phung, V.T.; Duong, T.H.; Vo, V.G. Anticancer Activity of Novel Plant Extracts and Compounds from *Adenosma bracteosum* (Bonati) in Human Lung and Liver Cancer Cells. *Molecules* **2020**, *25*, 2912. [CrossRef]

52. Anwar, F.; Latif, S.; Ashraf, M.; Gilani, A. Moringa oleifera: A food plant with multiple medicinal uses. *Phytother. Res.* **2006**, *21*, 17–25. [CrossRef]

53. Luetragoon, T.; Sranujit, R.P.; Noysang, C.; Thongsri, Y.; Potup, P.; Suphrom, N.; Nuengchamnong, N.; Usuwanthim, K. Anti-Cancer Effect of 3-Hydroxy-β-Ionone Identified from *Moringa oleifera* Lam. Leaf on Human Squamous Cell Carcinoma 15 Cell Line. *Molecules* **2020**, *25*, 3563. [CrossRef]

54. Jabeen, A.; Sharma, A.; Gupta, I.; Kheraldine, H.; Vranic, S.; Al Moustafa, A.-E.; Al Farsi, H.F. *Elaeagnus angustifolia* Plant Extract Inhibits Epithelial-Mesenchymal Transition and Induces Apoptosis via HER2 Inactivation and JNK Pathway in HER2-Positive Breast Cancer Cells. *Molecules* **2020**, *25*, 4240. [CrossRef] [PubMed]

55. Roda, E.; De Luca, F.; Locatelli, C.A.; Ratto, D.; Di Iorio, C.; Savino, E.; Bottone, M.G.; Rossi, P. From a Medicinal Mushroom Blend a Direct Anticancer Effect on Triple-Negative Breast Cancer: A Preclinical Study on Lung Metastases. *Molecules* **2020**, *25*, 5400. [CrossRef] [PubMed]

56. Huang, X.F.; Sheu, G.T.; Chang, K.F.; Huang, Y.C.; Hung, P.H.; Tsai, N.M. *Pogostemon cablin* Triggered ROS-Induced DNA Damage to Arrest Cell Cycle Progression and Induce Apoptosis on Human Hepatocellular Carcinoma In Vitro and In Vivo. *Molecules* **2020**, *25*, 5639. [CrossRef] [PubMed]

57. Rotondo, R.; Oliva, M.A.; Staffieri, S.; Castaldo, S.; Giangaspero, F.; Arcella, A. Implication of Lactucopicrin in Autophagy, Cell Cycle Arrest and Oxidative Stress to Inhibit U87Mg Glioblastoma Cell Growth. *Molecules* **2020**, *25*, 5843. [CrossRef] [PubMed]

58. Vilariño, M.; García-Sanmartín, J.; Ochoa-Callejero, L.; López-Rodríguez, A.; Blanco-Urgoiti, J.; Martínez, A. Macrocybin, a Natural Mushroom Triglyceride, Reduces Tumor Growth In Vitro and In Vivo Through Caveolin-Mediated Interference with the Actin Cytoskeleton. *Molecules* **2020**, *25*, 6010. [CrossRef]

59. Grymel, M.; Pastuch-Gawołek, G.; Lalik, A.; Zawojak, M.; Boczek, S.; Krawczyk, M.; Erfurt, K. Glycoconjugation of Betulin Derivatives Using Copper-Catalyzed 1,3-Dipolar Azido-Alkyne Cycloaddition Reaction and a Preliminary Assay of Cytotoxicity of the Obtained Compounds. *Molecules* **2020**, *25*, 6019. [CrossRef] [PubMed]

60. Ferraz da Costa, D.C.; Rangel, L.P.; Quarti, J.; Santos, R.A.; Silva, J.L.; Fialho, E. Bioactive Compounds and Metabolites from Grapes and Red Wine in Breast Cancer Chemoprevention and Therapy. *Molecules* **2020**, *25*, 3531. [CrossRef] [PubMed]

61. Muniyandi, K.; George, B.; Parimelazhagan, T.; Abrahamse, H. Role of Photoactive Phytocompounds in Photodynamic Therapy of Cancer. *Molecules* **2020**, *25*, 4102. [CrossRef] [PubMed]

62. Verebová, V.; Beneš, J.; Staničová, J. Biophysical Characterization and Anticancer Activities of Photosensitive Phytoanthraquinones Represented by Hypericin and Its Model Compounds. *Molecules* **2020**, *25*, 5666. [CrossRef] [PubMed]

63. Hoenke, S.; Serbian, I.; Deigner, H.P.; Csuk, R. Mitocanic Di- and Triterpenoid Rhodamine B Conjugates. *Molecules* **2020**, *25*, 5443. [CrossRef] [PubMed]

 *molecules*

Article

Polish Propolis—Chemical Composition and Biological Effects in Tongue Cancer Cells and Macrophages

Joanna Wezgowiec [1,*], Anna Wieczynska [2,3], Wlodzimierz Wieckiewicz [4,*], Julita Kulbacka [5], Jolanta Saczko [5], Natalia Pachura [6], Mieszko Wieckiewicz [1], Roman Gancarz [2] and Kazimiera A. Wilk [2]

[1] Department of Experimental Dentistry, Wroclaw Medical University, 50-425 Wroclaw, Poland; m.wieckiewicz@onet.pl
[2] Department of Engineering and Technology of Chemical Processes, Wroclaw University of Science and Technology, 50-370 Wroclaw, Poland; anna.wieczynska@pwr.edu.pl (A.W.); roman.gancarz@pwr.edu.pl (R.G.); kazimiera.wilk@pwr.edu.pl (K.A.W.)
[3] Institute of Genetics and Microbiology, University of Wroclaw, 51-148 Wroclaw, Poland
[4] Department of Prosthetic Dentistry, Wroclaw Medical University, 50-425 Wroclaw, Poland
[5] Department of Molecular and Cellular Biology, Wroclaw Medical University, 50-556 Wroclaw, Poland; julita.kulbacka@umed.wroc.pl (J.K.); jolanta.saczko@umed.wroc.pl (J.S.)
[6] Department of Chemistry, Wroclaw University of Environmental and Life Sciences, 50-375 Wroclaw, Poland; natalia.pachura@upwr.edu.pl
* Correspondence: joanna.wezgowiec@umed.wroc.pl (J.W.); wlodzimierz.wieckiewicz@umed.wroc.pl (W.W.)

Academic Editors: José Antonio Lupiáñez, Amalia Pérez-Jiménez and Eva E. Rufino-Palomares
Received: 14 April 2020; Accepted: 20 May 2020; Published: 22 May 2020

Abstract: The purpose of this study was to compare the chemical composition and biological properties of Polish propolis. Ethanol, ethanol-hexane, hexane and hexane-ethanol extracts of propolis from three different regions of Poland were prepared. On the basis of the evaluation of their chemical composition as well as the extraction yield and free radical scavenging activity, the ethanol and hexane-ethanol extractions were proposed as the most effective methods. Subsequently, the biological properties of the extracts were evaluated to investigate the selectivity of an anticancer effect on tongue cancer cells in comparison to normal gingival fibroblasts. The obtained products demonstrated anticancer activity against tongue cancer cells. Additionally, when the lowest extract concentration (100 µg/mL) was applied, they were not cytotoxic to gingival fibroblasts. Finally, a possible anti-inflammatory potential of the prepared products was revealed, as reduced mitochondrial activity and proliferation of macrophages exposed to the extracts were observed. The results obtained indicate a potential of Polish propolis as a natural product with cancer-selective toxicity and anti-inflammatory effect. However, further studies are still needed to thoroughly explain the molecular mechanisms of its action and to obtain the promising health benefits of this versatile natural product.

Keywords: total phenolic content; total flavonoid content; GC-MS; DPPH; antioxidant; anticancer agent; anti-inflammatory agent; gingival fibroblasts; oral cancer; natural extract

---

1. Introduction

Nature, as an immemorial source of diverse active molecules, continues to serve as a major inspiration for drug development. Therapeutic applications of natural products offer great opportunities for modern medicine, while being simultaneously a huge challenge due to the problem of standardization procedures and the chemical complexity of these substances. On the other hand, such complexity is inevitable and a final therapeutic effect of a whole extract in general is better than effects of individual compounds since it results from the synergistic activity of the extract components [1].

One of the most attractive natural products is propolis—the resinous substance collected by bees from plants and mixed with wax and enzymes. It is then used to strengthen and protect their hives as well as to prevent decomposition of intruders' carcasses. People have also widely used propolis in folk medicine, as it is known for a broad spectrum of biological properties including antibacterial, antifungal, antiviral, anti-inflammatory, antioxidant and anticancer activity [2]. Nowadays, it is used in the cosmetics industry, i.e., as a component of anti-acne creams and products for oral hygiene [3]. However, the therapeutic potential of propolis is still untapped and many research groups continue investigation of a chemical composition and biological properties of this material. The studies revealed a variability in propolis composition depending on the geographical region of collection and the plant sources. For instance, bud exudates of different poplar buds are the main source of propolis collected in the temperate zone, including Europe [4]. Silva-Carvalho et al. reported that poplar propolis is mainly composed of flavonoids, phenolic acids and its esters [3].

In particular, contemporary oral medicine may benefit from the wide spectrum of propolis activities. Many dental specialties which make use of this natural product have been reported [5]. Research on Polish propolis is mainly focused on its antimicrobial properties [6–11]. Interestingly, no research on the use of Polish propolis against oral cancer has been published so far. There is little research concerning the antiproliferative effect of Polish propolis on glioblastoma cells [12], colon, lung and breast cancer cells [13], as well as prostate cancer cells [14]. On the other hand, taking into account global data, the problem of oral cancer treatment is still unsolved. In 2018, new cases of oral cancer occurred globally in approximately 355,000 people and caused 177,000 deaths. The most common oral cancer type is tongue squamous cell carcinoma (TSCC), characterized by high lymphatic metastasis, recurrence and drug resistance. The current treatment approaches include surgery, which may be followed by radiotherapy and/or chemotherapy. However, there is still no effective therapeutic strategy and the death toll linked to this disease is still increasing [15].

The purpose of this study was to evaluate the anticancer properties of three different types of propolis from different regions of Poland on the in vitro model of tongue cancer cells. For this reason, ethanol, ethanol-hexane, hexane and hexane-ethanol extracts of Polish propolis were prepared. Normal human gingival fibroblasts were used as a control group of non-cancer cells and a murine macrophage-like cell line was used to evaluate anti-inflammatory potential of the prepared products. Additionally, chemical composition and antioxidant activity of the prepared extracts were compared.

2. Results

2.1. Extraction Yield

The extraction yields of the propolis extracts were calculated and are presented in Table 1. The extraction yield values of the ethanol extracts of propolis (EEP) were higher than the hexane extracts of propolis (HEP) and the highest values of extraction yield were obtained for propolis from Masovia (P2) and West Pomerania Province (P3). Therefore, the results indicated that ethanol was a better solvent than *n*-hexane. In addition, the hexane-ethanol extracts of propolis (HEEP) had the second highest extraction yields among all the propolis extracts analyzed.

2.2. Total Polyphenol Content

The total polyphenol content (TPC) was determined with the Folin–Ciocalteu method (Table 2). The statistical analysis revealed that there was not strong variation between the TPC of all propolis harvested in different provinces of Poland ($F(2, 109) = 0.86794$; $p = 0.42270$). However, regardless of a type of propolis, there were statistically significant differences of TPC ($F(3, 108) = 1178.4$; $p = 0.0000$) between different extracts, such as ethanol extract of propolis (EEP), ethanol-hexane extract of propolis (EHEP), hexane extract of propolis (HEP) and hexane-ethanol extract of propolis (HEEP). The TPC for

EEP and HEEP was above 220 mg GAE (gallic acid equivalent)/g of the propolis extract, while the TPC for EHEP and HEP was below 50 mg GAE/g. Tukey's post-hoc test revealed that all differences of TPC between each type of propolis extract tested were statistically significant at $p < 0.05$. Interestingly, when only EEP and HEEP were considered, the strong differences among TPC of the propolis harvested in different provinces were observed ($F(2, 51) = 31.058$; $p = 0.00000$). Thus, the propolis extracts from West Pomerania Province (P3) had the highest TPC, while the lowest TPC was obtained for propolis extracts from Podlasie (P1).

Table 1. Extraction yields of the prepared extracts; P1—propolis from Podlasie, P2—propolis from Masovia, P3—propolis from West Pomerania Province, EEP—ethanol extract of propolis, EHEP—ethanol-hexane extracts of propolis, HEP—hexane extract of propolis, HEEP—hexane-ethanol extracts of propolis.

Symbol	Sequence of Solvents	Extraction Yield [%]		
		P1	P2	P3
EEP	ethanol	33.4	57.5	63.7
EHEP	ethanol–hexane	24.2	8.4	13.3
HEP	hexane	28.2	17.5	14.5
HEEP	hexane–ethanol	32.9	42.7	47.9

Table 2. Total polyphenol content and total flavonoid content of the prepared extracts; the results are expressed as mean ± SD; P1—propolis from Podlasie, P2—propolis from Masovia, P3—propolis from West Pomerania Province, EEP—ethanol extract of propolis, EHEP—ethanol-hexane extracts of propolis, HEP—hexane extract of propolis, HEEP—hexane-ethanol extracts of propolis, GAE—gallic acid equivalent, QE—quercetin equivalent.

Propolis Extract	P1	P2	P3
	Total Polyphenol Content [mg GAE/g]		
EEP	222.05 ± 14.29	259.63 ± 11.73	275.79 ± 13.42
EHEP	16.36 ± 1.12	19.60 ± 1.07	18.02 ± 1.09
HEP	20.45 ± 4.08	45.02 ± 7.22	38.84 ± 6.40
HEEP	249.92 ± 8.64	277.19 ± 14.28	308.92 ± 15.85
	Total Flavonoid Content [mg QE/g]		
EEP	18.76 ± 0.66	22.19 ± 0.44	19.79 ± 0.19
EHEP	11.10 ± 0.06	10.87 ± 0.03	12.99 ± 0.07
HEP	12.23 ± 0.21	13.49 ± 0.13	14.45 ± 0.19
HEEP	19.00 ± 0.57	22.46 ± 0.40	21.63 ± 0.25

2.3. Total Flavonoid Content

The total flavonoid content (TFC), evaluated via aluminum chloride method, was presented in Table 2. Similarly to the measurement results of TPC, this analysis also revealed no statistically significant differences among propolis of different origin ($F(2, 133) = 3.3270$; $p = 0.03891$). On the other hand, differences among various extracts—EEP, EHEP, HEP and HEEP—were statistically significant ($F(3, 132) = 360.77$; $p = 0.0000$). Tukey's post-hoc test revealed that all differences of TFC between each extract type tested were statistically significant at $p < 0.05$, except for the differences between EEP and HEEP samples ($p = 0.122385$). For all the ethanol and hexane-ethanol extracts (EEP and HEEP) analyzed, TFC was above 18.76 mg QE (quercetin equivalent)/g of the propolis extract, while ethanol-hexane and hexane extracts (EHEP and HEP) were characterized by significantly lower TFC. The highest TFC among all samples tested was found for propolis extracts from Masovia (EEP_P2 and HEEP_P2).

2.4. GC-MS Analysis

The chemical composition of EEP from different regions of Poland (Podlasie, Masovia and West Pomerania Province) was determined using gas chromatography–mass spectrometry (GC-MS) and is presented in Appendix A, Table A1. Briefly, the analysis of EEP revealed the presence of seventy-two components, out of which sixty-two were identified. The main components of the material analyses were TMS derivatives of 4-coumaric acid, D-fructose, D-glucose, D-mannopyranose, benzoic acid, lignoceric acid, ferulic acid and naringenin. GC-MS analysis of the ethanol extracts of propolis from Podlasie (EEP_P1) and Masovia (EEP_P2) showed a higher concentration of aromatic acids than the ethanol extract of propolis from West Pomerania Province (EEP_P3). The concentration of the compounds selected is presented in Table A1. The results indicate that the highest concentration of 4-coumaric acid and caffeic acid was measured in EEP_P2, while the lowest one was found in EEP_P3. Furthermore, the highest concentration of ferulic acid and benzoic acid was measured in EEP_P1, while the lowest one was found in EEP_P3.

The chemical composition of HEP from different regions of Poland is presented in Appendix A, Table A2. The profile of the compounds of the n-hexane extracts of propolis, determined by GC-MS, contains forty-one compounds (out of which forty were identified). The results showed domination of waxes and fatty acids derivatives of TMS. The main compounds of HEP_P1 and HEP_P2 were methyl triacontyl ether, heptacosane, pentacosane and lignoceric acid. The main compounds of HEP_P3 were heptacosane, lignoceric acid, 13-octadecanoic acid and methyl triacontyl ether. In addition, HEPs contained around two times more (HEP_P1: 8.10%, HEP_P2: 8.47%) or four times more (HEP_P3: 13.19%) benzoic acid than EEP.

The chemical composition of HEEP harvested from different regions of Poland is presented in Appendix A, Table A3. The profile of the compounds of the hexane-ethanol extracts of propolis contained sixty-five compounds (out of which sixty-two were identified). High similarity of the content of EEP and HEEP was observed. The dominant compounds in HEEP were 4-coumaric acid, D-fructose, D-glucose, D-mannopyranose and ferulic acid.

Fatty Acids Composition

The complete chemical composition of fatty acids in HEP from different regions of Poland is presented in Appendix A, Table A4. Fourteen compounds were identified when analyzing fatty acids contained in propolis of different origins. The main components found in the HEP fraction are: hexadecanoic acid methyl ester, heptadecanoic acid methyl ester, oleic acid methyl ester and tetracosanoic acid methyl ester.

2.5. DPPH Free Radical Scavenging Activity

A 2,2-diphenyl-1-picrylhydrazyl (DPPH) assay was used to measure antioxidant activity of the propolis extracts, with the results presented in Figure 1.

As shown in the Figure 1a–c, all the tested ethanol-hexane and hexane extracts (EHEP and HEP) obtained from propolis harvested in different regions of Poland (P1, P2, P3) had only minimal DPPH free radical scavenging activity compared to the standard. Therefore, they were assumed as having no effect at all. In contrast, for EEP and HEEP, the free radical scavenging activity increased with the increase of the extracts' concentration from 0 to 200 μg/mL. For these active extracts their IC_{50} were calculated (the concentration of extracts that inhibits the formation of DPPH free radicals by 50%) and showed in Table 3. Statistically significant differences among IC_{50} of propolis from different regions of Poland were demonstrated ($F(2, 55) = 43.365$; $p = 0.00000$). Tukey's post-hoc test revealed that all differences of IC_{50} between each type of propolis tested (P1, P2, P3) were statistically significant at $p < 0.05$. Regardless of the type of propolis studied, the type of extract did not significantly influence the obtained values of IC_{50} ($F(1, 56) = 0.09896$; $p = 0.75425$). The lowest IC_{50} values were calculated for propolis extracts from West Pomerania Province (P3), indicating the highest antioxidant potential of these preparations among all the extracts tested.

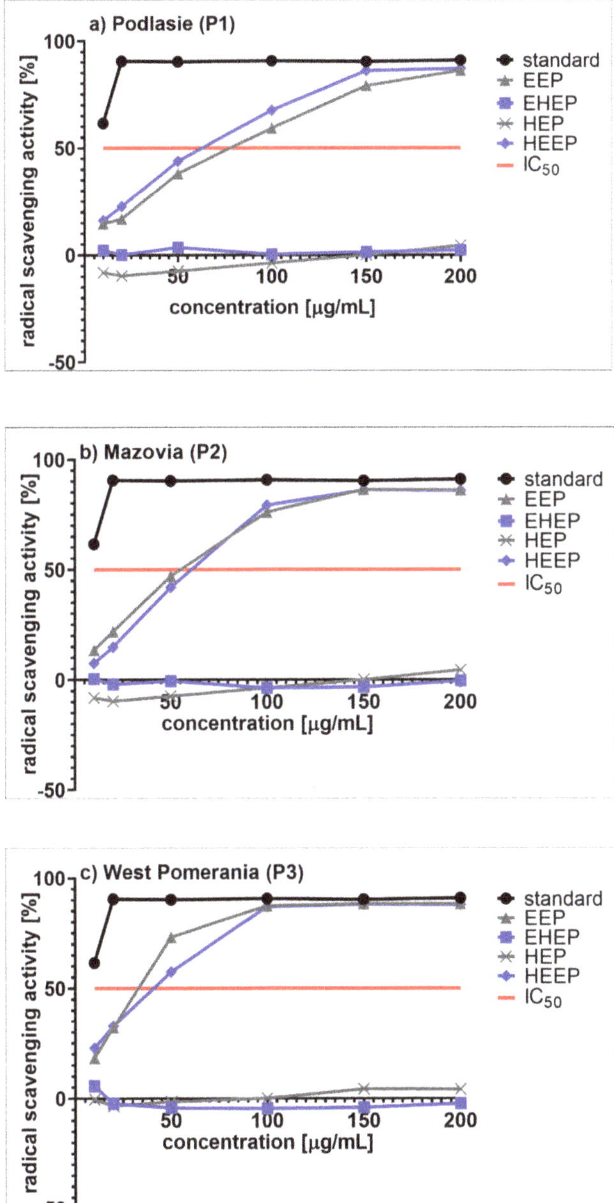

Figure 1. DPPH free radical scavenging activity of the prepared propolis extracts from: (**a**) Podlasie (P1); (**b**) Masovia (P2); (**c**) West Pomerania Province (P3); EEP— ethanol extract of propolis, EHEP—ethanol-hexane extracts of propolis, HEP—hexane extract of propolis, HEEP—hexane-ethanol extracts of propolis.

Table 3. The concentration of extracts that inhibits the formation of DPPH free radicals by 50% (IC$_{50}$); P1—propolis from Podlasie, P2—propolis from Masovia, P3—propolis from West Pomerania Province, EEP—ethanol extract of propolis, HEEP—hexane-ethanol extracts of propolis.

Propolis Extract	IC$_{50}$ [µg/mL]		
	P1	**P2**	**P3**
EEP	78.02 ± 4.86	55.07 ± 7.39	33.01 ± 2.73
HEEP	62.84 ± 14.59	60.72 ± 2.89	40.92 ± 7.55

2.6. Anticancer Activity

The anticancer activity of the selected Polish propolis extracts was evaluated on human squamous cell carcinoma derived from tongue (SCC-25) after incubation for 5 min and 24 h. For this purpose, 3-(4,5-dimethyl-2-thiazolyl)-2,5-diphenyl-2H-tetrazolium bromide assay (MTT assay) and sulforhodamine B assay (SRB assay) were performed. In addition, for both methods, 24 h incubation with human gingival fibroblasts (HGFs) was used as a control model to investigate the effects of propolis in normal, i.e., non-cancer cells. The cytotoxicity values of EEP and HEEP harvested from three different regions in Poland and applied at three concentrations (100, 500 and 1000 µg/mL) are presented in Figure 2 (MTT assay results) and in Figure 3 (SRB assay results).

2.6.1. MTT Assay

When 5 min of incubation with the propolis extracts was applied, mitochondrial activity of SCC-25 cells was only slightly reduced (Figure 2a). Moreover, when concentrations of all the extracts tested were increased, the mitochondrial activity was still above 80% compared to the control. However, the prolonged 24 h incubation period affected the cell viability significantly (Figure 2b). Three-way ANOVA revealed that for all tested extracts of Polish propolis the differences between groups based on the propolis type or extraction type were not statistically significant ($p = 0.093920$ and $p = 0.493920$, respectively). The only factor determining significant differences between groups was the extract concentration ($p = 0.000000$). For tongue cancer cells, incubation with each of the tested propolis extract at a concentration of 500 and 1000 µg/mL resulted in a decrease of mitochondrial activity to ca. 20% of the control. When the concentration of the propolis extracts applied was 100 µg/mL, the mitochondrial activity was most reduced for EEP_P3 (33% of the control) and least reduced for EEP_P1 (59% of the control), therefore, EEP_P1 was less active. The results obtained for HGFs treated with propolis indicated that the tested propolis extracts impaired also the viability of normal cells (Figure 2c). Incubation of HGFs with each tested propolis extract at a concentration of 500 and 1000 µg/mL reduced the mitochondrial activity to ca. 40% compared to the control. In addition, when the propolis extract concentration of 100 µg/mL was applied, EEP_P3 was the most active propolis extract, which reduced the mitochondrial activity to 52% of the control. Furthermore, HEEP_P2 reduced the mitochondrial activity to 76% compared to the control and therefore it was the least active propolis extract. Three-way ANOVA results for HGFs revealed that all the factors studied (type of propolis, type of extract and extract concentration) were source of significant variation at $p < 0.05$.

2.6.2. SRB Assay

When SCC-25 cells were incubated with Polish propolis for 5 min, for all the concentrations of all the extracts tested the total protein content of cells was above 93% compared to the control (Figure 3a). Therefore, no cytotoxic effect of propolis extracts after a short-time incubation was revealed. However, the prolonged incubation, i.e., 24 h, affected the cellular proliferation significantly (Figure 3b). The results showed that 24 h incubation of tongue cancer cells with increasing concentration of propolis extract resulted in a decrease of total protein content. For example, when the concentrations of 100 and 500 µg/mL of all Polish propolis extracts were applied, cellular protein content was reduced to ca. 55% of the control. Notably, the least activity was observed at 100 µg/mL of EEP_P1, that reduced the cellular protein content to 72% compared to the control. However, for 500 µg/mL of HEEP_P1 the cellular protein content was reduced to 45% of the control. Finally, when tongue cancer cells were incubated with each of the Polish propolis extracts tested at a concentration of 1000 µg/mL, it resulted in a decrease in total protein content to ca. 45% compared to the control. The results for HGFs indicated that the analyzed propolis extracts at concentrations of 500 and 1000 µg/mL impaired the proliferation of normal cells (Figure 3c). Incubation of HGFs with 1000 µg/mL propolis extracts reduced total protein content to ca. 30% compared to the control. Incubation of HGFs with 1000 µg/mL of HEEP_P2 resulted in the lowest level of protein content, reduced to 20% of the control. On the other hand, incubation of HGFs with 1000 µg/mL of EEP_P1 resulted in the highest level of total protein content, i.e., 40% of the control. Additionally, incubation of HGFs with 500 µg/mL propolis extracts also induced a significant cytotoxicity. In contrast, for the concentration of 100 µg/mL, the lowest level of total protein content, i.e., 79% of the control, was observed when HGFs were incubated with EEP_P3 and HEEP_P3. The three-way ANOVA of results both for SCC-25 and HGFs revealed that all the factors studied (type of propolis, type of extract and extract concentration) were sources of significant variation at $p < 0.05$.

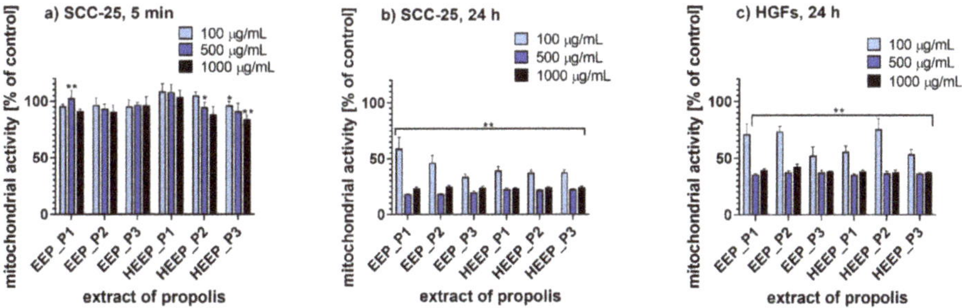

Figure 2. MTT assay results for: (**a**) tongue cancer cells (SCC-25) incubated for 5 min with propolis extracts; (**b**) tongue cancer cells (SCC-25) incubated for 24 h with propolis extracts; (**c**) human gingival fibroblasts (HGFs) incubated for 24 h with propolis extracts; * $p < 0.05$, ** $p < 0.005$; P1—propolis from Podlasie, P2—propolis from Masovia, P3—propolis from West Pomerania Province, EEP—ethanol extract of propolis, HEEP—hexane-ethanol extracts of propolis.

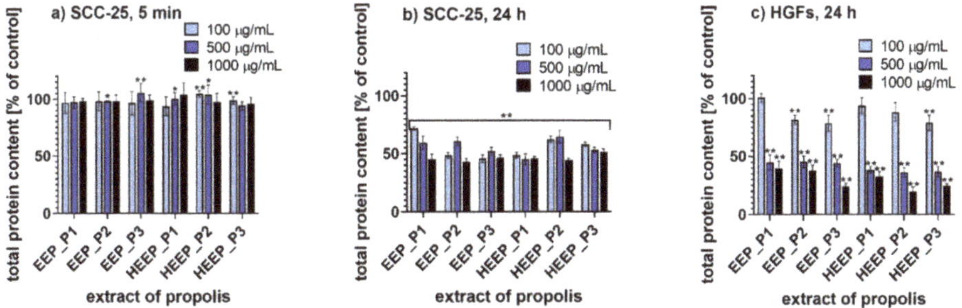

Figure 3. Sulforhodamine B (SRB) assay results for: (**a**) tongue cancer cells (SCC-25) incubated for 5 min with propolis extracts; (**b**) tongue cancer cells (SCC-25) incubated for 24 h with propolis extracts; (**c**) human gingival fibroblasts (HGFs) incubated for 24 h with propolis extracts; * $p < 0.05$, ** $p < 0.005$; P1—propolis from Podlasie, P2—propolis from Masovia, P3—propolis from West Pomerania Province, EEP—ethanol extract of propolis, HEEP—hexane-ethanol extracts of propolis.

2.7. Anti-Inflammatory Potential

Anti-inflammatory potential of the propolis extracts selected was evaluated on murine macrophage-like cell line (P388-D1) via MTT assay (Figure 4a) and SRB assay (Figure 4b) after 24 h of incubation. For all the analyzed concentrations of all the extracts tested it was observed that the prolonged 24 h incubation period affected the cellular mitochondrial activity and proliferation significantly. Incubation with each tested propolis extract at a concentration of 1000 µg/mL resulted in a decrease in mitochondrial activity of P388-D1 cells to ca. 19% of the control (Figure 4a) and a decrease in total protein content to ca. 38% of the control (Figure 4b). When the lowest concentration of extracts (100 µg/mL) was applied, the cellular mitochondrial activity was reduced to 48% and the cellular protein content to ca. 55% of the control.

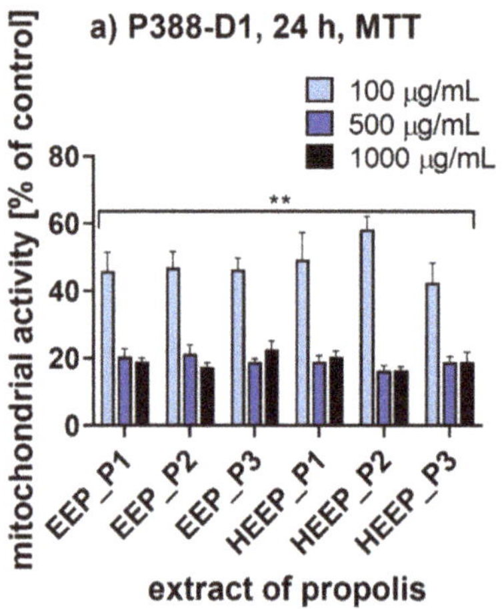

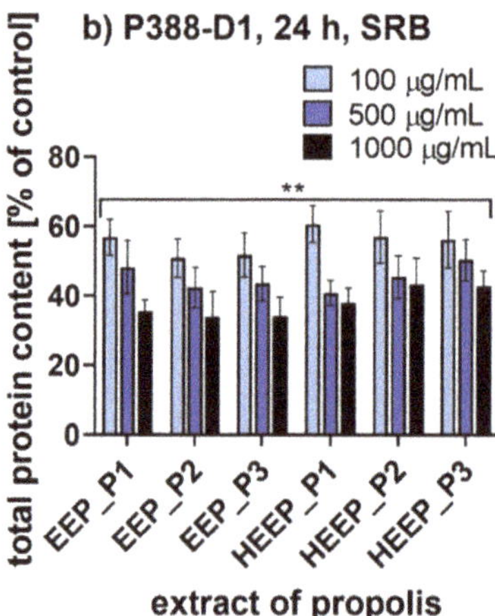

Figure 4. Results for murine macrophage cells (P388-D1) incubated for 24 h with propolis extracts; (**a**) MTT assay; (**b**) SRB assay; ** $p < 0.005$; P1—propolis from Podlasie, P2—propolis from Masovia, P3—propolis from West Pomerania Province, EEP—ethanol extract of propolis, HEEP—hexane-ethanol extracts of propolis.

3. Discussion

Propolis demonstrated antiproliferative activity on various cancer cell lines. It was reported that this natural product can block specific oncogene signaling pathways, leading to a decrease in cell proliferation. It can also increase apoptosis, exert antiangiogenic effects, and modulate the tumor microenvironment [3,16].

In spite of these beneficial properties, research on the anticancer activity of propolis on human tongue cancer cells is very limited. Antiproliferative activity of the ethanol extract of Chilean propolis on human mouth epidermoid carcinoma cells (KB) was demonstrated by Russo et al. [17]. Furthermore, Yen et al. and Chiu et al. showed an anti-inflammatory effect of various propolis extracts by inhibiting one of the inflammatory markers—COX-2—in KB cell line [18]. The study of Salehi et al. determined the chemopreventive effect of Iranian propolis on dysplastic changes in the rats' tongue epithelium after administration of carcinogens (DMBA). The results have showed that propolis can prevent DMBA-induced dysplasia of the oral mucosa in animal model [19]. A similar effect was obtained for hydroalcoholic extract of Brazilian red propolis (HERP) on oral squamous cell carcinoma (OSCC) in rodents. The research revealed that HERP inhibited tumor growth and progression [20].

The anticancer effect of propolis is often attributed to one of its active components—caffeic acid phenethyl ester (CAPE). It can be considered as a potential support for therapy of patients with oral squamous cell carcinoma due to the ability to inhibit cellular proliferation and to prevent cancer metastasis [21–23]. On the other hand, the other approach to the clarification of the natural drug's mechanisms of action is more comprehensive and takes into account a complexity of the product rather than the effect of its individual components. The study of Czyżewska et al. suggested that the synergistic effect of different polyphenols (chrysin, galangin, pinocembrin, caffeic acid, p-coumaric acid and ferulic acid) is responsible for the propolis' ability to inhibit the growth of human tongue cancer cells through apoptosis [24]. Another study indicated the synergistic effect of the main components of Iranian propolis on mouth epidermoid carcinoma (KB) cells. MTT assay revealed that IC_{50} values of EEP and its main component, quercetin (Q) were 40 μg/mL and 195 μg/mL respectively after 48 h of incubation [25].

In this study, the whole extracts of Polish propolis were evaluated in terms of the selectivity of their anticancer effect on the tongue cancer cells in comparison to the normal gingival fibroblasts. Chemical analyses revealed that ethanolic and hexane-ethanol extraction were the most effective methods of raw propolis extraction to receive the most chemically complex product. This conclusion confirms the findings of the other studies indicating ethanol extraction as the most common method of raw propolis processing [26–28]. The second proposed method—ethanol–hexane extraction—may be an interesting alternative allowing the wax content removal [29]. Both spectroscopic and chromatographic methods enabled determination of a chemical character of the extracts obtained. The chemical compounds identified in the prepared propolis extracts are analogous to the results described by Sahinler and Kaftanogl [30] as well as by Anjum et al. [31], showing high concentration of the aromatic acids, hydrocarbons, alcohols, polyphenols and fatty acids. The presence of phenolic compounds in the propolis extracts is particularly promising when its anticancer activity is considered [32].

The biological analysis of the selected systems showed that the prolonged 24 h incubation of cells with propolis significantly affected the cell viability measured via MTT and SRB assays. Differences between groups, based on the propolis type or extraction type, were not statistically significant. This may confirm the hypothesis that differences in the chemical composition of the extracts obtained did not influence the general biological effect induced by them. It should be emphasized that higher concentrations of the propolis extracts (500 and 1000 μg/mL) significantly affected the viability of normal HGFs as well. For this reason, only the extract concentration of 100 μg/mL could be considered as effective selectively in cancer cells. Similar results demonstrating the cytotoxic effect of propolis on normal human fibroblasts were obtained by Tyszka-Czochara et al. [33], Popova et al. [13] and in our previous study [10]. Moreover, the study presented by Popova et al. revealed the similar chemical profile of the propolis sample (mainly flavanones and dihydroflavonols, as well as a series of esters of p-coumaric acid, ferulic acid, benzoic acid and fatty acids (palmitic acid, linoleic acid, oleic acid) compared to the extracts analyzed in our study [13]).

Additionally, due to the polyphenolic content of propolis, the anti-inflammatory activity of the extracts prepared was verified on macrophage models commonly used in case of natural compounds [34,35]. Szliszka et al. suggested that phenolic compounds may be responsible for a crucial contribution of Brazilian green propolis in the modulation of chemokine-mediated inflammation [34]. In our study, the impairment of the cellular proliferation and mitochondrial activity observed in macrophage-like cell line (P388-D1) suggested a possible anti-inflammatory activity of the prepared extracts. Here we have observed that the effect was dependent on the cytotoxic effect of propolis extracts applied.

In the future, the preliminary results reported in this research should be used to select the ethanol and hexane-ethanol extraction as the most effective methods of propolis extraction to obtain chemically complex and biologically active products. The prepared extracts should become a subject of an in-depth analysis aimed at the identification of the most active components and at the investigation of a precise molecular mechanism of their anticancer and anti-inflammatory action. In addition, the selected natural extracts could be combined with conventional chemotherapeutic regimens in order to propose safer and more effective treatment of cancer [36]. Finally, functional polymer microparticles for encapsulation of biologically active compounds could be designed and manufactured [37].

4. Materials and Methods

4.1. Material

The research materials were propolis samples originating from three different regions in Poland (Table 4). Raw propolis was collected from beehives manually. Before processing it was stored at room temperature under dark conditions.

Table 4. Geographical origin of the Polish propolis examined.

Symbol	Region of Origin	The Most Abundant Plants in the Region	Bee Species
P1	Podlasie (Hajnowka)	spruce (*Picea abies L.*)—30%, pine (*Pinus sylvestris L.*)—27%, alder (*Alnus glutinosa L.*)—20%, sessile oak (*Quercus petraea L.*)—10%, silver birch (*Betula pendula L.*)—7%	*Apis mellifera carnica x Apis mellifera caucasica*
P2	Mazovia (Ciechanow)	pine (*Pinus sylvestris L.*)—70%, alder (*Alnus glutinosa L.*)—10%, sessile oak (*Quercus petraea L.*)—10%, silver birch (*Betula pendula L.*)—7%	*Apis mellifera carnica*
P3	West Pomerania (Miedzyzdroje)	pine (*Pinus sylvestris L.*)—75%, alder (*Alnus glutinosa L.*)—5%, beech (*Fagus sylvatica L.*)—5%, sessile oak (*Quercus petraea L.*)—5%, silver birch (*Betula pendula L.*)—4%	*Apis mellifera mellifera*

4.2. Extraction

Ethanol, ethanol-hexane, hexane and hexane-ethanol extracts of Polish propolis were prepared according to the procedure illustrated in Figure 5. For this purpose, 5 g of raw propolis was cut into small pieces, dissolved in 50 mL of 70% ethanol (POCH, Poland) or 50 mL of hexane (POCH, Poland) and stirred for 48 h at room temperature under dark conditions, using a magnetic stirrer (Big-squid, IKA, Germany). Subsequently, the samples were centrifuged at 10,500 rpm for 10 min at room temperature, using a 5804 centrifuge (Eppendorf, Germany). The supernatant obtained was named ethanol extract of propolis (EEP) and hexane extract of propolis (HEP). Then, the residue was extracted one more time with ethanol or hexane to obtain EEP_II or HEP_II, respectively. Subsequently, the residue left after ethanol extraction was treated twice with hexane to obtain ethanol-hexane extracts (EHEP and EHEP_II). The residue left after hexane extraction was dissolved twice with 70% ethanol to obtain hexane-ethanol extracts (HEEP and HEEP_II). Non-dissolved residues were discarded.

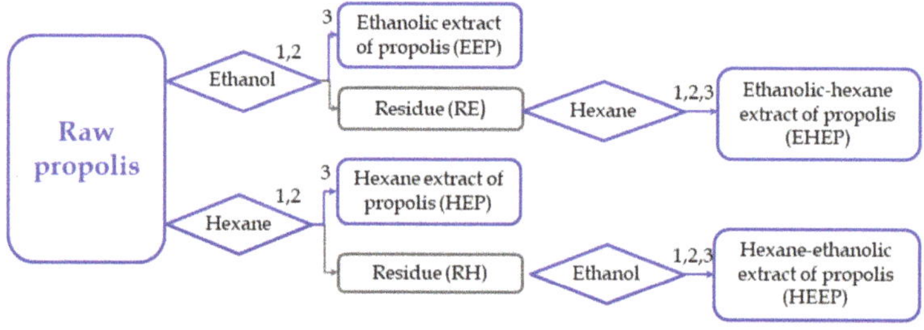

Figure 5. Schematic illustration of procedure for propolis extract preparation; EEP—ethanol extract of propolis, EHEP—ethanol-hexane extracts of propolis, HEP—hexane extract of propolis, HEEP—hexane-ethanol extracts of propolis, RE—residue after extraction with ethanol, RH—residue after extraction with hexane.

The extracts were evaporated to dryness at 40 °C using a RV 10 rotary vacuum evaporator (IKA, Germany) and stored at 4 °C under dark conditions. After evaporation, the samples obtained were weighted using an analytical balance: WPS 510/C/2 (Radwag, Poland); extraction yields were expressed in percentage as a ratio of the mass of the sample after evaporation to the mass of the propolis material before extraction. The samples obtained after second extraction with the same solvent (EEP_II, HEP_II, EHEP_II, HEEP_II) were not subjected to further analysis due to their small quantity. Then, the samples were dissolved in methanol (POCH, Poland) at a concentration of 1 mg/mL (for chemical studies) or in DMSO (POCH, Poland) at a concentration of 100 mg/mL (for biological studies).

4.3. Total Polyphenol Content

The total soluble phenolic compounds in the samples were determined using the Folin–Ciocalteu colorimetric method [38]. For this purpose, 100 µL of analyzed propolis extract was dissolved in methanol (1 mg/mL) and then mixed with 900 µL of distilled water and 100 µL of Folin and Ciocalteu's phenol reagent (Sigma-Aldrich, Poland). After 5 min of incubation, 1 mL of 7% Na_2CO_3 (POCH, Poland) and 400 µL of distilled water were added. Subsequently, the mixture was incubated for 2 h and the absorbance was measured at 765 nm using a UV-Vis spectrophotometer: SP 8001 (Metertech, Norway). Gallic acid (Sigma-Aldrich, Poland) was used as a standard. The results were expressed in mg of gallic acid equivalent per g of propolis extract (mg GAE/g). The minimum number of measurements for each extract was n = 9.

4.4. Total Flavonoid Content

The total flavonoid contents in the samples were determined using an aluminum chloride method [39]. Briefly, 100 µL of propolis extract dissolved in methanol (1 mg/mL) was mixed with 100 µL of 2% $AlCl_3$ (Sigma-Aldrich, Poland). After 15 min of incubation, the absorbance was measured at 435 nm using a UV-Vis spectrophotometer: SP 8001 (Metertech, Norway). Quercetin (Sigma-Aldrich, Poland) was used as a standard. The results were expressed in mg of quercetin equivalent per g of propolis extract (mg QE/g). The minimum number of measurements for each extract was n = 9.

4.5. GC-MS Analysis

The propolis extracts obtained (EEP, HEP and HEEP) were evaluated in terms of a low-molecular-weight compound content by means of derivatization with N,O-bis (trimethylsilyl) trifluoroacetamide (BSTFA) silylation approach on gas chromatography, coupled with mass spectrometry (Shimadzu GC-MS QP 2020, Shimadzu, Kyoto, Japan). Each of the extracts was evaporated under reduced pressure. Then, 500 µL of pyridine and 50 µL of BSTFA were added to all samples. The mixture was placed in a vial and heated for 15 min at 70 °C. Separation was achieved using Zebron ZB-5 capillary column with a length of 30 m, inner diameter of 0.25 mm, and film thickness of 0.25 µm (Phenomenex, Torrance, CA, USA). The GC-MS analysis was performed according to the following parameters: scan mode with mass range from 40 to 1050 *m/z* in electronic impact (EI) mode at 70 eV; mode at 10 scan s^{-1} mode. Analyses were conducted using helium as a carrier gas at a flow rate of 1.0 mL min^{-1} in a split ratio of 1:20 and the following program: (a) 100 °C for 1 min; (b) rate of 2.0 °C min^{-1} from 100 to 190 °C; (c) rate of 5 °C min^{-1} from 190 to 300 °C. An injector was held at 280 °C, respectively. Compounds were identified by using two different analytical methods that compare: retention times with authentic chemicals (Supelco C7-C40 Saturated Alkanes Standard), and obtained mass spectra with available library data (Willey NIST 17, match index >90%).

Fatty Acids Composition

The lipid fraction was obtained according to the previously described method [40]. In the next step, the extracted nonpolar fraction, approx. 30 mg, was saponified (10 min at 75 °C) with 2 mL of 0.5 M KOH/MeOH solution and subjected to methylation (10 min at 75 °C) using 2 mL of 14% (*v/v*) BF3/MeOH (Sigma-Aldrich, St. Louis, MO, USA). Subsequently, water was added to reaction mixture and methyl esters of fatty acids were extracted with 10 mL of hexane (UQF Wroclaw, Poland), then washed with 10 mL 10% sodium bicarbonate (UQF Wroclaw, Poland) and desiccated with anhydrous sodium sulphate. The organic phase was evaporated under reduced pressure and stored at −27 °C until chromatographical analysis. The FAME profile was assessed using gas chromatograph coupled with a mass spectrometer (Shimadzu GCMS QP 2020, Shimadzu, Kyoto, Japan). Separation was achieved using Zebron ZB-FAME capillary column with a length of 60 m, inner diameter of 0.20 mm, and film thickness of 0.20 μm (Phenomenex, Torrance, CA, USA). The GC-MS analysis was according to the following parameters: scan mode with mass range from 40 to 400 *m/z* in electronic impact (EI) mode at 70 eV; mode at 3 scan s^{-1} mode. Analyses were conducted using helium as a carrier gas at a flow rate of 1.8 mL min^{-1} in a split ratio of 1:10 and the following program: (a) 80 °C for 2 min; (b) rate of 3.0 °C min^{-1} from 80 to 180 °C; (c) rate of 8 °C min^{-1} from 180 to 240 °C. An injector was held at 280 °C, respectively. Compounds were identified by using two different analytical methods that compare: retention times with authentic chemicals (Supelco 37 Component FAME Mix), and obtained mass spectra with available library data (Willey NIST 17, match index >90%).

4.6. DPPH Free Radical Scavenging Activity

The 2,2-diphenyl-1-picrylhydrazyl (DPPH) free radical scavenging activity was determined using a method described by Yang et al. [39]. For this purpose, 100 μL of propolis extract dissolved in methanol (10, 20, 50, 100, 150, 200 μg/mL) was placed into a 96 well plate (Nunc, Denmark) and 100 μL of 0.2 mM DPPH solution (Sigma-Aldrich, Poland) was added. After 15 min of incubation, the absorbance was measured at 517 nm using a multiwell-plate reader (EnSpire Multimode Reader, Perkin Elmer, USA). Ascorbic acid (P.P.H. STANLAB Sp.J., Poland) was used as a standard. The percentage inhibition capacity was calculated from the following equation:

$$\text{percentage inhibition} = (A_0 - A_1)/(A_0 \times 100),$$

where A_0 is the absorbance of the control group and A_1 is the absorbance of the extracts.

4.7. Biological Characterisation

Taking into account the results of the chemical analyses, hexane extracts (HEP) and ethanol-hexane extracts (EHEP) were excluded from further studies. Biological analyses were conducted only for ethanol extracts (EEP) and hexane-ethanol extracts (HEEP), which were characterized by the highest TPC, TFC and DPPH free radical scavenging activity.

4.7.1. Cell Culture

Human squamous cell carcinomas derived from tongue (SCC-25 cell line, ATCC CRL-1628, ATCC, USA) were cultured in a 1:1 mixture of Dulbecco's Modified Eagle's Medium (DMEM) and Ham's F12 medium (Lonza, Switzerland) supplemented with 10% fetal bovine serum (FBS, Sigma-Aldrich, Poland), and antibiotics: penicillin/streptomycin (Sigma-Aldrich, Poland), as recommended by ATCC.

Human gingival fibroblasts (HGFs) were mechanically isolated from a fragment of gingival tissue (1–2 mm) in healthy patients, according to the procedure described by Dominiak and Saczko [41]. The biopsies were provided by the Department of Dental Surgery at the Wroclaw Medical University in accordance with the requirements of the Bioethics Commission of Wroclaw Medical University (Bioethical Committee approval, No.: KB-8/2010). The fragment of tissue was taken by a scalpel and immediately placed on Petri dishes (60 mm, Nunc, Denmark) with DMEM (Sigma-Aldrich, Poland) containing 10% FBS (Sigma-Aldrich, Poland) and antibiotics: penicillin/streptomycin (Sigma-Aldrich, Poland).

Murine macrophage-like cells (P388-D1 cell line, ATCC CCL-46, ATCC, USA) were cultured in a 1:1 mixture of DMEM and RPMI 1640 medium (Lonza, Switzerland) supplemented with 10% FBS (Sigma-Aldrich, Poland) and antibiotics: penicillin/streptomycin (Sigma-Aldrich, Poland).

All cell lines were incubated in a humidified atmosphere at 37 °C and 5% CO_2. After trypsinization with 0.25% trypsin-EDTA (Sigma-Aldrich, Poland), the cells were passaged and grown in 25 cm^2 flasks (Equimed, Poland). In order to evaluate cytotoxicity of the extracts tested, cells were seeded into a 96-well plate (Nunc, Denmark). After 24 h, the culture medium was removed and then propolis extracts, diluted with an appropriate culture medium (100, 500 and 1000 μg/mL), were added for 5 min or 24 h. MTT and SRB assays were performed 24 h later. All results were referred to the untreated control cells.

4.7.2. MTT Assay

To evaluate cytotoxicity of propolis extracts (EEP and HEEP) on the basis of differences in mitochondrial function, 3-(4,5-dimethyl-2-thiazolyl)-2,5-diphenyl-2*H*-tetrazolium bromide (MTT) assay was performed. Cells were incubated for 90 min with 100 μL of the MTT reagent (Sigma-Aldrich, Poland) at 37 °C. Then, formazan crystals were dissolved by addition of 100 μL of acidic isopropanol and by mixing. The absorbance was measured at 570 nm using a multiwell plate reader (EnSpire Multimode Reader, Perkin Elmer, USA). The results were expressed as the percentage of treated cells with altered mitochondrial function in relation to untreated control cells with normal mitochondrial activity, considered as 100%.

4.7.3. SRB Assay

To evaluate cytotoxicity of propolis extracts on the basis of differences in total protein content in cells, sulforhodamine B (SRB) assay was performed. The protocol was based on the procedure described in [42]. Cell monolayers were fixed with 10% (vol/vol) trichloroacetic acid (Roth, Poland) for 1 h at 4 °C, subsequently washed (five times) in cold water and desiccated. Cell staining was performed for 30 min using 0.4% SRB (Sigma-Aldrich, Poland) in 1% acetic acid (Sigma-Aldrich, Poland) at room temperature. After incubation, the excess of dye was removed by means of washing with 1% (*v/v*) acetic acid (four times). Plates were desiccated and the protein-bound dye was dissolved in 10 mM Tris base solution (pH 10.5) (BioShop, Canada). The absorbance was measured at 490 nm using a multiwell plate reader (GloMax Discover, Promega, USA). The results were expressed as the percentage of total protein content in treated cells in relation to untreated control cells.

4.8. Statistical Analysis

The results are presented as means ± standard deviation (SD) values for minimum n = 9 repeats. The results were analyzed by one-way ANOVA and $\alpha = 0.05$ using Statistica ver. 13.3 software (StatSoft, Poland). *F*-values and *p*-values were determined, the values $p \leq 0.05$ were considered as statistically significant. Tukey's HSD test was performed when ANOVA indicated statistically significant results. Additionally, for the MTT and SRB assays, the statistical significance of the differences between mean values of different groups and the untreated control group was evaluated by Student's *t*-test. The values $p \leq 0.05$ were marked with an asterisk and considered as statistically significant. Finally, for MTT and SRB assay results, three–way ANOVA test was performed to indicate, which factor (type of propolis, type of extract, extract concentration) determines significant differences between groups, $p \leq 0.05$ were considered as statistically significant.

5. Conclusions

This study has revealed differences in chemical composition and antioxidant activity of the extracts of three different types of Polish propolis obtained after extraction with ethanol, hexane and combinations of both. The products selected (EEP and HEEP) demonstrated anticancer activity in the tongue cancer cells and cytotoxicity towards murine macrophages. In addition, EEP and HEEP did not have any cytotoxic effect in the normal gingival fibroblasts when the lowest concentration was applied.

The following conclusions can be drawn on the basis of the results obtained:

- The highest total extraction yields were obtained for ethanol and hexane-ethanol extracts (EEP and HEEP);

- Total polyphenol content (TPC) and total flavonoid content (TFC) of ethanol and hexane-ethanol extracts (EEP and HEEP) were much higher than TPC and TFC of ethanol-hexane and hexane extracts (EHEP and HEP);

- Antioxidant potential of ethanol and hexane-ethanol extracts (EEP and HEEP) was much higher than that of ethanol-hexane and hexane extracts (EHEP and HEP);

- The extracts selected (EEP and HEEP) demonstrated anticancer activity in the tongue cancer cells; 24 h incubation affected cell viability and cellular proliferation significantly;

- The propolis extracts tested at higher concentrations (500 and 1000 µg/mL) impaired the proliferation of normal cells as well;

- The observed cytotoxicity of the extracts prepared towards murine macrophages requires further investigation to evaluate their possible anti-inflammatory potential.

As a final conclusion, we can select the minimal dose of 100 µg/mL of the extracts applied, which caused anticancer effect on human tongue cancer cells with limited cytotoxic effect on normal mucosal cells and simultaneous anti-inflammatory potential. However, further studies on Polish propolis are still necessary in order to thoroughly explain the molecular mechanisms of its action and to obtain promising health benefits of this versatile natural product.

Author Contributions: Conceptualization, J.W., W.W. and K.A.W.; methodology, J.W. and A.W.; validation, J.K., J.S., R.G. and K.A.W.; formal analysis, J.W., A.W. and N.P.; investigation, J.W., A.W. and N.P.; resources, J.W., W.W., J.K., J.S., M.W. and K.A.W.; writing—original draft preparation, J.W.; writing—review and editing, A.W., J.K., J.S., M.W., R.G. and K.A.W.; visualization, J.W.; supervision, J.K., J.S., R.G. and K.A.W.; project administration, J.W.; funding acquisition, J.W., W.W., M.W. and K.A.W. All authors have read and agreed to the published version of the manuscript.

Funding: This research was funded by the Wroclaw Medical University, grant number STM. B022.17.013 (J. Weżgowiec), and by a statutory activity subsidy from the Polish Ministry of Science and Higher Education for the Faculty of Chemistry at the Wroclaw University of Science and Technology (extracts preparation).

Acknowledgments: A. Szumny from Department of Chemistry, Wroclaw University of Environmental and Life Sciences, is acknowledged for his supervision of GC-MS analysis and for a critical revision of the manuscript.

Conflicts of Interest: The authors declare no conflict of interest.

Appendix A

Table A1. GC-MS profile of ethanol extracts of propolis (EEP) from different regions of Poland: *EEP_P1*—Podlasie (Hajnowka), *EEP_P2*—Masovia (Ciechanow), *EEP_P3*—West Pomerania Province (Miedzyzdroje).

	Substances	RT	RI $_{exp}$	RI $_{lit}$	EEP_P1[%]	EEP_P2[%]	EEP_P3[%]
1	Benzyl alcohol, TMS derivative	6.110	1155	1152	0.12	0.15	0.1
2	Benzoic Acid, TMS derivative	8.795	1246	1249	4.35	3.8	3.15
3	Glycerol, TMS	10.162	1287	1289	0.78	0.99	1.55
4	Butanedioic acid, 2TMS derivative	11.445	1320	1321	0.12	0.04	0.13
5	1-Monoacetin, 2O-TMS	11.740	1326	1324	0.13	0.07	0.08
6	4-Hydroxybenzaldehyde, TMS derivative	13.737	1373	1383	0.22	0.16	0.08
7	Hydroquinone, 2TMS derivative	15.308	1409	1408	0.24	0.2	0.08
8	Malic acid, 3TMS derivative	20.278	1511	1497	0.09	0.06	1.1
9	5-Oxoproline, TMS derivative	20.680	1521	1527	0.1	0.09	-
10	Vanillin, TMS derivative	21.622	1536	1530	1.48	0.5	0.37
11	Cinnamic acid, TMS derivative	22.048	1545	1542	0.12	0.14	0.25
12	4-Hydroxybenzoic acid, 2TMS derivative	26.623	1634	1635	0.27	0.22	0.08
13	Dodecanoic acid, TMS	27.858	1658	1655	0.13	0.09	-
14	β-D-Xylopyranose, 4TMS derivative	34.330	1784	1777	0.17	-	-
15	o-Coumaric acid, 2TMS derivative	34.975	1797	1815	0.26	0.26	0.09
16	D-Psicofuranose	37.465	1848	1837	1.28	1.2	0.76
17	D-Fructose, 5TMS derivative	37.855	1856	1867	7.4	9.55	8.51
18	D-Sorbitol, 6TMS derivative	41.075	1922	1920	0.22	0.09	7.15
19	D-Glucose, 5TMS derivative	41.635	1934	1928	4.82	8.5	-

Table A1. Cont.

	Substances	RT	RI $_{exp}$	RI $_{lit}$	EEP_P1[%]	EEP_P2[%]	EEP_P3[%]
20	4-Coumaric acid, 2TMS derivative	42.223	1947	1949	10.74	13.68	1.34
21	D-Glucitol, 6TMS derivative	43.902	1982	1980	0.44	0.34	0.18
22	Gallic acid, 4TMS derivative	44.118	1986	1987	0.39	0.1	0.21
23	Salicylic acid, trimethylsilyl ether, benzyl ester	45.910	2028	2025	0.55	0.49	0.11
24	D-Mannopyranose, 5TMS derivative	46.347	2038	2037	4.87	8.84	7.8
25	D-Gluconic acid, 6TMS derivative	46.875	2052	2043	0.16	0.16	0.09
26	Palmitic Acid, TMS derivative	47.070	2057	2050	0.85	0.7	0.69
27	Isoferulic acid, 2TMS derivative	48.435	2090	2081	0.17	0.67	2.7
28	Ferulic acid, 2TMS derivative	48.950	2103	2103	4.79	2.95	2.69
29	Myo-Inositol, 6TMS derivative	49.822	2132	2129	0.18	0.1	1.17
30	Phtalic acid derivative*	50.387	2150	-	1.07	1.19	4.02
31	Caffeic acid, 3TMS derivative	50.548	2157	2155	1.06	2.15	-
32	Unknown	51.908	2202	-	1.54	2.48	0.19
33	13-Octadecenoic acid, (E)-, TMS derivative	52.383	2222	2228	0.59	0.72	0.88
34	3,7,11,15-Tetramethyl-2,6,10,14-hexadecatetraene-1-ol trimethylsilyl ether	52.697	2236	2234	1.65	2.7	0.25
35	Tricosane	54.272	2300	2300	1.14	0.42	0.08
36	Unknown	55.733	2370	-	-	-	1.21
37	Unknown	56.810	2424	-	-	-	2.4
38	Pterostilbene, trimethylsilyl ether	57.507	2462	-	0.39	0.48	0.82
39	Pentacosane	58.223	2501	2506	0.83	0.26	0.64
40	Unknown	58.548	2519	-	4.41	4.32	1.5
41	Ethyl trans-caffeate, bis(tert-butyldimethylsilyl) ether	58.675	2527	2547	0.11	0.08	0.16
42	Bisphenol C*	59.085	2551	-	1.47	3.58	5.2
43	1-Docosanol, TMS derivative	59.220	2558	2557	0.25	0.48	0.67
44	Unknown	59.577	2579	-	0.3	0.19	0.45
45	Butanoic acid, 4-methoxy-2-nitro-, 2,6-bis(1,1-dimethylethyl)-4-methoxyphenyl ester	59.990	2604	2595	0.6	0.25	3.25
46	Unknown	60.465	2635	-	0.45	0.71	0.56
47	Behenic acid, TMS derivative	60.650	2645	2644	1.05	0.37	-
48	Unknown	60.938	2664	-	0.82	1.22	2.59
49	Unknown	61.132	2675	-	0.27	0.59	1.51
50	Unknown	61.280	2685	-	2.12	0.96	0.9
51	Maltose, 8TMS derivative, isomer 2	61.415	2693	2693	1.56	2.96	5.33
52	n-Heptacosane	61.527	2700	2700	0.82	0.2	-
53	Sucrose, 8TMS derivative	61.700	2712	2712	2.38	0.43	0.23
54	D-(+)-Turanose, octakis(trimethylsilyl) ether, methyloxime (isomer 1)	61.925	2727	2724	0.53	1.96	4.92
55	Maltose, OTMS	62.035	2735	2733	0.38	0.18	0.11
56	D-Cellobiose, (isomer 2), 8TMS derivative	62.258	2749	2762	1.7	2.64	0.43
57	Naringenin, O,O'-bis(trimethylsilyl)-	62.612	2772	2778	1.07	3.48	6.06
58	Unknown	63.180	2813	-	-	-	2.08
59	Isosakuranetin, TMS derivative	63.345	2821	2818	0.74	1.43	0.2
60	Lignoceric acid, TMS derivative	63.648	2842	2838	7.82	3.45	0.35
61	Sakuranetin, TMS derivative	64.200	2882	2877	0.96	0.58	0.26
62	Catechine, 5TMS derivative	64.872	2932	2938	0.17	0.13	-
63	Gettibiose, TMS derivative	65.712	2991	2989	0.39	0.38	0.29
64	Triacontane	65.933	3009	3003	0.44	0.36	0.27
65	Pectolinaringenin, TMS derivative	66.108	3021	3037	0.54	0.41	0.38
66	Hexacosanoic acid, TMS derivative	66.367	3041	3039	0.7	0.18	1.84
67	Nonacosan-10-ol, O-TMS	66.840	3078	3048	2.59	0.64	-
68	Nonacosan-9-ol, O-TMS	66.925	3085	3053	2.13	1.06	0.91
69	Hentriacontane	67.132	3100	3103	0.36	0.2	0.83
70	Kaempferol, 4TMS	67.298	3114	3112	0.29	0.86	0.77
71	Trimethylsilyl octacosanoate, TMS derivative	69.220	3256	3229	0.52	0.2	0.52
72	Methyl triacontyl ether	69.512	3275	3233	8.38	0.97	0.44

RI (retention time); *RI$_{exp}$* and *RI$_{lit}$* indicate retention indices based on experiments and literature, respectively.

Table A2. GC-MS profile of hexane extracts of propolis (HEP) from different regions of Poland: *HEP_P1*—Podlasie (Hajnowka), *HEP_P2*—Masovia (Ciechanow), *HEP_P3*—West Pomerania Province (Miedzyzdroje).

	Substances	RT	RI $_{exp}$	RI $_{lit}$	HEP_P1[%]	HEP_P2[%]	HEP_P3[%]
1	Benzoic Acid, TMS derivative	8.765	1246	1249	8.1	8.47	13.19
2	Glycerol, TMS	10.137	1287	1289	0.82	0.47	0.6
3	Decanoic acid, TMS derivative	17.830	1460	1450	0.15	1.5	0.14
4	Vanillin, TMS derivative	21.590	1536	1530	0.79	0.36	0.27
5	Cinnamic acid, TMS derivative	21.998	1545	1542	0.17	0.26	0.87
6	Dodecanoic acid, TMS	27.828	1658	1655	0.31	0.37	0.36
7	β-D-Xylopyranose, 4TMS derivative	34.300	1784	1777	0.25	1.36	-
8	D-Fructose, 5TMS derivative	37.770	1856	1867	0.36	0.36	0.38
9	4,7,10-Hexadecatrienoic acid, methyl ester	40.008	1899	1902	0.36	0.13	-
10	D-Sorbitol, 6TMS derivative	41.032	1922	1920	0.2	-	-
11	4-Coumaric acid, 2TMS derivative	42.122	1947	1949	0.33	0.19	0.21
12	Salicylic acid, trimethylsilyl ether, benzyl ester	45.865	2028	2025	0.87	1.84	0.63
13	Palmitic Acid, TMS derivative	47.035	2057	2050	2.48	3.06	2.96
14	Ferulic acid, 2TMS derivative	48.887	2103	2103	1.2	0.76	3.01
15	Phtalic acid derivative*	50.357	2150		1.04	1.31	0.21
16	Methyl caffeate, 2TMS derivative	51.882	2201	1997	1.02	1.67	0.07
17	13-Octadecenoic acid, (E)-, TMS derivative	52.365	2222	2228	1.7	4.3	8.05
18	3,7,11,15-Tetramethyl-2,6,10,14-hexadecatetraene-1-ol trimethylsilyl ether	52.677	2236	2234	0.77	1.21	-
19	Stearic acid, TMS derivative	53.047	2249	2246	0.55	0.89	1.15
20	Tricosane	54.255	2300	2300	3.55	2.67	3.05
21	Arachidic acid, TMS derivative	57.212	2446	2447	0.43	0.4	0.63
22	Pterostilbene, trimethylsilyl ether	57.723	2462	-	0.34	0.28	0.13
23	Pentacosane	58.200	2501	2506	5.24	4.6	4.00
24	Ethyl trans-caffeate, bis(tert-butyldimethylsilyl) ether	58.518	2527	2547	0.84	1.54	1.2
25	Bisphenol C*	59.043	2551		0.3	0.52	1.28
26	Behenic acid, TMS derivative	60.617	2643	2644	2.11	1.42	2.19
27	Unknown	61.255	2683	-	2.02	1.56	1.77
28	n-Heptacosane	61.512	2700	2700	14.11	14.31	12.45
29	Maltose, OTMS	62.007	2735	2733	0.38	0.19	0.1
30	D-Cellobiose, (isomer 2), 8TMS derivative	62.290	2749	2762	0.37	1.16	0.55
31	Octacosane	63.003	2798	2800	0.56	1.09	0.68
32	Lignoceric acid, TMS derivative	63.620	2842	2838	7.72	5.88	12.02
33	Sakuranetin, TMS derivative	64.170	2882	2877	0.53	-	0.23
34	Nonacosane	64.438	2899	2900	7.15	9.24	7.34
35	Hexacosanoic acid, TMS derivative	66.345	3041	3039	0.63	0.67	2.21
36	Nonacosan-10-ol, O-TMS	66.817	3078	3048	5.01	3.31	3.27
37	Nonacosan-9-ol, O-TMS	66.907	3085	3053	4.36	3.24	2.7
38	Hentriacontane	67.115	3100	3103	4.25	6.66	4.62
39	Myristic acid, 9-hexadecenyl ester, (Z)-	68.128	3177	3130	0.6	0.34	0.17
40	Trimethylsilyl octacosanoate, TMS derivative	69.188	3256	3229	1.21	1.01	0.5
41	Methyl triacontyl ether	69.493	3276	3233	16.8	11.36	6.82

RI (retention time); $RI_{exp.}$ and $RI_{lit.}$ indicate retention indices based on experiments and literature, respectively.

Table A3. GC-MS profile of hexane-ethanol extracts of propolis (HEEP) from different regions of Poland: *HEEP_P1*—Podlasie (Hajnowka), *HEEP_P2*—Masovia (Ciechanow), *HEEP_P3*—West Pomerania Province (Miedzyzdroje).

	Substances	RT	RI $_{exp}$	RI $_{lit}$	HEEP_P1[%]	HEEP_P2[%]	HEEP_P3[%]
1	Benzyl alcohol, TMS derivative	6.112	1155	1152	0.09	0.04	0.08
2	Benzoic Acid, TMS derivative	8.797	1246	1249	3.16	1.24	1.27
3	Cinnamaldehyde	9.905	1272	1274	0.04	0.08	0.06
4	Glycerol, TMS	10.170	1287	1289	0.87	1.05	1.8
5	Butanedioic acid, 2TMS derivative	11.465	1320	1321	0.12	0.05	0.18
6	1-Monoacetin, 2O-TMS	11.748	1326	1324	0.14	0.06	0.09
7	4-Hydroxybenzaldehyde, TMS derivative	13.748	1373	1383	0.26	0.19	0.14
8	Hydroquinone, 2TMS derivative	15.325	1409	1408	0.4	0.22	0.13
9	Cinnamyl alcohol, trimethylsilyl ether	16.273	1422	1428	-	0.02	0.03
10	Malic acid, 3TMS derivative	20.282	1511	1497	0.12	0.12	0.81
11	5-Oxoproline, TMS derivative	20.693	1521	1527	0.12	0.06	0.05
12	Vanillin, TMS derivative	21.640	1537	1530	1.82	0.46	0.27
13	Cinnamic acid, TMS derivative	22.043	1545	1542	0.07	0.05	0.15
14	3,4-Dihydroxybenzaldehyde,	26.045	1622	1612	-	0.13	0.27

Table A3. *Cont.*

	Substances	RT	RI $_{exp}$	RI $_{lit}$	HEEP_P1[%]	HEEP_P2[%]	HEEP_P3[%]
15	4-Hydroxybenzoic acid, 2TMS derivative	26.638	1634	1635	0.48	0.22	0.18
16	Dodecanoic acid, TMS	27.668	1658	1655	0.14	0.02	0.08
17	β-D-Xylopyranose, 4TMS derivative	34.563	1784	1777	0.13	0.03	0.03
18	o-Coumaric acid, 2TMS derivative	34.968	1797	1815	0.28	0.25	0.09
19	4-Methoxycinnamic acid, TMS derivative	36.602	1830	1833	-	0.08	0.26
20	D-Psicofuranose	37.475	1848	1837	1.46	1.59	2.24
21	D-Fructose, 5TMS derivative	37.865	1856	1866	12.58	11.92	2.03
22	D-Sorbitol, 6TMS derivative	41.092	1922	1920	0.16	0.1	-
23	D-Glucose, 5TMS derivative	41.658	1934	1928	7.14	11.9	8.27
24	4-Coumaric acid, 2TMS derivative	42.252	1947	1949	16.74	13.52	10.74
25	D-Glucitol, 6TMS derivative	43.917	1982	1980	0.43	0.25	0.44
26	Gallic acid, 4TMS derivative	44.125	1987	1987	0.68	0.13	-
27	Salicylic acid, trimethylsilyl ether, benzyl ester	45.932	2028	2025	0.1	0.08	0.2
28	D-Mannopyranose, 5TMS derivative	46.353	2038	2037	6.33	13.13	9.13
29	D-Gluconic acid, 6TMS derivative	46.885	2052	2043	0.15	0.2	0.08
30	Palmitic Acid, TMS derivative	47.083	2057	2050	0.49	0.18	0.09
31	Isoferulic acid, 2TMS derivative	48.453	2090	2081	0.09	0.78	2.95
32	Ferulic acid, 2TMS derivative	48.965	2103	2103	7.93	3.02	3.86
33	Myo-Inositol, 6TMS derivative	49.845	2132	2129	0.09	0.13	1.3
34	Phtalic acid derivative*	50.403	2150	-	0.7	0.24	0.03
35	Caffeic acid, 3TMS derivative	50.565	2157	2155	0.89	0.83	0.03
36	Linoleic acid, TMS	51.922	2203	2212	1.31	2.48	4.6
37	13-Octadecenoic acid, (E)-, TMS derivative	52.392	2222	2228	0.17	1.77	0.06
38	3,7,11,15-Tetramethyl-2,6,10,14-hexadecatetraene-1-ol trimethylsilyl ether	52.703	2236	2234	1.01	0.07	0.07
39	2′,6′-Dihydroxy 4′-methoxydihydrochalcone, trimethylsilyl ether	52.717	2418	2405	-	1.99	0.1
40	Pterostilbene, trimethylsilyl ether	57.520	2462	-	0.43	0.77	2.58
41	Pentacosane	58.563	2501	2506	5.85	0.44	0.87
42	Ethyl trans-caffeate, bis(tert-butyldimethylsilyl) ether	58.560	2527	2547	-	3.55	2.22
43	Bisphenol*	59.088	2551		1.65	0.13	0.12
44	1-Docosanol, TMS derivative	59.223	2558	2557	0.19	4.51	4.76
45	Butanoic acid, 4-methoxy-2-nitro-, 2,6-bis(1,1-dimethylethyl)-4-methoxyphenyl ester	59.998	2604	2595	0.72	0.32	0.67
46	Unknown	60.475	2635		0.99	0.84	3.44
47	Behenic acid, TMS derivative	60.550	2645	2644	0.31	0.94	0.83
48	Unknown	60.947	2664		0.95	1.73	2.77
49	Unknown	61.285	2685		3.22	0.69	1.67
50	Maltose, 8TMS derivative, isomer 2	61.422	2693	2693	1.83	3.42	5.26
51	n-Heptacosane	61.537	2700	2700	0.21	0.11	0.31
52	Sucrose, 8TMS derivative	61.705	2712	2712	5.88	0.52	2.61
53	D-(+)-Turanose, octakis(trimethylsilyl) ether, methyloxime (isomer 1)	61.937	2727	2724	0.71	2.66	4.98
54	Maltose, OTMS	62.040	2735	2733	0.29	0.13	0.12
55	D-Cellobiose, (isomer 2), 8TMS derivative	62.262	2749	2762	1.51	0.32	1.3
56	Naringenin, O,O′-bis(trimethylsilyl)-	62.617	2772	2778	1.14	4.1	5.95
57	Isosakuranetin, TMS derivative	63.352	2821	2818	0.94	1.37	2.13
58	Lignoceric acid, TMS derivative	63.650	2842	2838	1.22	1.68	0.44
59	Sakuranetin, TMS derivative	64.215	2882	2877	1.75	0.64	0.11
60	Catechine, 5TMS derivative	64.880	2932	2938	0.33	0.11	0.07
61	Gettibiose, TMS derivative	65.720	2991	2989	0.76	0.36	0.66
62	Triacontane	65.948	3009	3003	0.78	0.4	0.35
63	Hexacosanoic acid, TMS derivative	66.375	3041	3039	0.5	0.69	2.04
64	Unknown	66.975	3085	-	0.66	0.61	0.74
65	Hentriacontane	67.137	3100	3103	0.49	0.23	0.77

RI (retention time); *RI$_{exp}$.* and *RI$_{lit}$.* indicate retention indices based on experiments and literature, respectively.

Table A4. Percentage of fatty acids in hexane extracts of propolis (HEP) from different regions of Poland: *HEP_P1*—Podlasie (Hajnowka), *HEP_P2*—Masovia (Ciechanow), *HEP_P3*—West Pomerania Province (Miedzyzdroje).

	Substances	RT	RI_{exp}	RI_{lit}	HEP_P1[%]	HEP_P2[%]	HEP_P3[%]
1	Benzoic acid, methyl ester	15.055	1623	1612	0.69	2.73	0.92
2	Lauric acid, methyl ester	18.820	1877	1804	1.85	2.32	1.28
3	cis-9-Tetradecenoic acid, methyl ester	25.410	2116	2026	5.16	1.94	2.30
4	Pentadecanoic acid, methyl ester	27.530	2194	2108	1.75	0.79	0.70
5	Hexadecanoic acid, methyl ester	29.590	2271	2208	31.17	27.63	20.54
6	Heptadecanoic acid, methyl ester	32.790	2666	2309	9.68	7.67	6.34
7	Octadecanoic acid, methyl ester	34.345	2394	2418	6.40	5.28	4.23
8	Oleic acid, methyl ester	35.050	2488	2434	18.18	25.30	25.90
9	cis-11-Octadecenoic acid, methyl ester	35.565	2512	2468	1.19	0.48	0.81
10	Linolenic acid, methyl ester	37.775	2625	2571	0.95	2.38	5.86
11	Eicosanoic acid, methyl ester	38.180	2651	2639	1.02	1.86	1.89
12	Docosanoic acid, methyl ester	40.790	2844	2835	4.70	5.71	5.00
13	Me. C20:4n3; Eicosa-(8,11,14,17)-tetraenoate <methyl>	41.060	2866	2865	2.48	2.37	2.75
14	Tetracosanoic acid, methyl ester	42.805	3067	3039	14.79	13.91	21.49

RI (retention time); RI_{exp}. and RI_{lit}. indicate retention indices based on experiments and literature, respectively.

References

1. Thomford, N.E.; Senthebane, D.A.; Rowe, A.; Munro, D.; Seele, P.; Maroyi, A.; Dzobo, K. Natural Products for Drug Discovery in the 21st Century: Innovations for Novel Drug Discovery. *Int. J. Mol. Sci.* **2018**, *19*, 1578. [CrossRef]

2. Sforcin, J.M. Biological Properties and Therapeutic Applications of Propolis. *Phytotherapy Res.* **2016**, *30*, 894–905. [CrossRef] [PubMed]

3. Carvalho, R.; Baltazar, F.; Aguiar, C.A.A. Propolis: A Complex Natural Product with a Plethora of Biological Activities That Can Be Explored for Drug Development. *Evid.-Based Complement. Altern. Med.* **2015**, *2015*, 1–29. [CrossRef] [PubMed]

4. Giannopoulou, E.; De Castro, S.L.; Marcucci, M.C. Propolis: Recent advances in chemistry and plant origin. *Apidologie* **2000**, *31*, 3–15. [CrossRef]

5. Więckiewicz, W.; Miernik, M.; Więckiewicz, M.; Morawiec, T. Does Propolis Help to Maintain Oral Health? *Evid.-Based Complement. Altern. Med.* **2013**, *2013*, 1–8. [CrossRef] [PubMed]

6. Przybyłek, I.; Karpiński, T.M. Antibacterial Properties of Propolis. *Molecules* **2019**, *24*, 2047. [CrossRef]

7. Grecka, K.; Kuś, P.M.; Okińczyc, P.; Worobo, R.; Walkusz, J.; Szweda, P. The Anti-Staphylococcal Potential of Ethanolic Polish Propolis Extracts. *Molecules* **2019**, *24*, 1732. [CrossRef]

8. Pobiega, K.; Kraśniewska, K.; Derewiaka, D.; Gniewosz, M. Comparison of the antimicrobial activity of propolis extracts obtained by means of various extraction methods. *J. Food Sci. Technol.* **2019**, *56*, 5386–5395. [CrossRef]

9. Piekarz, T.; Mertas, A.; Wiatrak, K.; Rój, R.; Kownacki, P.; Śmieszek-Wilczewska, J.; Kopczyńska, E.; Wrzoł, M.; Cisowska, M.; Szliszka, E.; et al. The Influence of Toothpaste Containing Australian Melaleuca alternifolia Oil and Ethanolic Extract of Polish Propolis on Oral Hygiene and Microbiome in Patients Requiring Conservative Procedures. *Molecules* **2017**, *22*, 1957. [CrossRef]

10. Wieczynska, A.; Wezgowiec, J.; Wieckiewicz, W.; Czarny, A.; Kulbacka, J.; Nowakowska, D.; Gancarz, R.; Wilk, K.A. Antimicrobial Activity, Cytotoxicity and Total Phenolic Content of Different Extracts of Propolis from the West Pomeranian Region in Poland. *Acta Pol. Pharm.-Drug Res.* **2017**, *74*, 715–722.

11. Wojtyczka, R.D.; Dziedzic, A.; Idzik, D.; Kępa, M.; Kubina, R.; Kabała-Dzik, A.; Smoleń-Dzirba, J.; Stojko, J.; Sajewicz, M.; Wąsik, T.J. Susceptibility of Staphylococcus aureus Clinical Isolates to Propolis Extract Alone or in Combination with Antimicrobial Drugs. *Molecules* **2013**, *18*, 9623–9640. [CrossRef]

12. Borawska, M.; Naliwajko, S.; Moskwa, J.; Markiewicz-Żukowska, R.; Puścion-Jakubik, A.; Soroczyńska, J. Anti-proliferative and anti-migration effects of Polish propolis combined with Hypericum perforatum L. on glioblastoma multiforme cell line U87MG. *BMC Complement. Altern. Med.* **2016**, *16*, 367. [CrossRef] [PubMed]

13. Popova, M.; Giannopoulou, E.; Skalicka-Woźniak, K.; Graikou, K.; Widelski, J.; Giannopoulou, E.; Kalofonos, H.P.; Sivolapenko, G.; Gaweł-Bęben, K.; Antosiewicz, B.; et al. Characterization and Biological Evaluation of Propolis from Poland. *Molecules* **2017**, *22*, 1159. [CrossRef] [PubMed]

14. Szliszka, E.; Sokół-Łętowska, A.; Kucharska, A.Z.; Jaworska, D.; Czuba, Z.; Krol, W. Ethanolic Extract of Polish Propolis: Chemical Composition and TRAIL-R2 Death Receptor Targeting Apoptotic Activity against Prostate Cancer Cells. *Evid.-Based Complement. Altern. Med.* **2013**, *2013*, 1–12. [CrossRef]

15. Yu, X.; Li, Z. MicroRNA expression and its implications for diagnosis and therapy of tongue squamous cell carcinoma. *J. Cell. Mol. Med.* **2015**, *20*, 10–16. [CrossRef] [PubMed]

16. Chan, G.C.; Cheung, K.-W.; Sze, D.M.-Y. The Immunomodulatory and Anticancer Properties of Propolis. *Clin. Rev. Allergy Immunol.* **2012**, *44*, 262–273. [CrossRef]

17. Russo, A.; Cardile, V.; Sánchez, F.; Troncoso, N.; Vanella, A.; Garbarino, J. Chilean propolis: Antioxidant activity and antiproliferative action in human tumor cell lines. *Life Sci.* **2004**, *76*, 545–558. [CrossRef]

18. Yen, C.-H.; Chiu, H.-F.; Wu, C.-H.; Lu, Y.-Y.; Han, Y.-C.; Shen, Y.; Venkatakrishnan, K.; Wang, C.-K. Beneficial efficacy of various propolis extracts and their digestive products by in vitro simulated gastrointestinal digestion. *LWT* **2017**, *84*, 281–289. [CrossRef]

19. Salehi, M.; Motallebnejad, M.; Moghadamnia, A.A.; Seyemajidi, M.; Khanghah, S.N.; Ebrahimpour, A.; Molania, T. An Intervention Airing the Effect of Iranian Propolis on Epithelial Dysplasia of the Tongue: A Preliminary Study. *J. Clin. Diagn. Res.* **2017**, *11*, ZC67–ZC70. [CrossRef]

20. Ribeiro, D.R.; Alves Ângela, V.F.; Dos Santos, E.P.; Padilha, F.F.; Gomes, M.Z.; Rabelo, A.S.; Cardoso, J.; Massarioli, A.P.; Alencar, S.M.; De Albuquerque-Júnior, R.L.C.; et al. Inhibition of DMBA-induced Oral Squamous Cells Carcinoma Growth by Brazilian Red Propolis in Rodent Model. *Basic Clin. Pharmacol. Toxicol.* **2015**, *117*, 85–95. [CrossRef]

21. Lee, Y.-T.; Don, M.-J.; Hung, P.-S.; Shen, Y.-C.; Lo, Y.-S.; Chang, K.-W.; Chen, C.-F.; Ho, L.-K. Cytotoxicity of phenolic acid phenethyl esters on oral cancer cells. *Cancer Lett.* **2005**, *223*, 19–25. [CrossRef] [PubMed]

22. Kuo, Y.-Y.; Jim, W.-T.; Su, L.-C.; Chung, C.-J.; Lin, C.-Y.; Huo, C.; Tseng, J.-C.; Huang, S.-H.; Lai, C.-J.; Chen, B.-C.; et al. Caffeic Acid Phenethyl Ester Is a Potential Therapeutic Agent for Oral Cancer. *Int. J. Mol. Sci.* **2015**, *16*, 10748–10766. [CrossRef] [PubMed]

23. Kuo, Y.-Y.; Lin, H.-P.; Huo, C.; Su, L.-C.; Yang, J.; Hsiao, P.-H.; Chiang, H.-C.; Chung, C.-J.; Wang, H.-D.; Chang, J.-Y.; et al. Caffeic Acid Phenethyl Ester Suppresses Proliferation and Survival of TW2.6 Human Oral Cancer Cells via Inhibition of Akt Signaling. *Int. J. Mol. Sci.* **2013**, *14*, 8801–8817. [CrossRef] [PubMed]

24. Czyżewska, U.; Siemionow, K.; Zaręba, I.; Miltyk, W. Proapoptotic Activity of Propolis and Their Components on Human Tongue Squamous Cell Carcinoma Cell Line (CAL-27). *PLOS ONE* **2016**, *11*, e0157091. [CrossRef]

25. Asgharpour, F.; Moghadamnia, A.A.; Zabihi, E.; Kazemi, S.; Namvar, A.E.; Gholinia, H.; Motallebnejad, M.; Nouri, H.R. Iranian propolis efficiently inhibits growth of oral streptococci and cancer cell lines. *BMC Complement. Altern. Med.* **2019**, *19*, 266–268. [CrossRef]

26. Sawaya, A.C.; Cunha, I.B.D.S.; Marcucci, M.C. Analytical methods applied to diverse types of Brazilian propolis. *Chem. Central J.* **2011**, *5*, 1–10. [CrossRef]

27. Rassu, G.; Cossu, M.; Langasco, R.; Carta, A.; Cavalli, R.; Giunchedi, P.; Gavini, E. Propolis as lipid bioactive nano-carrier for topical nasal drug delivery. *Colloids Surfaces B Biointerfaces* **2015**, *136*, 908–917. [CrossRef]

28. Kubilienė, L.; Laugaliene, V.; Pavilonis, A.; Maruska, A.; Majiene, D.; Barčauskaitė, K.; Kubilius, R.; Kasparaviciene, G.; Savickas, A. Alternative preparation of propolis extracts: Comparison of their composition and biological activities. *BMC Complement. Altern. Med.* **2015**, *15*, 156. [CrossRef]

29. Maciejewicz, W. Isolation of Flavonoid Aglycones from Propolis by A Column Chromatography Method and Their Identification By Gc-Ms And Tlc Methods. *J. Liq. Chromatogr. Relat. Technol.* **2001**, *24*, 1171–1179. [CrossRef]

30. Anjum, S.I.; Ullah, A.; Khan, K.A.; Attaullah, M.; Khan, H.; Ali, H.; Bashir, M.A.; Tahir, M.; Rana, R.M.; Ghramh, H.A.; et al. Composition and functional properties of propolis (bee glue): A review. *Saudi J. Boil. Sci.* **2019**, *26*, 1695–1703. [CrossRef] [PubMed]

31. Sahinler, N.; Kaftanoglu, O. Natural product propolis: Chemical composition. *Nat. Prod. Res.* **2005**, *19*, 183–188. [CrossRef] [PubMed]

32. Roleira, F.M.F.; Da Silva, E.J.T.; Varela, C.; Costa, S.; Silva, T.; Garrido, J.; Borges, F. Plant derived and dietary phenolic antioxidants: Anticancer properties. *Food Chem.* **2015**, *183*, 235–258. [CrossRef] [PubMed]

33. Tyszka-Czochara, M.; Paśko, P.; Reczyński, W.; Szlósarczyk, M.; Bystrowska, B.; Opoka, W. Zinc and propolis reduces cytotoxicity and proliferation in skin fibroblast cell culture: Total polyphenol content and antioxidant capacity of propolis. *Boil. Trace Element Res.* **2014**, *160*, 123–131. [CrossRef] [PubMed]

34. Szliszka, E.; Kucharska, A.Z.; Sokół-Łętowska, A.; Mertas, A.; Czuba, Z.; Krol, W. Chemical Composition and Anti-Inflammatory Effect of Ethanolic Extract of Brazilian Green Propolis on Activated J774A.1 Macrophages. *Evid.-Based Complement. Altern. Med.* **2013**, *2013*, 1–13. [CrossRef] [PubMed]

35. Bueno-Silva, B.; Rosalen, P.L.; Alencar, S.M.; Mayer, M.P.A. Anti-inflammatory mechanisms of neovestitol from Brazilian red propolis in LPS-activated macrophages. *J. Funct. Foods* **2017**, *36*, 440–447. [CrossRef]

36. Parashar, K.; Sood, S.; Mehaidli, A.; Curran, C.; Vegh, C.; Nguyen, C.; Pignanelli, C.; Zhang, Z.; Liang, G.; Wang, Y.; et al. Evaluating the Anti-cancer Efficacy of a Synthetic Curcumin Analog on Human Melanoma Cells and Its Interaction with Standard Chemotherapeutics. *Molecules* **2019**, *24*, 2483. [CrossRef]

37. Tsirigotis-Maniecka, M.; Szyk-Warszynska, L.; Michna, A.; Warszyński, P.; Wilk, K.A. Colloidal characteristics and functionality of rationally designed esculin-loaded hydrogel microcapsules. *J. Colloid Interface Sci.* **2018**, *530*, 444–458. [CrossRef]

38. Makhafola, T.J.; Elgorashi, E.; McGaw, L.J.; Verschaeve, L.; Jn, E. The correlation between antimutagenic activity and total phenolic content of extracts of 31 plant species with high antioxidant activity. *BMC Complement. Altern. Med.* **2016**, *16*, 490. [CrossRef]

39. Yang, H.; Dong, Y.; Du, H.; Shi, H.; Peng, Y.; Li, X. Antioxidant Compounds from Propolis Collected in Anhui, China. *Molecules* **2011**, *16*, 3444–3455. [CrossRef]

40. Zali, A.G.; Ehsanzadeh, P.; Szumny, A.; Matkowski, A. Genotype-specific response of Foeniculum vulgare grain yield and essential oil composition to proline treatment under different irrigation conditions. *Ind. Crop. Prod.* **2018**, *124*, 177–185. [CrossRef]

41. Dominiak, M.; Saczko, J. Method of Primary Culture of Human Fibroblasts for Autologous Augmentation. Patent No. PL209784B1, 31 October 2011.

42. Houghton, P.; Fang, R.; Techatanawat, I.; Steventon, G.; Hylands, P.J.; Lee, C. The sulphorhodamine (SRB) assay and other approaches to testing plant extracts and derived compounds for activities related to reputed anticancer activity. *Methods* **2007**, *42*, 377–387. [CrossRef] [PubMed]

Sample Availability: Samples of the extracts of propolis prepared are available from the authors.

Article

Anticancer Activity of Novel Plant Extracts and Compounds from *Adenosma bracteosum* (Bonati) in Human Lung and Liver Cancer Cells

Ngoc Hong Nguyen [1], Qui Thanh Hoai Ta [2], Quang Thang Pham [3], Thi Ngoc Han Luong [3], Van Trung Phung [4], Thuc-Huy Duong [5] and Van Giau Vo [6,7,*]

[1] CirTech Institute, Ho Chi Minh City University of Technology (HUTECH), Ho Chi Minh City 700000, Vietnam; nn.hong@hutech.edu.vn

[2] Institute of Research and Development, Duy Tan University, Danang 550000, Vietnam; tathoaiqui@duytan.edu.vn

[3] Institute of Applied Science, Ho Chi Minh City University of Technology (HUTECH), Ho Chi Minh City 700000, Vietnam; pquangthang1@gmail.com (Q.T.P.); ngochanlt96@gmail.com (T.N.H.L.)

[4] Center for Research and Technology Transfer, Vietnam Academy of Science and Technology, Hanoi 100000, Vietnam; pvtrung@ict.vast.vn

[5] Department of Organic Chemistry, University of Education, Ho Chi Minh City 700000, Vietnam; huydt@hcmue.edu.vn

[6] Bionanotechnology Research Group, Ton Duc Thang University, Ho Chi Minh City 700000, Vietnam

[7] Faculty of Pharmacy, Ton Duc Thang University, Ho Chi Minh City 700000, Vietnam

* Correspondence: vovangiau@tdtu.edu.vn

Academic Editors: José Antonio Lupiáñez, Amalia Pérez-Jiménez and Eva E. Rufino-Palomares

Received: 30 May 2020; Accepted: 16 June 2020; Published: 24 June 2020

Abstract: Cancer is the second leading cause of death globally, and despite the advances in drug development, it is still necessary to develop new plant-derived medicines. Compared with using conventional chemical drugs to decrease the side effects induced by chemotherapy, natural herbal medicines have many advantages. The present study aimed to discover the potential cytotoxicity of ethanol extract and its derived fractions (chloroform, ethyl acetate, butanol, and aqueous) of *Adenosma bracteosum* Bonati. (*A. bracteosum*) on human large cell lung carcinoma (NCI-H460) and hepatocellular carcinoma (HepG2). Among these fractions, the chloroform showed significant activity in the inhibition of proliferation of both cancerous cells because of the presence of bioactive compounds including xanthomicrol, 5,4'-dihydroxy-6,7,8,3'-tetramethoxyflavone, and ursolic acid which were clearly revealed by nuclear magnetic resonance spectroscopy (^{1}H-NMR, ^{13}C-NMR, Heteronuclear Multiple Bond Coherence, and Heteronuclear Single Quantum Coherence Spectroscopy) analyses. According to the radical scavenging capacity, the 5,4'-dihydroxy-6,7,8,3'-tetramethoxyflavone compound (AB2) exhibited the highest anticancer activity on both NCI-H460 and HepG2 with IC_{50} values of 4.57 ± 0.32 and 5.67 ± 0.09 µg/mL respectively, followed by the ursolic acid with the lower percent inhibition at 13.05 ± 0.55 and 10.00 ± 0.16 µg/mL, respectively ($p < 0.05$). Remarkably, the AB2 compound induced to significant increase in the production of reactive oxygen species accompanied by attenuation of mitochondrial membrane potential, thus inducing the activation of caspase-3 activity in both human lung and liver cancer cells. These results suggest that *A. bracteosum* is a promising source of useful natural products and AB2 offers opportunities to develop the novel anticancer drugs.

Keywords: *Adenosma bracteosum*; extract; anti-cancer; cell line; isolated compounds; caspase-3

---

1. Introduction

A major problem of public health, cancer is one of the main causes of death globally. The prevalence of this disease is rising, however, more rapidly in Africa, Asia, and Central and South America that make up about 70% of cancer deaths in the world [1]. Many studies have been focusing on the development of agent for cancer therapies [2–4]. The chemotherapy is one of the ways to treat this disease and the advances in anticancer drugs have improved patient care. Unfortunately, the conventional chemical drugs also cause adverse side effects on normal cells/tissue, such as bone marrow function inhibition, nausea, vomiting, and alopecia [5,6]. On the other hand, natural antioxidants and many phytochemicals have been recently suggested as anti-cancer adjuvant therapies because of their anti-proliferative and pro-apoptotic properties [3,4]. Hence, the continuing search for anticancer agents/compounds from plants played a critical role to find the possible ways to have safe and to decrease the side effects induced by chemotherapy since natural herbal medicines have many advantages [7–10].

Over several decades, around 200 new chemical compounds have been approved to fight cancer, 50% of that come from structurally originally natural products and their modifications to be safe and have many advantages [11,12]. Owing to their structural diversity, organic molecules (e.g., terpenes, flavonoids, alkaloids, lignans, saponins, vitamins, glycosides, oils, and other secondary metabolites) play a vital role in selective inhibition of proliferation and induction of cancerous cell death [13,14]. Among methoxylated flavones, xanthomicrol was first identified [15] and isolated from *Dracocephalum kotschyii* Boiss [16] which was able to inhibit proliferation of a number of malignant cells [16,17] because of its inhibition of endothelial cell proliferation via decreased vascular endothelial growth factor activity [18]. In terms of ursolic acid's anti-cancer effect, many studies reported that the underlying mechanisms were the inhibition of tumorigenesis and cancer cell proliferation, as well as apoptosis modulation, cell cycle arrest prevention, and autophagy promotion through in vitro and in vivo models [19–23]. As part of the continuing investigation on natural antitumor from herbs, Bai et al. 2010 first isolated 5,4'-dihydroxy-6,7,8,3'-tetramethoxyflavone from the aerial parts of *Rabdosia rubescens*, which was able to exhibit cytotoxicity in a various of cancer cell lines, however, the exact mechanisms of its health benefits are unclear [24].

Adenosma bracteosum Bonati (*A. bracteosum*) belongs to the family Scrophulariaceae, and is used in the treatment of liver diseases because it contains around twelve compounds of essential oil such as thymol (25.6%), linalool (13.1%), and (E)-β-farnesene (9.5%), etc., [25,26]. Although this plant is reported as an efficient medicinal plant, the cytotoxicity against lung and liver cancers and the response of cell lines to the plant extract have not been described. The present study aimed to discover cytotoxic and apoptotic potential of ethanol extract and its derived fractions (chloroform, ethyl acetate, butanol, and aqueous) as well as isolated compounds of *A. bracteosum* on human large cell lung carcinoma (NCI-H460) and hepatocellular carcinoma (HepG2).

2. Materials and Methods

2.1. Chemicals and Reagents

Camptothecin, DMSO (dimethyl sulfoxide), ethanol, FBS (Fetal bovine serum), HEPES, glucose, L-glutamine, phenol red, sodium bicarbonate, sulforhodamine-B, streptomycin were purchased from Sigma-Aldrich (St. Louis, MO, USA). n-hexane, ethyl acetate, chloroform, silicagel, methanol, trichloroacetic acid, and TLC were purchased from Merck (Darmstadt, Germany).

2.2. Plant Material, Extraction, and Isolation

The aerial part of *A. bracteosum* was collected in December 2018 from Ba Den Mountain, Tay Ninh province, Vietnam. A voucher specimen (No. UH-B02) was deposited in the herbarium of the CirTech Institute, Ho Chi Minh City University of Technology (Vietnam). Total of 7 kg of the shade-dried and powdered aerial part of *A. bracteosum* (7 kg) was mashed in ethanol 70% for 2 days. To obtain 800 g of crude ethanol extract, the solution was filtered, followed by removing of solvent under reduced pressure. To obtain aqueous, n-hexane, chloroform (CHCl3), ethyl acetate (EtOAc), and n-butanol fractions, these extracts were then suspended in DW and successively partitioned with the solvents, respectively. The chloroform extract (104.6 g) was applied to normal phase silica gel CC, and eluted with a gradient of *n*-hexane-EtOAc (2:8 to 0:10, *v/v*) then eluted with EtOAc-CH$_3$OH (1:1, *v/v*) to give 10 fractions (**A1–10**). Fraction **A7** (35.5 g) was fractionated in the same manner as mentioned previously to afford eight subfractions **A7.1-8**. Purifying the subfraction **A7.4** (5.5 g) by CC provided seven fractions (**A7.4.1-7**). Fraction **A7.4.1.4** was rechromatographed by CC using the mixture of *n*-hexane-CHCl$_3$ (2:8, *v/v*) as mobile phase to afford compound **AB1** (135 mg). Likewise, multiple purification of fraction **A7.4.1.5** by CC yielded compound **AB2** (172 mg) and **AB3** (127 mg).

2.3. Free Radical Scavenging Activity Assay

The samples were measured using the stated method for the 2,2-diphenyl-1-picrylhydrazyl (DPPH) assay [27]. A total of 50 µL of sample solutions were used to react with 2 mL of DPPH• of 6–5 M in 80% methanol. All samples were scanned at the absorbance wavelength (515 nm) after 16 min. The % of DPPH• radical inhibition was calculated using the following equation: % Inhibition = [($A_{C(0)}$ − $A_{S(t)}$) / $A_{C(0)}$)] × 100, where $A_{C(0)}$ and $A_{C(t)}$ observed at t = 0 and 16 min, respectively.

2.4. Assessment of ABTS Radical Scavenging Activity

A previous procedure [28] was used to test the radical scavenging activity of the samples. Briefly, ABTS—2,2′-azino-bis(3-ethylbenzothiazoline-6-sulfonic acid) was dissolved in water, reacted with potassium persulfate (2.45 mM) to obtain the working solution. A 1.0 mL of the working solution was mixed with 10 mL samples and measured at 734 nm after 6 min. Ascorbic acid was used as positive control.

2.5. Brine Shrimp Lethality Bioassay

The cytotoxic activity of the samples (the ethanol and aqueous crude extracts, chloroform, ethyl acetate, n-butanol, and aqueous fractions and the isolated compound) was determined using the Brine shrimp lethality bioassay [29,30]. Briefly, brine shrimp eggs (Artemia salina) were incubated in a vessel containing sterile artificial seawater produced by dissolving 38 g of table salt in 1 L of distilled water at 28–30 °C with good aeration (using air pump), under a continuous light condition (60 W lamp) for 48 h. After hatching, nauplii were collected with a Pasteur pipette, ten brine shrimps were moved into each well containing the seawater.

Sample stock solutions were made by dissolving 5 mg of each sample in 1 mL of DMSO. Test solutions of different concentrations (1000, 800, 400, 200, 100, 50, 25, 12.5, and 6.25 µg/mL) were collected using the serial dilution technique with seawater. The solutions were transferred to individual vials and 5 mL of the seawater including 10 nauplii shrimp were taken into each vial. A control group comprising same volumes of DMSO (as in the sample vials) and 10 nauplii in 5 mL of artificial seawater was used. After 24 h, the vials were examined with a magnifying glass and the amount of nauplii survived in each vial was counted. The percent mortality of the brine shrimp nauplii was determined for control and increasing concentration from the data and the values of the LC$_{50}$ were determined. Potassium dichromate was used as a reference standard.

2.6. Cell Culture and Sulforhodamine B Assay

The NCI-H460 and HepG2 were grown in Dulbecco's Modified Eagle Medium (DMEM) containing (1% penicillin/streptomycin, 1% L-glutamine, and 10% FBS) and were incubated consecutively in culture flasks (37 °C and 5% CO$_2$). The sulforhodamine B (SRB) assay was performed to determine the cytotoxicity effect of the extract, fractions and compounds as previously described [31,32]. Briefly, the cells at 7.5 × 10^3 cells/mL were seeded in a transparent 96-well plate (Falcon, Franklin, NJ, USA) and incubated (37 °C and 5% CO$_2$) for 24 h. Then, the old DMEM medium was replaced by 100 µL of the crude extract and its fractions or camptothecin at different concentrations (100, 75, 50, 30, 20, 10, 5, 2.5, 1.0, and 0.5 µg/mL), followed by incubation at 37 °C for 48 h. Subsequently, the treated and non-treated cells were fixed by adding 50 µL of cold trichloroacetic acid 50% and incubated for 1 h at 4 °C. The plates were washed with distilled water, air dried, and stained with 0.2% SRB (20–30 min, room temperature). The plates were then washed with 1% acetic acid to remove unbound dye. After air-drying, 200 µL Tris-base 10 mM was added to each well. The plates were shaken in ELISA photometer for 20 min and absorbance was measured at wavelength of 492 nm and a reference wavelength of 620 nm. The effect on the cell growth was calculated as:

$$I\% = 1 - [OD\ (492 - 620)\ sample\ /\ OD\ (492 - 620)\ blank]) \times 100\%$$

where $I\%$ = % growth inhibition; OD = optical density. In addition, the IC_{50} values (the concentration corresponding to 50% cell-growth-inhibition rate) were also determined on the basis of linear-regression analyses.

2.7. DNA Fragmentation and Apoptosis Induced by AB2

In order to analyze DNA fragmentation, NCI-H460 and HepG2 cells were induced apoptosis by treating with AB2 at IC_{50} and $2 \times IC_{50}$. DNA purification kit was applied to extract DNA, according to the manufacturer's instructions (Thermo Fisher Scientific, CA, USA). Passing quantification, 2 µg of each DNA sample was loaded to electrophoresis on a 1.5% agarose gel, then the gel was photographed under ultraviolet illumination after staining with ethidium bromide (10 µg/mL).

The effect of AB2 on cell apoptosis was evaluated by flow cytometer with Annexin V-FITC/PI staining kit (Thermo Fisher Scientific, CA, USA), according to the manufacturer's instructions. Briefly, the cells were treated with AB2 compound at IC50 and $2 \times$ IC50 concentrations for 24 h. After harvesting, cells were suspended in 300 µL binding buffer, and then stained with Annexin-FITC and/or propium iodide. Positive controls for apoptosis were stained with only Annexin-FITC. Positive controls for necrosis were stained with only propium iodide. At least 10^4 cells were analyzed by flow cytometer (BD, Biosciences, San Jose, CA, USA) and data were analyzed using FlowJo vX.0.7 (Tree Star, Inc., Ashland, OR, USA).

2.8. Detection of Mitochondrial Membrane Potential

The changes in mitochondrial membrane potential were determined using TMRE (tetramethylrhodamine, ethyl ester) mitochondrial membrane potential assay kit (Abcam, Cambridge, UK), according to the manufacturer's instructions. Briefly, NCI-H460 and HepG2 cells in 96-well plate at 1×10^4 cells per well were incubated for 24 h. Cells were treated with AB2 compound at IC_{50} and $2 \times IC_{50}$ concentrations for 24 h. After incubation time periods, the cells were harvested, washed twice in PBS, re-suspended in media supplemented with TMRE (200 nM), incubated at 37 °C for 20 min in the dark. Then, the media was replaced once with 100 µL of PBS/0.2% BSA, and then the fluorescence of TMRE was measured at an excitation wavelength of 549 nm by using a microplate reader (Perkin Elmer, Victor X5, Norwalk, CT, USA). The carbonyl cyanide-p-trifluoromethoxyphenylhydrazone (FCCP)-treated cells were used as a positive control.

2.9. Fluorescent Assays for Measuring Caspase-3 Activity and Intracellular Reactive Oxygen Species (ROS) Generation

Caspase activity assays in multi-well plate formats represent powerful tools for understanding experimental modulation of the apoptotic response. Caspase-3, -8, -9 activities were measured using the activity assay kit (Abcam, Cambridge, UK)) were purchased by Abcam according to the manufacturer's guidelines. Briefly, NCI-H460 and HepG2 cells were cultured into 96-well plates at a density of 2×10^4 cells/well. Then, old media were replaced by fresh ones having the AB2 compound at IC_{50} and $2 \times IC_{50}$ concentrations, and cells were further incubated for 6 h. Total of 100 µL of Caspase reagent was added to each well. The fluorescence of each well was measured at excitation/emission (Ex/Em) = 535/620 nm, Ex/Em = 490/525 nm, and Ex/Em = 370/450 nm for Caspase 3, Caspase 8, and Caspase 9 respectively for detecting fluorescence intensity using a plate-reading fluorescence reader (Perkin Elmer, Victor X5, Norwalk, CT, USA).

ROS generation was tested by using the ROS assay kit (ab113851, Abcam) according to the manufacturer's instructions. Briefly, the cells (5×10^4 cells/well) were cultured in 96-well plates. DCFH-DA was added to the cells at 37 °C for 1 h in the dark. After incubation with AB2 compound at IC_{50} and $2 \times IC_{50}$ concentrations for 0, 4, 8, 16, 20, and 24 h, cells were rinsed with PBS. Cells were measured on a fluorescent plate reader (Perkin Elmer, Victor X5, Norwalk, CT, USA), and mean ± standard deviation was plotted for three replicates from each condition. Tert-butyl hydrogen peroxide (TBHP), which mimics ROS activity to oxidize DCFDA to fluorescent DCF, was used as positive control.

2.10. Western Blot Analysis

NCI-H460 and HepG2 cells were treated with AB2 compound at IC_{50} and $2 \times IC_{50}$ concentrations and Camptothecin (4 µg/mL), then total protein was extracted using radioimmunoprecipitation assay buffer. Following by centrifuging at 14,000 rpm for 20 min at 4 °C, total proteins were obtained, measured, and subjected onto 12% sodium dodecyl sulfate-polyacrylamide gel electrophoresis (SDS-PAGE), separated and transferred onto a nitrocellulose membrane. Then, 5% bovine serum albumin (BSA) was applied to block the membranes, followed by probing with anti-active caspase-3 and anti-β-actin primary antibodies with gentle agitation overnight at 4 °C. Afterward, the corresponding HRP-conjugated secondary antibody was applied and incubated for 1 h at room temperature after washing three times with TBST buffer (Tris-buffered saline, 0.1% Tween 20). The blots were visualized with enhanced chemiluminescence detection and quantified by densitometry using Image J version 1.47.

2.11. Statistical Analysis

The experiments were performed in triplicate. The data are expressed as mean ± standard deviation. Significant differences between groups were determined by using Student *t*-test. *p* value of less than 0.05 was considered significant.

3. Results and Discussion

3.1. Brine Shrimp Lethality Test

The brine shrimp assay is significantly associated with in vitro growth inhibition of human solid tumor cell lines which was demonstrated by the National Cancer Institute (NCI, USA) and it can show the value of this bioassay as a pre-screening tool for antitumor drug research [33]. This test is well correlated with antitumor activity (cytotoxicity) and can be used to monitor the activity of bioactive natural products [34]. As shown in Figure 1, the cytotoxicity demonstrates a relationship between the concentration of the samples and the degree of lethality, suggesting that the samples are biologically active. LC_{50} values were calculated using graph extrapolation as are shown in Table 1. Comparison to aqueous extract, the LC_{50} values for the ethanol extract at 24 h were 647.64 µg/mL, has revealed that it more exhibited toxic expressions (LC_{50} was less than 1.0 mg/mL) against the brine shrimp [35]. The ethanol extract was then used to guide the fractionation process of the plant extract to isolate potential anti-cancer compounds. The chloroform fraction from ethanol extract was the most active among all fraction. Ethyl acetate fraction, n-butanol fraction, and aqueous fraction gave $LC_{50} > 1000$ µg/mL, that are considered to be inactive. The variation in their results is possibly because of the different polarities of the solvents, the ethanol solvent is less polar than aqueous one and the phytochemicals of the ethanol extract contain specific molecules that have provided its cytotoxic activity against brine shrimp. As a result, the chloroform fraction of ethanol extract exhibited more toxicity against the brine shrimp at 205.58 µg/mL compared to that of other fractions, which was further used to isolate compounds.

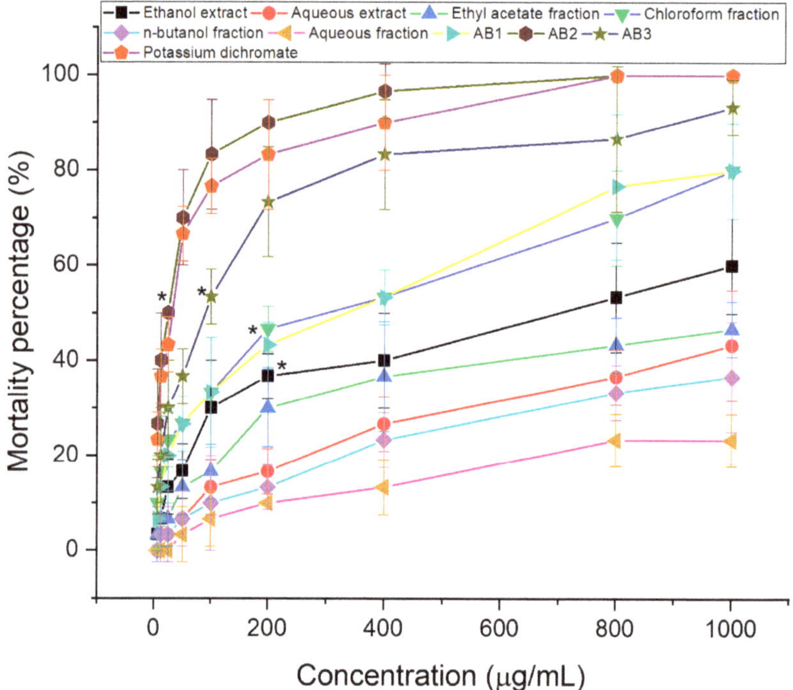

Figure 1. Brine shrimp lethality of extracts, fractions (A), and isolated compounds (B) at 24 h, * mean significant difference ($p < 0.05$) compared to control.

Table 1. Brine shrimp toxicity expressed as LC_{50} value.

Sample	Ethanol Extract	Aqueous Extract	Chloroform Fraction	Ethyl Acetate Fraction	*n*-Butanol Fraction	Aqueous Fraction	Compound			Potassium Dichromate
							AB1	AB2	AB3	
LC_{50} (µg/mL)	647.64	>1000	205.58	>1000	>1000	>1000	202.8	20.34	65.71	27.75

3.2. In Vitro Screening for Cytotoxic Activity of Extracts and Fractions

The cytotoxic effect of extracts and fractions from *A. bracteosum* on the growth of NCI-H460 and HepG2 cancer cell lines were investigated by SRB assay. The cell inhibition activity of ethanol and aqueous extracts, chloroform, ethyl acetate, *n*-butanol, and aqueous fractions of *A. bracteosum* showed in Figure 2. The ethanol extract had high activities on the human non-small cell lung cancer carcinoma NCI-H460 (90.46 ± 1.61%) and HepG2 (69.23 ± 4.71%) cell lines but at the same concentration of 100 µg/mL, aqueous had low activities (<25%). The chloroform fraction from ethanol extract had the highest activity as compared to other fractions that percentage of cytotoxicity on cell line NCI-H460 and HepG2 was 89.29 ± 0.63% and 76.40 ± 3.62%, respectively. As revealed in Figure 2, ethanol extract and chloroform fraction had significant cytotoxic activity against the cells in increasing dose concentration. The death of 50% of the tumor cells of ethanol extract on HepG2 and NCI-H460 was 39.15 ± 0.61 and 30.31 ± 1.60 µg/mL while IC_{50} of chloroform fraction was 36.34 ± 0.48 and 38.35 ± 2.04, respectively. HepG2 cell was more sensitive to chloroform fraction than ethanol extract. The chloroform fraction was therefore further studied for the isolation of pure compounds.

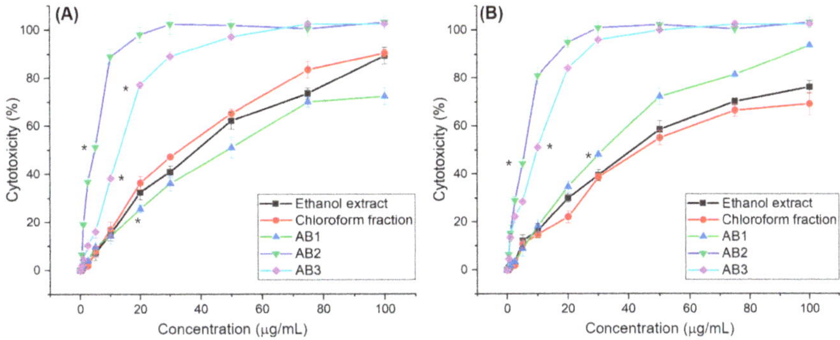

Figure 2. Cytotoxicity of extract, fraction, and compounds on NCI-H460 (**A**) and HepG2 (**B**) cells at different concentration, * mean significant difference ($p < 0.05$) compared to control.

3.3. Spectroscopic Data of Isolated Compounds

The NMR spectra were measured on Bruker 500 Avance spectrometer (Karlsruhe, Germany) (500 MHz for ^{1}H NMR and 125 MHz for ^{13}C NMR) spectrometers with tetramethylsilane (TMS) as internal standard. Chemical shifts are expressed in ppm with reference to the residual protonated solvent signals (DMSO-d_6 with δ_H 2.50 and δ_C 39.52, acetone-d_6 with δ_H 2.05 and δ_C 29.840). The EIMS were recorded on a HRESIMS Bruker MicroTOF (Billerica, MA, USA). TLC was carried out on precoated silica gel 60 F$_{254}$ and spots were visualized by UV$_{254nm}$, UV$_{365nm}$ lamp (Spectroline, Westbury, NY, USA). Gravity column chromatography was performed with silica gel 60 (0.040–0.063 mm).

Compound AB1: Yellow amorphous powder. ESI-MS *m/z*: 343.08 [M − H]$^-$. ^{1}H-NMR (500 MHz, CD$_3$OD): δ 7.85 (1H, *d*, *J* = 8.0 Hz, H-2′/H-6′), 6.94 (1H, *d*, *J* = 8.0 Hz, H-3′/H-5′), 6.60 (1H, s, H-3), 4.09 (3H, s, OCH$_3$), 3,97 (3H, s, OCH$_3$), 3.90 (3H, s, OCH$_3$). ^{13}C-NMR (125 MHz, Acetone–d$_6$): δ 183.1 (C-4), 152.0 (C-5), 164.3 (C-2), 161.3 (C-4′), 149.0 (C-8a), 136.0 (C-8), 133.0 (C-6), 128.5 (C-2′), 128.5 (C-6′), 122.3 (C-1′), 116.1 (C-3′), 116.1 (C-5′), 106.7 (C-4a), 102.8 (C-3), 61.4 (-OCH$_3$), 61.0 (-OCH$_3$), 60.1 (-OCH$_3$). These spectroscopic data were consistent with those of xanthomicrol [16] (Figure 3). 1D and 2D NMR spectra of **AB1** were also provided in Supplementary Materials file.

Compound AB2: Yellow amorphous powder. ESI-MS *m/z*: 373.10 [M-H]$^-$. ^{1}H-NMR (500 MHz, Acetone–d$_6$): δ 7.66 (1H, *dd*, *J* = 1.5, 8.0 Hz, H-2′), 7.65 (1H, *brs*, H-6′), 7.04 (1H, *d*, *J* = 8.0 Hz, H-3′), 6.75 (1H, s, H-3), 4.09 (3H, s, OCH$_3$), 3.97 (3H, s, OCH$_3$), 3.90 (3H, s, OCH$_3$), 3.88 (3H, s, OCH$_3$). ^{13}C-NMR (125 MHz, Acetone–d$_6$): 183.0 (C-4), 164.4 (C-2), 152.0 (C-7), 151.0 (C-4′), 148.0 (C-8a), 145.7 (C-5′), 136.0 (C-8), 133.0 (C-6), 122.4 (C-1′), 120.6 (C-2′), 115.6 (C-3′), 109.6 (C-6′) 106.7 (C-4a), 103.0 (C-3), 61.4 (-OCH$_3$), 61.0 (-OCH$_3$), 60.1 (-OCH$_3$), 55.6 (-OCH$_3$). These results were consistent with those of 5,4′-dihydroxy-6,7,8,3′-tetramethoxyflavone [24,36,37] (Figure 3). 1D and 2D NMR spectra of **AB2** were also provided in Supporting Materials file. NMR data assignments were completed through detailed analysis of HMBC experimental data (Figure 4).

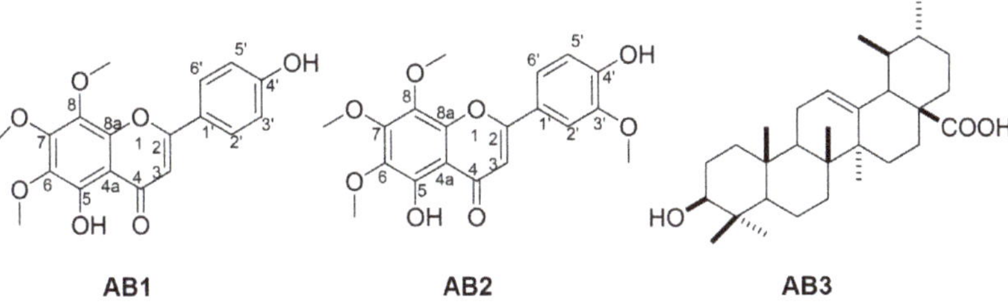

Figure 3. Chemical structures of AB1, AB2 and AB3.

Compound AB3: White amorphous powder. ESI-MS *m/z:* 455.37 [M-H]⁻ ¹H-NMR (500 MHz, DMSO–d₆): δ 0.66 ppm (1H, d, *J* = 11.3 Hz, H-5), 0.77 ppm (3H, s, 24-CH₃), 0.81 ppm (3H, s, 26-CH₃), 0.85 (3H, d, *J* = 6.4 Hz, 29-CH₃), 0.90 ppm (3H, d, *J* = 4.2 Hz, 25-CH₃), 0.92 (3H, s, 30-CH₃), 0.94 (3H, d, *J* = 6.2 Hz, 23-CH₃), 1.08 (3H, s, 27-CH₃), 1.99 ppm (1H, dd, *J* = 4.1, 13.4 Hz, H-6), 2.11 ppm (1H, d, *J* = 13.1 Hz, H-18), 3.00 ppm (1H, dd, *J* = 6.8, 9.9 Hz, H-3), 5.14 ppm (1H, dd, *J* = 3.4, 6.8 Hz, H-12). ¹³C-NMR (125 MHz, DMSO–d₆): δ 178.0 (C-28), 138.4 (C-13), 124.4 (C-12), 76.9 (C-3), 54.8 (C-5), 52.4 (C-18), 47.0 (C-17), 46.8 (C-9), 41.6 (C-14), 39.1 (C-8), 38.5 (C-4), 38.5 (C-19), 38.4 (C-20), 38.3 (C-1), 36.5 (C-10), 36.3 (C-22), 27.0 (C-2), 18.0 (C-6), 32.7 (C-7), 30.2 (C-21), 28.3 (C-23), 27.6 (C-15), 23.8 (C-16), 23.3 (C-27), 22.9 (C-11), 21.1 (C-30), 17.0 (C-26), 16.9 (C-29), 16.1 (C-24), 15.2 (C-25). This compound was identified as ursolic acid [19] (Figure 3) and also demonstrated to be effective for the treatment of a wide spectrum of diseases [38] including antioxidant [39], anti-inflammatory [40], antibacterial [41], antiviral [41], antifungal [41], antipyretic [38], anticancer [42], antitumor [42], antiwrinkle [43], anti-hypertension [44], and hepatoprotective activities [45].

Indeed, LC₅₀ values of AB1, AB2, and AB3 were observed to be 202.80, 20.34, and 65.71 µg/mL, respectively, whereas this figure was revealed for the positive control (potassium dichromate) at 24 h was 27.75 µg/mL, indicating these toxic compounds expressions well against the brine shrimp. Hence, further investigation of these compounds to their toxic expressions on cancer cell lines should be pursued.

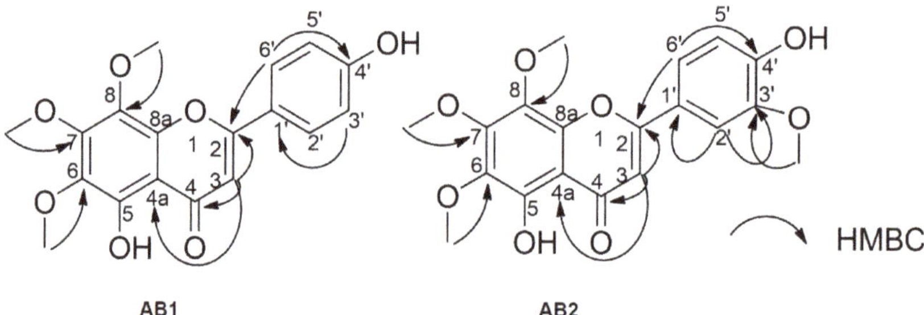

Figure 4. Selected heteronuclear multiple bond correlation (HMBC) correlations from **AB1** and **AB2**.

3.4. Bioactivity of Isolated Compounds

3.4.1. Antioxidant Activity Assessments

ABTS+ and 2,2-diphenyl-1-picrylhydrazile (DPPH) assays are widely used to evaluate compounds' ability to determine their antioxidant potential. The free radical inhibition of three compounds and reference antioxidant (ascorbic acid) at concentrations was shown in Figure 5A, the DPPH scavenging activity of AB1, AB2, and AB3 increased progressively with the concentration. AB2 was the most active radical scavenging which showed an IC₅₀ of 4.04 µg/mL and was 1.37 fold lower than that of ascorbic acid (IC₅₀ = 2.95 µg/mL), followed by AB1 and AB3 with the IC₅₀ values were 4.45 and 11.93 µg/mL, respectively ($p < 0.05$).

The ABTS+ assay is additionally an important method for quantifying radical scavenging activity, which can provide comparable results to those obtained in the DPPH assay. As indicated in Figure 5B, three compounds significantly exhibited ABTS-free radical scavenging activity. Three compounds had an antioxidant activity comparable to that of ascorbic acid. AB2 was a strong active radical scavenger, showing an IC₅₀ of 6.53 ± 0.16 µg/mL and 1.24 times lower than that of ascorbic acid, followed by AB1 and AB3 with the IC₅₀ values of 7.09 and 11.41 µg/mL, respectively ($p < 0.05$). In both the assays, AB2 exhibited the highest ABTS and DPPH radical scavenging activities. The correlation between DPPH and ABTS methods used to measure the antioxidant activity of the compounds has been examined. The DPPH radical activity also showed a strong correlation with ABTS (R^2 = 97.56%).

Reactive oxygen species (ROS) cause a variety of cancers in humans [46]. ROS can damage macro biomolecules such as proteins, lipids, and DNA and reduce the DNA repair capability that can result in normal cells being transformed into cancer cells by mutating key genes [47]. Research on potent antioxidants or scavengers thus contributes to the prevention of cancer. Some previous studies have shown that phenols and flavonoids have strong antioxidants and are effective anticancer agents through anti-angiogenic and apoptosis activities [48,49]. In fact, in this study, A2, a flavonoid showed good antioxidant and cytotoxic effects.

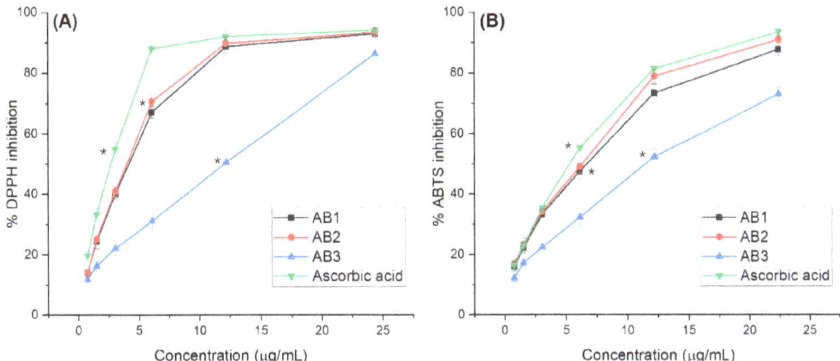

Figure 5. (**A**) 2,2-diphenyl-1-picrylhydrazyl (DPPH) and (**B**) ABTS+ radical scavenging activities. AB1: Xanthomicrol, AB2: 5,4'-dihydroxy-6,7,8,3'-tetramethoxyflavone, AB3: Ursolic acid, ABTS: 2,2'-azino-bis(3-ethylbenzothiazoline-6-sulfonic acid Values are given as mean ± standard deviation, * mean significant difference ($p < 0.05$) compared to control.

3.4.2. Brine Shrimp Bioassay

AB1, AB2, and AB3 were identified as confirmed by their cytotoxic effects through brine shrimp bioassay (Figure 1). LC_{50} values of AB1, AB2, and AB3 were 202.80, 20.34, and 65.71 μg/mL, respectively, which demonstrated (possessed) significant activity against the brine shrimp. The positive control LC_{50} values were 27.75 μg/mL, indicating toxic expressions against the brine shrimp. AB2 had a higher LC_{50} than the positive control, highlighting the cytotoxic effect.

3.5. In Vitro Screening for Cytotoxic Activity of Compounds

The cell inhibition activity of compounds including AB1, AB2, and AB3 is presented in Figure 2. Calculation of the dose of AB1, AB2, and AB3 revealed the following IC_{50} values for NCI-H460 and HepG2 cells: 32.5 ± 0.41 and 49.2 ± 0.81 μg/mL; 4.57 ± 0.32 and 5.67 ± 0.09 μg/mL; 13.05 ± 0.55 and 10.00 ± 0.16 μg/mL, respectively. According to the criteria of the National Cancer Institute and Geran protocol, extracts with $IC_{50} \leq 20$ μg/mL = highly cytotoxic, IC_{50} ranged between 21 and 200 μg/mL = moderately cytotoxic, IC_{50} ranged between 201 and 500 μg/mL = weakly cytotoxic, and $IC_{50} > 501$ μg/mL = no cytotoxicity [50–52]. These results clearly reveal that test of the ethanol extract, chloroform fraction and AB1 on the two cell lines revealed less cytotoxicity. Interestingly, the compound AB2 (5,4'-dihydroxy-6,7,8,3'-tetramethoxyflavone) and AB3 (ursolic acid) have excellent cytotoxic activity for two types of cancer cells, in which the highest percent anticancer activity was observed in AB2, followed by the ursolic acid with the lower percent inhibition at the same concentration. These compounds AB2 and AB3 are greatly considered to have in vitro cytotoxic activity with an IC_{50} value ≤10 μg/mL for the cells.

Among the isolated compounds, ursolic acid (AB3) has been well documented as a naturally synthesized pentacyclic triterpenoid, widely distributed in different fruits and vegetables [53–55] and has been known as an excellent anticancer agent [8,11] by inducing apoptosis in several human cancer cells [56,57]. Remarkably, the compound 5,4'-dihydroxy-6,7,8,3'-tetramethoxyflavone (AB2) was isolated for the first time in the *A. bracteosum*. Further research investigating the cellular and molecular mechanisms underlying the effects of 5,4′-dihydroxy-6,7,8,3′-tetramethoxyflavone is required for the development of new therapeutic agents.

3.6. DNA Fragmentation and Apoptosis Induced by 5,4′-Dihydroxy-6,7,8,3′-Tetramethoxyflavone

To seek the mechanism of cell death mediated by 5,4′-dihydroxy-6,7,8,3′-tetramethoxyflavone (AB2), we first performed DNA fragmentation assay, which is characteristic for apoptosis. We treated the cells with 5,4′-dihydroxy-6,7,8,3′-tetramethoxyflavone at its IC_{50} for 24 h, and DNA was then isolated and separated by agarose gel electrophoresis. As shown in Figure 6A, the HepG2 cells treated with 5 μg/mL of AB2 showed significant fragmentation after 24 h of AB2 treatment compared to the treatment of camptothecin at 4 μg/mL. Also, an oligo nucleosomal ladder of fragmented DNA was obtained after the treatment in NCI-H460 cells (Figure 6B), whereas no DNA fragments were observed when untreated in both cells. These data strongly suggest that 5,4′-dihydroxy-6,7,8,3′-tetramethoxyflavone is a potent inducer of apoptosis in the cells.

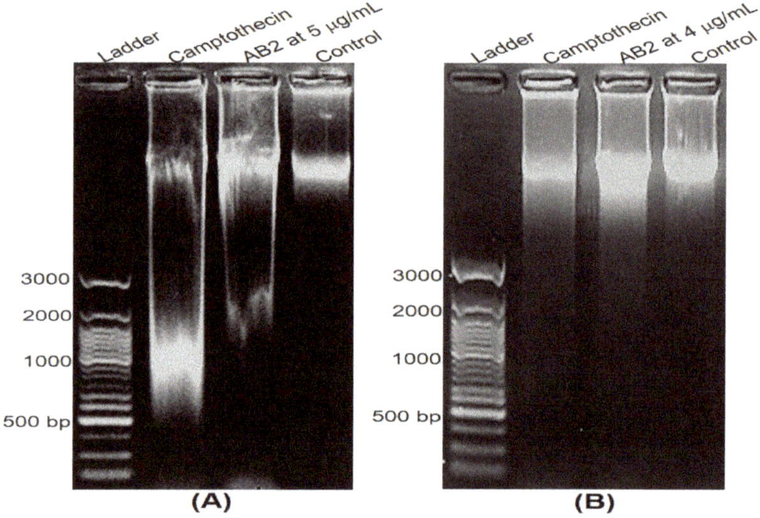

Figure 6. Apoptosis induced by 5,4′-dihydroxy-6,7,8,3′-tetramethoxyflavone at its IC_{50} in HepG2 (**A**) and NCI-H460 (**B**) cells.

In addition, to further evaluate the potential mechanism of the compound 5,4′-dihydroxy-6,7,8,3′-tetramethoxyflavone-induced inhibition of human lung and hepatocyte carcinoma cell proliferation, we continuously examined the mode of cell death on NCI-H460 and HepG2 cells Annexin V-FITC/propidium iodide (PI) staining caused by the compound. Meanwhile, PI was used to only stain necrotic cells because of its increased membrane permeability and eventual propidium uptake in the cells [58]. In this study, Figure 7 reveals total apoptosis percent from treated and untreated NCI-H460 and HepG2 cells which were determined through the flow-cytometric outcomes. As shown in Figure 7A, NCI-H460 cells were treated with AB2 at 4 μg/mL ($\sim IC_{50}$) caused the lowest apoptosis (42.1%) within 24 h of treatment, relative to the control cells ($\sim$5.1%), whereas the figures seemed to be similar for HepG2 cells (Figure 7B). Remarkably, AB2 exposure, (8 μg/mL$\sim$2 × IC_{50}) and (10 μg/mL$\sim$2 × IC_{50}), caused a significant increase in apoptosis and necrosis in both NCI-H460 and HepG2 cells, which were about 2.3-fold and 2.2-fold higher than that of at IC_{50} doses, respectively. With respect to the cytotoxicity effect, AB2 could lead to DNA damage, apoptosis in terms of the number of cell death events, because it contains a number ROS radicals [59].

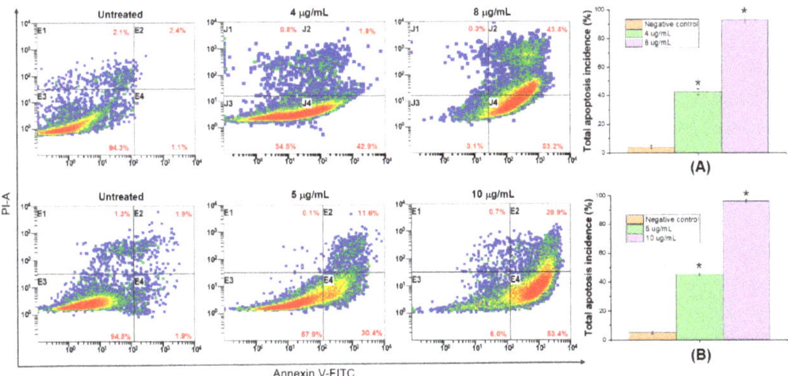

Figure 7. The total percentages of AB2-induced apoptosis in NCI-H460 (**A**) and HepG2 (**B**) cells, * mean significant difference (*p* < 0.05) compared to control.

3.7. 5,4′-Dihydroxy-6,7,8,3′-Tetramethoxyflavone Triggers Apoptosis Sensitivity by ROS- Caspase-3 Mediated in Human Hepatocyte Carcinoma Cell

The disruption of mitochondrial membrane potential may result in apoptosis [60]; thus, we next evaluated, by TMRE assay, whether AB2 mediated the cells. FCCP treatment as in the positive control group caused a significant loss of mitochondrial membrane potentials in comparison to the control group (*p* < 0.05). Interestingly, the same figure was observed following the treatment (IC$_{50}$ and 2 × IC$_{50}$) with AB2 (*p* < 0.05), strongly revealing the presence of AB2-mediated perturbation of mitochondrial metabolic activity in NCI-H460 and HepG2 cells (Figure 8A).

ROS is known to play a dual role; either that may be harmful depending on their accumulation levels, generally contributing to cell death either by apoptosis or necrosis at the levels beyond the cellular antioxidant defense mechanisms [60–62]. Generally, higher mitochondrial membrane potential results in greater adenosine triphosphate (ATP) production and greater ROS production [63]. Thus, the intracellular ROS levels were determined by measuring the intensity of a highly fluorescent derivative 2′,7′-dichlorofluorescein (DCF) which is generated from an externally applied non-fluorescent substance, DCFH-DA by the cellular redox reactions. As shown in Figure 8B–C, AB2 dose-dependently increased the ratio of green fluorescence in NCI-H460 and HepG2 cells suggesting that AB2 excitants production of intracellular ROS in the cells compared to the control group, camptothecin (*p* < 0.05). The antioxidant from AB2 succeeded in inverting the cytotoxic effect of both cells. These figures strongly indicate that AB2 treatment in a concentration- and time- dependent manner led to ROS-dependent and independent cell death in human lung and hepatocellular carcinoma cells.

During apoptosis, the permeabilization of the mitochondrial outer membrane caused the release of cytochrome c [64], which induces caspase activation to orchestrate the death of the cell due to its loss of mitochondrial function and generation of ROS [65]. Overproduction of ROS could be associated with cell homeostasis imbalance, mitochondrial damage, and apoptosis [60,66]. Activation of apoptosis pathways through the mitochondrial pathway could be is one of the key steps in apoptosis [67,68], which is generally related to recruitment of caspase family proteins including caspase-8 and/or caspase-3 [69]. Thus, the caspase activities pathway is checked to evaluate whether or not AB2 affects the induction of apoptosis. As shown in Figure 8D–E, the cells treated with AB2 resulted in increased percentage of caspase-3, -8, -9 activities compared to the negative control. However, AB2 did not affect much the caspase-8 and -9 activities and resulted in slightly increased expression in the cells in comparison to caspase-3 outcomes. Indeed, there was a significantly increased apoptotic cell frequency in the cells by ~4.5 and ~7.2-fold by treatment with 5,4′-dihydroxy-6,7,8,3′-tetramethoxyflavone at IC$_{50}$ and 2 × IC$_{50}$ respectively, as compared with the untreated groups, suggesting that AB2 stimulates caspase-3 in human lung and hepatocellular carcinoma cells.

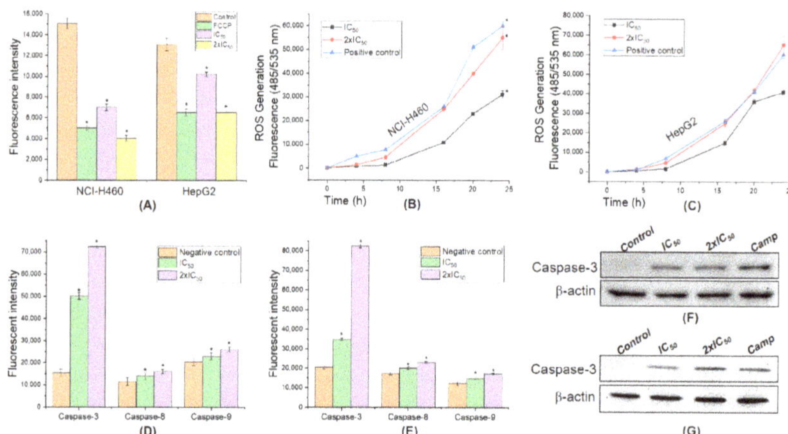

Figure 8. Potential mechanism of action of 5,4′-dihydroxy-6,7,8,3′-tetramethoxyflavone on human liver and lung cancer cells. (**A**) Mitochondrial membrane potential; ROS production in NCI-H460 (**B**) and HepG2 (**C**) cells by DCFH-DA essay. The caspase activity in the NCI-H460 (**D**) and HepG2 (**E**) cells treated with AB2. Expression of active caspase-3 protein on NCI-H460 (**F**) and HepG2 (**G**) cells was confirmed by Western blot analysis, * mean significant difference ($p < 0.05$) compared to control.

Finally, the expression level of apoptosis-related active protein (cleaved) caspase-3 was confirmed by Western blot assay. The expression of active caspase-3 protein was clearly revealed by only IC_{50} and $2xIC_{50}$ of AB2 treatment compared to the positive control group (Figure 8F–G). This figure is consistent in comparison with the significantly increased levels of active caspase-3 from fluorescence assay. Our results revealed that the role of 5,4′-dihydroxy-6,7,8,3′-tetramethoxyflavone-induced apoptosis in both human lung and hepatocellular carcinoma cells was dependent on the activation caspase-3.

These findings revealed that these extracts and compounds are very potent antioxidants because of their radical scavenging capacity and their capacity to stimulate ROS-mediated mitochondrial pathway because of the activation of caspase-3. The interactions may result from new anticancer molecules in which antioxidants can also be enhanced to eliminate cancer cells through the apoptosis pathway, where antioxidants from medicinal plants have shown a great cytotoxic potential [70,71]. This has been demonstrated through caspase-independent cell death from *H. speciosa* [72] and *J. Decurrens* [73], whereas others such as *C. adamantium* [74], *S. velutina* [72], and *S. Adstringens* [75] destroyed malignant haematologic cells or melanoma cells through apoptosis. The search for new antioxidants containing toxicity profile is desirable, and the *A. bracteosum* (Bonati) demonstrated here may represent interesting targets for this purpose.

4. Conclusions

The current investigation strongly demonstrate that *A. bracteosum* could significantly inhibit the growth of human NCI-H460 and HepG2 cells as well as the brine shrimp. Importantly, a new compound, 5,4′-dihydroxy-6,7,8,3′-tetramethoxyflavone was investigated, and demonstrated to be probably by ROS-mediated mitochondrial pathway, followed by activated caspase-3, associated with the apoptosis cells death. These results suggest that *A. bracteosum* is a promising source of useful natural products and the new compound offers opportunities to develop novel anticancer drug after its full apoptosis activity has been clinically addressed.

Supplementary Materials: The following are available online at http://www.mdpi.com/1420-3049/25/12/2912/s1. Figure S1: ESI mass spectrum of AB1. Figure S2: The ^{1}H NMR spectrum of AB1 in methanol-d_4. Figure S3: The ^{13}C NMR spectrum of AB1 in acetone-d_6. Figure S4: HSQC spectrum of AB1 in acetone-d_6. Figure S5: HMBC spectrum of AB1 in Acetone-d_6. Figure S6: EI mass spectrum of AB2. Figure S7: The ^{1}H NMR spectrum of AB2 in acetone-d_6. Figure S8: The ^{13}C NMR spectrum of AB2 in acetone-d_6. Figure S9: HSQC spectrum of AB2 in acetone-d_6. Figure S10: HMBC spectrum of AB2 in Acetone-d_6. Figure S11: The ^{1}H NMR spectrum of AB3 in DMSO-d_6. Figure S12: The ^{13}C NMR spectrum of AB3 in DMSO-d_6. Table S1: NMR spectral data of AB1 and AB2 compounds. Table S2: Compare AB3 compound spectral data and reference.

Author Contributions: N.H.N., Q.T.H.T., Q.T.P., T.N.H.L., V.T.P., T.-H.D., and V.G.V. conceived, designed, performed the experiments; Q.T.T.H., Q.T.P., T.N.H.L., V.T.P., and T.H.D. performed compound isolation and bioactivity essays; N.H.N. substantive supervision and project administration; N.H.N. and V.G.V. wrote and revised the whole manuscript. All authors have read and agreed to the published version of the manuscript.

Funding: This research received no external funding.

Conflicts of Interest: No conflict of interest associated with this work.

References

1. Siegel, R.L.; Miller, K.D.; Jemal, A. Cancer statistics, 2020. *CA Cancer J. Clin.* **2020**, *70*, 7–30. [CrossRef]
2. Pucci, C.; Martinelli, C.; Ciofani, G. Innovative approaches for cancer treatment: Current perspectives and new challenges. *Ecancermedicalscience* **2019**, *13*, 961. [CrossRef]
3. Chikara, S.; Nagaprashantha, L.D.; Singhal, J.; Horne, D.; Awasthi, S.; Singhal, S.S. Oxidative stress and dietary phytochemicals: Role in cancer chemoprevention and treatment. *Cancer Lett.* **2018**, *413*, 122–134. [CrossRef]
4. Singh, S.; Sharma, B.; Kanwar, S.S.; Kumar, A. Lead Phytochemicals for Anticancer Drug Development. *Front. Plant Sci.* **2016**, *7*, 8973. [CrossRef]
5. Sak, K. Chemotherapy and Dietary Phytochemical Agents. *Chemother. Res. Pract.* **2012**, *2012*, 1–11. [CrossRef]
6. Baskar, R.; Dai, J.; Wenlong, N.; Yeo, R.; Yeoh, K.-W. Biological response of cancer cells to radiation treatment. *Front. Mol. Biosci.* **2014**, *1*, 24. [CrossRef]
7. Wood, D.M.; Athwal, S.; Panahloo, A. The advantages and disadvantages of a 'herbal' medicine in a patient with diabetes mellitus: A case report. *Diabet. Med.* **2004**, *21*, 625–627. [CrossRef]
8. Yin, S.-Y.; Wei, W.-C.; Jian, F.-Y.; Yang, N.-S. Therapeutic applications of herbal medicines for cancer patients. *Evid. Based Complement. Altern. Med.* **2013**, *2013*, 302426. [CrossRef]
9. Nguyen, N.H.; Nguyen, T.T.; Ma, P.C.; Ta, Q.T.H.; Duong, T.-H.; Van Giau, V. Potential Antimicrobial and Anticancer Activities of an Ethanol Extract from Bouea macrophylla. *Molecules* **2020**, *25*, 1996. [CrossRef]
10. Duong, T.H.; Beniddir, M.A.; Trung, N.T.; Phan, C.D.; Vo, V.G.; Nguyen, V.K.; Le, Q.L.; Nguyen, H.D.; Le Pogam, P. Atypical Lindenane-Type Sesquiterpenes from Lindera myrrha. *Molecules* **2020**, *25*, 1830. [CrossRef]
11. Iqbal, J.; Abbasi, B.A.; Ahmad, R.; Mahmood, T.; Kanwal, S.; Ali, B.; Khalil, A.T.; Shah, A.; Alam, M.M.; Badshah, H. Ursolic acid a promising candidate in the therapeutics of breast cancer: Current status and future implications. *Biomed. Pharmacother.* **2018**, *108*, 752–756. [CrossRef]
12. Agarwal, G.; Carcache, P.J.B.; Addo, E.M.; Kinghorn, A.D. Current status and contemporary approaches to the discovery of antitumor agents from higher plants. *Biotechnol. Adv.* **2020**, *38*, 107337. [CrossRef]
13. Iqbal, J.; Abbasi, B.H.; Batool, R.; Mahmood, T.; Ali, B.; Khalil, A.T.; Kanwal, S.; Shah, A.; Ahmad, R. Potential phytocompounds for developing breast cancer therapeutics: Nature's healing touch. *Eur. J. Pharmacol.* **2018**, *827*, 125–148. [CrossRef]
14. Avtanski, D.; Poretsky, L. Phyto-polyphenols as potential inhibitors of breast cancer metastasis. *Mol. Med.* **2018**, *24*, 29. [CrossRef]
15. Power, F.B.; Salway, A.H. Chemical Examination of Micromeria Chamissonis. *J. Am. Chem. Soc.* **1908**, *30*, 251–265. [CrossRef]
16. Jahaniani, F.; Ebrahimi, S.A.; Rahbar-Roshandel, N.; Mahmoudian, M. Xanthomicrol is the main cytotoxic component of Dracocephalum kotschyii and a potential anti-cancer agent. *Phytochemistry* **2005**, *66*, 1581–1592. [CrossRef]
17. Fattahi, M.; Cusido, R.M.; Khojasteh, A.; Bonfill, M.; Palazon, J. Xanthomicrol: A comprehensive review of its chemistry, distribution, biosynthesis and pharmacological activity. *Mini-Rev. Med. Chem.* **2014**, *14*, 725–733. [CrossRef] [PubMed]
18. Abbaszadeh, H.; Ebrahimi, S.A.; Akhavan, M.M. Antiangiogenic Activity of Xanthomicrol and Calycopterin, Two Polymethoxylated Hydroxyflavones in Both In Vitro and Ex Vivo Models. *Phytother. Res.* **2014**, *28*, 1661–1670. [CrossRef]

19. Harmand, P.-O.; Delage, C.; Simon, A.; Duval, R.E. Ursolic acid induces apoptosis through mitochondrial intrinsic pathway and caspase-3 activation in M4Beu melanoma cells. *Int. J. Cancer* **2005**, *114*, 1–11. [CrossRef] [PubMed]

20. Chen, C.-J.; Shih, Y.-L.; Yeh, M.-Y.; Liao, N.-C.; Chung, H.-Y.; Liu, K.-L.; Lee, M.-H.; Chou, P.-Y.; Hou, H.-Y.; Chou, J.-S.; et al. Ursolic Acid Induces Apoptotic Cell Death Through AIF and Endo G Release Through a Mitochondria-dependent Pathway in NCI-H292 Human Lung Cancer Cells In Vitro. *In Vivo* **2019**, *33*, 383–391. [CrossRef]

21. Woźniak, Ł.; Skąpska, S.; Marszałek, K. Ursolic Acid—A Pentacyclic Triterpenoid with a Wide Spectrum of Pharmacological Activities. *Molecules* **2015**, *20*, 20614–20641. [CrossRef]

22. Weng, H.; Tan, Z.-J.; Hu, Y.-P.; Shu, Y.; Bao, R.-F.; Jiang, L.; Wu, X.-S.; Li, M.; Ding, Q.; Wang, X.-A.; et al. Ursolic acid induces cell cycle arrest and apoptosis of gallbladder carcinoma cells. *Cancer Cell Int.* **2014**, *14*, 96. [CrossRef] [PubMed]

23. Kim, K.H.; Seo, H.S.; Choi, H.S.; Choi, I.; Shin, Y.C.; Ko, S.-G. Induction of apoptotic cell death by ursolic acid through mitochondrial death pathway and extrinsic death receptor pathway in MDA-MB-231 cells. *Arch. Pharm. Res.* **2011**, *34*, 1363–1372. [CrossRef]

24. Bai, N.; He, K.; Zhou, Z.; Lai, C.-S.; Zhang, L.; Quan, Z.; Shao, X.; Pan, M.-H.; Ho, C.-T. Flavonoids from Rabdosia rubescens exert anti-inflammatory and growth inhibitory effect against human leukemia HL-60 cells. *Food Chem.* **2010**, *122*, 831–835. [CrossRef]

25. Tsankova, E.T.; Kuleva, L.V.; Thanh, L.T. Composition of the Essential Oil of Adenosma bracteosum Bonati. *J. Essent. Oil Res.* **1994**, *6*, 305–306. [CrossRef]

26. Nguyen, N.H.; Pham, Q.T.; Luong, T.N.H.; Le, H.K.; Van Giau, V. Potential Antidiabetic Activity of Extracts and Isolated Compound from Adenosma bracteosum (Bonati). *Biomolecules* **2020**, *10*, 201. [CrossRef] [PubMed]

27. Von Gadow, A.; Joubert, E.; Hansmann, C.F. Comparison of the Antioxidant Activity of Aspalathin with That of Other Plant Phenols of Rooibos Tea (Aspalathus linearis), α-Tocopherol, BHT, and BHA. *J. Agric. Food Chem.* **1997**, *45*, 632–638. [CrossRef]

28. Re, R.; Pellegrini, N.; Proteggente, A.; Pannala, A.; Yang, M.; Rice-Evans, C. Antioxidant activity applying an improved ABTS radical cation decolorization assay. *Free Radic. Boil. Med.* **1999**, *26*, 1231–1237. [CrossRef]

29. Naher, S.; Aziz, A.; Akter, M.I.; Rahman, S.M.M.; Sajon, S.R.; Mazumder, K. Anti-diarrheal activity and brine shrimp lethality bioassay of methanolic extract of Cordyline fruticosa (L.) A. Chev. leaves. *Clin. Phytosci.* **2019**, *5*, 15. [CrossRef]

30. Syahmi, A.R.M.; Vijayarathna, S.; Sasidharan, S.; Latha, L.Y.; Kwan, Y.P.; Lau, Y.L.; Shin, L.N.; Chen, Y.; Vijayaratna, S. Acute Oral Toxicity and Brine Shrimp Lethality of Elaeis guineensis Jacq., (Oil Palm Leaf) Methanol Extract. *Molecules* **2010**, *15*, 8111–8121. [CrossRef]

31. Orellana, E.A.; Kasinski, A.L. Sulforhodamine B (SRB) Assay in Cell Culture to Investigate Cell Proliferation. *Bio-Protocol* **2016**, *6*, 1984. [CrossRef] [PubMed]

32. Skehan, P.; Scudiero, M.; Vistica, D.; Bokesch, H.; Kenney, S.; Storeng, R.; Monks, A.; McMahon, J.; Warren, J.T.; Boyd, M.R. New Colorimetric Cytotoxicity Assay for Anticancer-Drug Screening. *J. Natl. Cancer Inst.* **1990**, *82*, 1107–1112. [CrossRef] [PubMed]

33. Anderson, J.E.; Goetz, C.M.; McLaughlin, J.L.; Suffness, M. A blind comparison of simple bench-top bioassays and human tumour cell cytotoxicities as antitumor prescreens. *Phytochem. Anal.* **1991**, *2*, 107–111. [CrossRef]

34. Arcanjo, D.; Albuquerque, A.; Melo-Neto, B.; Santana, L.; Medeiros, M.; Citó, A.; Arcanjo, D.D.R. Bioactivity evaluation against Artemia salina Leach of medicinal plants used in Brazilian Northeastern folk medicine. *Braz. J. Boil.* **2012**, *72*, 505–509. [CrossRef]

35. Meyer, B.; Ferrigni, N.; Putnam, J.; Jacobsen, L.; Nichols, D.; McLaughlin, J. Brine Shrimp: A Convenient General Bioassay for Active Plant Constituents. *Planta Med.* **1982**, *45*, 31–34. [CrossRef]

36. Song, M.; Charoensinphon, N.; Wu, X.; Zheng, J.; Gao, Z.; Xu, F.; Wang, M.; Xiao, H. Inhibitory Effects of Metabolites of 5-Demethylnobiletin on Human Nonsmall Cell Lung Cancer Cells. *J. Agric. Food Chem.* **2016**, *64*, 4943–4949. [CrossRef]

37. Song, M.; Wu, X.; Charoensinphon, N.; Wang, M.; Zheng, J.; Gao, Z.; Xu, F.; Li, Z.; Li, F.; Zhou, J.; et al. Dietary 5-demethylnobiletin inhibits cigarette carcinogen NNK-induced lung tumorigenesis in mice. *Food Funct.* **2017**, *8*, 954–963. [CrossRef]

38. Seo, D.Y.; Lee, S.R.; Heo, J.-W.; No, M.-H.; Rhee, B.D.; Ko, K.S.; Kwak, H.-B.; Han, J. Ursolic acid in health and disease. *Korean J. Physiol. Pharmacol.* **2018**, *22*, 235–248. [CrossRef]

39. Liobikas, J.; Majiene, D.; Trumbeckaite, S.; Kuršvietienė, L.; Masteikova, R.; Kopustinskiene, D.M.; Savickas, A.; Bernatoniene, J. Uncoupling and Antioxidant Effects of Ursolic Acid in Isolated Rat Heart Mitochondria. *J. Nat. Prod.* **2011**, *74*, 1640–1644. [CrossRef]

40. Kashyap, D.; Tuli, H.S.; Punia, S.; Sharma, A.K.; Sharma, A. Ursolic Acid and Oleanolic Acid: Pentacyclic Terpenoids with Promising Anti-Inflammatory Activities. *Recent Pat. Inflamm. Allergy Drug Discov.* **2016**, *10*, 21–33. [CrossRef]

41. Jesus, J.A.; Lago, J.H.G.; Laurenti, M.D.; Yamamoto, E.; Passero, L.F.D. Antimicrobial Activity of Oleanolic and Ursolic Acids: An Update. *Evid.-Based Complement. Altern. Med.* **2015**, *2015*, 1–14. [CrossRef] [PubMed]

42. Shishodia, S.; Majumdar, S.; Banerjee, S.; Aggarwal, B.B. Ursolic acid inhibits nuclear factor-kappaB activation induced by carcinogenic agents through suppression of IkappaBalpha kinase and p65 phosphorylation: Correlation with down-regulation of cyclooxygenase 2, matrix metalloproteinase 9, and cyclin D1. *Cancer Res.* **2003**, *63*, 4375–4383. [PubMed]

43. Sultana, N. Clinically useful anticancer, antitumor, and antiwrinkle agent, ursolic acid and related derivatives as medicinally important natural product. *J. Enzym. Inhib. Med. Chem.* **2011**, *26*, 616–642. [CrossRef]

44. Jayaprakasam, B.; Olson, L.K.; Schutzki, R.E.; Tai, M.-H.; Nair, M.G. Amelioration of Obesity and Glucose Intolerance in High-Fat-Fed C57BL/6 Mice by Anthocyanins and Ursolic Acid in Cornelian Cherry (Cornus mas). *J. Agric. Food Chem.* **2006**, *54*, 243–248. [CrossRef] [PubMed]

45. Sundaresan, A.; Radhiga, T.; Pugalendi, K.V. Effect of ursolic acid and Rosiglitazone combination on hepatic lipid accumulation in high fat diet-fed C57BL/6J mice. *Eur. J. Pharmacol.* **2014**, *741*, 297–303. [CrossRef]

46. Aggarwal, V.; Tuli, H.S.; Varol, A.; Thakral, F.; Yerer, M.B.; Sak, K.; Varol, M.; Jain, A.; Khan, A.; Sethi, G. Role of Reactive Oxygen Species in Cancer Progression: Molecular Mechanisms and Recent Advancements. *Biomolecules* **2019**, *9*, 735. [CrossRef] [PubMed]

47. Ziech, D.; Franco, R.; Georgakilas, A.G.; Georgakila, S.; Malamou-Mitsi, V.; Schoneveld, O.; Pappa, A.; Panagiotidis, M. The role of reactive oxygen species and oxidative stress in environmental carcinogenesis and biomarker development. *Chem. Interact.* **2010**, *188*, 334–339. [CrossRef] [PubMed]

48. Sharma, N.; Dobhal, M.P.; Joshi, Y.C.; Chahar, M.K. Flavonoids: A versatile source of anticancer drugs. *Pharmacogn. Rev.* **2011**, *5*, 1–12. [CrossRef]

49. Ghasemzadeh, A.; Jaafar, H. Profiling of phenolic compounds and their antioxidant and anticancer activities in pandan (Pandanus amaryllifolius Roxb.) extracts from different locations of Malaysia. *BMC Complement. Altern. Med.* **2013**, *13*, 341. [CrossRef]

50. Grever, M.R.; Schepartz, S.A.; Chabner, B.A. The National Cancer Institute: Cancer drug discovery and development program. *Semin. Oncol.* **1992**, *19*, 622–638. [PubMed]

51. Goldin, A.; Venditti, J.M.; Macdonald, J.S.; Muggia, F.M.; Henney, J.E.; Devita, V.T., Jr. Current results of the screening program at the Division of Cancer Treatment, National Cancer Institute. *Eur. J. Cancer* **1981**, *17*, 129–142. [CrossRef]

52. Protocols for screening chemical agents and natural products against animal tumors and other biological systems: Drug Evaluation Branch. *Nat. Cancer Inst.* **1962**, *3*, 1–107.

53. Yin, R.; Li, T.; Tian, J.X.; Xi, P.; Liu, R.H. Ursolic acid, a potential anticancer compound for breast cancer therapy. *Crit. Rev. Food Sci. Nutr.* **2017**, *58*, 568–574. [CrossRef]

54. Yamaguchi, H.; Noshita, T.; Kidachi, Y.; Umetsu, H.; Hayashi, M.; Komiyama, K.; Funayama, S.; Ryoyama, K. Isolation of Ursolic Acid from Apple Peels and Its Specific Efficacy as a Potent Antitumor Agent. *J. Health Sci.* **2008**, *54*, 654–660. [CrossRef]

55. Jäger, S.; Trojan, H.; Kopp, T.; Laszczyk, M.N.; Scheffler, A. Pentacyclic Triterpene Distribution in Various Plants—Rich Sources for a New Group of Multi-Potent Plant Extracts. *Molecules* **2009**, *14*, 2016–2031. [CrossRef]

56. Liu, P.; Du, R.; Yu, X. Ursolic Acid Exhibits Potent Anticancer Effects in Human Metastatic Melanoma Cancer Cells (SK-MEL-24) via Apoptosis Induction, Inhibition of Cell Migration and Invasion, Cell Cycle Arrest, and Inhibition of Mitogen-Activated Protein Kinase (MAPK)/ERK Signaling Pathway. *Med. Sci. Monit.* **2019**, *25*, 1283–1290. [PubMed]

57. Feng, X.; Su, X. Anticancer effect of ursolic acid via mitochondria-dependent pathways. *Oncol. Lett.* **2019**, *17*, 4761–4767. [CrossRef]

58. Rieger, A.M.; Nelson, K.L.; Konowalchuk, J.D.; Barreda, D.R. Modified annexin V/propidium iodide apoptosis assay for accurate assessment of cell death. *J. Vis. Exp.* **2011**, *2011*, e2597. [CrossRef] [PubMed]

59. Redza-Dutordoir, M.; Averill-Bates, D.A. Activation of apoptosis signalling pathways by reactive oxygen species. *Biochim. Biophys. Acta (BBA) Mol. Cell Res.* **2016**, *1863*, 2977–2992. [CrossRef] [PubMed]

60. Vo, V.G.; An, S.S.A.; Hulme, J.P. Mitochondrial therapeutic interventions in Alzheimer's disease. *J. Neurol. Sci.* **2018**, *395*, 62–70.

61. Fiers, W.; Beyaert, R.; Declercq, W.; Vandenabeele, P. More than one way to die: Apoptosis, necrosis and reactive oxygen damage. *Oncogene* **1999**, *18*, 7719–7730. [CrossRef] [PubMed]

62. Mates, J.M.; Sanchez-Jimenez, F.M. Role of reactive oxygen species in apoptosis: Implications for cancer therapy. *Int. J. Biochem. Cell Biol.* **2000**, *32*, 157–170. [CrossRef]

63. Suski, J.M.; Lebiedzińska-Arciszewska, M.; Bonora, M.; Pinton, P.; Duszynski, J.; Wieckowski, M.R. Relation Between Mitochondrial Membrane Potential and ROS Formation. *Methods Mol. Biol.* **2011**, *810*, 183–205.

64. Waterhouse, N.J.; Goldstein, J.C.; Von Ahsen, O.; Schuler, M.; Newmeyer, D.D.; Green, D.R. Cytochrome C Maintains Mitochondrial Transmembrane Potential and Atp Generation after Outer Mitochondrial Membrane Permeabilization during the Apoptotic Process. *J. Cell Biol.* **2001**, *153*, 319–328. [CrossRef] [PubMed]

65. Ricci, J.E.; Gottlieb, R.A.; Green, D.R. Caspase-mediated loss of mitochondrial function and generation of reactive oxygen species during apoptosis. *J. Cell Biol.* **2003**, *160*, 65–75. [CrossRef]

66. Xie, L.-L.; Shi, F.; Tan, Z.; Li, Y.; Bode, A.M.; Cao, Y. Mitochondrial network structure homeostasis and cell death. *Cancer Sci.* **2018**, *109*, 3686–3694. [CrossRef]

67. Zhou, W.; Yuan, X.; Zhang, L.; Su, B.; Tian, D.; Li, Y.; Zhao, J.; Wang, Y.-M.; Peng, S. Overexpression of HO-1 assisted PM2.5-induced apoptosis failure and autophagy-related cell necrosis. *Ecotoxicol. Environ. Saf.* **2017**, *145*, 605–614. [CrossRef]

68. Eisenberg-Lerner, A.; Bialik, S.; Simon, H.-U.; Kimchi, A. Life and death partners: Apoptosis, autophagy and the cross-talk between them. *Cell Death Differ.* **2009**, *16*, 966–975. [CrossRef]

69. Peixoto, M.S.; Galvão, M.F.D.O.; De Medeiros, S.R.B. Cell death pathways of particulate matter toxicity. *Chemosphere* **2017**, *188*, 32–48. [CrossRef]

70. Greenwell, M.; Rahman, P.K. Medicinal Plants: Their Use in Anticancer Treatment. *Int. J. Pharm. Sci. Res.* **2015**, *6*, 4103–4112.

71. Seca, A.M.L.; Pinto, D.C.G.A. Plant Secondary Metabolites as Anticancer Agents: Successes in Clinical Trials and Therapeutic Application. *Int. J. Mol. Sci.* **2018**, *19*, 263. [CrossRef]

72. Santos, U.P.; Campos, J.F.; Torquato, H.F.V.; Paredes-Gamero, E.J.; Carollo, C.A.; Estevinho, L.M.; Souza, K.D.P.; Dos Santos, U.P. Antioxidant, Antimicrobial and Cytotoxic Properties as Well as the Phenolic Content of the Extract from Hancornia speciosa Gomes. *PLoS ONE* **2016**, *11*, e0167531. [CrossRef] [PubMed]

73. Casagrande, J.C.; Macorini, L.F.B.; Antunes, K.Á.; Dos Santos, U.P.; Campos, J.F.; Dias-Júnior, N.M.; Sangalli, A.; Cardoso, C.A.L.; Vieira, M.D.C.; Rabelo, L.A.; et al. Antioxidant and Cytotoxic Activity of Hydroethanolic Extract from Jacaranda decurrens Leaves. *PLoS ONE* **2014**, *9*, e112748. [CrossRef]

74. Campos, J.F.; Espindola, P.P.D.T.; Torquato, H.F.V.; Vital, W.; Justo, G.Z.; Silva, D.B.; Carollo, C.A.; Souza, K.D.P.; Paredes-Gamero, E.J.; Dos Santos, E.L. Leaf and Root Extracts from Campomanesia adamantium (Myrtaceae) Promote Apoptotic Death of Leukemic Cells via Activation of Intracellular Calcium and Caspase-3. *Front. Pharmacol.* **2017**, *8*, 466. [CrossRef] [PubMed]

75. Baldivia, D.S.; Leite, D.F.; De Castro, D.T.H.; Campos, J.F.; Dos Santos, U.P.; Paredes-Gamero, E.J.; Carollo, C.A.; Silva, D.B.; Souza, K.D.P.; Dos Santos, E.L. Evaluation of In Vitro Antioxidant and Anticancer Properties of the Aqueous Extract from the Stem Bark of Stryphnodendron adstringens. *Int. J. Mol. Sci.* **2018**, *19*, 2432. [CrossRef] [PubMed]

Sample Availability: Samples of the compounds are available from the authors.

Article

Anti-Cancer Effect of 3-Hydroxy-β-Ionone Identified from *Moringa oleifera* Lam. Leaf on Human Squamous Cell Carcinoma 15 Cell Line

Thitiya Luetragoon [1,2], Rungnapa Pankla Sranujit [3], Chanai Noysang [3], Yordhathai Thongsri [1], Pachuen Potup [1], Nungruthai Suphrom [4], Nitra Nuengchamnong [5] and Kanchana Usuwanthim [1,*]

[1] Cellular and Molecular Immunology Research Unit, Faculty of Allied Health Sciences, Naresuan University, Phitsanulok 65000, Thailand; nok.hong.yok49@gmail.com (T.L.); yordhathai.k@gmail.com (Y.T.); pachuenp@nu.ac.th (P.P.)

[2] Faculty of Medicine, School of Human Development and Health, University of Southampton, Southampton SO16 6YD, UK

[3] Thai Traditional Medicine College, Rajamangala University of Technology Thanyaburi, Pathum Thani 12130, Thailand; rungnapa_s@rmutt.ac.th (R.P.S.); chanai_n@rmutt.ac.th (C.N.)

[4] Department of Chemistry, Faculty of Science and Center of Excellence for Innovation in Chemistry, Naresuan University, Phitsanulok 65000, Thailand; suphrom.n1@gmail.com

[5] Science Laboratory Centre, Faculty of Science, Naresuan University, Phitsanulok 65000, Thailand; Nitran@nu.ac.th

* Correspondence: kanchanau@nu.ac.th; Tel.: +66-55-966-411; Fax: +66-55-966-234

Academic Editors: José Antonio Lupiáñez, Amalia Pérez-Jiménez, Eva E. Rufino-Palomares and José Rubén Tormo
Received: 9 May 2020; Accepted: 1 August 2020; Published: 5 August 2020

Abstract: Squamous cell carcinoma is the most common type of head and neck cancer worldwide. Radiation and chemotherapy are general treatments for patients; however, these remedies can have adverse side effects and tumours develop drug resistance. Effective treatments still require improvement for cancer patients. Here, we investigated the anti-cancer effect of *Moringa oleifera* (MO) Lam. leaf extracts and their fractions, 3-hydroxy-β-ionone on SCC15 cell line. SCC15 were treated with and without MO leaf extracts and their fractions. MTT assay was used to determine cell viability on SCC15. Cell cycle and apoptosis were evaluated by the Muse™ Cell Analyser. Colony formation and wound closure analysis of SCC15 were performed in 6-well plates. Apoptosis markers were evaluated by immunoblotting. We found that Moringa extracts and 3-HBI significantly inhibited proliferation of SCC15. Moreover, they induced apoptosis and cell cycle arrest at G2/M phase in SCC15 compared to the untreated control. MO extracts and 3-HBI also inhibited colony formation and cell migration of SCC15. Furthermore, we observed the upregulation of cleaved caspase-3 and Bax with downregulation of anti-apoptotic Bcl-2, indicating the induction of cancer cell apoptosis. Our results revealed that MO extracts and 3-HBI provided anti-cancer properties by inhibiting progression and inducing apoptosis of SCC15.

Keywords: squamous cell carcinoma; *Moringa oleifera*; anti-cancer; 3-hydroxy-β-ionone

---

1. Introduction

Cancer is a noncommunicable disease and the leading cause of death worldwide. Around 18.1 million new cases of cancer and 9.6 million deaths from the disease were reported in 2018 [1]. Global incidences of head and neck squamous cell carcinoma (HNSCC) were reported at more than 830,000 cases with 430,000 deaths each year [2]. Major risk factors of HNSCC are high smoking levels and alcohol consumption [3]. HNSCC arises from the mucosal surfaces at various sites including skin, nasal cavity, paranasal sinuses, oral cavity, salivary glands, pharynx and larynx [4]. Treatment for HNSCC usually involves therapy with surgery, radiation, chemotherapy, targeted therapy, immunotherapy and combination therapy [2]. Nevertheless, these treatments result in adverse effects including nausea, vomiting, fatigue, mucositis, dysphagia and dermatitis [5]. Major hallmarks during cancer development include the ability to proliferate, evade apoptosis, uncontrolled replicative potential, induction of angiogenesis and tissue invasion and metastasis to other organs. Hence, many drugs and treatment methods have been developed to interfere with each step and impede tumour growth and progression [6]. The goal of medical scientists and researchers is to improve the best treatment for good quality of life for cancer patients. Therefore, targeted therapy and alternative low toxic treatments have received intense focus. Several studies have examined the effects of natural products against tumours by inducing apoptosis via the P53 tumour suppressor and reactive oxygen species production [7–9].

Moringa oleifera Lam. (MO) is known as the miracle tree and is widely cultivated in Asia and Africa. All the different parts of MO have been reported to have medicinal use [10]. The leaves of this plant have been intensively studied and contain high amounts of vitamins, carotenoids, polyphenols, phenolic acids, flavonoids, alkaloids, glucosinolates, isothiocyanates, tannins and saponins [11]. Many studies have reported on the biological activities of Moringa leaf such as antidiabetic [12], antioxidant [13], antibacterial [14] and kidney and hepatic protective effect [15,16]. Our previous study demonstrated that MO leaf extract provides anti-inflammatory potential by reducing the production of pro-inflammatory mediators such as interleukin-6, tumour necrosis factor-α and cyclooxygenase-2 via inactivation of NF-κB, inhibiting both IκB-α degradation and nuclear translocation of p65 [17,18]. 3-hydroxy-β-ionone (3-HBI) derived from MO leaf extract (Figure 1) had potent anti-inflammatory effects [17], while *in vitro* studies showed that soluble extract from MO leaf induced apoptosis and inhibited tumour cell growth in human non-small cell lung cancer A549 and human hepatocellular carcinoma HepG2 cells [19,20]. Another *in vitro* study reported that MO leaf extract and its compounds including eugenol, isopropyl isothiocyanate, D-allose and hexadeconoic acid ethyl ester decreased cell motility and colony formation, inhibited cell growth and triggered cell apoptosis against breast cancer and colorectal cancer cell lines [21]. Astragalin and isoquercetin from bioactive fractions of *M. oleifera* leaf extract suppressed proliferation of HCT116 colon cancer cells by downregulation of ERK1/2 phosphorylation [22]. In addition, glucomoringin from *Moringa oleifera* induced oxidative stress and apoptosis via p53 and Bax activation and Bcl-2 inhibition in human astrocytoma grade IV CCF-STTG1 cells [23], and also promoted apoptosis of SH-SY5Y human neuroblastoma cells through the modulation of NF-κB and apoptotic factors [24]. However, the effect of MO leaf extract on squamous cell carcinoma (SCC) 15 cell line remains unknown.

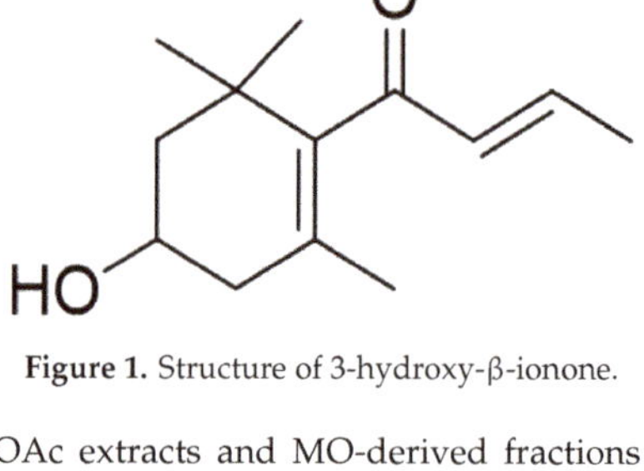

Figure 1. Structure of 3-hydroxy-β-ionone.

In this study, crude EtOAc extracts and MO-derived fractions were tested for anti-SCC15 activities. Active MO-derived fractions were fraction no. 6, sub-fraction no. 6.17.2, LC-MS base peak chromatogram no. 6 identified as 3-HBI (BPC6) and 3-HBI. Our findings revealed that MO leaf extract and its active compound, 3-HBI strongly inhibited tumour cell growth and triggered apoptosis via over-expression of cleaved caspase-3 and Bax, while down-regulating the anti-apoptotic Bcl-2.

2. Results

2.1. Cellular Cytotoxicity of MO Extract, Fraction, and Sub-Fraction on Human Monocyte-Derived Macrophages and SCC15

MTT was developed to evaluate the optimal concentration of antiproliferative effect of Moringa extracts, compound, and drugs. Human monocyte-derived macrophages (MDMs) and SCC15 were treated with different concentrations of extracts, compound, and cisplatin for 24 h. The half maximal inhibitory concentrations (IC_{50}) values of Moringa extracts, compound and drug are shown in Figure 2A,B. A five percent inhibitory concentration (IC_5) results of 3-HBI and cisplatin on MDMs were 18.46 µg/mL and 5.32 µg/mL, respectively (Figure 2A). While IC_5 of crude ethyl acetate (EtOAc) and fraction no. 6 were 26.84 µg/mL and 84.89 µg/mL [17]. These concentrations were used as the non-toxic optimal concentration for cell culture treatment. IC_{50} values of 3-HBI and cisplatin were 487.53 µg/mL and 21.33 µg/mL, respectively (Figure 2A). The IC_{50} values of crude EtOAc, fraction no. 6, 3-HBI and cisplatin on SCC15 cell line were 214.28, 114.55, 243.22 and 28.44 µg/mL, respectively (Figure 2B). Crude EtOAc, fraction no. 6 and 3-HBI showed a strong effect in inhibiting the proliferation of SCC15 cancer cells with lower IC_{50} values compared to primary MDM.

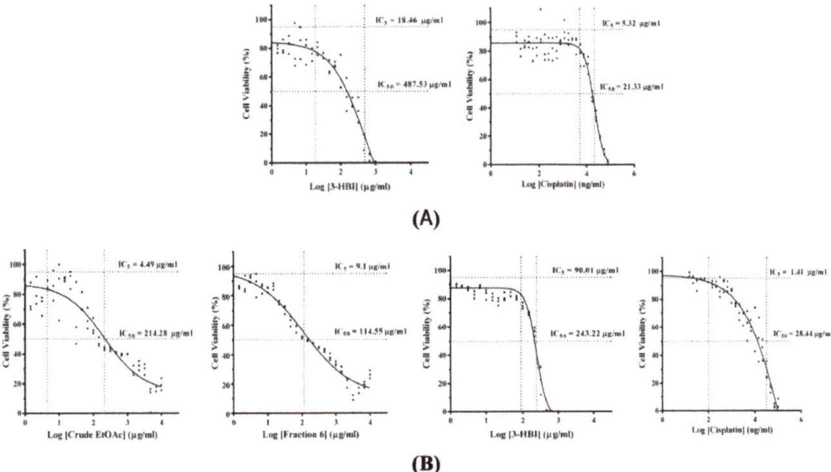

Figure 2. Investigation of the effect of MO crude extracts, fraction no. 6, 3-HBI and cisplatin on cell viability of MDMs and SCC15 cell lines. MTT assay was performed after cell treatment for 24 h. (**A**) IC_5 and IC_{50} results of MDMs after treatment with different concentrations of crude extract, fraction no. 6, 3-HBI and cisplatin. (**B**) IC_{50} results for antiproliferative effect of crude extract, fraction no.6, 3-HBI and cisplatin on SCC15. IC: inhibitory concentration; 3-HBI: 3-hydroxy-β-ionone.

2.2. Effect of MO Extract, Fraction, and Sub-Fraction on Cell Cycle of SCC15

Cell cycle assay was assessed using Muse™ Cell Cycle Kit. The nuclear DNA of the cell line was intercalated with propidium iodide (PI). Cells were discriminated at different phases of the cell cycle based on differential DNA content including G0/G1, S and G2/M phase. To investigate the effect of cisplatin, crude MO extract, fraction no. 6, sub-fraction no. 6.17.2, BPC6 (LC-MS base peak chromatogram no. 6 identified as 3-HBI), and 3-HBI on cell cycle progression, both untreated and treated SCC15 were investigated by Muse™ Cell Analyser. Figure 3A shows the DNA content index histogram of the control and treated cells in each cell cycle phase (G0/G1, S and G2/M). The bar graphs demonstrate that the percentage of cells in G2/M stages. After treatment with cisplatin, crude EtOAc, fraction no. 6, sub-fraction no. 6.17.2, BPC6, and 3-HBI, the G2/M enrichment phase significantly increased in SCC15 compared to the untreated control (Figure 3B). This result indicates that crude

EtOAc, fraction no. 6, sub-fraction no. 6.17.2, BPC6, and 3-HBI significantly increased cell cycle arrest at G2/M phase in SCC15.

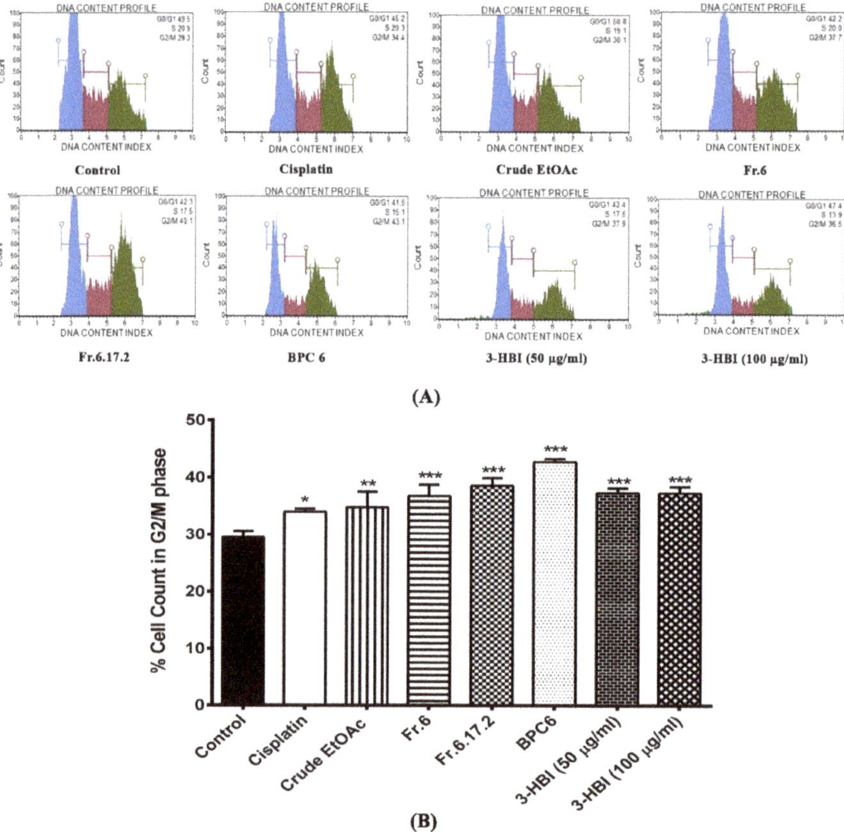

Figure 3. Efficacy of MO extract, fraction no. 6, sub-fraction no. 6.17.2, BPC6, and 3-HBI on distribution of SCC15 in the cell cycle analysed by Muse™ Cell Analyser. (**A**) DNA content index histogram of cell populations in each phase of the cell cycle in SCC15 cell line after treatment with cisplatin, crude EtOAc, Fr.6, Fr.6.17.2, BPC6, and 3-HBI for 24 h. (**B**) Bar graphs of percentage of cell in G2/M phase for SCC15 cell line. Data are presented as means ± SEM. * $p \leq 0.05$, ** $p \leq 0.01$, and *** $p \leq 0.001$ compared to control. Control: untreated SCC15; crude EtOAc: crude ethyl acetate; Fr.6: fraction no. 6; Fr.6.17.2: sub-fraction no. 6.17.2; BPC6: base peak chromatogram no. 6; 3-HBI: 3-hydroxy-β-ionone.

2.3. MO Extract, Fraction, and Sub-Fraction Induce Apoptosis in SCC15

We next evaluated the induction of apoptosis in SCC15 cell line after 24 h of treatment with MO extract, fraction no. 6, sub-fraction no. 6.17.2, BPC6, and 3-HBI. Apoptosis assay was performed by Muse™ Cell Analyser using the Muse™ Annexin V & Dead Cell Kit procedure with dot plots of SCC15 cell line stained with annexin V and 7-AAD (7-amino-actinomycin D). The first and second quadrants represent dead cells and late apoptotic cells, respectively, while the third and fourth quadrants represent live cells and early apoptotic cells, respectively, as shown in Figure 4A. These dot plots reveal that crude EtOAc, fraction no. 6 and sub-fraction no. 6.17.2 triggered late apoptosis and cell death in SCC15 cell line. Interestingly, BPC6 and 3-HBI induced early and late apoptotic cells similar to cisplatin. Figure 4B shows the statistical analysis of total apoptotic cells represented in the form of bar graphs. We observed a significantly increased percentage of total apoptotic cells (early and late apoptosis) after treatment with crude EtOAc, fraction no. 6, sub-fraction no. 6.17.2, BPC6, and 3-HBI compared to the untreated control ($p < 0.001$). The average percentage of total apoptotic cells increased from 4.41% in the control to 31.62% in cisplatin treatment. Moreover, treatment with crude EtOAc, fraction no. 6, sub-fraction no. 6.17.2, BPC6, 3-HBI (50 µg/mL), and 3-HBI (100 µg/mL) enhanced apoptotic cells to 17.85, 26.69, 21.08, 27.60, 14.76 and 17.0%, respectively. Our findings suggested that treatment with crude EtOAc, fraction no. 6, sub-fraction no. 6.17.2, BPC6, and 3-HBI strongly enhanced apoptosis in SCC15 cell line.

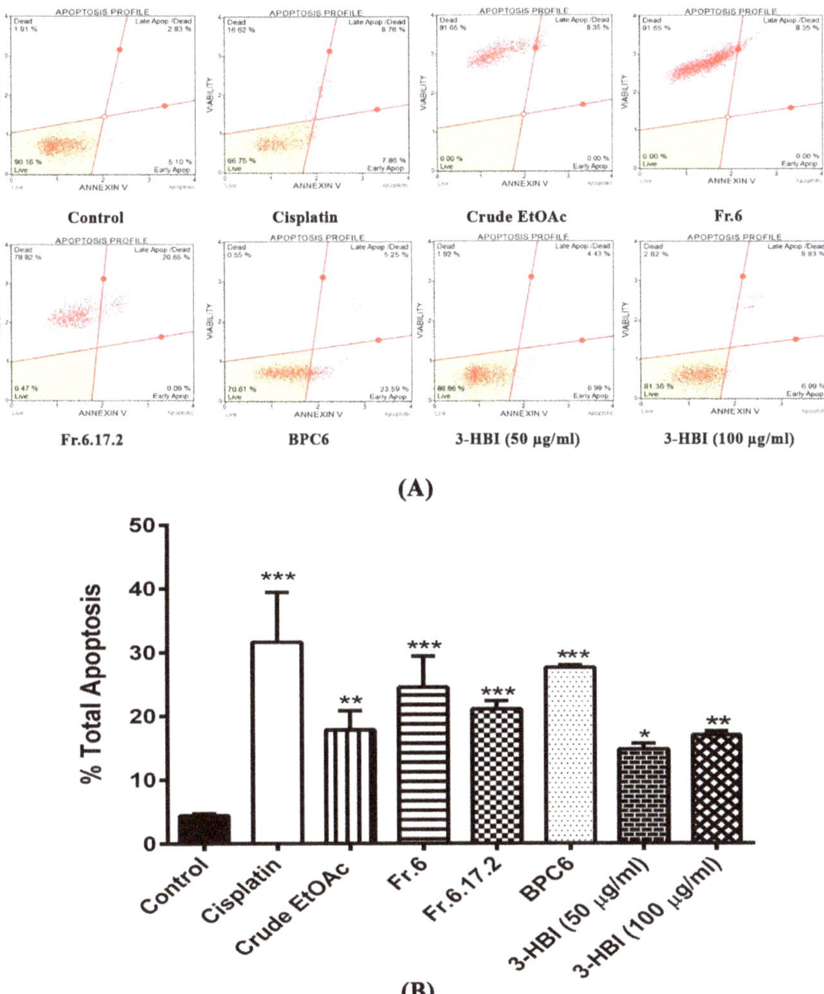

Figure 4. Determination of apoptosis in SCC15 cell line after treatment with MO extract, fraction no. 6, sub-fraction no. 6.17.2 BPC6, and 3-HBI for 24 h. (**A**) Dot plots of annexin V and 7-AAD dual staining showing the percentage of cell populations in each of the four quadrants. (**B**) Bar graphs showing quantitative data of percentage of total apoptotic cells. Results are presented as means ± SEM. * $p \le 0.05$, ** $p \le 0.01$, and *** $p \le 0.001$ compared to control. Control: untreated SCC15; crude EtOAc: crude ethyl acetate; Fr.6: fraction no. 6; Fr.6.17.2: sub-fraction no. 6.17.2; BPC6: base peak chromatogram no. 6; 3-HBI: 3-hydroxy-β-ionone.

2.4. MO Extract and 3-HBI Inhibit Colony Formation of SCC15

The ability of SCC15 cell line to form colonies in the presence or absence of MO extract, fraction no. 6, sub-fraction no. 6.17.2, BPC 6 and 3-HBI for 24 h was studied. SCC15 cell lines were seeded into 6-well plates and incubated for one week. Cells were fixed and stained with 0.5% crystal violet. We found that MO extract, fraction no. 6 and sub-fraction no. 6.17.2 strongly inhibited colony formation of SCC15, similar to cisplatin treatment (Figure 5A,B). Furthermore, colony formation was significantly reduced in a concentration-dependent manner by 3-HBI compared to the control, as shown in Figure 5B. A similar result was observed in SCC15 treated with BPC6. These data confirmed that MO extract, fraction no. 6, sub-fraction no. 6.17.2, BPC6 and 3-HBI showed potential to inhibit the formation of colonies in SCC15 cell lines as an important factor in cancer survival and progression.

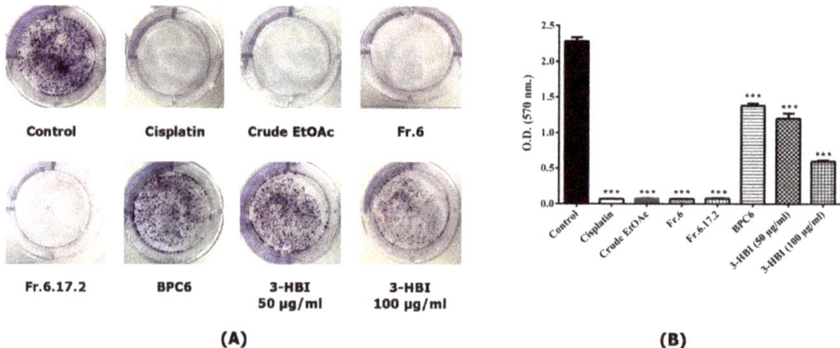

Figure 5. Colony formation of SCC15 cell line performed in 6-well plates with cells stained by crystal violet. (**A**) Untreated control and cells treated with MO extract, fraction no. 6, sub-fraction no. 6.17.2, BPC 6 and 3-HBI for 24 h. (**B**) Colony quantification measured by a microplate reader at OD 570 nm. Data are presented as means ± SEM. *** $p \leq 0.001$, compared to control. Control: untreated SCC15; crude EtOAc: crude ethyl acetate; Fr.6: fraction no. 6; Fr.6.17.2: sub-fraction no. 6.17.2; BPC6: base peak chromatogram no. 6; 3-HBI: 3-hydroxy-β-ionone.

2.5. MO Extract and 3-HBI Inhibit Migration of SCC15

Wound closure assay is a method of *in vitro* study for analysing the migration of cell populations. We evaluated the migrative ability of SCC15 cell line after treatment with and without MO extract, fraction no. 6, sub-fraction no. 6.17.2, BPC6 and 3-HBI for 36 h. Cell monolayers were scratched with a similar size at baseline. We observed a significant inhibition of cell migration after treatment for 6 h. The percentage of wound area for cells treated with cisplatin, crude extract, fraction no. 6, sub-fraction no. 6.17.2, BPC6 and 3-HBI was significantly higher when compared to the control. At 36 h after wound induction, the wound area of the untreated control was almost closed. On the other hand, for the wound areas of cells treated with drugs, all extracts and compounds were still widely open (Figure 6A,B).

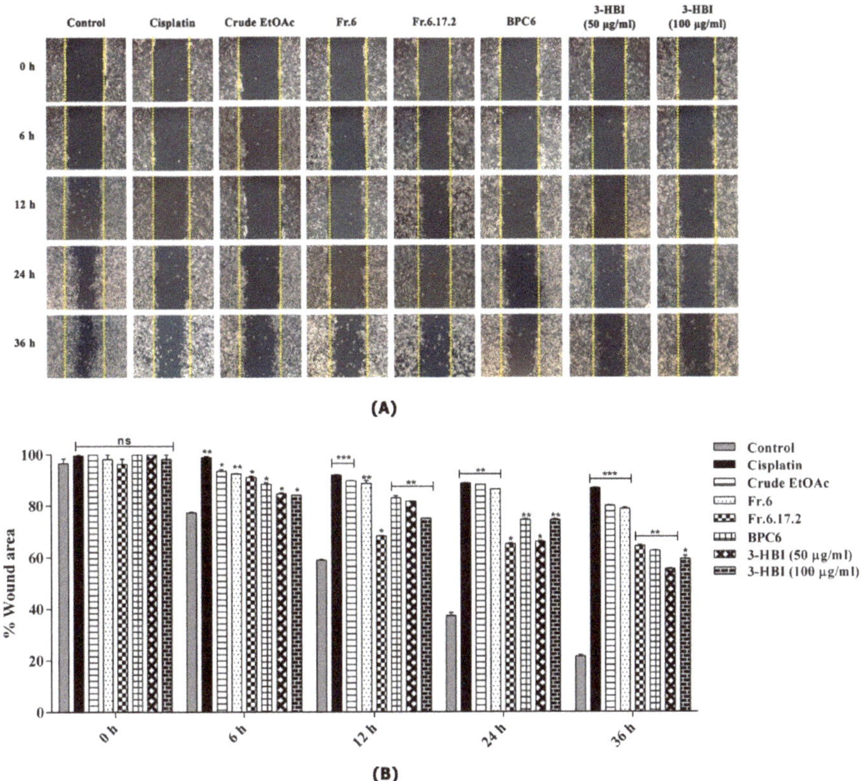

Figure 6. Wound closure analysis of SCC15 cell line for untreated control and cells treated with MO extract, fraction no. 6, sub-fraction no. 6.17.2, BPC 6 and 3-HBI for 36 h. (**A**) Pictures of wound area were taken at 0, 6, 12, 24 and 36 h with an inverted microscope. (**B**) Size of wound area was measured using ImageJ system software. Percentage of wound area was calculated, and bar graphs were represented as means ± SEM. ns = not significant, * $p \leq 0.05$, ** $p \leq 0.01$, and *** $p \leq 0.001$ compared to control. Control: untreated SCC15; crude EtOAc: crude ethyl acetate; Fr.6: fraction no. 6; Fr.6.17.2: sub-fraction no. 6.17.2; BPC6: base peak chromatogram no. 6; 3-HBI: 3-hydroxy-β-ionone.

2.6. Effect of MO Extracts and Their Fractions on Apoptosis Signaling Pathway in SCC15 Cell Line

Western blotting was performed to evaluate apoptosis in SCC15 cells after treatment with MO extract and its fractions. The expression of housekeeping protein β-actin was considered as an equal amount of protein was loaded. The pro-apoptotic Bax was significantly upregulated by treatment with cisplatin, MO crude extract, fraction no. 6, sub-fraction no. 6.17.2, BPC6 and 3-HBI. Moreover, they significantly decreased anti-apoptotic Bcl-2 compared to the control ($p \leq 0.001$). Interestingly, MO extracts and their fractions reduced pro-caspase-3 expression, as well as significantly inducing the activation of cleaved caspase-3 (Figure 7A–E). Our results demonstrated that MO extracts and their fractions showed the potential to induce apoptosis of SCC15 cell line by inducing the activation of cleaved caspase-3 and Bax. Furthermore, these extracts and compound significantly decreased anti-apoptotic Bcl-2, which showed strong efficacy similar to the positive drug control.

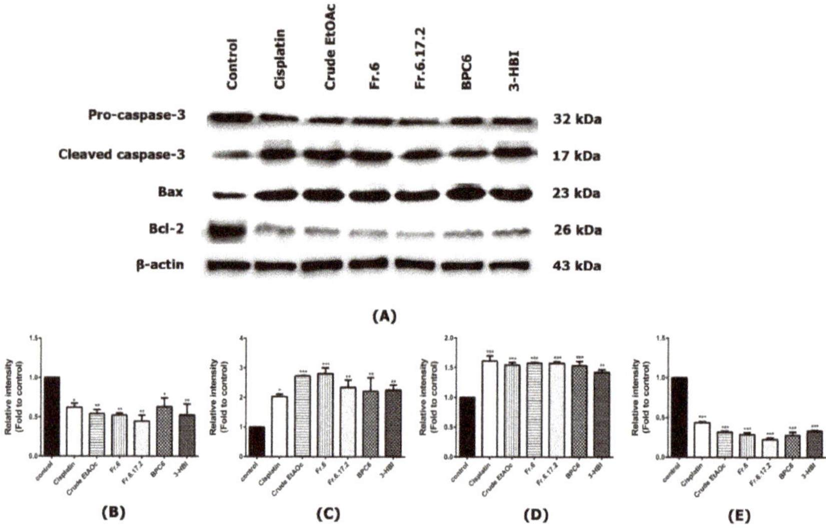

Figure 7. The inhibitory effect of MO extract, fraction no. 6, sub-fraction no. 6.17.2, BPC 6 and 3-HBI on the protein expressions of SCC15 cell line. Cells were treated under different conditions for 24 h. Total proteins were extracted and determined by Western blot analysis. (**A**) Band intensity of total protein levels of pro-caspase-3, cleaved caspase-3, Bax and Bcl-2. β-actin was used as a loading control. (**B**) Relative intensity of pro-caspase-3, (**C**) cleaved caspase-3, (**D**) Bax and (**E**) Bcl-2 were quantified by scanning densitometry and normalised to control. Data are shown as mean ± SEM. * $p \leq 0.05$, ** $p \leq 0.01$, and *** $p \leq 0.001$ compared to control. Control: untreated SCC15; Crude EtOAc: crude ethyl acetate; Fr.6: fraction no. 6; Fr.6.17.2: sub-fraction no. 6.17.2; BPC6: base peak chromatogram no. 6; 3-HBI: 3-hydroxy-β-ionone.

3. Discussion

Cancer cells lack a regulatory system that prevents cell overgrowth. Cancer progression involves a multi-step process including self-sufficiency in proliferative signalling, uncontrolled growth, evading programmed cell death, induction of angiogenesis and inducing invasion and metastasis [6]. Several studies have examined the effects of natural compounds against tumours by inhibiting cancer proliferation and inducing apoptosis. Chikusetsu isolated from *Aralia taibaiensis* induced apoptosis in human prostate cancer [7]. Capsaicin (trans-8-methyl-N-vanillyl-6-nonenamide) is a derivative from chili peppers inhibited the migration of cholangiocarcinoma cells by downregulating metalloproteinase-9 expression [25] and promoting apoptosis by stimulated p53 and Bax expression in HCT116 human colon carcinomas [26]. Arabinogalactan and curcumin have been extensively studied for their anti-cancer properties and both natural products decreased cell growth and significantly increased Bax/Bcl2 ratio as well as cleaved-caspase3 level in MDA-MB-231 human breast cancer cells [8]. Aloe-emodin derived from *Rheum undulatum* L. inhibited proliferation and induced apoptosis via activation of caspase-9 and caspase-3 in SCC15 cells [27]. Our results showed that 3-HBI derived from MO leaf extract inhibited SCC15 cell growth and triggered apoptosis via over-expression of cleaved caspase-3 and Bax, which down-regulated the anti-apoptotic Bcl-2. Additionally, our previous finding indicated that 3-HBI of MO leaf had potent biological anti-inflammatory effects by inhibiting NF-κB translocation in LPS-treated MDMs, leading to down-regulation of pro-inflammatory mediators [17,18]. Thus, 3-HBI exhibited both anti-cancer and anti-inflammatory activities and might be a novel effective therapeutic drug for head and neck cancer since NF-κB signalling pathways are targeted for therapeutic applications in many cancers including HNSCC. Accordingly, the major class of cellular targets controlling NF-κB consists of chemokines, regulators of apoptosis, cell proliferation and cell cycle regulators [28–30]. Curcumin has also been shown to inhibit NF-κB. This regulates several cellular processes including cell growth and survival by suppressing Bcl-2 and cyclin D1, IL-6, COX-2, and MMP-9 protein expression in HNSCC [31]. Erlotinib and EGCG of green tea extract synergistically inhibited HNSCC growth via inhibiting NF-κB in a p53-dependent manner [32].

MO leaf has various medicinal properties including antioxidant, anti-inflammatory, antiulcer, hepatoprotective activities, antibacterial and antifungal activities [33]. Cancer research involving Moringa leaf has been conducted in both *in vitro* and *in vivo* such as Moringa in MDA-MB-231 breast cancer cells, human HCT8 colon cancer cells and mouse melanoma [19,21,34]. This study evaluated the anti-cancer properties of Moringa leaf extract and 3-HBI bioactive compound including cell viability, cell cycle, apoptosis, migration, and colony formation of SCC15 cell line. The MTT assay was performed to evaluate cell viability of human MDMs and SCC15 cell line in the presence of MO leaf extract and its fractions. Our results showed that concentrations of the extract and compound caused cytotoxicity of SCC15 but non-toxicity in normal cells. Interestingly, this result concurred with our findings, indicating cell cycle arrest and an increase in cell apoptosis. We found that MO extracts and fractions caused a significant increase in cell population at the G2/M phase compared to the untreated control ($p \leq 0.001$). A previous study by Al-Asmari et al. (2015), showed similar findings with cell cycle arrest at the G2/M phase in MDA-MB-231 and HCT-8 cancer cell lines after treatment with Moringa leaf extract [21]. Cell migration and colony formation are hallmarks of tumour progression. Wound closure assays allow the observation of cell migration in confluent monolayer cell cultures, while colony formation assay is an *in vitro* technique for studying the survival and proliferation of cancer cells based on the ability of single cells to grow into colonies [35]. In this study, MO leaf fractions were able to significantly inhibit colony formation and cell migration of SCC15 cell line.

Apoptosis is programmed cell death that generally occurs in tissue during development as a homeostatic mechanism. Apoptosis is activated via two pathways, commonly known as intrinsic and extrinsic. The intrinsic pathway is initiated by pro-apoptotic proteins such as Bax and Bad that damage the mitochondrial membrane, leading to release of cytochrome C. Then, the formation of apoptosome complex activates procaspase-9 and stimulates caspase 3-6-7, resulting in apoptosis. The extrinsic pathway is initiated by death receptors at the cell surface to the intracellular signalling pathways. Then, caspase 8 is activated leading to stimulate downstream caspase 3-6-7 [36,37]. In various tumours, pro-apoptotic members are normally downregulated while anti-apoptotic factors are upregulated [6]. Our study showed that cell apoptosis in SCC15 cell line was strongly induced by MO leaf extracts and their fractions. Our results were further confirmed through activating the apoptosis signalling pathway by significantly increasing pro-apoptotic (BAX) and cleaved caspase-3 while suppressing the expression of anti-apoptotic protein Bcl-2 compared to the untreated control. These results indicated that 3-HBI bioactive compound of MO leaf showed strong anti-cancer activity by inducing apoptosis in SCC15 cell line via a caspase-dependent mechanism.

4. Materials and Methods

4.1. Preparation of Moringa Oleifera Leaf Extracts and Compound Identification

Moringa leaves were ground to a powder and extracted at room temperature with EtOAc as described in our previous study [17]. Then, 128 g of crude EtOAc extract was obtained after evaporation of the solvent. Fractions and sub-fractions were separated from the crude extract using flash column chromatography (Merck, Darmstadt, Germany). Gradient elution was performed by solvent system with increasing the polarity gradually including hexane, hexane-EtOAc and EtOAc-Methanol (MeOH). Apocarotenoid monoterpene namely 3-hydroxy-β-ionone, an active compound was identified from the Moringa sub-fraction by LC-ESI-QTOF-MS/MS (Agilent Technologies, Inc., Singapore). The crude EtOAc extracts and MO-derived fractions (fraction no.6, sub-fraction no. 6.17.2, BPC6 and 3-HBI) were tested for anti-cancer activities.

4.2. Monocyte Isolation

Human MDMs were used as a primary normal cell control. Buffy coat was obtained from the Blood Bank, Naresuan University Hospital, Phitsanulok, Thailand. Ethics approval was obtained from the Human Ethics Committee of Naresuan University (IRB no. 1013/60). Buffy coat was diluted with Hank's balanced salt solution (HBSS) and then overlaid on 5 mL Lymphoprep (Stemcell Technologies, Singapore) and centrifuged at 2000 rpm for 30 min. The mononuclear cell layer was collected and washed twice with HBSS buffer. Peripheral blood mononuclear cells (PBMCs) were collected and suspended in 5 mL of RPMI medium. Then, the monocytes were separated by size sedimentation centrifugation using Percoll (GE Healthcare Bio-Sciences AB, Uppsala, Sweden). The PBMC suspension was carefully overlaid on 10 mL of 46% Percoll solution and centrifuged for 30 min at 2000 rpm. Monocytes between Percoll were collected and washed with HBSS, followed by centrifugation for 10 min at 1300 rpm. Isolated monocytes were cultured in RPMI 1640 and supplemented with 10% fetal bovine serum (FBS) and 1% Antibiotic-Antimycotic purchased from Gibco™ (Thermo Fisher Scientific, Inc., New York, NY, USA). Cells were incubated at 37 °C with 5% CO_2 for 2 weeks with media replacement every 3 days.

4.3. Cell Line and Culture Conditions

Squamous cell carcinoma 15 (ATCC® CRL-1623™) was purchased from ATCC. Cells were cultured in Dulbecco's modified Eagle's medium (DMEM) and Ham's F12 medium (Caisson Labs Inc., Smithfield, UT, USA) containing 10% FBS. Cells were incubated at 37 °C with 5% CO_2 with medium renewal every 2–3 days. SCC15 cells were seeded in 24-well plates at a density of 5×10^4 cells/well and incubated for 24 h. The cells were treated with MO crude extracts, their fractions and 3-HBI (Santa Cruz Biotechnology, Inc., Dallas, TX, USA) for 24 h. Untreated and Cisplatin (Sigma-Aldrich, St. Louis, MO, USA) treated SCC15 cells were used as control conditions.

4.4. Cell Viability Assay

MTT assay was used to determine the growth inhibitory role of MO leaf extract. MDMs and SCC15 were seeded in a 96-well plate at a density of 1×10^4 cells/well and treated with various concentrations of extract, compound, and cisplatin. Cells were incubated at 37 °C for 24 h. Then, 50 µL of MTT (0.5 mg/mL) (Invitrogen, Carlsbad, CA, USA) in medium-free serum was added in all samples and they were incubated at 37 °C for 3 h. MTT reagent was removed and formazan crystals were dissolved in 100 µL of DMSO. The absorbance of formazan solution was measured at 590 nm by a microplate reader (PerkinElmer, Inc., Waltham, MA, USA). This method followed previous protocol [38]. IC_{50} of cisplatin, crude EtOAc extract and 3-HBI were calculated by concentration response relationships/sigmoidal curve fitting analysis. IC_5 was selected as non-toxic for cellular experiments.

4.5. Cell Cycle Analysis

To confirm the growth inhibitory role of MO extracts and derivative compounds, cell cycle assay was analysed using Muse™ Cell Cycle Kit (Merck, Darmstadt, Germany) following the manufacturer's protocol. Experimental conditions of SCC15 were harvested using trypsin/EDTA solution (Thermo Fisher Scientific) and incubated at 37 °C for 5 min. Then, 200 µL of completed DMEM HamF12 were added to stop the reaction of trypsin. Cells were aspirated and centrifuged at 1500 rpm for 5 min, then fixed with 70% ethanol and incubated for at least 3 h at −20 °C. Cells were washed twice with cold PBS (phosphate buffer saline), resuspended in 200 µL of Muse™ Cell cycle reagent, mixed gently and incubated for 30 min at room temperature in the dark. Cell cycle stage was then analysed by Muse™ Cell Analyser.

4.6. Cell Apoptosis Analysis

Muse™ Annexin V & Dead Cell Kit (Merck, Darmstadt, Germany) was used for the apoptosis study. SCC15 cells from all experimental conditions were harvested by trypsin/EDTA solution as described in cell cycle assay. Cells were washed in PBS and resuspended in medium with 1% FBS. Then, 100 µL of Muse™ Annexin V & Dead Cell Reagent were added, mixed gently, and incubated for 20 min at room temperature in the dark. Cell apoptosis was measured using Muse™ Cell Analyser following the manufacturer's protocol.

4.7. Colony Formation Assay

Colony formation assay was used to study the potentiality of a single cell forming colonies. This assay followed the previous description [35]. SCC15 cell line was seeded into 6-well plates at a density of 500 cells/well and incubated for 24 h in standard culture conditions at 37 °C. Cells were then treated with drugs and Moringa extracts including crude EtOAc, fraction no. 6, sub-fraction no. 6.17.2, BPC6 and 3-HBI for 24 h. Cells were incubated for 1 week with media replacement every 3 days and then fixed with 10% neutral buffer formalin solution for 30 min. The fixative reagent was removed, and the cells were stained with 2 mL of 0.5% crystal violet and incubated for 60 min at room temperature on a rotator. Cells were washed 4 times in a stream of tap water and the plate was air dried for at least 2 h at room temperature. Then, 2 mL of methanol were added to each well and the plate was incubated for 20 min at room temperature on a rotator. Optical density of each well was measured at 570 nm with a microplate reader (PerkinElmer, Inc.).

4.8. Wound Closure Assay

Cell migration of SCC15 cell line was evaluated by wound closure assay modified from a previous method [39]. SCC15 cells were seeded into 6-well plates at a density of 1×10^6 cells/mL and incubated at 37 °C until reaching 80% confluence as a monolayer. The cell monolayer was scraped in a straight line with a SPLScar Scratcher (SPL Life Sciences, Gyeonggi-do, Korea). Detached cells were removed and washed twice with 1 mL of medium. Cells were treated as described in the colony formation assay and incubated for 36 h. Snapshots were taken of the experimental cell plates at several time points including 6, 12, 24 and 36 h using an inverted microscope (Carl Zeiss Microscopy GmbH, Jena, Germany). The distance of the wound area was analysed by ImageJ system software.

4.9. SDS-PAGE and WESTERN BLOT ANALYSIS

Total proteins of SCC15 cell line from each condition were extracted by ice-cold RIPA lysis buffer (Bio Basic Inc., New York, NY, USA) in the presence of Halt Protease/Phosphatase Inhibitor Cocktails (Thermo Fisher Scientific) and centrifuged at 12,000 rpm for 15 min at 4 °C. Quantification of total protein concentration was performed by Bradford Coomassie-binding, colourimetric method. Protein extract was mixed with an equal volume of 4X Laemmli loading buffer and heated to denature at 95 °C for 5 min. Samples were loaded into wells of 12% SDS-polyacrylamide gel electrophoresis (PAGE) with proteins separated according to molecular weight and transferred to a polyvinylidene fluoride membrane (Bio-Rad Laboratories, Inc., Hercules, CA, USA). For Western blot analysis, the membrane was blocked for 2 h at room temperature with blocking buffer containing 5% bovine serum albumin (Capricorn Scientific GmbH, Ebsdorfergrund, Germany) in Tris-buffered saline with Tween 20 (TBST) buffer. The membrane was blotted using primary antibodies specific to cleaved-caspase 3 (Asp175, p17) (Affinity Biosciences, Cincinnati, OH, USA), β-actin, pro-caspase 3, Bax and Bcl-2 (Santa Cruz Biotechnology, Inc., Dallas, TX, USA) overnight at 4 °C on a rotator. The membrane was washed with TBST and incubated with horseradish peroxidase-conjugated goat anti-mouse IgG (H + L) secondary antibody (Thermo Fisher Scientific) for one hour at room temperature. The membrane was observed by soaking in chemiluminescence substrate for 5 min and placed in a ChemiDoc XRS+

Imaging System (Bio-Rad Laboratories, Inc., Hercules, CA, USA). The chemiluminescence signal of the blotted membrane was detected by Image Studio Lite software (LI-COR Corporate, Lincoln, NE, USA).

4.10. Statistical Analysis

All experiments were performed in three independent batches of experiments to provide accurate results. One-way ANOVA and the Bonferroni multiple comparisons test were used for data analysis with GraphPad Prism software. A confidence interval of 95% ($p = 0.05$) was used in all statistical analyses.

5. Conclusions

Moringa extract and its compound, 3-HBI suppressed cell proliferation and induced apoptosis in SCC15 cell line through the activation of cleaved caspase-3 and Bax as well as suppressing anti-apoptotic factor, Bcl-2. Moreover, treatment with MO extract and 3-HBI significantly increased the G2/M phase arrest of cell cycle progression in SCC15 (Figure 8). We observed a significant inhibition of cell migration as well as colony formation in SCC15 cells after treatment with crude extract and 3-HBI. Our findings suggest that MO extract and 3-HBI have potential as an anti-cancer treatment. This is the first report concerning MO extract and 3-HBI activity against SCC15 cell line.

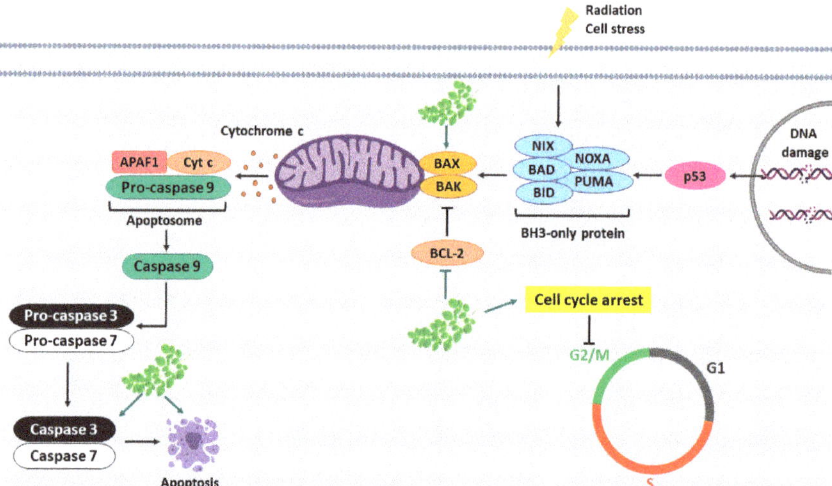

Figure 8. Apoptosis signalling pathway and cell cycle arrest in response to Moringa extract and 3-HBI. MO crude extract and 3-HBI induced G2/M cell cycle arrest and apoptosis in SCC15 cell line by increasing the Bax and cleaved caspase-3 expression and downregulating the levels of Bcl-2.

Author Contributions: This research was contributed by authors "conceptualization, K.U., R.P.S., C.N. and Y.T.; methodology, T.L. and K.U.; software, N.S., N.N. and T.L.; validation, P.P. and K.U.; formal analysis, T.L.; investigation, T.L.; resources, N.S. and N.N.; data curation, K.U., R.P.S. and Y.T.; writing—original draft preparation, T.L.; writing—review and editing, K.U., R.P.S., C.N., N.S. and N.N.; visualization, K.U. and T.L.; supervision, K.U.; project administration, K.U.; funding acquisition, K.U., R.P.S., C.N., Y.T. and P.P." All authors have read and agreed to the published version of the manuscript.

Funding: This study was supported by the Thailand Science Research and Innovation (PHD60I0017 and RSA6080042) and National Research Council of Thailand (NRCT).

Acknowledgments: We would like to thank Khaolaor Laboratories Co., Ltd., Samutprakan, Thailand for providing *Moringa oleifera* dried powder, Assoc. Prof Kornkanok Ingkaninan from Bioscreening Unit, Faculty of Pharmaceutical Sciences and Center of Excellence for Innovation in Chemistry, Naresuan University and Assoc. Prof Chavi Yenjai from Natural Products Research Unit, Department of Chemistry and Center of Excellence for Innovation in Chemistry, Faculty of Science, Khon Kaen University, Thailand for helpful giving reagents and suggestions.

Conflicts of Interest: The authors have declared that there is no conflict of interest.

References

1. Bray, F.; Ferlay, J.; Soerjomataram, I.; Siegel, R.L.; Torre, L.A.; Jemal, A. Global cancer statistics 2018: GLOBOCAN estimates of incidence and mortality worldwide for 36 cancers in 185 countries. *CA Cancer J. Clin.* **2018**, *68*, 394–424. [CrossRef]
2. Cramer, J.D.; Burtness, B.; Le, Q.T.; Ferris, R.L. The changing therapeutic landscape of head and neck cancer. *Nat. Rev. Clin. Oncol.* **2019**, *16*, 669–683. [CrossRef]
3. Castellanos, M.R.; Pan, Q. Novel p53 therapies for head and neck cancer. *World J. Otorhinolaryngol. Head Neck Surg.* **2016**, *2*, 68–75. [CrossRef] [PubMed]
4. Walden, M.J.; Aygun, N. Head and neck cancer. *Semin. Roentgenol.* **2013**, *48*, 75–86. [CrossRef] [PubMed]
5. Cooper, J.S.; Pajak, T.F.; Forastiere, A.A.; Jacobs, J.; Campbell, B.H.; Saxman, S.B.; Kish, J.A.; Kim, H.E.; Cmelak, A.J.; Rotman, M.; et al. Postoperative concurrent radiotherapy and chemotherapy for high-risk squamous-cell carcinoma of the head and neck. *N. Engl. J. Med.* **2004**, *350*, 1937–1944. [CrossRef] [PubMed]
6. Hanahan, D.; Weinberg, R.A. Hallmarks of cancer: The next generation. *Cell* **2011**, *144*, 646–674. [CrossRef] [PubMed]
7. Zhu, W.B.; Tian, F.J.; Liu, L.Q. Chikusetsu (CHI) triggers mitochondria-regulated apoptosis in human prostate cancer via reactive oxygen species (ROS) production. *Biomed. Pharmacother.* **2017**, *90*, 446–454. [CrossRef] [PubMed]
8. Moghtaderi, H.; Sepehri, H.; Attari, F. Combination of arabinogalactan and curcumin induces apoptosis in breast cancer cells in vitro and inhibits tumor growth via overexpression of p53 level in vivo. *Biomed. Pharmacother.* **2017**, *88*, 582–594. [CrossRef]
9. Usuwanthim, K.; Wisitpongpun, P.; Luetragoon, T. Molecular Identification of Phytochemical for Anticancer Treatment. *Anticancer Agents Med. Chem.* **2020**, *20*, 651–666. [CrossRef]
10. Abd Rani, N.Z.; Husain, K.; Kumolosasi, E. Moringa Genus: A Review of Phytochemistry and Pharmacology. *Front. Pharmacol.* **2018**, *9*, 108. [CrossRef]
11. Vergara-Jimenez, M.; Almatrafi, M.M.; Fernandez, M.L. Bioactive Components in Moringa Oleifera Leaves Protect against Chronic Disease. *Antioxidants* **2017**, *6*, 91. [CrossRef] [PubMed]
12. Leone, A.; Bertoli, S.; Di Lello, S.; Bassoli, A.; Ravasenghi, S.; Borgonovo, G.; Forlani, F.; Battezzati, A. Effect of Moringa oleifera Leaf Powder on Postprandial Blood Glucose Response: In Vivo Study on Saharawi People Living in Refugee Camps. *Nutrients* **2018**, *10*, 1494. [CrossRef] [PubMed]
13. Okumu, M.O.; Ochola, F.O.; Mbaria, J.M.; Kanja, L.W.; Gakuya, D.W.; Kinyua, A.W.; Okumu, P.O.; Kiama, S.G. Mitigative effects of Moringa oleifera against liver injury induced by artesunateamodiaquine antimalarial combination in wistar rats. *Clin. Phytosci.* **2017**, *3*, 18. [CrossRef]
14. Kalpana, S.; Moorthi, S.; Kumari, S. Antimicrobial activity of different extracts of leaf of Moringa oleifera (Lam) against gram positive and gram negative bacteria. *Int. J. Curr. Microbiol. Appl. Sci.* **2013**, *2*, 514–518.
15. Omodanisi, E.I.; Aboua, Y.G.; Chegou, N.N.; Oguntibeju, O.O. Hepatoprotective, Antihyperlipidemic, and Anti-inflammatory Activity of Moringa oleifera in Diabetic-induced Damage in Male Wistar Rats. *Pharmacogn. Res.* **2017**, *9*, 182–187.
16. Adeyemi, O.S.; Elebiyo, T.C. Moringa oleifera Supplemented Diets Prevented Nickel-Induced Nephrotoxicity in Wistar Rats. *J. Nutr. Metab.* **2014**, *2014*, 958621. [CrossRef]
17. Luetragoon, T.; Pankla Sranujit, R.; Noysang, C.; Thongsri, Y.; Potup, P.; Suphrom, N.; Nuengchamnong, N.; Usuwanthim, K. Bioactive Compounds in Moringa oleifera Lam. Leaves Inhibit the Pro-Inflammatory Mediators in Lipopolysaccharide-Induced Human Monocyte-Derived Macrophages. *Molecules* **2020**, *25*, 191. [CrossRef]
18. Kooltheat, N.; Sranujit, R.P.; Chumark, P.; Potup, P.; Laytragoon-Lewin, N.; Usuwanthim, K. An ethyl acetate fraction of Moringa oleifera Lam. Inhibits human macrophage cytokine production induced by cigarette smoke. *Nutrients* **2014**, *6*, 697–710. [CrossRef]
19. Jung, I.L. Soluble extract from Moringa oleifera leaves with a new anticancer activity. *PLoS ONE* **2014**, *9*, e95492. [CrossRef]
20. Jung, I.L.; Lee, J.H.; Kang, S.C. A potential oral anticancer drug candidate, Moringa oleifera leaf extract, induces the apoptosis of human hepatocellular carcinoma cells. *Oncol. Lett.* **2015**, *10*, 1597–1604. [CrossRef]

21. Al-Asmari, A.K.; Albalawi, S.M.; Athar, M.T.; Khan, A.Q.; Al-Shahrani, H.; Islam, M. Moringa oleifera as an Anti-Cancer Agent against Breast and Colorectal Cancer Cell Lines. *PLoS ONE* **2015**, *10*, e0135814. [CrossRef]

22. Tragulpakseerojn, J.; Yamaguchi, N.; Pamonsinlapatham, P.; Wetwitayaklung, P.; Yoneyama, T.; Ishikawa, N.; Ishibashi, M.; Apirakaramwong, A. Anti-proliferative effect of Moringa oleifera Lam (Moringaceae) leaf extract on human colon cancer HCT116 cell line. *Trop. J. Pharm. Res.* **2017**, *16*, 371–378. [CrossRef]

23. Rajan, T.S.; De Nicola, G.R.; Iori, R.; Rollin, P.; Bramanti, P.; Mazzon, E. Anticancer activity of glucomoringin isothiocyanate in human malignant astrocytoma cells. *Fitoterapia* **2016**, *110*, 1–7. [CrossRef]

24. Cirmi, S.; Ferlazzo, N.; Gugliandolo, A.; Musumeci, L.; Mazzon, E.; Bramanti, A.; Navarra, M. Moringin from Moringa Oleifera Seeds Inhibits Growth, Arrests Cell-Cycle, and Induces Apoptosis of SH-SY5Y Human Neuroblastoma Cells through the Modulation of NF-kappaB and Apoptotic Related Factors. *Int. J. Mol. Sci.* **2019**, *20*, 1930. [CrossRef]

25. Lee, G.R.; Jang, S.H.; Kim, C.J.; Kim, A.R.; Yoon, D.J.; Park, N.H.; Han, I.S. Capsaicin suppresses the migration of cholangiocarcinoma cells by down-regulating matrix metalloproteinase-9 expression via the AMPK-NF-kappaB signaling pathway. *Clin. Exp. Metastasis* **2014**, *31*, 897–907. [CrossRef] [PubMed]

26. Kim, M.Y.; Trudel, L.J.; Wogan, G.N. Apoptosis induced by capsaicin and resveratrol in colon carcinoma cells requires nitric oxide production and caspase activation. *Anticancer Res.* **2009**, *29*, 3733–3740. [PubMed]

27. Li, Q.; Wen, J.; Yu, K.; Shu, Y.; He, W.; Chu, H.; Zhang, B.; Ge, C. Aloe-emodin induces apoptosis in human oral squamous cell carcinoma SCC15 cells. *BMC Complement. Altern. Med.* **2018**, *18*, 296. [CrossRef]

28. Rahman, M.A.; Amin, A.R.; Shin, D.M. Chemopreventive potential of natural compounds in head and neck cancer. *Nutr. Cancer* **2010**, *62*, 973–987. [CrossRef]

29. Perkins, N.D. Integrating cell-signalling pathways with NF-kappaB and IKK function. *Nat. Rev. Mol. Cell Biol.* **2007**, *8*, 49–62. [CrossRef]

30. Tergaonkar, V. NFkappaB pathway: A good signaling paradigm and therapeutic target. *Int. J. Biochem. Cell Biol.* **2006**, *38*, 1647–1653. [CrossRef]

31. Aggarwal, S.; Takada, Y.; Singh, S.; Myers, J.N.; Aggarwal, B.B. Inhibition of growth and survival of human head and neck squamous cell carcinoma cells by curcumin via modulation of nuclear factor-kappaB signaling. *Int. J. Cancer* **2004**, *111*, 679–692. [CrossRef] [PubMed]

32. Amin, A.R.; Khuri, F.R.; Chen, Z.G.; Shin, D.M. Synergistic growth inhibition of squamous cell carcinoma of the head and neck by erlotinib and epigallocatechin-3-gallate: The role of p53-dependent inhibition of nuclear factor-kappaB. *Cancer Prev. Res.* **2009**, *2*, 538–545. [CrossRef] [PubMed]

33. Anwar, F.; Latif, S.; Ashraf, M.; Gilani, A.H. Moringa oleifera: A food plant with multiple medicinal uses. *Phytother. Res.* **2007**, *21*, 17–25. [CrossRef] [PubMed]

34. Purwal, L.; Pathak, A.K.; Jain, U.K. In vivo anticancer activity of the leaves and fruits of moringa oleifera on mouse melanoma. *Pharmacologyonline* **2010**, *1*, 655–665.

35. Franken, N.A.; Rodermond, H.M.; Stap, J.; Haveman, J.; van Bree, C. Clonogenic assay of cells in vitro. *Nat. Protoc.* **2006**, *1*, 2315–2319. [CrossRef] [PubMed]

36. Elmore, S. Apoptosis: A review of programmed cell death. *Toxicol. Pathol.* **2007**, *35*, 495–516. [CrossRef]

37. Creagh, E.M. Caspase crosstalk: Integration of apoptotic and innate immune signalling pathways. *Trends Immunol.* **2014**, *35*, 631–640. [CrossRef]

38. van Meerloo, J.; Kaspers, G.J.; Cloos, J. Cell sensitivity assays: The MTT assay. *Methods Mol. Biol.* **2011**, *731*, 237–245.

39. Liang, C.C.; Park, A.Y.; Guan, J.L. In vitro scratch assay: A convenient and inexpensive method for analysis of cell migration in vitro. *Nat. Protoc.* **2007**, *2*, 329–333. [CrossRef]

Sample Availability: 3-hydroxy-β-ionone is available from the authors.

Article

Unveiling the Differential Antioxidant Activity of Maslinic Acid in Murine Melanoma Cells and in Rat Embryonic Healthy Cells Following Treatment with Hydrogen Peroxide

Khalida Mokhtari [1,2], Amalia Pérez-Jiménez [3], Leticia García-Salguero [1], José A. Lupiáñez [1] and Eva E. Rufino-Palomares [1,*]

[1] Department of Biochemistry and Molecular Biology I, Faculty of Sciences, University of Granada, Avenida Fuentenueva, s/n, 18071 Granada, Spain; khalidafadoua@hotmail.com (K.M.); elgarcia@ugr.es (L.G.-S.); jlcara@ugr.es (J.A.L.)

[2] Laboratory of Bioresources, Biotechnologies, Ethnopharmacology and Health, Department of Biology, Faculty of Sciences, Mohammed I University of Oujda, Oujda 60000, Morocco

[3] Department of Zoology, Faculty of Sciences, University of Granada, Avenida Fuentenueva, s/n, 18071 Granada, Spain; calaya@ugr.es

* Correspondence: evaevae@ugr.es; Tel.: +34-958-243252

Academic Editor: Domenico Trombetta
Received: 3 August 2020; Accepted: 31 August 2020; Published: 3 September 2020

Abstract: Maslinic acid (MA) is a natural triterpene from *Olea europaea* L. with multiple biological properties. The aim of the present study was to examine MA's effect on cell viability (by the MTT assay), reactive oxygen species (ROS levels, by flow cytometry) and key antioxidant enzyme activities (by spectrophotometry) in murine skin melanoma (B16F10) cells compared to those on healthy cells (A10). MA induced cytotoxic effects in cancer cells (IC_{50} 42 µM), whereas no effect was found in A10 cells treated with MA (up to 210 µM). In order to produce a stress situation in cells, 0.15 mM H_2O_2 was added. Under stressful conditions, MA protected both cell lines against oxidative damage, decreasing intracellular ROS, which were higher in B16F10 than in A10 cells. The treatment with H_2O_2 and without MA produced different responses in antioxidant enzyme activities depending on the cell line. In A10 cells, all the enzymes were up-regulated, but in B16F10 cells, only superoxide dismutase, glutathione S-transferase and glutathione peroxidase increased their activities. MA restored the enzyme activities to levels similar to those in the control group in both cell lines, highlighting that in A10 cells, the highest MA doses induced values lower than control. Overall, these findings demonstrate the great antioxidant capacity of MA.

Keywords: antioxidant activity; antioxidant enzymes; anti-proliferative activity; maslinic acid; melanoma; *Olea europaea* L.; ROS levels

---

1. Introduction

An imbalance between pro-oxidant and antioxidant molecules can lead to an oxidative stress situation that modifies normal cell physiology due to protein, lipid, carbohydrate and nucleic acid damage [1]. Many cellular processes depend on the variations in the levels of ROS and NADPH that take place during their development and that, fundamentally, are determined by the activity of the different production systems for this reduced coenzyme, especially those belonging to the pentose phosphate pathway (PPP), glucose-6-phosphate dehydrogenase (G6PDH), 6-phosphogluconate dehydrogenase (6PGDH) and, also, NADP-dependent isocitrate dehydrogenase (ICDH-NADP). The cellular redox state is key to interpreting the behavior of most of these key cellular processes for the vital development of organisms, such as cell differentiation [2–4], cellular growth [5–8], cell nutrition [6,9–11] and cell aging [12].

Reactive oxygen species (ROS) and reactive nitrogen species (RNS) are the most important endogenous pro-oxidants produced by normal cellular metabolism [13]. Under a normal physiological situation, an antioxidant defense system neutralizes ROS. This antioxidant system involves enzymes such as catalase (CAT), which reduces hydrogen peroxide; superoxide dismutase (SOD), which detoxifies the superoxide radical; glutathione peroxidase (GPX), which reduces hydrogen peroxide and other organic peroxides; S-transferase glutathione (GST), which detoxifies harmful molecules; glutathione reductase (GR), which regenerates glutathione (GSH) from its oxidized form (GSSG) by an NADPH-dependent pathway; and G6PDH, which produces NADPH for GR's mechanism. Beside enzymes, other molecules can also act as antioxidants, such as NADPH, GSH and different vitamins, among other molecules [14–16]. Despite this antioxidant system, an excessive production of ROS can lead to an oxidative stress situation inducing several types of damage to different biomolecules [17]. In cancer processes, an excessive production of ROS has been related to genomic instability due to the induction of DNA damage and to the alterations in signaling pathways involved in survival, proliferation and apoptosis resistance. Moreover, it has been demonstrated that ROS induce vascular endothelial growth factor (VEGF) expression, producing neovascularization and fast expansion of the cancer, modifying angiogenesis and metastasis, which induce the malignancy of cancer cells [17]. The use of natural compounds with antioxidant capacity that reduce ROS levels could decrease metastatic progression [15,17]. Furthermore, the immune defense system can be altered by ROS, since high levels of these molecules produced by NADPH oxidase can inhibit monocytes and macrophages and weaken the response of T-cells due to the down-regulation of several cytokines [17].

Hydrogen peroxide (H_2O_2) is originated from O_2 by SOD. It is not a free radical as such, but it is a reactive molecule, since it has the capacity to generate the hydroxyl radical in the presence of metals such as iron [1]. H_2O_2 is an important metabolite that arises mainly during aerobic metabolism, although it can derive from other sources [18]. H_2O_2 is a messenger molecule that diffuses through cells and tissues, inducing different effects that include deformations of the shapes of cells, the initiation of proliferation and the recruitment of immune cells. If not controlled, an excess of H_2O_2 can cause uncontrolled oxidative stress in cells, producing irreparable damage [19]. Therefore, cell death and survival processes can be modulated by controlling intracellular H_2O_2 levels by the use of antioxidants. Natural products are gaining interest due to their cost-effectiveness, few side effects and high availability. As such, products such as maslinic acid (MA) have received heightened interest.

MA, $C_{30}H_{48}O_4$ ($2\alpha,3\beta$-dihydroxyolean-12-en-28-oic acid) is a natural pentacyclic triterpene, also known as crataegolic acid and derived from its structural analogue, oleanolic acid (Figure 1). It is formed by 30 carbon atoms grouped into five cycles that, as substituents, have two methyl groups each on the C-4 and C-21 carbons; single methyl groups on the C-8, C-10 and C-15 carbons; two hydroxyl groups each on the C-2 and C-3 carbons; one carboxyl group on the C-28 carbon; and a double bond on the C-12 and C-13 [20–23].

Maslinic acid is present in various plants, many of them common in the Mediterranean diet, such as eggplants, spinach, lentils, chickpeas, and even different aromatic herbs [24]. Moreover, it is especially abundant in *Crataegus oxyacantha*, in the surface wax of the fruits and leaves of *Olea europaea* L. and in the solid waste from olive oil extraction [25]. Furthermore, MA is among the main triterpenes present in olives and olive oil. Its concentration in the oil depends on the type of olive oil and the variety of olive tree [23].

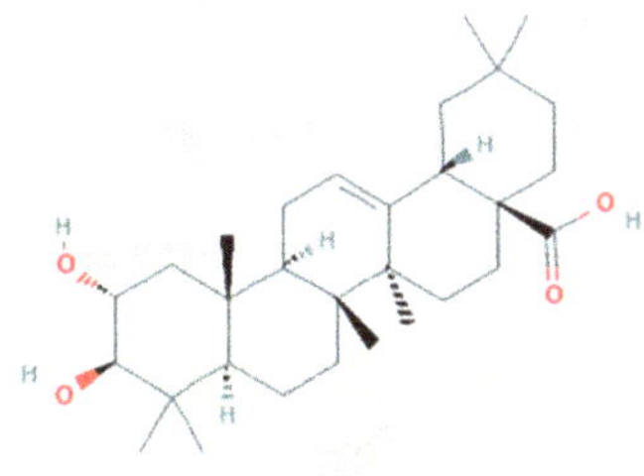

Maslinic acid

Figure 1. Schedule chemical structure of maslinic acid from pomace olive (*Olea europaea* L.) (PubChem).

MA enjoys high pharmacological interest, owing to its anti-tumor effect in certain types of cancer besides its anti-inflammatory, antioxidant, anti-diabetogenic, anti-hypertensive, anti-viral and cardioprotective effects, among others [1,23]. Among the bioactivities attributed to MA, its antioxidant effect is the most contradictory. It has been observed that the oxidative status induced by CCl_4 was decreased by MA treatment, which reduced lipid peroxides in the plasma and the susceptibility of lipids to peroxidation [26]. In the same way, the LDL oxidation in the plasma of rabbits produced by $CuSO^{4-}$ was decreased by MA obtained from *Punica granatum* [27]. In human plasma, MA showed peroxyl-radical scavenging and chelating capacities for copper but did not show a positive effect for the prevention of LDL oxidation [28]. Research in macrophages revealed that MA can act in a similar way to catalase, decreasing the generation of H_2O_2, but does not produce any direct inhibitory effects on NO and superoxide formation [29]. In a previous study, we concluded that MA produced an increase in the ROS level under stress conditions caused by the absence of FBS in melanoma cells [15]. Furthermore, in this same study, MA had an antioxidant effect at lower assayed levels; however, at higher dosages, MA induced cellular damage by apoptosis.

The aim of the present study was to evaluate MA's antioxidant effects in murine skin melanoma (B16F10) cells and in a healthy cell line derived from the thoracic aorta of an embryonic rat (A10), by analyzing the proliferation, ROS production and the activity of the main antioxidant enzymes under cellular stress conditions induced by H_2O_2.

2. Results

2.1. MA Decreases Proliferation of B16F10 Cells by a Dose-Dependent Mechanism

We evaluated MA's effect on B16F10 melanoma and A10 cell line proliferation using the MTT assay (Figure 2A,B). B16F10 and A10 cells were exposed to different doses of MA (0 to 212 μM) for 24 h. Then, cell survival was compared with that of untreated control cells. The percentage of living cells decreased as the dose of MA increased in cancer cells (B16F10), while in healthy cells (A10), MA did not produce any cytotoxic effect, even at the highest doses used (210 μM). In B16F10 cells, the growth inhibition values (IC_{50}) in response to MA were 42 μM after 24 h of addition of this compound. Based on this value, the rest of studies were performed with an exposure of cells to 10.6 μM ($IC_{50/4}$), 21.2 μM ($IC_{50/2}$), 42.3 μM (IC_{50}) and 84.6 μM ($2 \cdot IC_{50}$) of MA for 24 h.

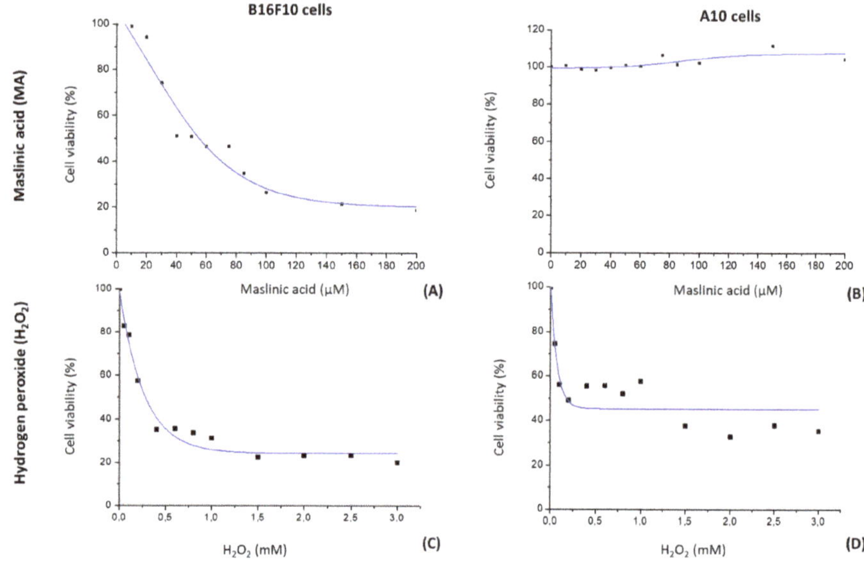

Figure 2. Cytotoxicity curves of maslinic acid (MA) for B16F10 murine melanoma cells (**A**) and A10 rat embryonic healthy cells (**B**) and of hydrogen peroxide (H_2O_2) for B16F10 murine melanoma cells (**C**) and A10 rat embryonic healthy cells (**D**). Cell proliferation was determined by the MTT assay. Values are expressed as means $\pm$ SEM ($n = 9$).

2.2. H_2O_2 Modifies Cell Viability

We examined H_2O_2's effects (0 to 3 mM of H_2O_2 for 24 h) on the cell viability of B16F10 and A10 cells to determine the optimal dose capable of producing stress without inducing cell death (Figure 2C,D). Cell survival was compared with that of untreated controls. The percentage of living cells decreased as the dose of H_2O_2 increased in both cell lines up to the concentration of 1.5 mM, beyoond which H_2O_2 had no cell viability effect. In both cell lines, we noticed an IC_{50} of 0.2 mM. In no case, with the doses used and the incubation time employed, was any mortality higher than 80% reached. Based on these results, the concentration of 0.15 mM of H_2O_2 was used to perform the rest of the studies.

2.3. Maslinic Acid's Influence on Mitochondrial-Membrane Potential

An assay of ROS production was performed to test the ROS levels that occurred over time in the presence of 0.150 mM H_2O_2 and different MA levels ($IC_{50/4}$, $IC_{50/2}$, IC_{50} and $2 \cdot IC_{50}$). The results are shown in Figure 3. After 24 h of H_2O_2 treatment, ROS levels increased significantly in both cell lines, compared to those in the controls. The increment observed upon H_2O_2 addition was offset by MA supplementation, which resulted in decreased intracellular ROS levels at any MA level used ($IC_{50/4}$, $IC_{50/2}$, IC_{50} and $2 \cdot IC_{50}$) in B16F10 and A10 cells. Moreover, in A10, ROS levels decreased below the value of control cells for all concentrations of MA tested.

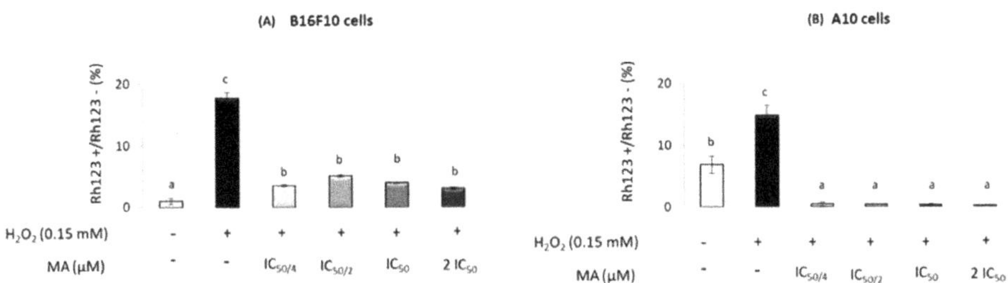

Figure 3. Positive fluorescent Rh123 on B16F10 cells (**A**) and A10 cells (**B**) with (+) or without (−) H_2O_2 and MA treatment at different doses: $IC_{50/4}$, $IC_{50/2}$, IC_{50} and $2 \cdot IC_{50}$ (10.6, 21.6, 42.3 and 84.6 µM, respectively). Values are expressed as means ± SEM (n = 9). Different letters indicate significant differences (p < 0.05).

2.4. MA Exerts Antioxidant Activity, Modulating Enzymatic Defense System

The results obtained for SOD in B16F10 cells showed that H_2O_2 increased its activity and MA (at any level) decreased it, but without significant differences induced by different MA concentrations. In A10, H_2O_2 increased the SOD activity, an increment that was mitigated with MA addition. Moreover, 42.3 and 84.6 µM MA (IC_{50} and $2 \cdot IC_{50}$, respectively) resulted in SOD activity levels lower than those found in the absence of H_2O_2 and MA (Figure 4A,B).

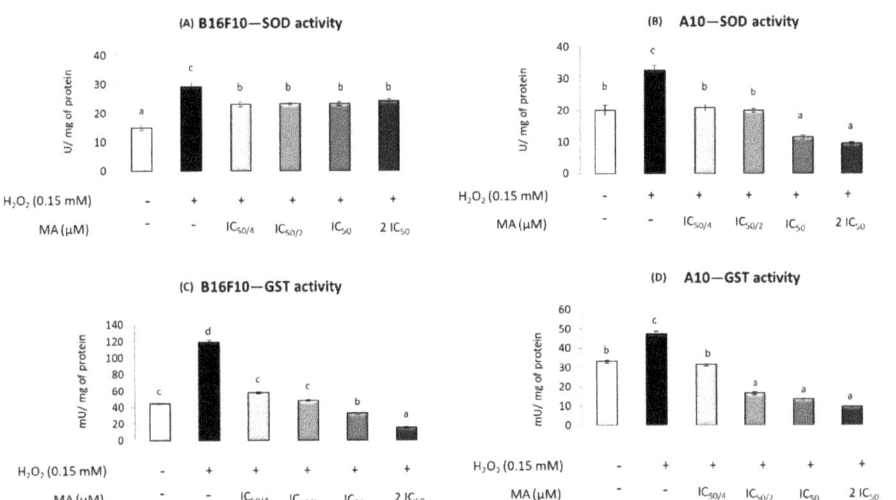

Figure 4. Effect of different maslinic acid doses—$IC_{50/4}$, $IC_{50/2}$, IC_{50} and $2 \cdot IC_{50}$ (10.6, 21.6, 42.3 and 84.6 µM, respectively)—on the specific activity of superoxide dismutase (SOD) in cancer cells, B16F10 (**A**), and normal cells, A10 (**B**), and glutathione S-transferase (GST) in cancer cells, B16F10 (**C**), and normal cells, A10 (**D**), subjected to the presence of hydrogen peroxide. Symbols (+) and (−) indicate the presence or absence of incubation with H_2O_2, respectively. Enzyme activities (U or mU × mg protein^{-1}) are expressed as means ± SEM (n = 9). Different letters indicate significant differences (p < 0.05).

In B16F10 cells, GST activity increased in the presence of H_2O_2, whereas the incubation with the $IC_{50/4}$ and $IC_{50/2}$ of MA decreased the levels, making them equal to those in the control without H_2O_2 and MA. The use of the IC_{50} and $2 \cdot IC_{50}$ MA concentrations induced GST activities lower than those in the control. Similar results were found in A10 for GST activity, with the exception of MA at $IC_{50/2}$, which also induced lower levels than control (Figure 4C,D).

In B16F10 cells, CAT activity decreased in the presence of H_2O_2 compared to that in the control, whereas MA increased the activity, inducing values higher than or similar to control when $IC_{50/4}$ or $IC_{50/2}$ were used. In A10 cells, H_2O_2 produced an increase in CAT activity, and its activity was not recovered until the highest doses of MA were used (IC_{50} and $2 \cdot IC_{50}$) (Figure 5A,B).

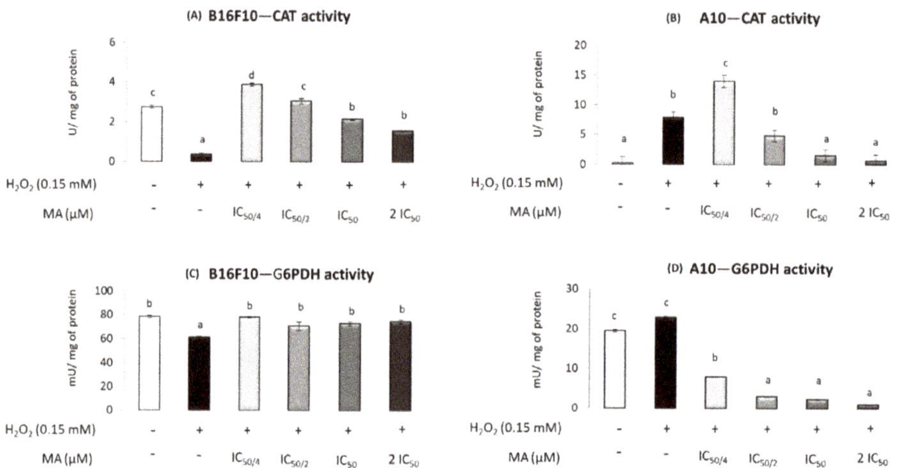

Figure 5. Effect of different maslinic acid doses—$IC_{50/4}$, $IC_{50/2}$, IC_{50} and $2 \cdot IC_{50}$ (10.6, 21.6, 42.3 and 84.6 µM, respectively)—on the specific activity of catalase (CAT) in cancer cells, B16F10 (**A**), and normal cells, A10 (**B**), and glucose 6-phosphate dehydrogenase (G6PDH) in cancer cells, B16F10 (**C**), and normal cells, A10 (**D**), subjected to the presence of hydrogen peroxide. Symbols (+) and (−) indicate the presence or absence of incubation with H_2O_2, respectively. Enzyme activities (U or mU × mg protein^{-1}) are expressed as means ± SEM ($n = 9$). Different letters indicate significant differences ($p < 0.05$).

Regarding G6PDH, the results indicated its activity decreased in the presence of H_2O_2 in B16F10 and MA at all the concentrations tested, with similar values to the control without H_2O_2 and MA. In A10 cells, no changes were induced by H_2O_2 in G6PDH, whereas MA decreased its activity at any tested level (Figure 5C,D).

GPX increased in the presence of H_2O_2 but no change was observed when MA was administrated in B16F10. In A10 cells, all concentrations of MA decreased the GPX activity increased by the H_2O_2, even below that in the control without H_2O_2 and MA (Figure 6A,B).

In B16F10, GR activity was decreased by H_2O_2 and no changes were observed at any MA level, with lower levels of activity maintained compared to the control. In A10, H_2O_2 increased GR activity, an increment that was mitigated with MA addition. Moreover, 42.3 and 84.6 µM MA induced values of GR activity lower than those found in the absence of H_2O_2 and MA (Figure 6C,D).

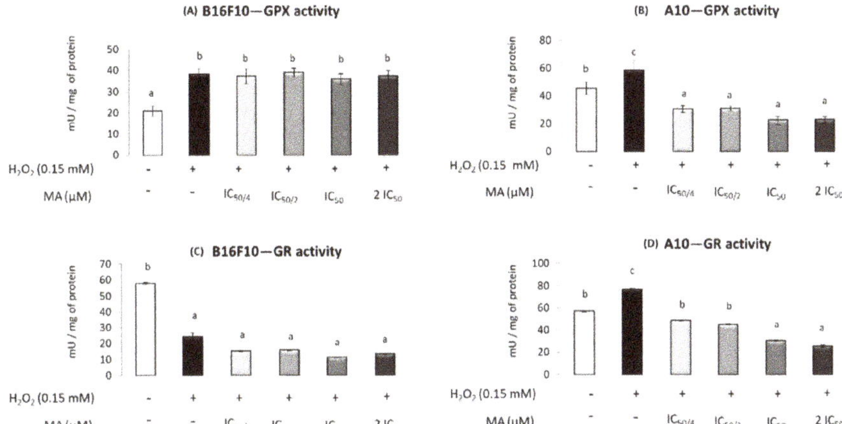

Figure 6. Effect of different maslinic acid doses—$IC_{50/4}$, $IC_{50/2}$, IC_{50} and $2 \cdot IC_{50}$ (10.6, 21.6, 42.3 and 84.6 µM, respectively)—on the specific activity of glutathione peroxidase (GPX) in cancer cells, B16F10 (**A**), and normal cells, A10 (**B**), and glutathione reductase (GR) in cancer cells, B16F10 (**C**), and normal cells, A10 (**D**), subjected to the presence of hydrogen peroxide. Symbols (+) and (−) indicate the presence or absence of incubation with H_2O_2, respectively. Enzyme activity (mU × mg protein^{-1}) is expressed as means ± SEM ($n = 9$). Different letters indicate significant differences ($p < 0.05$)

3. Discussion

Maslinic acid (MA) is a pentacyclic triterpene abundant in the surface wax of fruits and leaves of *Olea europaea* L. with many demonstrated biological activities [23,30,31]. For this reason, MA is appreciated as a chemopreventive agent in different diseases such as cancer [23,28], cardiovascular and neurodegenerative pathologies, etc. [26].

In different cell lines, MA's cytotoxic effect has been studied, including in both cancer and healthy cells. In this sense, it has been shown that this triterpene affects cells in different ways depending on the kind of cells and the conditions of the experiment. In the present study, MA showed cytotoxic effects only in B16F10 cells, and this effect was important, as the IC_{50} of MA was 42 µM. In another study performed in melanoma cells by Kim et al. [32], the authors observed a lower MA cytotoxicity effect. Thus, they concluded that the IC_{50} for MA in SK-MEL-3 was also 42 µM, but in their study, the incubation period was 48 h versus the 24 h used in the present study. Other results obtained in our research group showed that in colon cancer cells, the IC_{50} value for MA was 30 µM in HT29 cells after 72 h of incubation [31,33,34], showing a higher cytotoxic effect in Caco-2 cells, in which the IC_{50} of MA was 10.82 µM, also after 72 h [35]. Other authors observed IC_{50} values for MA ranging between 32 and 64 µM in different cancer cells such as lung (A549), ovary (SK-OV-3), colon (HCT15) and glioma (XF498) cells after 48 h of incubation [32]. In several bladder cancer cell lines incubated with MA for 48 h, other authors found IC_{50} values between 20 and 300 µM [36]. Notwithstanding this, compared to in melanoma cells, a non-cytotoxic effect was found in A10 healthy cells in the present study. These results are relevant since they demonstrate the selective cytotoxic effect of this compound. Studies focused on the selective effect of MA are scarce. Reyes et al. [22] observed that epithelial intestinal cells incubated with 30 µM MA for 72 h (IC_{50} value for HT29) exhibited a survival rate of 78% for IEC-6 cells and 68% for IEC-18.

An intracellular redox balance is crucial to ensure viability, growth and the diversity of cell functions [15]. An excess of ROS is related to pathological processes producing oxidative damage when the antioxidant defense system is not able to counteract it. Hence, there is a growing interest of the scientific community and industry in obtaining substances of natural origin aimed at preventing these pathologies and alterations caused by ROS. In this context, it is important to characterize the protective biochemical functions of natural antioxidants and to study their intracellular pathway signaling. A large number of plant constituents, such as MA, have antioxidant properties [37].

The present study examined the antioxidant effects of MA in skin melanoma cancer cells (B16F10) and thoracic aorta of embryonic rat cells (A10). The major findings were that MA improves the oxidative stress caused by a H_2O_2 excess, decreasing the intracellular ROS levels in both cell lines. H_2O_2 was used in this study because it is known that when present in excess, it is one of the major compounds that can damage cells [38]. Other authors have clearly shown the effects of H_2O_2 on the viability and ROS production of different cell types that were subsequently treated with other antioxidant natural compounds that reversed the oxidative damage caused by the H_2O_2 [39,40]. Furthermore, studies similar to ours with other ROS-producing molecules (i.e., CCl_4) observed that MA counteracted the lipid peroxidation generated in the nucleus [26]. Similarly, MA prevented the $CuSO_4^-$-induced oxidation of rabbit plasma LDL [41]. Moreover, Yang et al. [42] evaluated the antioxidant effects of MA derivatives that showed radical-scavenging activity and inhibition of NO production in RAW 264.7 cells.

Regarding MA's effect on antioxidant enzymes in the B16F10 cell line, as expected, SOD, GST and GPX activities were increased in response to the H_2O_2 addition. However, H_2O_2 decreased CAT, G6PDH and GR activities. A H_2O_2 excess is responsible for cellular damage that includes an imbalance in membranes and biomolecule alterations. This fact supposes an extra energetic cost for the maintenance of cellular homeostasis in order to repair the damage produced. Both cellular damage and energy requirements could affect G6PDH activity, by either the impaired glucose transporters or enhanced glycolysis pathway, decreasing the glucose available for the pentose phosphate pathway. The regulation of NADPH levels is essential for understanding the behavior of numerous physiological processes, being especially important for growth and cell differentiation [2] but also for antioxidant processes, among others. NAPDH is mainly generated by the G6PDH enzyme and is required for the GPX and CAT activity involved in the H_2O_2 removal pathway. Thus, CAT is protected from inactivation by NADPH. Moreover, this reduction equivalent is used by GR to regenerate the oxidized glutathione to its reduced form required for GPX activity [43,44]. The lower G6PDH activity observed when H_2O_2 was added could result in a decrease in available NADPH levels, which might justify the reduction in CAT and GR activities. Similar results have been observed in previous studies in B16F10 cells subjected to stress conditions induced by FBS absence [15]. Mokhtari et al. [15] reported low activity levels for CAT that were produced by low NADPH levels due to an imbalance in cellular homeostasis.

It has been established that MA is a compound with antioxidant capacity [15,26,28]. When MA was added to the B16F10 cells, the changes induced by H_2O_2 excess in SOD and GST activities were counteracted. In this sense, the scavenger MA's effects reduced ROS levels and, subsequently, the need for the action of these antioxidant defenses. Moreover, the observed decrease in ROS levels due to MA would result in less cellular damage, which would reverse the possible mechanisms responsible for the decrease in G6PDH when H_2O_2 is added, raising its activity up to the control values. This recovered G6PDH activity would produce the NADPH level required to prevent and reverse the down-regulation of CAT [43]. Finally, a dose-response effect was observed for both GPX and CAT, highlighting the antioxidant behavior of MA in these cells.

The results found in this work, in A10 healthy cells, revealed that all the enzyme activities increased in the stressful conditions originated by the H_2O_2 addition, except for G6PDH, whose activity was slightly higher. MA, in these cells, restored the antioxidant enzyme activities, inducing values similar to those in the control group in cells treated with the lowest MA levels ($IC_{50/4}$). Furthermore, when cells were treated with the two highest doses of MA (IC_{50} and $2 \cdot IC_{50}$), the activity of all antioxidant enzymes, with the exception of CAT, decreased below control levels. This fact confirms the relevant antioxidant effect of MA [15,28,43].

4. Materials and Methods

4.1. Compounds

Maslinic acid was obtained from olive pomace and kindly donated by Biomaslinic S.A., Granada, Spain. The extract is a chemically pure white powder composed of 98% maslinic acid and stable when stored at 4 °C. It was dissolved before use at 10 mg/mL in 50% dimethyl sulfoxide (DMSO) and 50% phosphate buffer solution (PBS). This solution was diluted in cell culture medium for assay purposes.

4.2. Cell Lines and Cultures

The mouse melanoma cell line B16F10 is a variant of the murine melanoma cell line B16. These cells show a higher metastatic potential than B16 cells [45]. A10 is a cell line derived from the thoracic aorta of an embryonic mouse, and it is used as a study model of smooth muscle cells (SMC). This cell type shows a great proliferative capacity and may be subcultured several times, allowing the rapid attainment of cell mass [46]. Both cell lines used were provided by the cell bank of the University of Granada (Spain). The cell lines were cultured in Dulbecco's modified Eagle's medium (DMEM) containing glucose (4.5 g/L) and L-glutamine (2 mM) from the commercial PAA brand, 10% heat-inactivated fetal bovine serum (FBS), and 0.5% gentamicin for B16F10 cells and 10,000 units/mL penicillin and 10 mg/mL streptomycin for A10 cells. The cell lines were maintained in a humidified atmosphere with 5% CO_2 at 37 °C. The cells were passaged at preconfluent densities by the use of a solution containing 0.05% trypsin and 0.5 mM EDTA. The cells were seeded in the culture dishes at the desired density. After 24 h, when the cells were attached to the dish, the cells were incubated with 0.15 mM hydrogen peroxide (H_2O_2), in order to produce a stress situation in the cells. Following that, the cells were incubated with several concentrations of MA as indicated below.

4.3. MTT Assay

The MTT assay was performed as described by Mokhtari et al. [15]. Briefly, a 200 µL cell suspension of B16F10 or A10 (1.5×10^3 cells/well) was cultured in 96-well plates. Subsequent to the adherence of the cells, different MA dilutions on a scale of 10 to 210 µM were added separately. The incubation times were 24 h for all cases. MTT was dissolved in the medium and added to the wells at a final concentration of 0.5 mg/mL. Following 2 h of incubation, the generated formazan was dissolved in DMSO. Absorbance was measured at 570 nm in a multiplate reader (Bio-tek®). The absorbance was proportional to the number of viable cells. The MA concentration leading to 50% inhibition of cell proliferation (IC_{50}) was determined. The results are expressed as the percentage of live cells compared with the control considered as 100% cell viability. Cell viability in B16F10 and A10 cells was also studied in the presence of H_2O_2 by the MTT assay. H_2O_2 was dissolved in the culture medium of the cell lines. The concentrations used were made extemporaneously and protected from light before use to prevent degradation of the compounds. After 24 h of incubation, the medium was removed; fresh medium was added with different concentrations of H_2O_2 per well—0, 0.05, 0.1, 0.2, 0.4, 0.6, 0.8, 1.1, 1.5, 2.0, 2.6 and 3 mM—to a final volume of 200 µL for 24 h.

4.4. Flow-Cytometry Analysis of the Mitochondrial-Membrane Potential

Changes in the mitochondrial-membrane potential can be examined by monitoring the cell fluorescence after double staining with rhodamine 123 (Rh123) and propidium iodide (PI). Rh123 is a membrane-permeable fluorescent cationic dye that is selectively taken up by mitochondria in direct proportion to the MMP (mitochondrial-membrane permeabilization) [47]. B16F10 and A10 cells (4×10^5 cells/well) were seeded on 12-well plates with 2 mL of medium and treated with 0.15 mM H_2O_2 for 24 h and MA at $IC_{50/4}$, $IC_{50/2}$, IC_{50} and $2 \cdot IC_{50}$ (10.6, 21.6, 42.3 and 84.6 µM, respectively) concentrations for 24 h more. Following the treatment, the medium was removed and fresh medium with dihydrorhodamine (DHR), at a final concentration of 5 µg/mL, was added. After 30 min of incubation, the medium was removed and the cells were washed and resuspended in PBS with 5 µg/mL of PI. The intensity of fluorescence from Rh123 and PI was determined using an ACS flow cytometer (Coulter Corporation, Hialeah, FL, USA), at the excitation and emission wavelengths of 500 and 536 nm, respectively. The experiments were performed three times with two replicates per assay.

4.5. Antioxidant Enzyme Assays

In order to prepare the samples for analytic procedures, cells were homogenized in RIPA buffer. Immediately, the cells were sonicated on ice for 5 min and maintained under moderate shaking at 4 °C for 1 h. Every 15 min, the samples were moderately shaken in a vortex. The lysates were spun in a centrifuge at $10,000\times g$ at 4 °C for 15 min. All the enzyme assays were carried out at 37 °C using a Power Wave X microplate scanning spectrophotometer (Bio-Tek Instruments, Winooski, VT, USA) and run in duplicate in 96-well microplates. The optimal substrate and protein concentrations for the measurement of maximal activity for each enzyme were established by preliminary assays. The enzymatic reactions were initiated by the addition of the cell extract, except for SOD, where xanthine oxidase was used. The millimolar extinction coefficients used for H_2O_2, NADH/NADPH and DTNB (5,5-dithiobis (2-nitrobenzoic acid)) were 0.039, 6.22 and 13.6, respectively. The assay conditions were as follows:

Superoxide dismutase (SOD; EC 1.15.1.1) activity was measured by the ferricytochrome c method using xanthine/xanthine oxidase as the source of superoxide radicals. The reaction mixture consisted of 50 mM potassium phosphate buffer (pH 7.8), 0.1 mM EDTA, 0.1 mM xanthine, 0.013 mM cytochrome c and 0.024 IU/mL xanthine oxidase. Activity is reported in units of SOD per milligram of protein. One unit of activity was defined as the amount of enzyme necessary to produce a 50% inhibition of the ferricytochrome c reduction rate [48].

Catalase (CAT; EC 1.11.1.6) activity was determined by measuring the decrease in hydrogen peroxide concentration at 240 nm according to Aebi [49]. The reaction mixture contained 50 mm potassium phosphate buffer (pH 7.0) and 10.6 mM freshly prepared H_2O_2.

Glucose-6-phosphate dehydrogenase (G6PDH; EC 1.1.1.49) activity was determined at pH 7.6 in a medium containing 50 mM HEPES buffer, 2 mM $MgCl_2$, 0.8 mM $NADP^+$ and glucose 6-phosphate used as substrate. The enzyme activity was determined by measuring the reduction of $NADP^+$ at 340 nm as previously described by Lupiáñez et al. [2] and Peragón et al. [50]. The change in absorbance at 340 nm was recorded and, after confirmation of no exogenous activity, the reaction started by the addition of substrate.

Glutathione S-transferase (GST; EC 2.5.1.18) activity was measured according to the method described by Habig et al. [51], using 1-chloro-2,4-dinitrobenzene as a substrate.

Glutathione peroxidase (GPX, EC 1.11.1.9) activity was determined using the method described by Flohe and Günzler [52], based on the oxidation of NADPH, which is used to regenerate the reduced glutathione (GSH) from oxidized glutathione (GS-SG) obtained by the action of glutathione peroxidase.

Glutathione reductase (GR, EC 1.8.1.7) was determined by the modified method of Carlberg and Mannervik [53]. We measured the decrease in absorbance produced by the oxidation of NADPH used by GR in the passage of oxidized glutathione (GS-SG) to reduced glutathione (GSH).

All enzyme activities (except for SOD) are expressed as units or milliunits per milligram of soluble protein (specific activity). One unit of enzyme activity was defined as the amount of enzyme required to transform 1 µmol of substrate per min under the above assay conditions.

Soluble protein concentrations were determined using the method of Bradford, with bovine serum albumin used as a standard.

4.6. Statistical Analysis

Data are shown as mean ± standard error mean (SEM). The statistical significance among different experimental groups was determined by one-way analysis of variance (ANOVA) tests. When F values ($p < 0.05$) were significant, means were compared using Tukey's HSD test. The SPSS version 15.0 for Windows software package was used for statistical analysis.

5. Conclusions

In conclusion, the results obtained in the present study demonstrate that MA exerts a selective anti-proliferative effect against the B16F10 cell line, whereas in A10 healthy cells, MA did not present a cytotoxic effect. This natural compound prevents the oxidative stress caused by H_2O_2 excess, decreasing the ROS production levels. Moreover, depending on the cell line, the antioxidant enzymatic responses were different in the presence of H_2O_2 and without MA. Thus, in healthy A10 cells, all enzymes up-regulated their activity, but in B16F10 cells, only SOD, GST and GPX increased it. In most cases, MA treatment restored the activities of enzymes to levels similar to those in the control group, highlighting that in A10 cells, the highest doses of MA (IC_{50} and $2 \cdot IC_{50}$) resulted in values below control. Overall, these findings demonstrate the great antioxidant capacity of maslinic acid.

Author Contributions: Conceptualization, J.A.L. and E.E.R.-P.; methodology, performing experiments, K.M., E.E.R.-P., A.P.-J. and L.G.-S.; data analysis, K.M. and E.E.R.-P.; results interpretation, E.E.R.-P. and A.P.-J.; writing—original draft preparation, K.M., E.E.R.-P. and A.P.-J.; writing—review and editing, J.A.L. and L.G.-S.; supervision, E.E.R.-P. and J.A.L.; project fund acquisition and administration, J.A.L. All authors have read and agreed to the published version of the manuscript.

Funding: This research was funded by the consolidated Research Group BIO-157, from the General Secretariat of Universities, Research and Technology of the Ministry of Economy, Innovation, Science and Employment Government of the Junta de Andalucía (Spain), and by the Research Contract no. C-3650-00 under the program FEDER-INNTERCONECTA from the Spanish Government.

Conflicts of Interest: The authors of this work declare that this research was carried out in the absence of any commercial or financial relationship and, therefore, does not present any conflict of interest now or in the future.

References

1. Rodríguez-Rodríguez, R. Oleanolic acid and related triterpenoids from olives on vascular function: Molecular mechanisms and therapeutic perspectives. *Curr. Med. Chem.* **2015**, *22*, 1414–1425. [CrossRef] [PubMed]
2. Lupiáñez, J.A.; Adroher, F.J.; Vargas, A.M.; Osuna, A. Differential behaviour of glucose 6-phosphate dehydrogenase in two morphological forms of *Trypanosoma cruzi*. *Int. J. Biochem.* **1987**, *19*, 1085–1089. [CrossRef]
3. Adroher, F.J.; Osuna, A.; Lupiáñez, J.A. Differential energetic metabolism during *Trypanosoma cruzi* differentiation. I: Citrate synthase, NADP-isocitrate and succinate dehydrogenases. *Arch. Biochem. Biophys.* **1988**, *267*, 252–261. [CrossRef]
4. Adroher, F.J.; Osuna, A.; Lupiáñez, J.A. Differential energetic metabolism during *Trypanosoma cruzi* differentiation. II. Hexokinase, phosphofructokinase and pyruvate kinase. *Mol. Cell. Biochem.* **1990**, *94*, 71–82. [CrossRef] [PubMed]
5. Peragón, J.; Barroso, J.B.; de la Higuera, M.; Lupiáñez, J.A. Relationship between growth and protein turnover rates and nucleic acids in the liver of rainbow trout (*Oncorhynchus mykiss*) during development. *Can. J. Fish. Aquat. Sci.* **1998**, *55*, 649–657. [CrossRef]
6. Peragón, J.; Barroso, J.B.; García-Salguero, L.; Aranda, F.; de la Higuera, M.; Lupiáñez, J.A. Selective changes in the protein-turnover rates and nature of growth induced in trout liver by long-term starvation followed by re-feeding. *Mol. Cell. Biochem.* **1999**, *201*, 1–10. [CrossRef]
7. Peragón, J.; Barroso, J.B.; García-Salguero, L.; de la Higuera, M.; Lupiáñez, J.A. Dietary alterations in protein, carbohydrates and fat increase liver protein-turnover rate and decrease overall growth rate in the rainbow trout (*Oncorhynchus mykiss*). *Mol. Cell. Biochem.* **2000**, *209*, 97–104. [CrossRef]
8. Peragón, J.; Barroso, J.B.; García-Salguero, L.; de la Higuera, M.; Lupiáñez, J.A. Dietary-protein effects on growth and fractional protein-synthesis and degradation rates in liver and white muscle of rainbow-trout (*Oncorhynchus mykiss*). *Aquaculture* **1994**, *124*, 35–46. [CrossRef]
9. Barroso, J.B.; Peragón, J.; García-Salguero, L.; de la Higuera, M.; Lupiáñez, J.A. Carbohydrate deprivation reduces NADPH-production in fish liver but not in adipose tissue. *Int. J. Biochem. Cell Biol.* **2001**, *33*, 785–796. [CrossRef]
10. Peragón, J.; Barroso, J.B.; García-Salguero, L.; de la Higuera, M.; Lupiáñez, J.A. Growth, protein-turnover rates and nucleic-acid concentrations in the white muscle of rainbow trout during development. *Int. J. Biochem. Cell Biol.* **2001**, *33*, 1227–1238. [CrossRef]

11. Sánchez-Muros, M.J.; García-Rejón, L.; Lupiáñez, J.A.; de la Higuera, M. Long-term nutritional effects on the primary liver and kidney metabolism in rainbow trout (*Oncorhynchus mykiss*). II. Adaptive response of glucose 6-phosphate dehydrogenase activity to high-carbohydrate/low-protein and high-fat/on-carbohydrate diets. *Aquacult. Nutr.* **1996**, *2*, 193–200. [CrossRef]

12. Nóbrega-Pereira, S.; Fernández-Marcos, P.J.; Brioche, T.; Gómez-Cabrera, M.C.; Salvador-Pascual, A.; Flores, J.M.; Vina, J.; Serrano, M. G6PD protects from oxidative damage and improves healthspan in mice. *Nat. Commun.* **2016**, *7*, 10894. [CrossRef] [PubMed]

13. Brennan, J.P.; Southworth, R.; Medina, R.A.; Davidson, S.M.; Duchen, M.R.; Shattock, M.J. Mitochondrial uncoupling, with low concentration FCCP, induces ROS-dependent cardioprotection independent of KATP channel activation. *Cardiovas. Res.* **2006**, *72*, 313–321. [CrossRef]

14. Kurze, A.K.; Buhs, S.; Eggert, D.; Oliveira-Ferrer, L.; Muller, V.; Niendorf, A.; Wagener, C.; Nollau, P. Immature O-glycans recognized by the macrophage glycoreceptor CLEC10A (MGL) are induced by 4-hydroxy-tamoxifen, oxidative stress and DNA-damage in breast cancer cells. *Cell Commun. Signal.* **2019**, *17*, 107. [CrossRef] [PubMed]

15. Mokhtari, K.; Rufino-Palomares, E.E.; Pérez-Jiménez, A.; Reyes-Zurita, F.J.; Figuera, C.; García-Salguero, L.; Medina, P.P.; Peragón, J.; Lupiáñez, J.A. Maslinic acid, a triterpene from olive, affects the antioxidant and mitochondrial status of B16F10 melanoma cells grown under stressful conditions. *Evid.-Based Complement. Altern. Med.* **2015**, *2015*, 272457. [CrossRef]

16. Ahmad, K.; Hafeez, Z.B.; Bhat, A.R.; Rizvi, M.A.; Thakur, S.C.; Azam, A.; Athar, F. Antioxidant and apoptotic effects of *Callistemon lanceolatus* leaves and their compounds against human cancer cells. *Biomed. Pharmacother.* **2018**, *106*, 1195–1209. [CrossRef]

17. De Santiago-Arteche, R. Efecto de la Quimioterapia Antineoplásica en Pacientes con Cancer Colorectal Sobre Biomarcadores del Estrés Oxidativo y del Estado Redox Plasmático. Ph.D. Thesis, University of Burgos, Burgos, Spain, 2010; p. 262.

18. Zhang, Y.S.; Ning, Z.X.; Yang, S.Z.; Wu, H. Antioxidation properties and mechanism of action of dihydromyricetin from *Ampelopsis grossedentata*. *Acta Pharm. Sin.* **2003**, *38*, 241–244.

19. Yang, W.F.; Zhao, W.L. Determination of ginsenosides Re, Rb1 in Panax quinquefolius by micellar electrokinetic chromatography. *China J. Chin. Mater. Med.* **2003**, *28*, 1135–1137.

20. Reyes-Zurita, F.J.; Medina-O'Donnell, M.; Ferrer-Martín, R.M.; Rufino-Palomares, E.E.; Martín-Fonseca, S.; Rivas, F.; Martínez, A.; García-Granados, A.; Pérez-Jiménez, A.; García-Salguero, L.; et al. The oleanolic acid derivative, 3-O-succinyl-28-O-benzyl oleanolate, induces apoptosis in B16-F10 melanoma cells via the mitochondrial apoptotic pathway. *RSC Adv.* **2016**, *6*, 93590–93601. [CrossRef]

21. Reyes-Zurita, F.J.; Rufino-Palomares, E.E.; Lupiáñez, J.A.; Cascante, M. Maslinic acid, a natural triterpene from *Olea europaea* L., induces apoptosis in HT29 human colon-cancer cells via the mitochondrial apoptotic pathway. *Cancer Lett.* **2009**, *273*, 44–54. [CrossRef] [PubMed]

22. Reyes, F.J.; Centelles, J.J.; Lupiáñez, J.A.; Cascante, M. (2Alpha,3beta)-2,3-dihydroxyolean-12-en-28-oic acid, a new natural triterpene from *Olea europea*, induces caspase dependent apoptosis selectively in colon adenocarcinoma cells. *FEBS Lett.* **2006**, *580*, 6302–6310. [CrossRef] [PubMed]

23. Rufino-Palomares, E.E.; Pérez-Jiménez, A.; Reyes-Zurita, F.J.; García-Salguero, L.; Mokhtari, K.; Herrera-Merchán, A.; Medina, P.P.; Peragón, J.; Lupiáñez, J.A. Anti-cancer and anti-angiogenic properties of various natural pentacyclic tri-terpenoids and some of their chemical derivatives. *Curr. Org. Chem.* **2015**, *19*, 919–947. [CrossRef]

24. Juan, M.E.; Planas, J.M. Bioavailability and metabolism of maslinic acid, a natural pentacyclic triterpene. In *Recent Advances in Pharmaceutical Sciences VI*; Publisher: Kerala, India, 2016; pp. 131–145.

25. Siewert, B.; Csuk, R. Membrane damaging activity of a maslinic acid analog. *Eur. J. Med. Chem.* **2014**, *74*, 1–6. [CrossRef] [PubMed]

26. Montilla, M.P.; Agil, A.; Navarro, M.C.; Jiménez, M.I.; García-Granados, A.; Parra, A.; Cabo, M.M. Antioxidant activity of maslinic acid, a triterpene derivative obtained from *Olea europaea*. *Planta Med.* **2003**, *69*, 472–474. [PubMed]

27. Barroso, J.B.; García-Salguero, L.; Peragón, J.; de la Higuera, M.; Lupiáñez, J.A. The influence of dietary-protein on the kinetics of NADPH production systems in various tissues of rainbow-trout (*Oncorhynchus mykiss*). *Aquaculture* **1994**, *124*, 47–59. [CrossRef]

28. Allouche, Y.; Warleta, F.; Campos, M.; Sánchez-Quesada, C.; Uceda, M.; Beltrán, G.; Gaforio, J.J. Antioxidant, antiproliferative, and pro-apoptotic capacities of pentacyclic triterpenes found in the skin of olives on MCF-7 human breast cancer cells and their effects on DNA damage. *J. Agric. Food Chem.* **2011**, *59*, 121–130. [CrossRef]

29. Chen, J.C.; Zhang, G.H.; Zhang, Z.Q.; Qiu, M.H.; Zheng, Y.T.; Yang, L.M.; Yu, K.B. Octanorcucurbitane and cucurbitane triterpenoids from the tubers of *Hemsleya endecaphylla* with HIV-1 inhibitory activity. *J. Nat. Prod.* **2008**, *71*, 153–155. [CrossRef]

30. Reyes-Zurita, F.J.; Rufino-Palomares, E.E.; García-Salguero, L.; Peragón, J.; Medina, P.P.; Parra, A.; Cascante, M.; Lupiáñez, J.A. Maslinic acid, a natural T¡triterpene, induces a death receptor-mediated apoptotic mechanism in Caco-2 p53-deficient colon adenocarcinoma cells. *PLoS ONE* **2016**, *11*, e0146178. [CrossRef]

31. Rufino-Palomares, E.E.; Reyes-Zurita, F.J.; García-Salguero, L.; Mokhtari, K.; Medina, P.P.; Lupiáñez, J.A.; Peragón, J. Maslinic acid, a triterpenic anti-tumoural agent, interferes with cytoskeleton protein expression in HT29 human colon-cancer cells. *J. Proteom.* **2013**, *83*, 15–25. [CrossRef]

32. Kim, Y.K.; Yoon, S.K.; Ryu, S.Y. Cytotoxic triterpenes from stem bark of *Physocarpus intermedius*. *Planta Med.* **2000**, *66*, 485–486. [CrossRef] [PubMed]

33. Reyes-Zurita, F.J.; Pachón-Pena, G.; Lizarraga, D.; Rufino-Palomares, E.E.; Cascante, M.; Lupiáñez, J.A. The natural triterpene maslinic acid induces apoptosis in HT29 colon cancer cells by a JNK-p53-dependent mechanism. *BMC Cancer* **2011**, *11*, 154. [CrossRef] [PubMed]

34. Sánchez-Tena, S.; Reyes-Zurita, F.J.; Díaz-Moralli, S.; Vinardell, M.P.; Reed, M.; García-García, F.; Dopazo, J.; Lupiáñez, J.A.; Gunther, U.; Cascante, M. Maslinic acid-enriched diet decreases intestinal tumorigenesis in Apc(Min/+) mice through transcriptomic and metabolomic reprogramming. *PLoS ONE* **2013**, *8*, e59392. [CrossRef] [PubMed]

35. Reyes-Zurita, F.J.; Rufino-Palomares, E.E.; Medina, P.P.; García-Salguero, L.; Peragón, J.; Cascante, M.; Lupiáñez, J.A. Antitumour activity on extrinsic apoptotic targets of the triterpenoid maslinic acid in p53-deficient Caco-2 adenocarcinoma cells. *Biochimie* **2013**, *95*, 2157–2167. [CrossRef] [PubMed]

36. Zhang, S.; Ding, D.; Zhang, X.; Shan, L.; Liu, Z. Maslinic acid induced apoptosis in bladder cancer cells through activating p38 MAPK signaling pathway. *Mol. Cell. Biochem.* **2014**, *392*, 281–287. [CrossRef]

37. Medina, I.; Lois, S.; Lizarraga, D.; Pazos, M.; Tourino, S.; Cascante, M.; Torres, J.L. Functional fatty fish supplemented with grape procyanidins. Antioxidant and proapoptotic properties on colon cell lines. *J. Agric. Food Chem.* **2006**, *54*, 3598–3603. [CrossRef] [PubMed]

38. Sroka, Z.; Cisowski, W. Hydrogen peroxide scavenging, antioxidant and anti-radical activity of some phenolic acids. *Food Chem. Toxicol.* **2003**, *41*, 753–758. [CrossRef]

39. Choi, B.S.; Kim, H.; Lee, H.J.; Sapkota, K.; Park, S.E.; Kim, S.; Kim, S.J. Celastrol from 'Thunder God Vine' protects SH-SY5Y cells through the preservation of mitochondrial function and inhibition of p38 MAPK in a rotenone model of Parkinson's disease. *Neurochem. Res.* **2014**, *39*, 84–96. [CrossRef]

40. Rafatian, G.; Khodagholi, F.; Farimani, M.M.; Abraki, S.B.; Gardaneh, M. Increase of autophagy and attenuation of apoptosis by Salvigenin promote survival of SH-SY5Y cells following treatment with H_2O_2. *Mol. Cell. Biochem.* **2012**, *371*, 9–22. [CrossRef]

41. Baricevic, D.; Sosa, S.; Della Loggia, R.; Tubaro, A.; Simonovska, B.; Krasna, A.; Zupancic, A. Topical anti-inflammatory activity of *Salvia officinalis* L. leaves: The relevance of ursolic acid. *J. Ethnopharmacol.* **2001**, *75*, 125–132. [CrossRef]

42. Yang, Z.G.; Li, H.R.; Wang, L.Y.; Li, Y.H.; Lu, S.G.; Wen, X.F.; Wang, J.; Daikonya, A.; Kitanaka, S. Triterpenoids from *Hippophae rhamnoides* L. and their nitric oxide production-inhibitory and DPPH radical-scavenging activities. *Chem. Pharm. Bull.* **2007**, *55*, 15–18. [CrossRef]

43. Kirkman, H.N.; Rolfo, M.; Ferraris, A.M.; Gaetani, G.F. Mechanisms of protection of catalase by NADPH—Kinetics and stoichiometry. *J. Biol. Chem.* **1999**, *274*, 13908–13914. [CrossRef] [PubMed]

44. Pérez-Jiménez, A.; Abellán, E.; Arizcun, M.; Cardenete, G.; Morales, A.E.; Hidalgo, M.C. Dietary carbohydrates improve oxidative status of common dentex (*Dentex dentex*) juveniles, a carnivorous fish species. *Comp. Biochem. Physiol. A* **2017**, *203*, 17–23. [CrossRef] [PubMed]

45. Wang, W.; Xu, Z.; Yang, M.; Liu, R.; Wang, W.; Liu, P.; Guo, D. Structural determination of seven new triterpenoids from *Kadsura heteroclita* by NMR techniques. *Magn. Reson. Chem.* **2007**, *45*, 522–526. [CrossRef] [PubMed]

46. Rao, M.S.; Subbarao, V. Effect of dexamethasone on ciprofibrate-induced cell proliferation and peroxisome proliferation. *Fundam. Appl. Toxicol.* **1997**, *35*, 78–83. [CrossRef]
47. Rothe, G.; Oser, A.; Valet, G. Dihydrorhodamine 123: A new flow cytometric indicator for respiratory burst activity in neutrophil granulocytes. *Die Nat.* **1988**, *75*, 354–355. [CrossRef]
48. McCord, J.M.; Fridovich, I. Superoxide dismutase. An enzymic function for erythrocuprein (hemocuprein). *J. Biol. Chem.* **1969**, *244*, 6049–6055.
49. Aebi, H. Catalase in vitro. *Methods Enzymol.* **1984**, *105*, 121–126.
50. Peragón, J.; Aranda, F.; García-Salguero, L.; Corpas, F.J.; Lupiáñez, J.A. Stimulation of rat-kidney hexose-monophosphate shunt dehydrogenase-activity by chronic metabolic-acidosis. *Biochem. Int.* **1989**, *18*, 1041–1050.
51. Habig, W.H.; Pabst, M.J.; Jakoby, W.B. Glutathione S-transferases. The first enzymatic step in mercapturic acid formation. *J. Biol. Chem.* **1974**, *249*, 7130–7139.
52. Flohe, L.; Gunzler, W.A. Assays of glutathione peroxidase. *Methods Enzymol.* **1984**, *105*, 114–121.
53. Carlberg, I.; Mannervik, B. Purification by affinity chromatography of yeast glutathione reductase, the enzyme responsible for the NADPH-dependent reduction of the mixed disulfide of coenzyme A and glutathione. *Biochim. Biophys. Acta* **1977**, *484*, 268–274. [CrossRef]

Sample Availability: Samples of the compound maslinic acid are available from the authors.

Article

Elaeagnus angustifolia Plant Extract Inhibits Epithelial-Mesenchymal Transition and Induces Apoptosis via HER2 Inactivation and JNK Pathway in HER2-Positive Breast Cancer Cells

Ayesha Jabeen [1,2,†], Anju Sharma [1,†], Ishita Gupta [1,2,†], Hadeel Kheraldine [1,2,3], Semir Vranic [1], Ala-Eddin Al Moustafa [1,2,*] and Halema F. Al Farsi [1,*]

[1] College of Medicine, QU Health, Qatar University, Doha P.O. Box 2713, Qatar; jabeen@qu.edu.qa (A.J.); anju.sharma7385@gmail.com (A.S.); ishugupta28@gmail.com (I.G.); hk1805332@student.qu.edu.qa (H.K.); svranic@qu.edu.qa (S.V.)
[2] Biomedical Research Centre, Qatar University, Doha P.O. Box 2713, Qatar
[3] College of Pharmacy, Qatar University, Doha P.O. Box 2713, Qatar
* Correspondence: aalmoustafa@qu.edu.qa (A.-E.A.M.); halfarsi@qu.edu.qa (H.F.A.F.); Tel.: +974-4403-7817 (A.-E.A.M.); +974-4403-7840 (H.F.A.F.)
† These authors contributed equally to this work.

Academic Editors: José Antonio Lupiáñez, Amalia Pérez-Jiménez and Eva E. Rufino-Palomares
Received: 18 August 2020; Accepted: 11 September 2020; Published: 16 September 2020

Abstract: *Elaeagnus angustifolia* (*EA*) is a medicinal plant used for treating several human diseases in the Middle East. Meanwhile, the outcome of *EA* extract on HER2-positive breast cancer remains nascent. Thus, we herein investigated the effects of the aqueous *EA* extract obtained from the flowers of *EA* on two HER2-positive breast cancer cell lines, SKBR3 and ZR75-1. Our data revealed that *EA* extract inhibits cell proliferation and deregulates cell-cycle progression of these two cancer cell lines. *EA* extract also prevents the progression of epithelial-mesenchymal transition (EMT), an important event for cancer invasion and metastasis; this is accompanied by upregulations of E-cadherin and β-catenin, in addition to downregulations of vimentin and fascin, which are major markers of EMT. Thus, *EA* extract causes a drastic decrease in cell invasion ability of SKBR3 and ZR75-1 cancer cells. Additionally, we found that *EA* extract inhibits colony formation of both cell lines in comparison with their matched control. The molecular pathway analysis of HER2 and JNK1/2/3 of *EA* extract exposed cells revealed that it can block HER2 and JNK1/2/3 activities, which could be the major molecular pathway behind these events. Our findings implicate that *EA* extract may possess chemo-preventive effects against HER2-positive breast cancer via HER2 inactivation and specifically JNK1/2/3 signaling pathways.

Keywords: *Elaeagnus angustifolia*; breast cancer; EMT; chemoprevention; apoptosis

1. Introduction

Breast cancer (BC) commonly affects women worldwide, comprising 25% of cancer cases [1]. There are several risk factors, both environmental and genetic, associated with the onset of breast cancer [2]. Gene-expression-profiling studies classified breast cancer into five molecular subtypes: Luminal (A and B), HER2, basal-like, and normal-like, using hierarchical cluster analysis [3]. Among all subtypes, HER2-positive breast cancer accounts for 20–25% and is associated with aggressive phenotype, poor prognosis, and survival rate, in addition to increased recurrence [4]. Systemic management modalities for HER2-positive breast cancer include chemotherapy, radiation, and targeted anti-HER2 treatment modalities [4–7]. Although the treatment is generally effective in the early stages of therapy, nevertheless, ~90% of primary and half of metastatic breast cancer cases resistance to therapy leads to treatment failure and mortality [8]. Thus, it is important to identify novel potent therapeutic agents that can inhibit cell proliferation of HER2+ breast cancer with minimal side-effects. An alternate to conventional therapy can be found in naturally present phytochemicals in foods such as vegetables, fruits, spices, and plant roots [9,10]. Traditionally, *Elaeagnus angustifolia* (*EA*) plant has been used extensively for centuries in the treatment of various diseases due to its antioxidant, anti-inflammatory, and antimicrobial properties [11], along with bioactive compounds (flavonoids and coumarins) [12] that regulate key events associated with cancer development, such as cell-signaling pathways, including Wnt-signaling, cell proliferation, cell-cycle progression, apoptosis, and epithelial-to-mesenchymal transition (EMT) [13,14].

Elaeagnus angustifolia (*EA*), commonly known as wild olive, oleaster, silver berry, or Russian olive [11,15], is a deciduous tree belonging to the family of *Elaeagnacea* (Araliaceae) widely distributed in the Middle East, as well as Mediterranean regions [12,16]. *EA* fruit is highly nutritious and contains vitamins (vitamin C, tocopherol, thiamine B1, and carotene), sugar, proteins, and several minerals, like potassium, iron, magnesium, and calcium [17–19]; the leaves and flower extract are rich in secondary metabolites such as coumarins, phenolcarboxylic acids, flavonoids, saponins, and tannins [16,20,21]. Previous studies have shown that *EA* exhibits anticancer effects as a result of its essential oils (ethyl cinnamate, 2-phenyl-ethyl benzoate, 2-phenyl-ethyl isovalerate, nerolidole, squalene, and acetaphenone), flavonoids (quercetin), and pro anthocyanosides [22,23]. In cancer, flavonoids are shown to enhance p53 expression and cause cell-cycle arrest in the G2/M phase [24]. Moreover, they are known to inhibit Ras protein expression and regulate heat-shock proteins in various cancers, mainly in leukemia and colorectal cancer [24]. One of the key flavonoid components of *EA* is Quercetin, which is an anti-proliferative agent [24]. Furthermore, quercetin also promotes TRAIL-induced apoptosis by enhancing the expression of Bax and inhibiting Bcl-2 protein [25–27]. Additionally, ethyl acetate has been shown to significantly reduce proliferation of Hela cells in vitro [22]. Apart from this, volatile oils present in the plant have medicinal properties and are also used in perfume industries [23,28]. *EA* possesses numerous therapeutic and pharmacological properties, including antifungal, antibacterial, antimutagenic, anti-inflammation, antioxidant, and gastroprotective effects [11,29–32]. Traditionally, *EA* is also used to cure other diseases, including osteoporosis, amoebic dysentery, jaundice, asthma, flu, cough, cold, nausea, diarrhea, sore throat, fever, tetanus, and female aphrodisiac [12,15,33,34]. However, there are limited studies regarding the role of *EA* extract on cancer. In this context, our group recently demonstrated that *EA* extract can reduce the progression of human oral cancer by the inhibition of angiogenesis and cell invasion via Erk1/Erk2 signaling pathways [12].

A previous study showed that hydroalcoholic extracts of *EA* flower significantly inhibit angiogenesis, one of the known hallmarks of cancer [35]. Nevertheless, there are no studies reported on the anticancer activity of *EA* in breast cancer, especially in HER2-positive type, and its mechanism of cancer inhibition. To investigate the potent therapeutic and antitumor properties of *EA* extract in human breast cancer and its underlying mechanism, we explored the effect of aqueous extract of *EA* flower on cell proliferation and cell-cycle progression, cell invasion, and colony formation in two HER2-positive human breast cancer cell lines (SKBR3 and ZR75-1).

2. Results

In order to determine the effects of *EA* extract on HER2-positive cell lines SKBR3 and ZR75-1, cells were treated with varying concentrations of *EA* extract (25, 50, 75, 100, 150, and 200 µL/mL) for 48 h. Treatment with EA extract reduced the number of proliferating HER2-positive breast cancer cells in a dose-dependent manner (Figure 1); notably, concentrations of 100 and 200 µL/mL showed a substantial decrease in cell viability of SKBR3 and ZR75-1 by 50% and 75%, respectively.

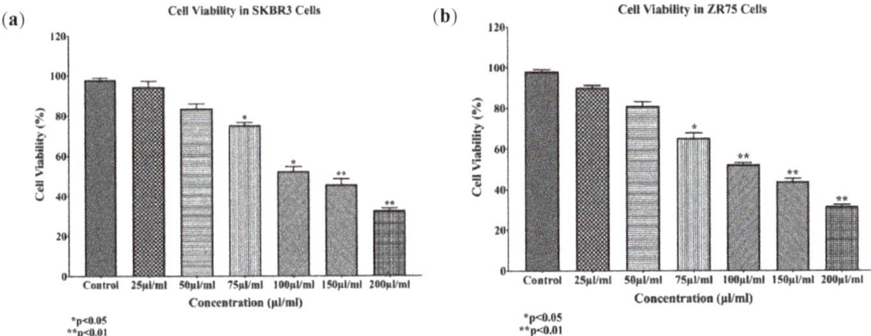

Figure 1. (**a**,**b**) The effects of different concentrations of *Elaeagnus angustifolia (EA)* plant extract on cell proliferation of HER2-positive breast cancer cell lines SKBR3 (**a**) and ZR75-1 (**b**) at 48 h. Data indicate an inverse relation between concentrations of *EA* extract and cell proliferation in both SKBR3 and ZR75-1 cell lines. Data are expressed as percent of growth ± SEM.

Meanwhile, and to examine whether the antiproliferative effect of the *EA* flower extract on SKBR3 and ZR75-1 cells is associated with cell-cycle deregulation, we analyzed cell-cycle phase distributions of *EA*-treated cells, using flow cytometric analysis. Our results showed that exposure to *EA* extract (100 and 200 µL/mL) for 48 h enhanced the G_0/G_1 phase, with a simultaneous decrease in S and G_2/M phases of both the breast cancer cell lines, thus indicating *EA*-induced cell-cycle inhibition (Figure 2). Furthermore, we observed a significant increase in the sub/G_1 phase of both cell lines, indicating that cells undergo apoptosis when treated with *EA* extract (Figure 2).

To confirm *EA*-induced apoptosis, Annexin V-FITC and 7-AAD staining by flow cytometry were performed. Therefore, the presence of apoptosis is clearly demonstrated in both cell lines (Figure 3).

Next, we examined the cell morphology of SKBR3 and ZR75-1 in addition to HNME-E6/E7 cell lines, using phase-contrast microscopy, under the effect of 100 and 200 µL/mL of *EA* extract. In the absence of treatment, SKBR3 and ZR75-1 cells displayed a round morphology and disorganized multilayered cells. In contrast, and as indicated in Figure 4a, treatment for 48 h with 100 and 200 µL/mL of *EA* plant extract led to a phenotypic conversion from round cells to epithelial-like phenotype. Clearly, cells became more flattened in appearance and showed an increase in cell-cell adhesion, in comparison with untreated cells (Figure 4a). However, at three days of treatment with 200 µL/mL of *EA* plant extract, cells started detaching from the surface of the tissue culture dish, indicating cell death in SKBR3 and ZR75-1 cells; however, this was not observed in the human normal immortalized mammary epithelial cell line (HNME-E6/E7), as shown in Figure 4b. Nevertheless, it is evident that *EA* extract inhibits cell proliferation HNME-E6/E7 cells, with a slight effect on their epithelial morphology (Figure 4b).

These results imply that moderate concentrations of 100 µL/mL of *EA* plant extract induce cell differentiation after 24 and/or 48 h, while higher concentrations (200 µL/mL of *EA* plant extract) can provoke apoptosis after 48 h of exposure.

Subsequently, and to analyze the anti-invasion effects of *EA* on HER2-positive breast cancer cells, Matrigel invasion assay was performed, using SKBR3 and ZR75-1 cells, upon *EA* treatment with 100 and 200 µL/mL concentrations; our data revealed that *EA* extract significantly inhibits cell invasion ability of both cell lines by ~70% to 88%, respectively, in comparison with control cells (Figure 5, $p < 0.05$). This suggests that *EA* plant extract can considerably downgrade cell invasion and metastasis of HER2-positive breast cancer.

On the other hand, we assessed the colony formation of SKBR3 and ZR75-1 cells, in soft agar, under the effect of *EA* plant extract at 100 and 200 µL/mL, for two weeks; we observed a significant decrease in the number of colonies for both cell lines treated with *EA* plant extract, compared with their matched control, as shown in Figure 6. SKBR3 sustained significant inhibition of colony formation by

60% ($p < 0.01$) and 80% ($p < 0.05$) when exposed to 100 and 200 μL/mL *EA* plant extract, in comparison to the control, respectively (Figure 6a). In parallel, ZR75-1 cell line also displayed a similar pattern after two weeks of treatment; the number of colonies decreased by 70% ($p < 0.05$) and 85% ($p < 0.01$) at 100 and 200 μL/mL concentrations, respectively (Figure 6b). This indicates that *EA* plant extract suppresses colony formation and probably tumor growth in vivo.

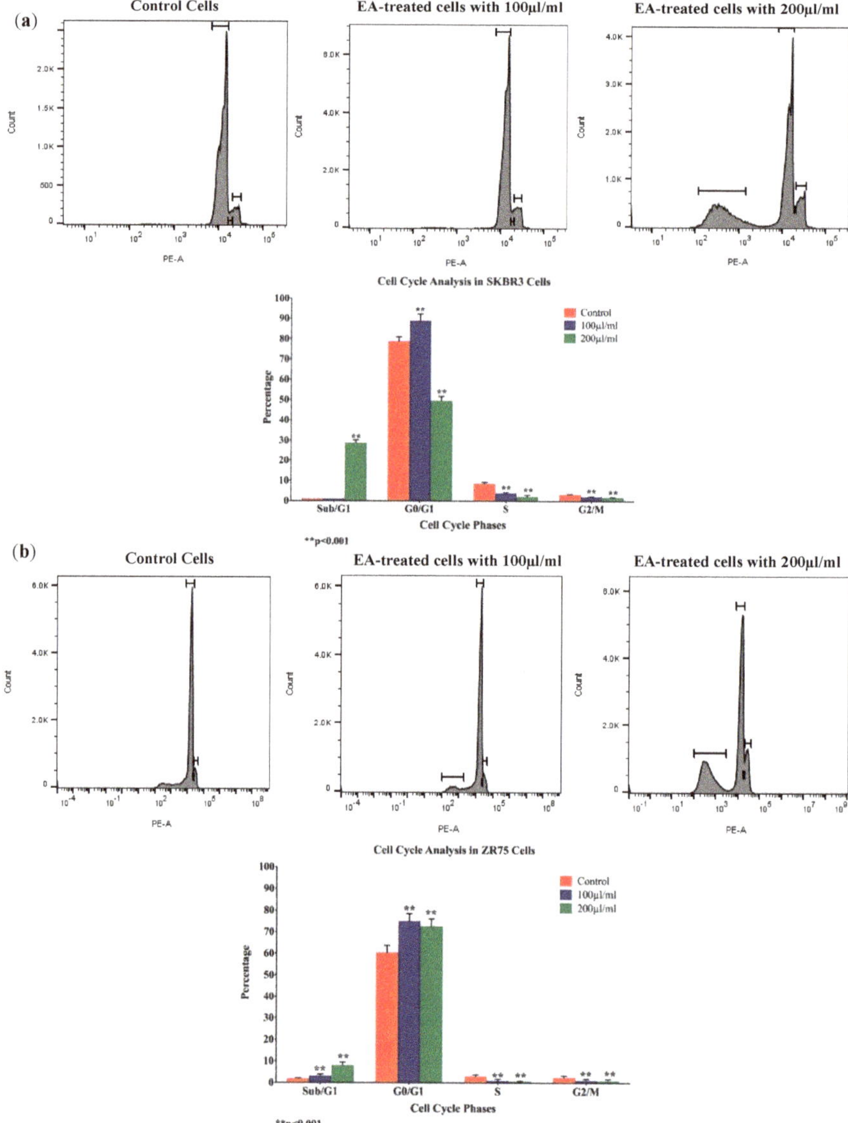

Figure 2. (**a,b**) Flow cytometry data analysis of SKBR3 and ZR75 cells after *EA*-treatment. Data demonstrate an increase in G_0/G_1 phase with simultaneous reduction in S and G_2/M phases in both cell lines. Meanwhile, there is a significant increase in cell apoptosis (Sub/G_1 phase) of SKBR3 cells treated with *EA*, and a small increase in cell apoptosis of treated ZR75 cells.

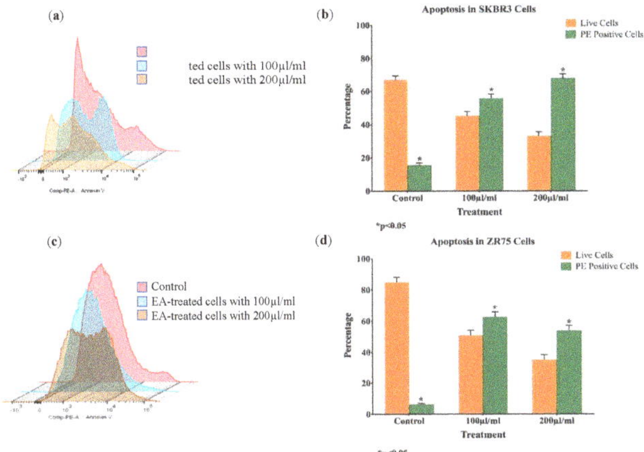

Figure 3. (**a**,**b**) Induction of apoptosis by *EA* extract in SKBR3 (**a**, **b**) and ZR75 (**c**, **d**) cells, as determined by Annexin V-FITC and 7-AAD apoptosis assay.

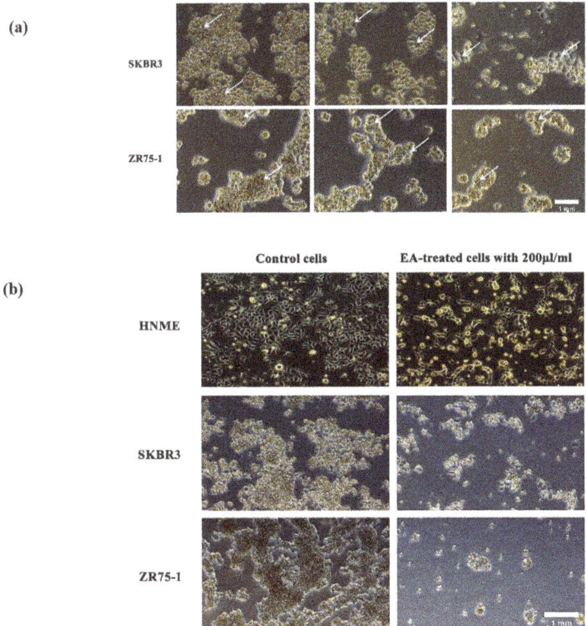

Figure 4. (**a**,**b**) *EA* plant extract induces morphological changes in HER2-positive cell lines, SKBR3 and ZR75-1. (**a**) We observe that treatment for 48 h with 100 and 200 µL/mL of *EA* extract induces epithelial transition and the formation of a monolayer of cells in both cell lines, in comparison with untreated (control) cells which display a round phenotype and form multilayers; arrows indicate epithelial morphology with clear cell-cell adhesion. (**b**) At three days of treatment of SKBR3, ZR75-1, and HNME-E6/E7 cell lines with 200 µL/mL of *EA* plant extract, the two cancer cell lines start detaching from the surface of the tissue culture dish, indicating cell death; this observation was not noted in the HNME-E6/E7 cells (images **a** and **b** at ×20 magnification).

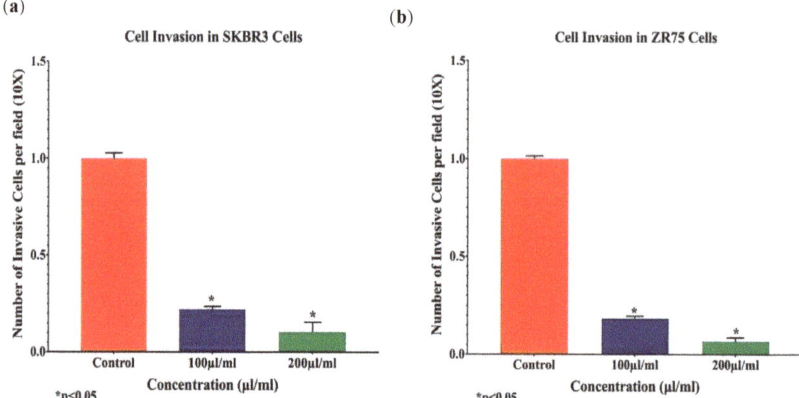

Figure 5. (**a,b**): The effects of *EA* flower extract on cell invasion of human HER2-positive breast cancer cells. *EA* extract inhibits cell invasion ability of SKBR3 (**a**) and ZR75-1 (**b**) cell lines by approximately 70% in comparison with their matched control cells (unexposed) ($p < 0.05$). Boyden chambers were used to assess cell-invasion ability of SKBR3 and ZR75-1 cell lines. Cancer cells treated for 24 h with 100 and 200 µL/mL *EA* plant extract showed a significant inhibition of cell invasion in both cell lines, when compared with their matched control ($p < 0.05$). Data are quantified by normalizing the number of invasive cells by their total number.

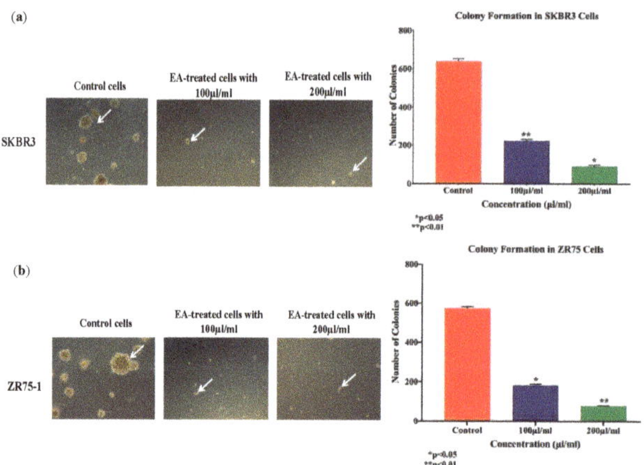

Figure 6. (**a,b**) Effect of *EA* flower extract on colony formation, in soft agar, in human HER2-positive cancer cell lines, SKBR3 (**a**) and ZR75-1 (**b**). *EA* extract inhibits colony formation of SKBR3 and ZR75-1, in comparison with their matched control cells (images of figure **a,b** at ×10 magnification). Colony formation in soft agar is a solid indicator of tumor formation in vivo. The colonies were counted manually and expressed as percentage of treatment relative to the control (mean ± SEM).

Based on the above data, we explored the expression patterns of key markers of EMT and cancer progression: E-cadherin, β-catenin, vimentin, and fascin; our data pointed out that *EA* extract enhances the expression of E-cadherin and β-catenin in SKBR3 and ZR75-1 cell lines, while the expression of vimentin and fascin are decreased in comparison to their control cells (Figures 7 and 8). In parallel, we examined the outcome of *EA* on pro-apoptotic proteins (caspase-3 and Bax) and anti-apoptotic protein (Bcl-2) with 100 and 200 µL/mL of *EA* plant extract after 24 and 48 h of exposure. We found enhanced expression of both pro-apoptotic proteins (Bax and caspase-3) in SKBR3 and ZR75-1 in *EA*-treated cells, compared to their control (Figures 7 and 8). In contrast, the expression of Bcl-2 was lost in SKBR3 and ZR75-1 (Figures 7 and 8). Our data suggest that high concentrations of *EA* induce apoptosis in HER2-positive cancer cells, which might be associated with the Bcl-2/Bax/caspase-3 signaling pathway.

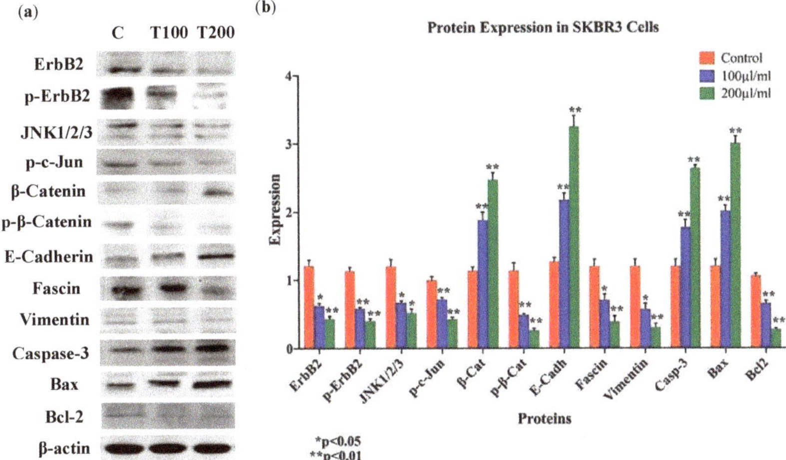

Figure 7. (**a,b**) Protein expression and molecular mechanisms of *EA* inhibitory actions in SKBR3 cell line. This plant extract induces an overexpression of E-cadherin, β-catenin, and downregulation of vimentin and fascin, while upregulating pro-apoptotic markers (Bax and Caspase-3), in comparison with their control and inhibiting anti-apoptotic markers (Bcl-2). Furthermore, *EA* plant extract inhibits the phosphorylation of ErbB2 and β-catenin, as well as the expression of JNK1/2/3. β-actin was used as a control for the proteins amount in this assay. Cells were treated with 100 and 200 µL/mL of *EA* extract for 48 h, as explained in the materials and methods and the results sections. (**a**) Blot image and (**b**) quantification of bands.

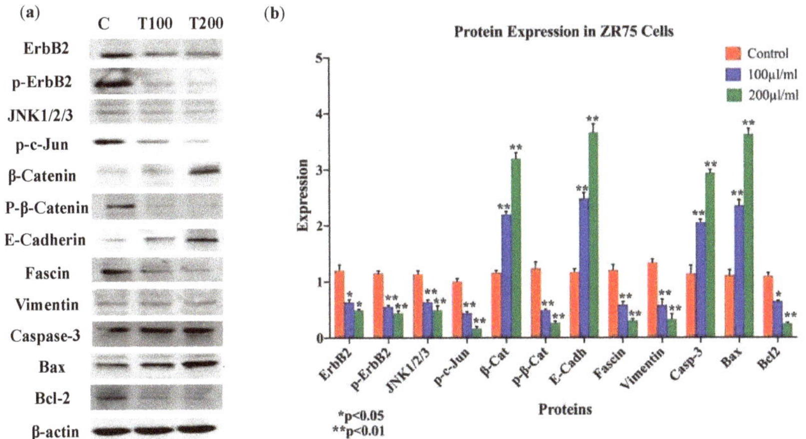

Figure 8. (**a,b**) Protein expression and molecular mechanisms of *EA* inhibitory actions in ZR75 cell line. This plant extract induces an overexpression of E-cadherin, β-catenin, and downregulation of vimentin and fascin; in addition, pro-apoptotic markers Bax and Caspase-3 are upregulated in comparison with their control, while anti-apoptotic marker Bcl-2 is inhibited. Furthermore, *EA* plant extract inhibits the phosphorylation of ErbB2 and β-catenin, as well as JNK1/2/3 expression. β-actin served as a control in this assay. Cells were treated with 100 and 200 µL/mL of *EA* extract for 48 h, as explained in the materials and Methods section. (**a**) Blot image and (**b**) quantification of bands.

Vis-à-vis the underlying molecular pathways of *EA* extract on cell proliferation, EMT progression, cell invasion, and colony formation of HER2-positive breast cancer cells, we assumed that HER2 activation, as well as c-Jun N-terminal kinase (JNK), could have major roles in regulating these events [36–39]; therefore, the expression patterns of HER2 and JNK1/2 were explored. We found that *EA* extract inhibits the phosphorylation of HER2 (with slight change in its expression level) and β-catenin, while it provokes a downregulation of JNK1/2 in SKBR3 and ZR75-1 upon treatment with *EA* plant extract after 24 and 48 h of exposure (Figures 7 and 8).

3. Discussion

In this study, we investigated the effect of *EA* extract in HER2-positive human breast cancer cell lines (SKBR3 and ZR75-1) with regard to certain parameters related to cell proliferation, cell cycle, morphological changes (round to epithelial-like transition: RELT), cell invasion, and colony formation. Additionally, we explored the molecular pathways behind these events. We report that *EA* plant extract can suppress cell proliferation, as well as dysregulate cell-cycle progression of SKBR3 and ZR75-1 cells, along with induction of RELT and inhibition of colony formation in both cell lines. *EA* plant is known for its antioxidant characteristics and has been used conventionally for the treatment of several diseases and inflammation [16,20,21]. Moreover, *EA* consists of bioactive compounds (flavonoids and neoclerodane diterpenoids), which can play a role in promoting apoptosis and cell-cycle progression, as well as inhibiting angiogenesis and EMT events, thus potentially preventing cancer development and progression [12,13,16,40]. Meanwhile, we herein demonstrate that *EA* plant extract inhibits cell proliferation and dysregulate cell-cycle and EMT progression of HER2-positive breast cancer cells.

Indeed, EMT is a crucial phenomenon in cancer progression, characterized by disruption of intracellular tight junctions and the loss of cell-cell contact and epithelial cell features, along with the gain of mesenchymal morphology [41]. On the other hand, it is well-known that cancer progression is characterized by loss of differentiation in human carcinomas, together with downregulation of E-cadherin, which is associated with the degree of tumor malignancy [33]. Moreover, previous studies on different types of human carcinomas have shown that loss of E-cadherin and β-catenin, in addition to enhanced expression of vimentin and fascin, can promote EMT, which is associated with cancer progression [41–44]. In this investigation, we analyzed the effect of *EA* on E-cadherin, β-catenin, vimentin, and fascin expression patterns in HER2-positive breast cancer cell lines. We observed two different major events that are provoked by *EA* treatment, namely induction of RELT "MET" and apoptosis. More specifically, we found that, upon *EA*-treatment for 24 and 48 h at both low and high concentrations (100 and 200 μL/mL), E-cadherin and β-catenin are upregulated, while vimentin, p-β-catenin, and fascin expressions are downregulated, which are important elements of mesenchymal-epithelial transition (MET), the opposite event of EMT, and RELT. Thus, *EA* plant extract can induce differentiation to an epithelial phenotype and consequently block cell invasion of the two HER2-positive human breast cancer cell lines. On the other hand, we found that *EA* extract inhibits colony formation of SKBR3 and ZR75-1 cell lines, which could be considered as an in vivo tumor formation.

On the other hand, regarding the molecular pathways of *EA* extract on our cell line models, we herein reported that, upon *EA*-treatment after 24 h at both low and high concentration, *EA* extract can inactivate HER2 receptor, as well as deregulate the expression patterns of JNK1/2, which can lead to increased expression of E-cadherin and β-catenin and decreased expression of vimentin and fascin, thus indicating restoration of cell-cell adhesion, especially E-cadherin/catenins complex. Furthermore, in accordance with our data, a study by Wang et al. showed overexpression of JNK to be linked with breast cancer cell migration and invasion, as well as EMT [45]. Therefore, in this present investigation, we show that *EA* plant extract can regulate the RELT/EMT event and inhibit cell invasion of the two human HER2-positive breast cancer cell lines. Moreover, these present data are concurrent with our recently published work regarding the outcome of *EA* plant extract on human oral cancer cells, where we have demonstrated that *EA* induces differentiation to an epithelial phenotype; and therefore, it causes a dramatic decrease in cell invasion and motility of human oral cancer cells, along with an upregulation of E-cadherin expression [12]. Additionally, our study of *EA* extract on oral cancer cells revealed that *EA* can inhibit the phosphorylation of Erk1/Erk2 and β-catenin, which could be behind the initiation of RELT/MET event and the overexpression of E-cadherin [12]. In the present work, we demonstrated that JNK1/2 pathway is one of the main molecular pathways of *EA* in HER2-positive human breast cancer cells.

JNK substrate proteins encompass several nuclear proteins, including transcription factors, as well as nuclear hormone receptors involved in maintaining various cellular activities comprising cell proliferation, differentiation, cell death, and cell survival [46]. JNKs phosphorylate and stimulate both, nuclear and non-nuclear proteins and form the transcription factor activator protein-1 (AP-1) by dimerization of the Jun proteins (c-Jun, JunB, and JunD) with the Fos proteins (c-Fos, FosB, Fra-1, and Fra-2); other downstream molecules include activating transcription factor 2 (ATF-2), c-Myc, p53, STAT1/3, Pax family of proteins, Elk1, NFAT, and Bcl-2 family (Bcl-2, Bcl-xl, Bad, Bim, and Bax) [47]. Of these nuclear substrates, c-jun is the most vital nuclear substrate; JNKs enhance c-jun transcription by binding and phosphorylating c-jun at Ser73 and Ser63 via Ha-Ras, c-Raf, and v-Src [48–50]. c-Jun, the downstream target of the JNK pathway, is necessary for Ras-induced carcinogenesis [51,52]. In vivo studies have indicated an oncogenic role of c-Jun in the liver [53–55], as well as intestinal cancers [54], thus indicating a pro-oncogenic role for the JNK/c-Jun axis. The activity of c-jun is essential for the Ha-Ras mediating carcinogenesis transformation. Our data indicate that *EA*-treatment reduced p-c-Jun expression, indicating an *EA*-tumor-suppressive role in cancer.

In parallel, it is evident that β-catenin signaling pathways are also involved in these events; this is based on the fact that β-catenin acts as a transcription regulator, as well as cell-cell adhesion molecule, which was elegantly reported by Kandouz et al., under the effect *Teucrium polium* on human prostate cancer cells [12,56]. Thus, it is possible that *EA* plant extract can have a similar effect on β-catenin pathways, especially since our data showed that *EA* extract inhibit β-catenin phosphorylation, consequently allowing it to translocate from the nucleus to undercoat membrane to act as a cell-cell adhesion protein leading to the inhibition of cell-invasion ability of SKBR3 and ZR75-1 cell lines. We surmise that *EA* exhibits anticancer activity due to the high levels of flavonoids, coumarins, and antioxidants [12,14,57].

Vis-à-vis the interaction between the activation of HER2 receptor and its downstream pathways, including JNK, it is well-established that HER2 overexpression causes homo- or heterodimerization, leading to phosphorylation of this receptor, which in turn triggers downstream signaling pathways responsible for important cellular functions, including cell proliferation, invasion, migration, angiogenesis, chemoresistance, and apoptosis [38,39]. We investigated the downstream target of HER2 stimuli, JNK, as JNK-dependent gene regulatory circuitry underlying cell-fate changes from epithelial to mesenchymal state. Our study demonstrates that *EA* slightly suppresses the expression of HER2 receptor, while mostly affecting its phosphorylation, as well as one of its main downstream targets JNK. More specifically, HER2 downregulation is associated with inhibition of proliferation and invasion of HER2-positive human breast cancer cells [58]; this correlates with our results of *EA*-induced decreased cell proliferation, cell invasion, and colony formation.

Regarding the outcome of high-concentration treatment of *EA* (200 μL/mL) after 48 h, we observed induction of apoptosis in *EA*-treated cells by analyzing the mitochondrial apoptosis regulators of Bcl-2 family (Bcl-2 and Bax), as well as Caspase-3 [59]. Bcl-2 homodimers have been shown to inhibit apoptosis; however, Bax homodimers initiates cell death [60]. Heterodimerization between Bax and Bcl-2 and its ratio of Bax to Bcl-2 determine the susceptibility of cells to apoptosis, whereas caspase-3 is known to act as a downstream target of Bax/Bcl-2 control and play a key role in the execution of apoptosis [60]. We herein report that *EA* can reduce the growth and provoke apoptosis of human HER2-positive breast cancer cells. This effect is associated with caspase-3 activation and reduced Bcl-2 expression. Moreover, mitochondrial Bax translocation and the expression of Bcl-2 slightly decreased upon *EA*-extract treatment, indicating that caspase-dependent pathways are involved in *EA*-induced apoptosis and Bcl2/Bax/Caspase-3-regulated cell death through JNK inactivation.

The JNK pathway is predominantly involved in the stimulation of the intrinsic apoptotic pathway facilitated by mitochondria [61]. However, the JNK pathway is also involved in TRAIL-induced apoptosis, autophagy, mitotic catastrophe, and immunogenic cell death [62–66]. Our data show upregulation of Bax and capsase-3 expression and downregulation of Bcl-2, indicating that apoptosis occurs via the extrinsic pathway as well [65]; a loss of JNK could primarily trigger extrinsic apoptosis. Moreover, Bax/Bcl-2/caspase-3 is also involved in other types of regulated cell death, including immunogenic cell death, mitotic catastrophe, and mitochondrial permeability transition (MPT)-driven necrosis [67], thus suggesting JNK inhibition to mediate Bax/Bcl-2/caspase-3 apoptosis. We herein showed loss of JNK, which is in concordance with a study by Wang et al., in breast cancer where overexpression of JNK did not cause apoptosis and correlated with poor prognosis [45]. Moreover, while activation of JNK results in loss of Bcl-2 expression [68–70], the mechanism is controversial as Bcl-2 phosphorylation enhances cell survival signaling [68,71–74], thus making the role of Bcl-2 phosphorylation in JNK-stimulated apoptosis nascent. Moreover, studies have demonstrated that JNK activation does not result in Bcl-2 phosphorylation [59,75], thus indicating that JNK might regulate another kinase or phosphatase resulting in Bcl-2 phosphorylation.

Although the roles of the Bcl-2 family of proteins in JNK-dependent apoptosis remain nascent, the results of the current study indicate that the proapoptotic Bax subfamily of Bcl-2-related proteins is not essential for JNK-dependent apoptosis. These data demonstrate a dual role of JNK in carcinogenesis which can be both oncogenic and tumor suppressive, as indicated previously. Alternatively, JNK activity can be tissue-specific and cell-type-dependent, differing based on tumor stage and status, as well as the presence of activated upstream and downstream molecules and stress signals [76–80]. Nevertheless, further work is needed to unravel the complexity of the interaction of JNK pathway and its molecules, to help pave the way for the development of anticancer therapeutic strategies. Moreover, in our laboratory, we are aiming to derive the active compounds of *EA* that can plausibly be involved in the inhibition of cancer progression.

4. Materials and Methods

4.1. Plant Collection and Extract Preparation

EA flowers were obtained during the second week of June, from Montreal, Quebec, Canada, and were dried and stored in a dark place, at room temperature, as previously described [12]. The extract was prepared by boiling 3 g of finely grounded dry *EA* flowers per 100 mL of autoclaved distilled water, at 150 °C, on a hot plate, for 20 min, with continuous stirring. The flower extract solution was then filtered, using a 0.45 μm filter unit, and stored at 4 °C until use. Dilutions were prepared in cell culture media for various applications. For each experiment, the extract was freshly prepared.

4.2. Cell Culture

Two different human HER2-positive breast cancer cell lines (SKBR3 and ZR75-1) derived from females were obtained from American Type Culture Collection (ATCC) (Rockville, MD, USA). Cell lines were grown and expanded in RPMI-1640 (Gibco, Life Technologies) supplemented with 10% fetal bovine serum (Gibco, Life Technologies, Massachusetts, MA, USA), 2 mm L-glutamine, 1% PenStrep antibiotic (Invitrogen, Life Technologies, Carlsbad, CA, USA) at 37 °C, and 5% CO_2 humidified atmosphere. Human normal mammary epithelial cells immortalized by E6/E7 of HPV type 16 (HNME-E6/E7) were used to assess plant extract toxicity [81]. Cells were maintained in Gibco® Keratinocyte-SFM (1X) media (Gibco, Life Technologies). All the experiments were carried out when cells were ~70–80% confluent.

4.3. Cell Viability Assay

HER-2-positive breast cancer cell lines, SKBR3 and ZR75-1, were seeded on clear bottom 96-well plates (10,000 cells/well) and cultured in RPMI-1640 supplemented with 10% fetal bovine serum (FBS) and 1% penicillin and streptomycin (100 µL/well).

Elaeagnus angustifolia (*EA*) solution was used to treat cells at different concentrations (25, 50, 75, 100, 150, and 200 µL/mL) for a period of 48 h. Control wells received 100 µL of media (control). The inhibition of cell viability was determined, using Alamar Blue Cell viability reagent (Invitrogen, Thermo Fisher Scientific, Waltham, MA, USA), according to the manufacturer's protocol. The shift in fluorescence was measured at 570 nm (excitation) and 600 nm (emission), in a fluorescent plate reader (Infinite M200, Tecan, Grödig, Austria), after 4 h of incubation with the dye. Relative cell proliferation was determined based on the fluorescence of *EA*-treated cells relative to that of control cells.

4.4. Cell Cycle and Apoptosis Assay

SKBR3 and ZR75-1 cells (1×10^6 cells/dish) were plated in 100 mm Petri dishes, with overnight incubation. The cells were then starved with serum-free RPMI-1640 medium for a period of 6–12 h to synchronize the cells into the G_0 phase of the cell cycle. Synchronized cells were then treated with *EA* extract (100 and 200 µL/mL) for 48 h. Cells were harvested, washed twice with PBS, fixed overnight in 70% ice-cold ethanol, and, subsequently, their DNA was stained with 50 µg/mL FXCycle PI/RNase staining solution (Invitrogen, Thermo Fisher Scientific) after RNase A treatment (50 µg/mL) (Thermo Fisher Scientific), at 37 °C, for 30 min, according to standard protocol [12]. Cell-cycle analysis was performed by flow-cytometry (BD Accuri C6, BD Biosciences, USA), and cells in G_0/G_1, S, G_2/M and the sub-G_0/G_1 (apoptotic) phases were quantified by using FlowJo software.

Furthermore, for apoptosis assay, the Annexin V-fluorescein isothiocyanate (FITC)/7-amino-actinomycin D (7-AAD) Apoptosis Kit-559763 (BD Biosciences, USA) was used as per the manufacturer's instructions. Briefly, cells (1×10^6 cells/dish) were seeded into 100 mm culture dishes and were maintained overnight in a medium containing 10% fetal bovine serum. The cells were collected by trypsinization and washed with phosphate buffered saline (PBS). Then, cells were resuspended in 200 µL of binding buffer. Annexin V staining was accomplished following product instructions (Clontech, Palo Alto, CA). In brief, 5 µL Annexin V-FITC and 5 µL 7-AAD were added to the samples for 15 min in the dark. However, for controls (unstained cells), they were stained with PE Annexin V (no 7-AAD) as well as with 7-AAD (no PE Annexin V). The cells were analyzed by flow cytometry (BD Accuri C6, BD Biosciences, San Jose, CA, USA). Data were presented as density plots of Annexin V-FITC and 7-AAD staining.

4.5. Cell Invasion Assay

Cell invasion assay was carried out in 24-well Biocoat Matrigel invasion chambers (pore size of 8 µm, Corning, USA) as per manufacturer's protocol. In brief, the bottom chamber was filled with RPMI-1640 medium, and the upper chamber was seeded with untreated, as well as treated, cells (5×10^4 cells), and then incubated at 37 °C. After 24 h incubation, non-invasive cells were scraped with a cotton swab, and cells that migrated to the lower surface of the membrane were fixed with methanol and stained with 0.4% crystal violet. For quantification, cells were counted under the Leica DMi1 inverted microscope (Leica Microsystems, Wetzlar, Germany) in five predetermined fields, as previously described [81]. Percentage inhibition of invasive cells was calculated with respect to untreated cells. Each experiment was carried out in triplicates.

4.6. Soft Agar Colony Formation Assay

Next, we determined the number of colonies formed prior and post-treatment, using soft agar growth assay. A total of 2×10^3 cells of SKBR3 and ZR75-1 were placed in their medium containing 0.2% agar with/without 100 and 200 µL/mL of *EA* extract (treated and control cells, respectively) and plated in a 6-well plate covered with a layer of 0.4% agar prepared in RPMI-1640 medium. Colony formation was examined every 2 days for a period of 2 weeks. Colonies in each well were counted, using the Leica SP8 UV/Visible Laser confocal microscope (Leica Microsystems, Wetzlar, Germany).

4.7. Western Blot Analysis

We analyzed the expression levels of proteins involved in the molecular pathways, such as apoptosis by Western blot analysis, as previously described by our group [81]. Briefly, SKBR3 and ZR75-1 cells (1×10^6 cells) were seeded and treated with *EA* extract (100 and 200 µL/mL) for 48 h. Cell lysates were collected, and equal amounts of protein (30 µg) were resolved on 10% polyacrylamide gels and electroblotted onto PVDF membranes. The PVDF membranes were probed with the following primary antibodies: anti-mouse E-cadherin (AbcamID#: ab1416), anti-rabbit β-catenin (CST 9562), anti-rabbit phosphorylated β-catenin (CST 4176), anti-rabbit Vimentin (Abcam: abID# 92547), anti-rabbit Fascin (AbcamID#: ab183891), anti-mouse Bax (ThermoFisher Scientific: MA5-14003), anti-mouse Bcl-2 (Abcam: abID# 692), anti-rabbit Caspase-3 (Abcam: abID# 13847), anti-mouse ErbB2 (Abcam: abID# 16901), anti-rabbit phosphorylated ErbB2 (Abcam: abID# 47262), anti-rabbit JNK1/JNK2/JNK3 (Abcam: abID# 179461), and anti-rabbit phosphorylated-c-Jun (Ser73) (Cell Signal Technologies, ID# 9164). To ensure equal loading of protein samples, the membranes were re-probed with anti-mouse β-actin (Abcam: abID# 6276).

Immunoreactivity was detected by using ECL Western blotting substrate (Pierce Biotechnology, Rockford, IL, USA), as described by the manufacturer.

In order to obtain a relative quantification of protein expressions, images acquired from Western blotting were analyzed, using ImageJ software. The intensity of the bands relative to the β-actin bands was used to calculate a relative expression of proteins in each cell line.

4.8. Statistical Analysis

The data were presented as mean ± SEM from three independent experiments performed in triplicates, and a t-test was used to compare the difference between treated and untreated cells. To evaluate significance for cell cycle, a Chi-square test was performed to compare significance between the different phases. Data were analyzed by using GraphPad Prism software (version 8.4.3), and differences with $p < 0.05$ were considered significant.

5. Conclusions

To the best of our knowledge, this is the first report, on the effect of *EA* in HER2-positive breast cancer and its underlying mechanism. Furthermore, this study brings about novel therapeutic potential by demonstrating the induced inhibition of HER2 and JNK activation by *EA* plant extract in human breast cancer cells. Our study points out that the downregulation of JNK can be one of the molecular pathways responsible of increasing E-cadherin and β-catenin and decreasing the expressions of vimentin and fascin. This is an interesting finding, since it can be potentially used as a target to inhibit cell invasion of HER2-positive breast cancer cells by reversing EMT or inducing RELT. In parallel, our data also demonstrate that high consecrations of *EA* trigger apoptosis, particularly in breast cancer cells, which is associated with Bcl-2/Bax/caspase-3 signaling pathway in HER2-positive cancer cells. We believe that *EA* might act as a candidate therapeutic agent based on its anticancer activity which can pave the way for potential more advanced therapeutic approaches in breast cancer management, especially HER2-positive cases.

Author Contributions: Conceptualization, A.-E.A.M. and H.F.A.F.; methodology, A.J., A.S., and H.K.; validation, A.J., A.S., and I.G.; resources, A.-E.A.M. and H.F.A.F.; data curation, A.J. and I.G.; writing—original draft preparation, A.S., I.G., and A.J.; writing—review and editing, I.G., S.V., and A.-E.A.M.; supervision, A.-E.A.M. and H.F.A.F.; funding acquisition, A.-E.A.M. and H.F.A.F. A.J., A.S., and I.G. contributed equally to this manuscript. All authors have read and agreed to the published version of the manuscript.

Funding: Our lab is supported by grants from Qatar University: # QUCP-CMED-2019-1, QUHI-CMED-19/20-1, and QUCG-CMED-20/21-2.

Acknowledgments: The authors would like to thank A. Kassab for her critical reading of the manuscript.

Conflicts of Interest: The authors declare no conflict of interest. The funders had no role in the design of the study; in the collection, analyses, or interpretation of data; in the writing of the manuscript; or in the decision to publish the results.

References

1. Ferlay, J.; Colombet, M.; Soerjomataram, I.; Mathers, C.; Parkin, D.M.; Piñeros, M.; Znaor, A.; Bray, F. Estimating the global cancer incidence and mortality in 2018: GLOBOCAN sources and methods. *Int. J. Cancer* **2019**, *144*, 1941–1953. [CrossRef] [PubMed]

2. Gupta, I.; Burney, I.; Al-Moundhri, M.S.; Tamimi, Y. Molecular genetics complexity impeding research progress in breast and ovarian cancers. *Mol. Clin. Oncol.* **2017**, *7*, 3–14. [CrossRef] [PubMed]

3. Perou, C.M.; Sørlie, T.; Eisen, M.B.; van de Rijn, M.; Jeffrey, S.S.; Rees, C.A.; Pollack, J.R.; Ross, D.T.; Johnsen, H.; Akslen, L.A.; et al. Molecular portraits of human breast tumours. *Nature* **2000**, *406*, 747–752. [CrossRef] [PubMed]

4. Sareyeldin, R.M.; Gupta, I.; Al-Hashimi, I.; Al-Thawadi, H.A.; Al Farsi, H.F.; Vranic, S.; Al Moustafa, A.-E. Gene Expression and miRNAs Profiling: Function and Regulation in Human Epidermal Growth Factor Receptor 2 (HER2)-Positive Breast Cancer. *Cancers* **2019**, *11*, 646. [CrossRef]

5. Giordano, S.H.; Temin, S.; Chandarlapaty, S.; Crews, J.R.; Esteva, F.J.; Kirshner, J.J.; Krop, I.E.; Levinson, J.; Lin, N.U.; Modi, S.; et al. Systemic Therapy for Patients with Advanced Human Epidermal Growth Factor Receptor 2–Positive Breast Cancer: ASCO Clinical Practice Guideline Update. *J. Clin. Oncol.* **2018**, *36*, 2736–2740. [CrossRef]

6. Ramakrishna, N.; Temin, S.; Chandarlapaty, S.; Crews, J.R.; Davidson, N.E.; Esteva, F.J.; Giordano, S.H.; Kirshner, J.J.; Krop, I.E.; Levinson, J.; et al. Recommendations on Disease Management for Patients with Advanced Human Epidermal Growth Factor Receptor 2–Positive Breast Cancer and Brain Metastases: ASCO Clinical Practice Guideline Update. *J. Clin. Oncol.* **2018**, *36*, 2804–2807. [CrossRef]

7. Vranic, S.; Beslija, S.; Gatalica, Z. Targeting HER2 expression in cancer: New drugs and new indications. *Bosn. J. Basic Med. Sci.* **2020**. [CrossRef]

8. Gonzalez-Angulo, A.M.; Morales-Vasquez, F.; Hortobagyi, G.N. Overview of Resistance to Systemic Therapy in Patients with Breast Cancer. In *Breast Cancer Chemosensitivity*; Yu, D., Hung, M.-C., Eds.; Springer: New York, NY, USA, 2007; pp. 1–22.

9. De Melo, F.H.M.; Oliveira, J.S.; Sartorelli, V.O.B.; Montor, W.R. Cancer Chemoprevention: Classic and Epigenetic Mechanisms Inhibiting Tumorigenesis. What Have We Learned So Far? *Front. Oncol.* **2018**, *8*. [CrossRef]

10. Gullett, N.P.; Amin, A.R.; Bayraktar, S.; Pezzuto, J.M.; Shin, D.M.; Khuri, F.R.; Aggarwal, B.B.; Surh, Y.-J.; Kucuk, O. Cancer Prevention with Natural Compounds. *Semin. Oncol.* **2010**, *37*, 258–281. [CrossRef]

11. Hamidpour, R.; Hamidpour, S.; Hamidpour, M.; Shahlari, M.; Sohraby, M.; Shahlari, N.; Hamidpour, R. Russian olive (Elaeagnus angustifolia L.): From a variety of traditional medicinal applications to its novel roles as active antioxidant, anti-inflammatory, anti-mutagenic and analgesic agent. *J. Tradit. Complement. Med.* **2016**, *7*, 24–29. [CrossRef]

12. Saleh, A.I.; Mohamed, I.; Mohamed, A.A.; Abdelkader, M.; Yalcin, H.C.; Aboulkassim, T.; Batist, G.; Yasmeen, A.; Al Moustafa, A.-E. Elaeagnus angustifolia Plant Extract Inhibits Angiogenesis and Downgrades Cell Invasion of Human Oral Cancer Cells via Erk1/Erk2 Inactivation. *Nutr. Cancer* **2018**, *70*, 297–305. [CrossRef] [PubMed]

13. Pudenz, M.; Roth, K.; Gerhauser, C. Impact of Soy Isoflavones on the Epigenome in Cancer Prevention. *Nutrients* **2014**, *6*, 4218–4272. [CrossRef] [PubMed]

14. Zhang, X.-J.; Jia, S.-S. Fisetin inhibits laryngeal carcinoma through regulation of AKT/NF-κB/mTOR and ERK1/2 signaling pathways. *Biomed. Pharmacother.* **2016**, *83*, 1164–1174. [CrossRef] [PubMed]

15. Amereh, Z.; Hatami, N.; Shirazi, F.H.; Gholami, S.; Hosseini, S.H.; Noubarani, M.; Kamalinejad, M.; Andalib, S.; Keyhanfar, F.; Eskandari, M.R. Cancer chemoprevention by oleaster (Elaeagnus angustifoli L.) fruit extract in a model of hepatocellular carcinoma induced by diethylnitrosamine in rats. *EXCLI J.* **2017**, *16*, 1046–1056. [PubMed]

16. Saboonchian, F.; Jamei, R.; Sarghein, S.H. Phenolic and flavonoid content of Elaeagnus angustifolia L. (leaf and flower). *Avicenna J. phytomedicine* **2014**, *4*, 231–238.

17. Boudraa, S.; Hambaba, L.; Zidani, S.; Boudraa, H. Mineral and vitamin composition of fruits of five underexploited species in Algeria: *Celtis australis* L., *Crataegus azarolus* L., *Crataegus monogyna* Jacq., *Elaeagnus angustifolia* L. and *Zizyphus lotus* L. *Fruits (Paris)* **2010**, *65*, 75–84. [CrossRef]

18. Fonia, A.; White, I.R.; White, J.M.L. Allergic contact dermatitis toElaeagnusplant (Oleaster). *Contact Dermat.* **2009**, *60*, 178–179. [CrossRef]

19. Taheri, J.B.; Anbari, F.; Maleki, Z.; Boostani, S.; Zarghi, A.; Pouralibaba, F. Efficacy of Elaeagnus angustifolia Topical Gel in the Treatment of Symptomatic Oral Lichen Planus. *J. Dent. Res. Dent. Clin. Dent. Prospect.* **2010**, *4*, 29–32.

20. Farzaei, M.H.; Bahramsoltani, R.; Abbasabadi, Z.; Rahimi, R. A comprehensive review on phytochemical and pharmacological aspects of E laeagnus angustifolia L. *J. Pharm. Pharmacol.* **2015**, *67*, 1467–1480. [CrossRef]

21. Niknam, F.; Azadi, A.; Barzegar, A.; Faridi, P.; Tanideh, N.; Zarshenas, M.M. Phytochemistry and Phytotherapeutic Aspects of Elaeagnus angustifolia L. *Curr. Drug Discov. Technol.* **2016**, *13*, 199–210. [CrossRef]

22. Torbati, M.; Asnaashari, S.; Afshar, F.H. Essential Oil from Flowers and Leaves of Elaeagnus Angustifolia (Elaeagnaceae): Composition, Radical Scavenging and General Toxicity Activities. *Adv. Pharm. Bull.* **2016**, *6*, 163–169. [CrossRef] [PubMed]

23. Ya, W.; Shang-Zhen, Z.; Chun-Meng, Z.; Tao, G.; Jian-Ping, M.; Ping, Z.; Qiu-xiu, R. Antioxidant and Antitumor Effect of Different Fractions of Ethyl Acetate Part from Elaeagnus angustifolia L. *Adv. J. Food Sci. Technol.* **2014**, *6*, 707–710. [CrossRef]

24. Kurdali, F.; Al-Shamma'A, M. Natural abundances of15N and13C in leaves of some N2-fixing and non-N2-fixing trees and shrubs in Syria. *Isot. Environ. Heal. Stud.* **2009**, *45*, 198–207. [CrossRef]

25. Murakami, A.; Ashida, H.; Terao, J. Multitargeted cancer prevention by quercetin. *Cancer Lett.* **2008**, *269*, 315–325. [CrossRef]

26. Zhang, H.; Zhang, M.; Yu, L.; Zhao, Y.; He, N.; Yang, X. Antitumor activities of quercetin and quercetin-5′,8-disulfonate in human colon and breast cancer cell lines. *Food Chem. Toxicol.* **2012**, *50*, 1589–1599. [CrossRef]

27. Duo, J.; Ying, G.G.; Wang, G.W.; Zhang, L. Quercetin inhibits human breast cancer cell proliferation and induces apoptosis via Bcl-2 and Bax regulation. *Mol. Med. Rep.* **2012**, *5*, 1453–1456. [CrossRef]

28. Kiseleva, T.I.; Chindyaeva, L.N. Biology of oleaster (Elaeagnus angustifolia L.) at the northeastern limit of its range. *Contemp. Probl. Ecol.* **2011**, *4*, 218–222. [CrossRef]

29. Faramarz, S.; Dehghan, G.; Jahanban-Esfahlan, A. Antioxidants in different parts of oleaster as a function of genotype. *BioImpacts* **2015**, *5*, 79–85. [CrossRef] [PubMed]

30. Panahi, Y.; Alishiri, G.; Bayat, N.; Hosseini, S.M.; Sahebkar, A. Efficacy of Elaeagnus Angustifolia extract in the treatment of knee osteoarthritis: A randomized controlled trial. *EXCLI J.* **2016**, *15*, 203–210. [PubMed]

31. Sahan, Y.; Dundar, A.N.; Aydın, E.; Kilci, A.; Dulger, D.; Kaplan, F.B.; Gocmen, D.; Celik, G. Characteristics of Cookies Supplemented with Oleaster (Elaeagnus angustifolia L.) Flour. I Physicochemical, Sensorial and Textural Properties. *J. Agric. Sci.* **2013**, *5*, 160. [CrossRef]

32. Tehranizadeh, Z.A.; Baratian, A.; Hosseinzadeh, H. Russian olive (Elaeagnus angustifolia) as a herbal healer. *BioImpacts* **2016**, *6*, 155–167. [CrossRef]

33. Asadiar, L.S.; Rahmani, F.; Siami, A. Assessment of genetic diversity in the Russian olive (Elaeagnus angustifolia) based on ISSR genetic markers. *Revista Ciência Agronômica* **2013**, *44*, 310–316. [CrossRef]

34. Natanzi, M.M.; Pasalar, P.; Kamalinejad, M.; Dehpour, A.R.; Tavangar, S.M.; Sharifi, R.; Ghanadian, N.; Balaei, M.R.; Gerayesh-Nejad, S. Effect of aqueous extract of Elaeagnus angustifolia fruit on experimental cutaneous wound healing in rats. *Acta medica Iran.* **2012**, *50*, 589–596.

35. Badrhadad, A.; Kh, P.; Mansouri, K. In vitro anti-angiogenic activity fractions from hydroalcoholic extract of Elaeagnus angustifolia L. flower and Nepeta crispa L. arial part. *J. Med. Plants Res.* **2012**, *6*, 4633–4639. [CrossRef]

36. Choi, Y.; Ko, Y.S.; Park, J.J.; Choi, Y.; Kim, Y.; Pyo, J.-S.; Jang, B.G.; Hwang, D.H.; Kim, W.H.; Lee, B.L. HER2-induced metastasis is mediated by AKT/JNK/EMT signaling pathway in gastric cancer. *World J. Gastroenterol.* **2016**, *22*, 9141–9153. [CrossRef]

37. Han, J.S.; Crowe, D.L. Jun amino-terminal kinase 1 activation promotes cell survival in ErbB2-positive breast cancer. *Anticancer. Res.* **2010**, *30*, 3407–3412.

38. Nahta, R. Molecular Mechanisms of Trastuzumab-Based Treatment in HER2-Overexpressing Breast Cancer. *ISRN Oncol.* **2012**, *2012*, 1–16. [CrossRef]

39. Wolf-Yadlin, A.; Kumar, N.; Zhang, Y.; Hautaniemi, S.; Zaman, M.; Kim, H.-D.; Grantcharova, V.; Lauffenburger, D.A.; White, F.M. Effects of HER2 overexpression on cell signaling networks governing proliferation and migration. *Mol. Syst. Biol.* **2006**, *2*, 54. [CrossRef]

40. Maggioni, D.; Biffi, L.; Nicolini, G.; Garavello, W. Flavonoids in oral cancer prevention and therapy. *Eur. J. Cancer Prev.* **2015**, *24*, 517–528. [CrossRef]

41. Wu, Y.; Sarkissyan, M.; Vadgama, J.V. Epithelial-Mesenchymal Transition and Breast Cancer. *J. Clin. Med.* **2016**, *5*, 13. [CrossRef]

42. Mao, X.; Duan, X.; Jiang, B. Fascin Induces Epithelial-Mesenchymal Transition of Cholangiocarcinoma Cells by Regulating Wnt/β-Catenin Signaling. *Med. Sci. Monit.* **2016**, *22*, 3479–3485. [CrossRef] [PubMed]

43. Satelli, A.; Li, S. Vimentin in cancer and its potential as a molecular target for cancer therapy. *Cell. Mol. Life Sci.* **2011**, *68*, 3033–3046. [CrossRef] [PubMed]

44. Xiao, C.; Wu, C.-H.; Hu, H.-Z. LncRNA UCA1 promotes epithelial-mesenchymal transition (EMT) of breast cancer cells via enhancing Wnt/beta-catenin signaling pathway. *Eur. Rev. Med. Pharmacol. Sci.* **2016**, *20*, 2819–2824. [PubMed]

45. Wang, J.; Kuiatse, I.; Lee, A.V.; Pan, J.; Giuliano, A.; Cui, X. Sustained c-Jun-NH2-kinase activity promotes epithelial-mesenchymal transition, invasion, and survival of breast cancer cells by regulating extracellular signal-regulated kinase activation. *Mol. Cancer Res.* **2010**, *8*, 266–277. [CrossRef]

46. Bubici, C.; Papa, S. JNK signalling in cancer: In need of new, smarter therapeutic targets. *Br. J. Pharmacol.* **2014**, *171*, 24–37. [CrossRef]

47. Bogoyevitch, M.A.; Kobe, B. Uses for JNK: The Many and Varied Substrates of the c-Jun N-Terminal Kinases. *Microbiol. Mol. Biol. Rev.* **2006**, *70*, 1061–1095. [CrossRef]

48. Leppä, S.; Saffrich, R.; Ansorge, W.; Bohmann, D. Differential regulation of c-Jun by ERK and JNK during PC12 cell differentiation. *EMBO J.* **1998**, *17*, 4404–4413. [CrossRef]

49. Li, L.; Porter, A.G.; Feng, Z. JNK-dependent Phosphorylation of c-Jun on Serine 63 Mediates Nitric Oxide-induced Apoptosis of Neuroblastoma Cells. *J. Biol. Chem.* **2004**, *279*, 4058–4065. [CrossRef]

50. Tournier, C. The 2 Faces of JNK Signaling in Cancer. *Genes Cancer* **2013**, *4*, 397–400. [CrossRef]

51. Lloyd, A.; Yancheva, N.; Wasylyk, B. Transformation suppressor activity of a Jun transcription factor lacking its activation domain. *Nature* **1991**, *352*, 635–638. [CrossRef]

52. Eferl, R.; Wagner, E.F. AP-1: A double-edged sword in tumorigenesis. *Nat. Rev. Cancer* **2003**, *3*, 859–868. [CrossRef]

53. Min, L.; Ji, Y.; Bakiri, L.; Qiu, Z.; Cen, J.; Chen, X.; Chen, L.; Scheuch, H.; Zheng, H.; Qin, L.; et al. Liver cancer initiation is controlled by AP-1 through SIRT6-dependent inhibition of survivin. *Nat. Cell Biol.* **2012**, *14*, 1203–1211. [CrossRef] [PubMed]

54. Nateri, A.S.; Spencer-Dene, B.; Behrens, A. Interaction of phosphorylated c-Jun with TCF4 regulates intestinal cancer development. *Nature* **2005**, *437*, 281–285. [CrossRef] [PubMed]

55. Maeda, S.; Karin, M. Oncogene at last—c-Jun promotes liver cancer in mice. *Cancer Cell* **2003**, *3*, 102–104. [CrossRef]

56. Kandouz, M.; Alachkar, A.; Zhang, L.; Dekhil, H.; Chehna, F.; Yasmeen, A.; Al Moustafa, A.-E. Teucrium polium plant extract inhibits cell invasion and motility of human prostate cancer cells via the restoration of the E-cadherin/catenin complex. *J. Ethnopharmacol.* **2010**, *129*, 410–415. [CrossRef] [PubMed]

57. Koirala, N.; Thuan, N.H.; Ghimire, G.P.; Van Thang, D.; Sohng, J.K. Methylation of flavonoids: Chemical structures, bioactivities, progress and perspectives for biotechnological production. *Enzym. Microb. Technol.* **2016**, *86*, 103–116. [CrossRef]

58. Roh, H.; Pippin, J.; Drebin, J.A. Down-regulation of HER2/neu expression induces apoptosis in human cancer cells that overexpress HER2/neu. *Cancer Res.* **2000**, *60*, 560–565.

59. Lei, K.; Nimnual, A.; Zong, W.-X.; Kennedy, N.J.; Flavell, R.A.; Thompson, C.B.; Bar-Sagi, D.; Davis, R.J. The Bax Subfamily of Bcl2-Related Proteins Is Essential for Apoptotic Signal Transduction by c-Jun NH2-Terminal Kinase. *Mol. Cell. Biol.* **2002**, *22*, 4929–4942. [CrossRef]

60. Gross, A.; McDonnell, J.M.; Korsmeyer, S.J. BCL-2 family members and the mitochondria in apoptosis. *Genes Dev.* **1999**, *13*, 1899–1911. [CrossRef]

61. Davis, R.J. Signal Transduction by the JNK Group of MAP Kinases. *Cell* **2000**, *103*, 239–252. [CrossRef]

62. Corazza, N.; Jakob, S.; Schaer, C.; Frese, S.; Keogh, A.; Stroka, D.; Kassahn, D.; Torgler, R.; Mueller, C.; Schneider, P.; et al. TRAIL receptor–mediated JNK activation and Bim phosphorylation critically regulate Fas-mediated liver damage and lethality. *J. Clin. Investig.* **2006**, *116*, 2493–2499. [CrossRef] [PubMed]

63. Lim, S.-C.; Jeon, H.J.; Kee, K.-H.; Lee, M.J.; Hong, R.; Han, S.I. Involvement of DR4/JNK pathway-mediated autophagy in acquired TRAIL resistance in HepG2 cells. *Int. J. Oncol.* **2016**, *49*, 1983–1990. [CrossRef] [PubMed]

64. Puduvalli, V.K.; Sampath, D.; Bruner, J.M.; Nangia, J.; Xu, R.; Kyritsis, A.P. TRAIL-induced apoptosis in gliomas is enhanced by Akt-inhibition and is independent of JNK activation. *Apoptosis* **2005**, *10*, 233–243. [CrossRef] [PubMed]

65. Recio-Boiles, A.; Ilmer, M.; Rhea, P.R.; Kettlun, C.; Heinemann, M.L.; Ruetering, J.; Vykoukal, J.; Alt, E. JNK pathway inhibition selectively primes pancreatic cancer stem cells to TRAIL-induced apoptosis without affecting the physiology of normal tissue resident stem cells. *Oncotarget* **2016**, *7*, 9890–9906. [CrossRef]

66. Reilly, E.O.; Tirincsi, A.; Logue, S.E.; Szegezdi, E. The Janus Face of Death Receptor Signaling during Tumor Immunoediting. *Front. Immunol.* **2016**, *7*. [CrossRef]

67. Galluzzi, L.; Vitale, I.; Aaronson, S.A.; Abrams, J.M.; Adam, D.; Agostinis, P.; Alnemri, E.S.; Altucci, L.; Amelio, I.; Andrews, D.W.; et al. Molecular mechanisms of cell death: Recommendations of the Nomenclature Committee on Cell Death 2018. *Cell Death Differ.* **2018**, *25*, 486–541. [CrossRef]

68. Deng, X.; Xiao, L.; Lang, W.; Gao, F.; Ruvolo, P.; May, W.S., Jr. Novel Role for JNK as a Stress-activated Bcl2 Kinase. *J. Biol. Chem.* **2001**, *276*, 23681–23688. [CrossRef]

69. Maundrell, K.; Antonsson, B.; Magnenat, E.; Camps, M.; Muda, M.; Chabert, C.; Gillieron, C.; Boschert, U.; Vial-Knecht, E.; Martinou, J.-C.; et al. Bcl-2 Undergoes Phosphorylation by c-Jun N-terminal Kinase/Stress-activated Protein Kinases in the Presence of the Constitutively Active GTP-binding Protein Rac1. *J. Biol. Chem.* **1997**, *272*, 25238–25242. [CrossRef]

70. Yamamoto, K.; Ichijo, H.; Korsmeyer, S.J. BCL-2 Is Phosphorylated and Inactivated by an ASK1/Jun N-Terminal Protein Kinase Pathway Normally Activated at G2/M. *Mol. Cell. Biol.* **1999**, *19*, 8469–8478. [CrossRef]

71. Breitschopf, K.; Haendeler, J.; Malchow, P.; Zeiher, A.M.; Dimmeler, S. Posttranslational Modification of Bcl-2 Facilitates Its Proteasome-Dependent Degradation: Molecular Characterization of the Involved Signaling Pathway. *Mol. Cell. Biol.* **2000**, *20*, 1886–1896. [CrossRef]

72. Dimmeler, S.; Breitschopf, K.; Haendeler, J.; Zeiher, A.M. Dephosphorylation Targets Bcl-2 for Ubiquitin-dependent Degradation: A Link between the Apoptosome and the Proteasome Pathway. *J. Exp. Med.* **1999**, *189*, 1815–1822. [CrossRef] [PubMed]

73. Ito, T.; Deng, X.; Carr, B.; May, W.S. Bcl-2 Phosphorylation Required for Anti-apoptosis Function. *J. Biol. Chem.* **1997**, *272*, 11671–11673. [CrossRef] [PubMed]

74. Ruvolo, P.P.; Deng, X.; May, W.S. Phosphorylation of Bcl2 and regulation of apoptosis. *Leukemia* **2001**, *15*, 515–522. [CrossRef]

75. Tournier, C.; Dong, C.; Turner, T.K.; Jones, S.N.; Flavell, R.A.; Davis, R.J. MKK7 is an essential component of the JNK signal transduction pathway activated by proinflammatory cytokines. *Genes Dev.* **2001**, *15*, 1419–1426. [CrossRef]

76. Cho, Y.-L.; Tan, H.W.S.; Saquib, Q.; Ren, Y.; Ahmad, J.; Wahab, R.; He, W.; Bay, B.; Shen, H.-M. Dual role of oxidative stress-JNK activation in autophagy and apoptosis induced by nickel oxide nanoparticles in human cancer cells. *Free. Radic. Biol. Med.* **2020**, *153*, 173–186. [CrossRef]

77. Dhanasekaran, D.N.; Reddy, E.P. JNK signaling in apoptosis. *Oncogene* **2008**, *27*, 6245–6251. [CrossRef]

78. Dou, Y.; Jiang, X.; Xie, H.; He, J.; Xiao, S.-S. The Jun N-terminal kinases signaling pathway plays a "seesaw" role in ovarian carcinoma: A molecular aspect. *J. Ovarian Res.* **2019**, *12*, 99. [CrossRef]

79. Gkouveris, I.; Nikitakis, N.G. Role of JNK signaling in oral cancer: A mini review. *Tumor Biol.* **2017**, *39*. [CrossRef]

80. Liu, J.; Lin, A. Role of JNK activation in apoptosis: A double-edged sword. *Cell Res.* **2005**, *15*, 36–42. [CrossRef] [PubMed]
81. Yasmeen, A.; Alachkar, A.; Dekhil, H.; Gambacorti-Passerini, C.; Al Moustafa, A.-E. Locking Src/Abl Tyrosine Kinase Activities Regulate Cell Differentiation and Invasion of Human Cervical Cancer Cells Expressing E6/E7 Oncoproteins of High-Risk HPV. *J. Oncol.* **2010**, *2010*, 1–10. [CrossRef] [PubMed]

Sample Availability: Samples of the aqueous *EA* extract compounds are available from the corresponding author per reasonable request.

 *molecules* 

Article

Antiproliferative and Pro-Apoptotic Effect of Uvaol in Human Hepatocarcinoma HepG2 Cells by Affecting G_0/G_1 Cell Cycle Arrest, ROS Production and AKT/PI3K Signaling Pathway

Gloria C. Bonel-Pérez [1,†], Amalia Pérez-Jiménez [2,†], Isabel Gris-Cárdenas [1], Alberto M. Parra-Pérez [1], José Antonio Lupiáñez [1], Fernando J. Reyes-Zurita [1], Eva Siles [3], René Csuk [4], Juan Peragón [3,*] and Eva E. Rufino-Palomares [1,*]

[1] Department of Biochemistry and Molecular Biology I, Faculty of Sciences, University of Granada, Avenida Fuentenueva, 1, 18071 Granada, Spain; gloria1797@correo.ugr.es (G.C.B.-P.); isagc@ugr.es (I.G.-C.); ampperez@correo.ugr.es (A.M.P.-P.); jlcara@ugr.es (J.A.L.); ferjes@ugr.es (F.J.R.-Z.)

[2] Department of Zoology, Faculty of Sciences, University of Granada, Avenida Fuentenueva, 1, 18071 Granada, Spain; calaya@ugr.es

[3] Department of Experimental Biology, University of Jaen, Campus Las Lagunillas s/n. 23071 Jaén, Spain; esiles@ujaen.es

[4] Berreich Organische Chemie, Martin Luther University Halle-Wittenberg, 06120 Halle (Saale), Germany; rene.csuk@chemie.uni-halle.de

* Correspondence: jperagon@ujaen.es (J.P.); evaevae@ugr.es (E.E.R.-P.); Tel.: +34-953-212523 (J.P.); +34-958-243252 (E.E.R.-P.)

† These authors contributed equally to this work.

Academic Editors: Enrique Barrajon-Catalan and Wolfgang Weigand

Received: 3 August 2020; Accepted: 15 September 2020; Published: 16 September 2020

Abstract: Natural products have a significant role in the development of new drugs, being relevant the pentacyclic triterpenes extracted from *Olea europaea L.* Anticancer effect of uvaol, a natural triterpene, has been scarcely studied. The aim of this study was to understand the anticancer mechanism of uvaol in the HepG2 cell line. Cytotoxicity results showed a selectivity effect of uvaol with higher influence in HepG2 than WRL68 cells used as control. Our results show that uvaol has a clear and selective anticancer activity in HepG2 cells supported by a significant anti-migratory capacity and a significant increase in the expression of HSP-60. Furthermore, the administration of this triterpene induces cell arrest in the G_0/G_1 phase, as well as an increase in the rate of cell apoptosis. These results are supported by a decrease in the expression of the anti-apoptotic protein Bcl2, an increase in the expression of the pro-apoptotic protein Bax, together with a down-regulation of the AKT/PI3K signaling pathway. A reduction in reactive oxygen species (ROS) levels in HepG2 cells was also observed. Altogether, results showed anti-proliferative and pro-apoptotic effect of uvaol on hepatocellular carcinoma, constituting an interesting challenge in the development of new treatments against this type of cancer.

Keywords: AKT/PI3K signaling pathway; apoptosis; human hepatocarcinoma HepG2 cells; migration activity; proliferation; oxidative stress; ROS level; uvaol

---

1. Introduction

Cancer is one of the leading causes of mortality worldwide, especially in developed countries, on account of the aging of the population [1]. Liver cancer is the fifth most common type, as well as the second type of tumor that causes more deaths globally. Specifically, hepatocellular carcinoma (HCC) accounts for 90% of primary liver neoplasms, whose incidence increases progressively with ageing, according to the EASL (European Association for the Study of the Liver) [2].

Hepatocarcinogenesis is a process initiated by different external stimuli that induce genetic changes in hepatocytes or hepatic stem cells, which can cause alterations in the processes of proliferation and apoptosis, by dysfunctions in the cell cycle and its regulation, which eventually lead to tissue dysplasia and can cause a neoplasm [3]. In response to this DNA damage, different control points can be activated throughout the cell cycle phases. Among all the proteins that regulate this process, p53 stands out [4]. The activity blockage of protein kinases regulated by p53 produces an inhibition of cell cycle progression, due to an arrest in G_1 phase [5]. It is known that p53 is frequently mutated in HCC, disrupting the correct role of this protein [6].

Another important pathway involved in hepatocarcinogenesis is the one controlled by c-*Myc* proto-oncogene. The elements of the *Myc* proto-oncogenes family have a role as strong transcription factors of other proteins that take part in the control of cell differentiation and proliferation, oncogenesis and apoptosis [7]. Studies show that c-*Myc* is overexpressed in HCC, in comparison with healthy patients [8].

One of the main characteristics of tumor cells is their resistance to cell death, so the apoptosis process is one of the most studied pathways. This programmed cell death can occur through two pathways: the mitochondrial or intrinsic and the one that involves death receptors, also named extrinsic. In response to various external stimuli, the mitochondria increases its permeability, releasing apoptotic mediators, among which proteins from the Bcl-2 family and cytochrome c stand out [9].

It is well established that NADPH molecules are essential in anabolic processes related to membranogenesis, as well as in nucleotide metabolism through ribose phosphate formation [10], and also plays a decisive role as a modulator of protein synthesis [11]. Therefore, the participation of NADPH in both metabolic aspects makes it especially important for growth and cell differentiation [12]. For these reasons, the regulation of NADPH levels is essential to understand the behavior of numerous physiological processes and, in this sense, nutritional conditions [13–16]; the presence of triterpenes [17,18]; and the redox state [19] modify significantly the levels of those reduction equivalents.

Reactive oxygen species (ROS) in non-pathological concentrations act as second messengers involved in several signal transduction pathways that regulate processes such as cell growth, proliferation and differentiation [20]. Therefore, cells have detoxification mechanisms that maintain a redox balance since, if they are altered, excessive production of ROS can lead to a situation of oxidative stress, which plays an important role in apoptosis and in the beginning of neoplasia development. Within these detoxification mechanisms, several enzymes stand out, such as superoxide dismutase (SOD), catalase (CAT) or glutathione peroxidase (GPX) [21].

Some pharmacological compounds used as HCC treatments, such as Sorafenib, inhibit VEGF angiogenic factor and MAPK pathway [22]. Since the efficacy of current therapies is low when advanced stages of HCC are considered, it is necessary to seek alternative treatments that could offer a better prognosis for patients [23]. Traditional medicine occupies an important place in the development of new drugs, since natural compounds are a recurrent source of molecules with bioactive properties [24].

The Mediterranean diet presents olive oil as its main exponent, which is obtained from the fruit of the *Olea europaea L.*. Its consumption is associated with various health benefits [24], among which there are a lower incidence of cardiovascular diseases, a lower production of ROS and even a reduction in the risk of cancer [25–27]. These properties are associated with their high content of monounsaturated fatty acids, with oleic acid being the most abundant, in addition to other minor compounds, such as tocopherols, polyphenols, triterpenes and squalene [28].

Uvaol (urs-12-ene-3,28-diol) is a natural alcohol pentacyclic triterpene whose formula and molecular weight is $C_{30}H_{50}O_2$ and 442.73 g/mol, respectively. It has a structural isomer, erythrodiol, from which it only differs in the position of a methyl group, since it is located in carbon 19, while uvaol presents it in carbon 20. The most notable feature of this compound is that it exhibits two hydroxyl groups in remote positions, specifically in carbons 3 and 28 (Figure 1A) [29]. Among the described set of triterpenes present in olive oil, uvaol has the lowest characterization of its bioactive properties [30]. Martín et al. [31] studied the effect of uvaol on the 1321N1 cell line and found that uvaol increased the rate of apoptotic cells, as well as producing an activation of the JNK. Allouche et al. [32] studied the effects of uvaol on MCF-7 cells observing a decrease in ROS production and cell viability.

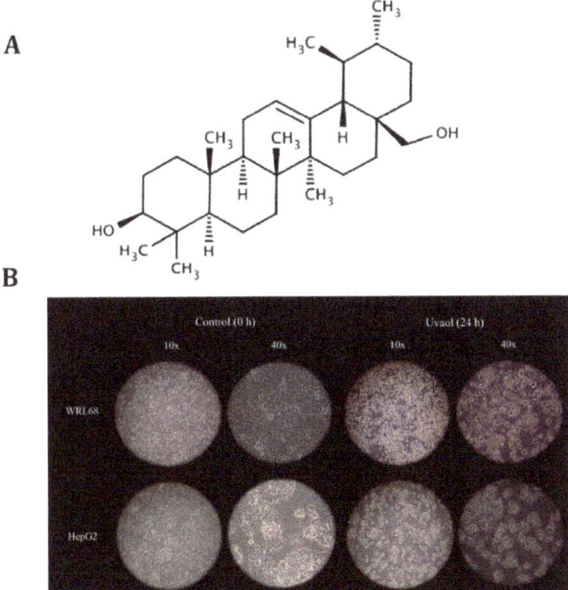

Figure 1. Chemical structure of uvaol (PubChem) (**A**). Morphological changes in response to uvaol treatment incubated at IC$_{50}$ after 24 h in the WRL68 (IC$_{50}$: 54.3 µg/mL) and HepG2 (IC$_{50}$: 25.2 µg/mL) cell lines (**B**).

Due to the promising results shown by the few studies carried out to date on the anticancer potential of uvaol, the aim of the present work was to evaluate its cytotoxic capacity, as well as its effect on proliferation, migration, morphology, cell cycle, apoptosis, oxidative stress, dual PI3K/MAPK signaling pathway and expression of protein markers involved in the processes described on the human control liver cell line WRL68 and the human hepatocellular carcinoma line HepG2.

2. Results

2.1. Uvaol Produces Morphological Changes in HepG2 Cells

Uvaol effect on cell morphology was analyzed using the IC$_{50}$ concentration of each cell line for 24 h. The images taken at 10 and 40 magnifications of the four experimental conditions described are shown in Figure 1B. Referring to the WRL68 line, we can observe that its growth on the support is homogeneous, with the cells being initially separated from each other and presenting a fusiform morphology at 10 magnifications. When the treatment described is applied, the cells undergo a structure change, since they become rounded and have cytoplasmic projections towards the environment, due to cell death induced by the administration of uvaol. In addition, this causes them to rise from the support on which they grow.

The morphology of HepG2 cells in a control situation is triangular and shows an islet-shaped growth, each of them being well delimited with respect to the rest of the surrounding islets. Additionally, its cytoplasm is much more irregular than the presented by WRL68 cells. After the compound's administration, as in the previous case, its characteristic morphology is blurred, becoming slightly more rounded, with very sharp projections and a loss of adhesion to the support.

2.2. Uvaol Has a Selectivity Cytotoxic Effect on the HepG2 Cell Lines

Uvaol effect was evaluated by the MTT assay on the WRL68 and HepG2 cell lines. The viability results obtained were represented in percentage (%) with respect to the different concentrations of compound administered (in µg/mL) for each cell line at 24, 48 and 72 h of exposure, showed using a sigmoidal adjustment (Figure 2A,B). Similarly, the IC_{20}, IC_{50} and IC_{80} values were calculated in the three times lapses described (Figure 2C). Figure 2A shows that, for the WRL68 line, uvaol reduced cell viability in a dose and time-dependent way, since the IC_{20}, IC_{50} and IC_{80} values decreased with this treatment in a progressive form for the three studied time lapses.

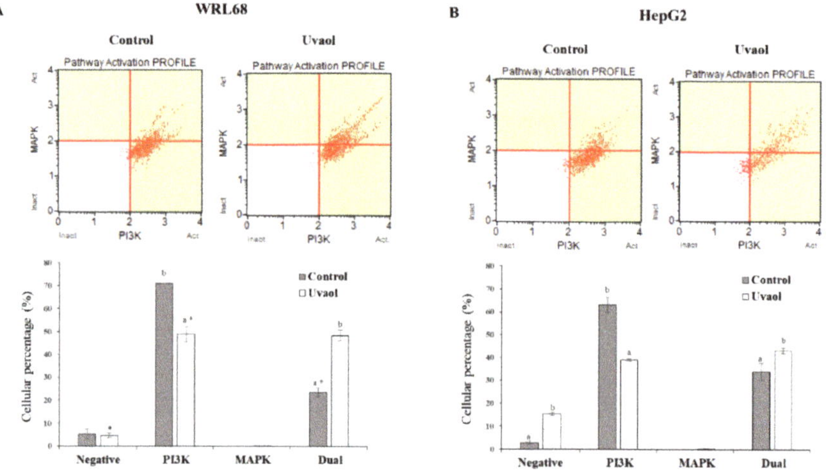

Figure 2. Cytotoxicity curves of uvaol during 24, 48 and 72 h in WRL68 (**A**) and HepG2 cells (**B**). IC_{20}, IC_{50} and IC_{80} values for the WRL68 and HepG2 lines after 24, 48 and 72 h of incubation with uvaol (**C**). These values are represented by mean ± SEM (n = 9).

In the HepG2 line there is also a progressive decrease in cell viability according to concentration and exposure time (Figure 2B). The IC_{50} value obtained was 54.3 µg/mL (12.3 µM) for the WRL68 line and 25.2 µg/mL (5.7 µM) for the HepG2 line after 24 h of uvaol exposure (Figure 2C). Therefore, the concentration needed to inhibit cell proliferation in tumor cells was half that the needed for the same purpose on the control line. Following tests were carried out using the mentioned IC_{50} concentrations and a period of 24 h uvaol incubation. Differences between the IC_{50} values of both cell lines at later time points (48 h and 72 h) were diminished, but still statistically significant.

2.3. Anti-Migratory Activity of Uvaol in HepG2 Cells

The wound healing assay (or scratch assay) aims to assess the ability of cancer cells to migrate in vitro. Figure 3 collects the several images taken and quantified in each experimental condition at 10 magnifications for 0, 6, 18, 30 and 42 h after the wound was made. Data in quantification for the WRL68 cell line (Figure 3) showed a progressive wound healing, occurring at 18 h, in the control situation, the junction of the separated areas. The complete closure occurred at 42 h. In the treated populations, the behavior of cells was different, since wound healing was maintained during more hours than the control group.

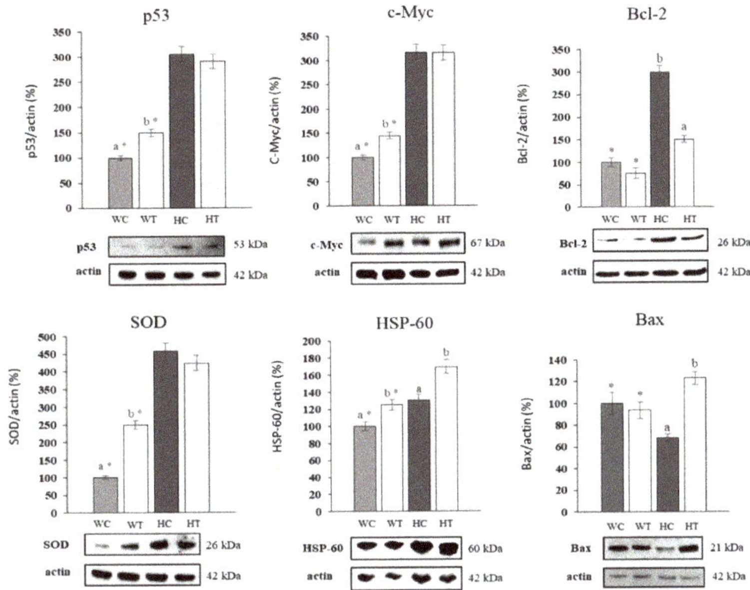

Figure 3. Representative image from wound healing assay of cells treated and no treated with uvaol at IC$_{50}$ concentration (0, 6, 18, 30 and 42 h) in **WRL68** and **HepG2** cells. In the graph, quantification of the cell-free region is showed during the time of wound healing assay (AU: arbitrary units).

For the HepG2 line (Figure 3), in the control populations the wound was not totally closed at the end, although the values of the quantification showed a progressive closure of the wound. In the treated populations, the wound remains until the end of the trial.

2.4. Uvaol Induces G$_0$/G$_1$ Arrest in HepG2 Cells

In order to analyze the possible effect of uvaol in the cell cycle for each line studied, the proportion of cells that were in each phase of the cycle (G$_0$/G$_1$, S and G$_2$/M) were measured in control and uvaol-incubation conditions. Figure 4 shows the images and quantifications of the cell cycle assay generated by the flow cytometer during its analysis for each experimental condition.

Uvaol produced a significant decrease in the percentage of cells in phase G$_0$/G$_1$ and phase S, while, on the contrary, induced an increase in the G$_2$/M phase for the WRL68 line (Figure 4A). In the case of the HepG2 cell line (Figure 4B), treatment with uvaol resulted in a statistically significant increase in the percentage of cells that were in the G$_0$/G$_1$ phase, while causing an equally significant reduction of cells in G$_2$/M phase. No differences were found in the S phase. When statistically comparing the results obtained between both lines, significant differences were observed in the behavior shown in each phase of the cell cycle after treatment with uvaol.

2.5. Apoptosis Is Enhanced in HepG2 Cells by Uvaol

Apoptosis assay provides information about the type of cell death that occurs in each line and for each situation studied: negative control, positive control (treatment with staurosporine with a 1 μg/mL concentration for 2 h) and treatment with uvaol. Figure 5 includes the images of each experimental condition generated by the flow cytometer during its analysis. The data obtained reflect the percentage of viable cells, those that suffer apoptosis and those presenting necrosis (Figure 5).

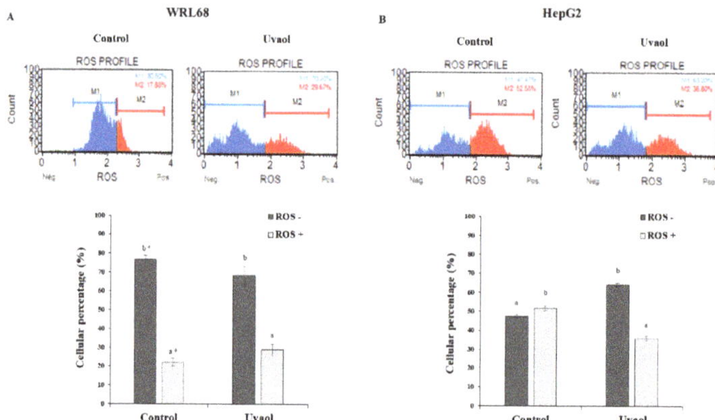

Figure 4. Cell cycle analysis obtained according to the Muse™ cell cycle kit. Panel (**A**) correspond to WRL68 cells and panel (**B**) to HepG2 cells. The cells were treated with IC_{50} of uvaol for 24 h. Top: histograms from a representative experiment show the effect of uvaol on cell cycle profile. Bottom: percentage of cells in each cellular cycle phase. Values are expressed as mean ± SEM (n = 12). Different letters indicate significant differences ($p < 0.05$) between control and uvaol treatment within each phases of the cell cycle for WRL168 or HepG2 cells. The inclusion of asterisks indicates significant differences ($p < 0.05$) between the different cells lines (WRL168 vs. HepG2), under the same treatment (control or uvaol) and phase of cell cycle.

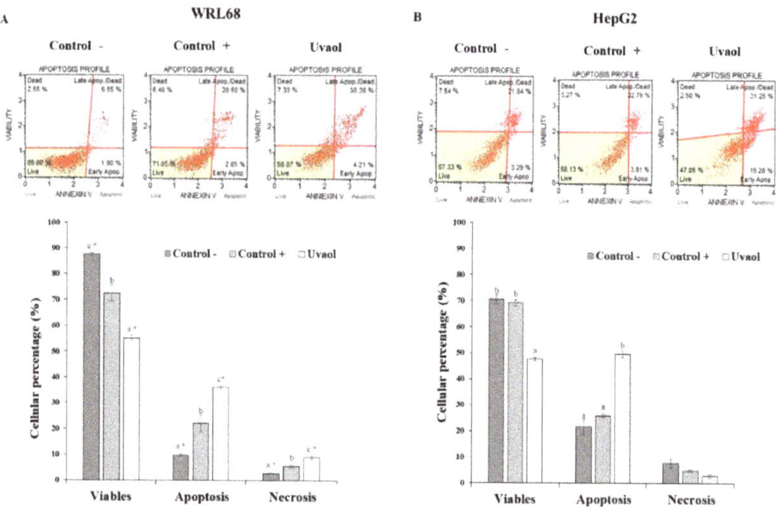

Figure 5. Apoptosis analysis obtained according to the Muse™ apoptosis kit. Panel (**A**) corresponds to the WRL68 cells and panel (**B**) to the HepG2 cells. Treatments included cells not treated (negative control) and cells incubated with staurosporine (1 µg/mL, positive control) or IC_{50} of uvaol for 24 h. Top: dot plots show a representative experiment of the different treatments. Bottom: percentage of lived, apoptotic and necrotic cells for each treatment. Values are expressed as mean ± SEM (n = 12). Different letters indicate significant differences ($p < 0.05$) between control and uvaol treatment within apoptosis stage (viable, apoptosis or necrosis) for WRL168 or HepG2 cells. The inclusion of asterisks indicates significant differences ($p < 0.05$) between different cells lines (WRL168 vs. HepG2), under the same treatment (control or uvaol) and apoptosis stage.

Figure 5A shows how both the addition of staurosporine (control +) and uvaol produced a significant decrease in the number of viable cells in the WRL68 line, which was accompanied by an equally significant increase in the number of apoptotic and necrotic cells in each situation with respect to control. In HepG2 cells (Figure 5B), treatment with staurosporine did not lead to an increase in apoptotic cells, while uvaol did induce a significant reduction in the percentage of viable cells, therefore, increasing the proportion of apoptotic cells. In none of the situations studied, changes in the rate of necrotic cells were generated.

When statistically comparing the results in both cell lines, significant differences were found in the death rate generated by uvaol in each case. The percentage of cells in apoptosis is higher in the HepG2 line, while, on the contrary, a higher rate of necrosis occurred in the WRL68 line, although significantly lower than the number of cells in apoptosis. Therefore, uvaol exerts a pro-apoptotic effect on both lines.

2.6. Uvaol Decreases ROS Production in HepG2 Cells

Intracellular ROS levels were measured by FACS under the four generic test conditions described above. Figure 6 includes the images of each experimental condition generated by the flow cytometer during its analysis. The results show the percentage of cells in each situation studied that express ROS (ROS+) and those that do not express them (ROS−).

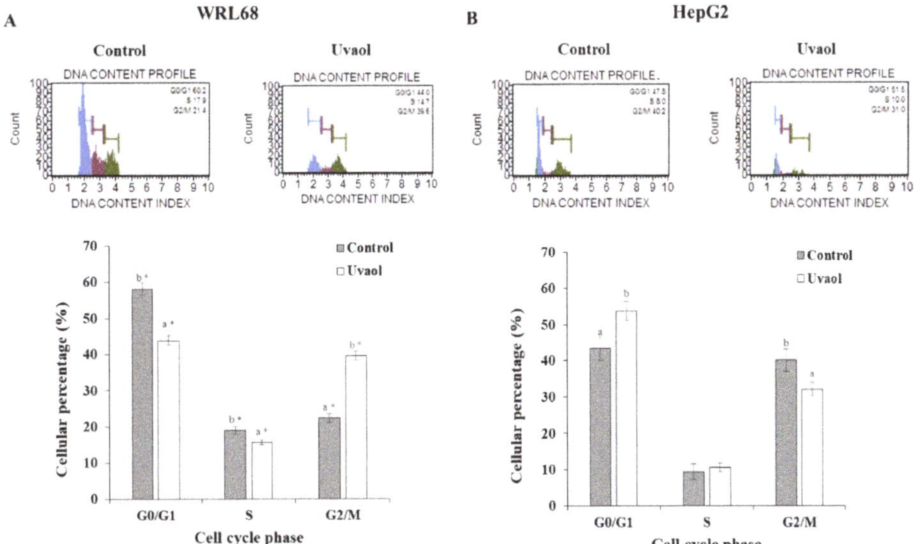

Figure 6. Reactive oxygen species (ROS) quantification was performed according to the Muse™ Oxidative stress kit. Panel (**A**) corresponds to the WRL68 cells and panel (**B**) to the HepG2 cells. Treatments included cells not treated and cells incubated with IC_{50} of uvaol for 24 h. Top: dot plots show a representative experiment of the different treatments. Bottom: percentage of ROS negative (ROS−) and ROS positive (ROS+) values observed for each treatment. Values are expressed as mean ± SEM (n = 12). Different letters indicate significant differences between ROS levels for each treatment and an asterisk indicates significant differences between cell lines for each treatment and ROS levels ($p < 0.05$).

Uvaol treatment maintained the ROS levels produced by the WRL68 cell line without significant differences with respect to the control situation (Figure 6A). The highest levels of ROS were produced in the control situation in the HepG2 cell line (Figure 6B). When treated with uvaol, a significant decrease in the percentage of cells expressing this type of reactive species was observed. When statistically comparing the results in both cell lines, no significant differences in ROS levels between the WRL68 and HepG2 lines after treatment were found. In control situation, significant differences were observed in ROS production levels, since HepG2 express higher levels of radicals.

2.7. Uvaol Produces Up and Down-Regulation of Target Proteins in HepG2 Cells

The expression of p53, c-Myc, Bcl-2, SOD, HSP-60 and Bax proteins for the WRL68 and HepG2 cell lines, in control situation and after treatment with the corresponding IC_{50} concentration of uvaol for 24 h, was measured through the Western blot technique. The level of expression obtained in each case was normalized with respect to the expression of actin (constitutive protein) and subsequently referred to that obtained in the control situation for the WRL68 line (Figure 7).

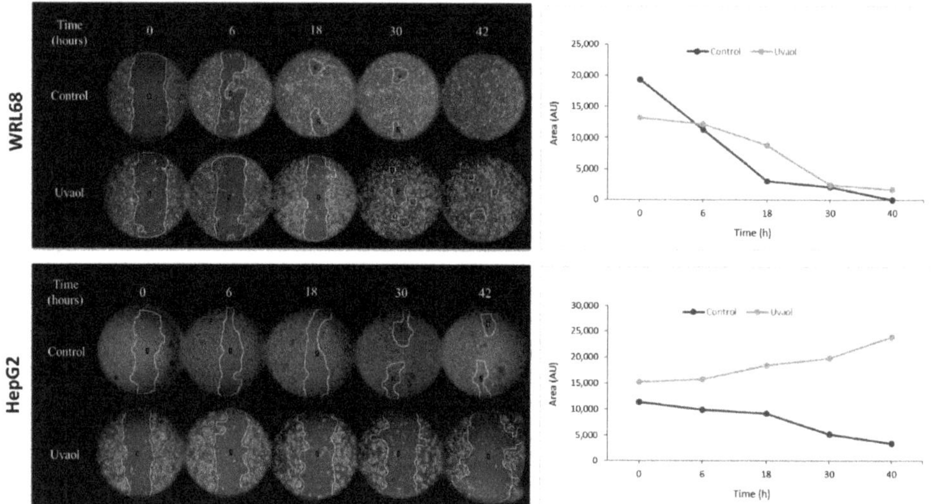

Figure 7. Western-blot analysis of **p53, c-Myc, Bcl-2, SOD, HSP-60** and **Bax** protein levels in WRL68 and HepG2 cells, untreated (WC and HC) and exposed to IC_{50} of uvaol for 24 h (WT and HT). The quantification of protein levels by densitometric analysis is shown in bar graphs. The results are the means ± SEM (n = 9) and are expressed as percentage of expression compared to actin. WRL68 values were used as 100% of expression and the rest of treatments were referred to them. Different letters indicate significant differences between treatments for each cell line and an asterisk indicates significant differences between cell lines for each treatment ($p < 0.05$).

The expression of p53 increased in the WRL68 line after treatment with uvaol regarding control, but remained unchanged in HepG2 cells, a situation that is repeated for SOD. In both cases, the percentage of control expression and after treatment for the HepG2 line was higher, with respect to expression after treatment in WRL68. In the case of c-Myc, an increase in its expression was observed in WRL68 with respect to the control, while it was maintained in HepG2. C-Myc levels in the control situation in HepG2 cells were higher than its expression after treatment in WRL68. The expression of HSP-60 increased in both lines when applying the treatment, being the percentage of expression after the addition of uvaol in WRL68 equal to the control situation in HepG2. Finally, the treatment did not produce changes in the expression of Bcl-2 either Bax in WRL68 cells, while their expression levels decreased and increased in HepG2 cells, respectively. In all the cases, the protein expressions were higher in the HepG2 cells than in the WRL68 cells when compared under the same experimental conditions.

2.8. Uvaol Modulate Dual PI3K/MAPK Signaling Pathway in HepG2 Cells

To analyze the mechanism by which uvaol exerts its anticancer activity, we studied whether such effect could be produced through the dual PI3K/MAPK signaling pathway. The results are shown in

Figure 8. Uvaol treatment produced a 22% decrease in cells that only activated the AKT/PI3K pathway, whereas the percentage of cells in which only ERK1/2/MAPK was activated, besides being too low, was not altered in the WRL68 cells. However, an increase of 25% was observed in cells that activated the dual pathway (MAPK and PI3K) (Figure 8A). When a net balance of activated signaling pathways is considered, no modification of the PI3K pathway was observed, but an activation of 25% took place in the ERK1/2/MAPK signaling pathway.

A   **B**

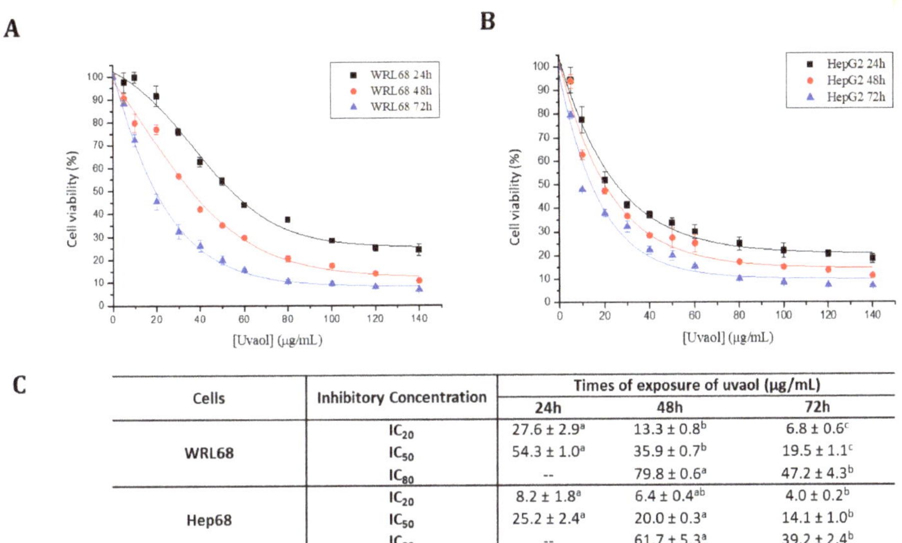

C

Cells	Inhibitory Concentration	Times of exposure of uvaol (µg/mL)		
		24h	48h	72h
WRL68	IC_{20}	27.6 ± 2.9^a	13.3 ± 0.8^b	6.8 ± 0.6^c
	IC_{50}	54.3 ± 1.0^a	35.9 ± 0.7^b	19.5 ± 1.1^c
	IC_{80}	--	79.8 ± 0.6^a	47.2 ± 4.3^b
Hep68	IC_{20}	8.2 ± 1.8^a	6.4 ± 0.4^{ab}	4.0 ± 0.2^b
	IC_{50}	25.2 ± 2.4^a	20.0 ± 0.3^a	14.1 ± 1.0^b
	IC_{80}	--	61.7 ± 5.3^a	39.2 ± 2.4^b

Figure 8. AKT/PI3K and ERK/1/2/MAPK dual pathway determination was performed according to the Muse™ PI3K/MAPK Dual Pathway activation kit. Panel (**A**) correspond to WRL68 cells and panel (**B**) to HepG2 cells. Treatments included cells not treated and cells incubated with IC_{50} of uvaol for 24 h. Top: dot plots show a representative experiment of the different treatments. Bottom: percentage of non-activated cells and cells with PI3K, MAPK and Dual pathway activated for each treatment. Values are expressed as mean ± SEM (n = 12). Different letters indicate significant differences between treatments for each activated pathway and an asterisk indicates significant differences between cell lines for each treatment and activated pathway ($p < 0.05$).

In HepG2 cells, uvaol effect on the percentage of cells that only activated AKT/PI3K induced a decrease of 25%. Similar to WRL68 cells, the percentage of cells that only activated ERK1/2/MAPK, besides being too low, was not altered in HepG2 cells. Regarding the activated dual pathway (MAPK and PI3K), uvaol only produced a 9% increase (Figure 8B). The net balance, in HepG2 cells, resulted in an inhibition of 16% in AKT/PI3K signaling pathway, whereas ERK1/2/MAPK signaling pathway was only increased in 9%.

3. Discussion

The Mediterranean diet is mainly characterized by the regular intake of olive oil, which presents in its composition different molecules with beneficial properties for health. Among them, pentacyclic triterpenes, such as uvaol, have been shown to have an anti-parasitic, anti-oxidant, anti-inflammatory and hepatoprotective activity, especially highlighting their potential as anticancer molecules [18]. Due to the lack of effective treatments against cellular hepatocarcinoma in advanced stages [33], and the fact that no researches about effect of uvaol have been performed in these type of cancer, this study was intended to characterize the anticancer activity of uvaol in the human cell lines WRL68 (hepatic control) and HepG2 (cellular hepatocarcinoma). For this purpose, its effect on the proliferation, migration, morphology, cell cycle, apoptosis, oxidative stress levels and protein markers of the processes described on the liver lines mentioned were determined.

The cytotoxicity assay is a useful test to preliminarily detect compounds that can affect the number of healthy cells in a total population. This trial showed that uvaol exerts an inhibition of cell proliferation in a dose and time dependent manner. Especially noteworthy is the result obtained for the 24 h IC_{50} concentration in both cell lines, being 25.2 µg/mL and 54.3 µg/mL for the HepG2 and WRL68 lines, respectively. As it can be seen, the concentration required to inhibit 50% of proliferation in control cells was more than twice the one used to inhibit the proliferation of tumor cells at the same rate, indicating greater susceptibility and specificity to the compound by hepatocarcinoma cells. To our knowledge, studies of the uvaol in cancer cells are limited. In this sense, similar results were observed in the human mammary tumor line MCF-7 in which uvaol produced anti-proliferative effects in a dose and time-dependent manner with values IC_{50} 11.06 µg/mL and 44.27 µg/mL [32]. Other triterpenes olive derived, such as maslinic acid, have been tested in HepG2 cells, in which the cytotoxic effect was lower to that observed in this study, with a IC_{50} of 47 µg/mL at 72 h [34]. These values demonstrate a greater cytotoxic effect of uvaol than maslinic acid. Moreover, oleanolic-type saponins lowered the growth and proliferation of HepG2 in transplanted tumor in mice [35].

Triterpenes present in olive oil, such as ursolic acid or oleanolic acid have demonstrated anti-migratory and anti-angiogenic activities on several tumor lines [18]. To study if uvaol also had this effect, a wound healing assay was performed in the HepG2 cell line, confirming that this compound has an anti-migratory effect. Moreover, this effect was corroborated by a simultaneous morphological characterization, since it was observed that, after the addition of the compound, cells presented alterations in their structure and adhesion ability. Uvaol induced these same changes on the human astrocytoma line 1321N1, making cells show a loss of adhesion [31]. Furthermore, the administration of oleanolic acid in HepG2 caused a cellular retraction and the appearance of cytoplasmic extensions characteristic of the apoptosis process [36].

Some natural triterpenes exert anti-proliferative effects, due to their interference with the progression of the cell cycle or by the induction of cell death through apoptosis [30]. The arrest of the cycle in some of its phases constitutes a defense mechanism that allows damage repair in DNA, characteristic of tumoral cells [4]. Likewise, apoptosis is an essential process in organisms, since it allows for the elimination of cells, in a controlled manner, whose behavior or characteristics are away from homeostasis [5]. For this reason, flow cytometry studies were conducted to determine the process involved in the cytotoxic effect of uvaol on the cell lines under study. The results obtained showed that this triterpene was capable of inducing cell death through apoptosis with the corresponding concentration of IC_{50} in both cell lines. Similar results were found in lymphoma cells (U937), in which 10 and 100 µM of uvaol induced apoptosis [32]. Notwithstanding, the same concentrations of uvaol did not show this effect in MDA-MB-231 (triple negative breast cancer) or in MCF-7 cells [32]. Uvaol-induced cell cycle arrest occurred in the G_0/G_1 phase for HepG2 cells, while in WRL68 it was in the G_2/M phase. The use of uvaol on the MDA-MB-231 line did not produce changes in the progression of the cycle [29], while a concentration of 44.27 µg/mL of uvaol on the MCF-7 line and of 22.11 µg/mL of oleanolic acid on the HepG2 line resulted in an arrest in the same phase of the cycle [32,36], results that match with those obtained. Therefore, the anti-proliferative effect of uvaol on HepG2 cells is due to, simultaneously, the induction of apoptosis and an arrest in the G_0/G_1 phase of the cell cycle.

In cancer initiation, ROS levels are usually increased, due to the generation of a pro-oxidant environment maintained over time. This situation of oxidative stress, generally induced by an exacerbated production of free radicals, or by an imbalance in endogenous cellular antioxidant systems, can lead to the production of DNA damage, an initial step in neoplastic development [21]. For this reason, controlling ROS levels inside a cancer cell could be a good strategy to mitigate the genetic material damage, or even prevent to it. Triterpenes are characterized by having an antioxidant activity [18], a quality manifested in the HepG2 results obtained in the present study. Similar results were obtained in the MCF-7 and MDA-MB-231 lines for uvaol and erythrodiol [29,32]. However, these compounds produced an increase in the intracellular levels of ROS in 1321N1 cells, which also resulted in the reduction of mitochondrial potential and in an induction of apoptosis via the JNK kinase pathway [31]. Therefore, the action of uvaol on the oxidative state of cells is specific to each tumor cell line. Nevertheless, during normal detoxification process of any foreign compound, such as uvaol inside the cell, ROS can be also produced [37]. An augmentation of ROS levels in response to uvaol treatment did not occur in WRL68 cells, probably due to the increase of SOD expression found in the present study.

p53 and c-Myc proteins are directly involved in the regulation of cell proliferation, as well as in the succession of the different phases of the cell cycle and in the apoptosis process [5,7]. For this reason, the expression profile of the mentioned proteins was characterized in the WRL68 and HepG2 cell lines. Contrary to expectations, p53 levels were higher in the HepG2 than in WRL68 cells. This may be due to the fact that, despite being described a decrease in p53 expression in patients with HCC, due to mutations at different points of the gene that encodes it (TP53) [8], the HepG2 line has this gene intact, being probably induced its expression by situations far from homeostasis in cancer cells, such as an increase in ROS intracellular levels, appreciated in our results. Due to the cellular arrest observed in the G_0/G_1 phase and the maintenance of p53 expression levels after treatment with uvaol in HepG2 cells, it is possible that the inhibition of cell cycle progression is on account of the alteration of the expression of other proteins involved in this process, such as the family of Cdks or their inhibitors, among which are p15, p16, p21 or p27 [5]. Similarly, the increase in c-Myc expression caused by treatment in the WRL68 line may be related to the higher rate of cells found in the G_2/M phase, since c-Myc presents an important proliferative function. One of the main activities of c-Myc is cell cycle control, since it has been descripted that c-Myc expression levels tightly correlate to cell proliferation. Indeed, c-Myc is in charge of inducing the expression of several positive regulators of the cell cycle, such as Cdk4/6, E2F, Cyclin E, Cyclin A, Cdk1/2, among other factors. C-Myc also represses several cell cycle inhibitors, such as p15, p16, p21 or p27 [38]. Overall, it can be stated that an overexpression of c-Myc correlates with an increase percentage of proliferating cells, that is, cells in G2/M phase, as observed in treated WRL68 cells. In the case of the HepG2 line, the increase found in the control situation for c-Myc falls within the expected range, since it is a cancer cell with a high proliferation capacity. This same expression profile was found by Koutb et al. [7] when analyzing the gene expression of the c-Myc gene in blood samples obtained from HCC patients.

The family of proteins that share BH domains contribute to the progression of the cell cycle and the induction of apoptosis, since they are involved in survival mechanisms and in pathways that prevent cell proliferation. The most important member of this family is Bcl-2, a protein with anti-apoptotic effects that is generally increased in certain types of neoplasms, among which is HCC, giving them greater invasive ability and a lower response to treatments [39]. In our study, we observed how Bcl-2 levels remained unchanged after treatment with uvaol in the WRL68 line. As expected, its expression was increased in untreated HepG2 cells with respect to the control line. The administration of uvaol produced a decrease in the levels of this anti-apoptotic protein in the cancer line, a result that is consistent with the induction of cell death due to apoptosis, and the arrest in the G_0/G_1 phase obtained in our findings. Other triterpenes have been reported to also induce apoptosis in HepG2 cells through a down-regulation in Bcl-2 [40]. Similar results have been also found in other cancer cell lines, such as HT29 [41]. Moreover, these authors also observed that Bax pro-apoptotic protein expression increased in response to treatment with maslinic acid [41]. These results are in concordance to the results observed in the present study in which uvaol treatment increased Bax in HepG2 cells.

Heat shock proteins (HSP) form a family with a fundamental role in the correct folding and functionalization of proteins synthesized inside the cell. Clinical studies in patients with HCC show that the expression of HSP-60 is diminished in tumor cells with respect to healthy livers. This finding is related to a greater invasive ability of this neoplasm and a lower survival rate in patients. The lower differentiation gives these cells greater mobility, resulting in a high rate of invasion and subsequent metastasis [42]. Our results show that uvaol treatment induces an increase in the expression of HSP-60 in the WRL68 control line, being this rise more acute for the HepG2 line. This effect of uvaol in the HCC lines is a promising result, since it is related to the results obtained by Zhang et al. [42]. These authors observed that after inducing an overexpression of the gene that codes for the HSP-60 in HepG2 cells, these cells developed a phenotype with lower migratory capacity and greater cell differentiation, thus, decreasing their metastatic ability [42]. Together with HSP-60 overexpression, wound healing results obtained in this study confirm the effect of uvaol as a potential anti-migratory compound.

The AKT/PI3K pathway is known to be one of the most important signaling routes, which participates in cell growth, proliferation, cellular apoptosis and cytoskeletal rearrangement [43,44]. AKT is the major downstream target of the AKT/PI3K pathway. In addition to its interaction with Bcl-2 family effectors, the survival signal to cells is transferred by phospho-AKT [45]. Many studies have demonstrated that this pathway is activated in several types of cancer [46]. The results of the present study showed that uvaol significantly decreased AKT/PI3K pathway in HepG2 cells, whereas no changes were observed in the WRL68 line. On the contrary, levels of ERK1/2/MAPK were increased in both lines, although this increment was significantly lower in hepatoma cells. This up-regulation on MAPK pathway was consequence of the down-regulation on PI3K, since PI3K and Ras pathways can intersect by cross-talk among their downstream effectors [47]. In other HCC studies (HepG2, Huh-7, Hep3B, and Sk-Hep-1 cell lines) using triterpenoids, such as ursolic acid or platycodin D, the results also showed an inhibition of AKT/PI3K signaling pathway [48–50]. These results support the hypothesis that apoptosis induced by uvaol is mediated by the modulation of the AKT/PI3K signaling pathway in HepG2 cells.

Considering the promising in vitro preliminary results of uvaol usage as an anticancer compound (anti-proliferative, pro-apoptotic and antioxidant) drawn from the present work, our future research will be focused on testing these same bioactive properties of uvaol in additional HCC cell lines, with the aim to finally move into in vivo preclinical models. Our next objective is to create HCC cell line-derived xenografts (CDX) and patient derived xenografts (PDX), to see if the observed antitumor properties of uvaol are maintained in vivo. These observations could provide enough evidence for considering a future usage of uvaol in humans as a nutraceutical compound in the food industry or even a therapeutic drug in the clinic; current milestones in cancer research.

4. Materials and Methods

4.1. Cell Cultures

The cell lines used were WRL68 (model liver cells) and HepG2 (hepatocellular carcinoma), provided by *"Centro de Instrumentación Científica"* (CIC) of the University of Granada. The cells were grown in Dulbecco's modified Eagle medium (DMEM) and minimum essential medium (MEM), respectively. In both cases, they were supplemented with 10% heat-inactivated foetal bovine serum (FBS) and 1% streptomycin/penicillin antibiotics. They were kept in a CO_2 incubator at 37 °C, 95% relative humidity and 5% CO_2. Cells were passaged at preconfluent densities by the use of a solution containing 0.05% trypsin and 0.5 mM EDTA. The cells were seeded in the culture dishes at the desired density with the appropriate culture medium.

4.2. Uvaol Solution

The compound tested in the experiments performed was uvaol, provided by Sigma® (St. Louis, MO, USA), with a purity greater than 95%. A stock solution of uvaol with a concentration of 1 mg/mL (in 40 µL of DMSO + 960 µL culture medium) was prepared and subsequently diluted in culture medium until reaching the concentrations required for each test.

4.3. Morphological Changes

A morphological characterization of both cell lines in triplicate was carried out in a control situation (cells growing attached in culture medium) and after uvaol treatment corresponding to the

IC_{50} calculated for 24 h. Optical microscopy (Olympus® model CX22LED, Tokyo, Japan) images of the cultures were obtained in the mentioned conditions at 10 and 40 magnifications for the cell line each after 24 h of the treatment's application.

4.4. MTT Assay

MTT assay was performed as described by Pérez-Jiménez et al. [51]. Briefly, samples containing 200 µL cell suspension (1×10^4 cells/well) were cultured in 96 well plates, in triplicate of three populations of the both cell lines. Subsequent to adherence of the cells within 12 h of incubation, Uvaol was added to the wells at a concentration between 0–140 µg/mL and maintained during 24, 48 and 72 h. MTT was dissolved in the medium and added to the wells at a final concentration of 0.5 mg/mL. Following 2 h of incubation, the generated formazan was dissolved in DMSO. Absorbance was measured at 570 nm in a multiplate reader (Bio-tek®, Winooski, VT, USA). The concentrations that caused 20%, 50% and 80% of inhibition of cell viability (IC_{20}, IC_{50} and IC_{80}) were calculated following the formula: % cell viability = (A0 − AT)/A0·100, where A0 is the control absorbance (100% of cell viability) and AT is de absorbance of the incubated cells with the different concentrations of uvaol. OriginPro 8 (OriginLab Corporation, Northampton, MA, USA) was used to performance a dose-response analysis by the following formula:

$$y = A_1 + \frac{A_2 - A_1}{1 + 10^{(Log\ x_0 - x)p}}$$

In which $Logx_0$ is the center of the curve, p the slope, A_1 the lower asymptote and A_2 the upper asymptote in the adjustment model described.

4.5. Scratch Assay

The wound healing or scratch assay was used to assess the migration capacity of the WRL68 and HepG2 cell lines. With this objective, 3×10^5 cells were seeded per well in a 6-well plate in triplicate for each cell line. When the populations reached a confluence of 80–90%, a sterile pipette tip was used to disrupt in a straight line (make a wound) the layer of cells attached to the culture surface. Subsequently, it was washed with PBS to prevent the cells that had been lifted from being able to rejoin. The control and monitoring of cell migration was carried out by imaging with optical microscopy the culture at different times (0 h, 6 h, 18 h, 30 h and 42 h) at 10 magnifications from the realization of the wound, to be able to identify the possible differences between the control cases (growing cells attached in culture medium) and after the addition of the IC_{50} concentration for 24 h in both lines. Quantification of wound healing was performed by using the MRI-Wound Healing Tool for ImageJ software (ImageJ 1.53a version).

4.6. Cell Sorting by Flow Cytometry

A total of 1.5×10^5 cells were seeded per well in 24-well plates for each cell line (quadruplicates for three different populations of each cell line). After 24 hallowing the cells to adhere, they were separated into four different groups: negative control WRL68 and HepG2 populations and WRL68 and HepG2 populations treated with the concentration corresponding to their uvaol IC_{50} for 24 h (in addition to other specific conditions of each test described in the corresponding section). Negative control populations were incubated only with the appropriate culture medium. The samples were analyzed in the Muse™ Cell Analyzer (Merck-Millipore®, Burlington, MA, USA).

4.6.1. Cell Cycle Assay

The method is based on the analysis of the cell cycle through the discrimination of three populations: cells in phase G_0/G_1, S or G_2/M. For this, the Muse™ Cell Cycle Kit purchased from Millipore (Billerica, MA, USA) was used following the specification of manufacturer.

4.6.2. Apoptosis Assay

Phosphatidylserine (PS) is a phospholipid that is usually found in the inner half of the cytoplasmic membrane, but is outsourced when loss of integrity occurs in the apoptosis process. The method is based on the recognition and binding to the PS exposed by annexin V protein, labeled with a fluorescent substance to detect the interaction and quantify it. We used the Muse[TM] Annexin V & Dead Cell kit purchased from Millipore (Billerica, MA, USA), which provides the percentage of viable cells, apoptotic and affected by necrosis. A positive control was carried out, incubating cells with staurosporine (1 µg/mL), a nonspecific inhibitor of kinase proteins isolated from the *Streptomyces staurosporeus* species, which shows the ability to induce the apoptotic pathway in a large number of tumor lines, being especially characterized this effect on the HepG2 cell line [35].

4.6.3. Measurement of ROS Production

The method is based on the use of dihydroetide (DHE), a reagent capable of crossing the cell membrane and interacting with superoxide anions, a type of ROS. In this process, DHE oxidizes and forms the DNA intercalating agent ethidium bromide, detectable by its fluorescence emission. This allows two cell populations to be distinguished, those that express remarkable levels of ROS (ROS+) and those that do not (ROS−). The Muse[TM] Oxidative Stress kit purchased from Millipore (Billerica, MA, USA) was used for this purpose.

4.7. Protein Extraction and Western Blot Analysis

Triplicates of three different populations of each cell line, WRL68 and HepG2, were seeded in 6-well plates with a density of 2.5×10^5 cells per well. The corresponding IC_{50} concentration for 24 h was added in the treated populations. For protein extraction, each sample was resuspended in 20 µL of RIPA buffer (150 mM NaCl, 1% Igepal, 0.5% deoxycholic acid, 0.1% SDS, 50 mM Tris HCl pH 7.5, 0.2 M PMSF, 700 mM OV_4 and Thermo Scientific® PierceTM inhibitor cocktail, Waltham, MA, USA), and held on ice for 15 min. They were centrifuged at $15,000 \times g$ for 10 min at 4 °C, the supernatant was collected and stored at −80 °C for further analysis. The quantification of the protein concentration was carried out through the Bradford method.

The methodology used for Western blot analysis is described by Mokhtari et al. [19]. Brie fly, the initial separation of the extracts was performed by a 12% polyacrylamide gel electrophoresis under denaturing conditions (SDS-PAGE). A standardized load of 30 µg of protein per well was used after denaturing the samples at 95 °C for 5 min and adding the loading buffer (0.25 mM Tris-HCl pH 6.8, 10% SDS, 1 M DTT and glycerol). The proteins were transferred, by the semi-dry transfer method, (60 mA per gel, 1 h) using the TransBlot Turbo system (BioRad®, Berkeley, CA, USA), to a PVDF membrane. The membranes were blocked with blocking buffer (TBS, 0.1% Tween-20 and 3% skimmed milk) for 1 h at room temperature, incubating with the specific primary antibodies against the proteins studied at 4 °C overnight. The primary antibodies (from Santa Cruz Biotechnology®, Dallas, TX, USA) were diluted in blocking buffer. The membranes were washed with TBS-T (TBS, 0.1% Tween-20) for 5 min three times under stirring, and incubated with the corresponding secondary antibody (from Sigma®, St. Louis, MO, USA) (Table 1) for 1 h at room temperature, repeating the previous washing process later. These secondary antibodies have the enzyme horseradish peroxidase (HRP) coupled, which allows chemiluminescence to be detected in the presence of the protein band due to its ability to oxidize the luminol solution (ECL-plus Western-blot detection system, GE Healthcare, Chicago, IL, USA). The reaction produced was captured inside a ChemiDoc Imaging System (Bio-Rad®, Berkely, CA, USA). The program used to analyze the images obtained was Image Lab Software (Bio-Rad®, for PC 6.1 version). The expression of each protein was normalized referring to actin levels and were represented based on the results obtained for the control situation (WRL68 line without treatment) as percentage of expression (%). The expression obtained in each protein was analyzed through the mentioned program in triplicate.

Table 1. Description of the polyclonal antibodies (IgG) used in the Western blot technique.

Protein	Primary Antibody	Secondary Antibody
p53	Rabbit anti-p53 antibody (1:500)	Anti-mouse IgG antibody (1:5000)
SOD	Rabbit anti-SOD antibody (1:250)	Anti-rabbit IgG antibody (1:5000)
c-Myc	Goat anti-c-Myc antibody (1:500)	Anti-goat IgG antibody (1:5000)
HSP-60	Mouse anti-HSP-60 antibody (1:250)	Anti-mouse IgG antibody (1:5000)
Bcl-2	Rabbit anti-Bcl-2 antibody (1:500)	Anti-rabbit IgG antibody (1:5000)
Bax	Mouse anti-Bax antibody (1:500)	Anti-mouse IgG antibody (1:5000)
Actin	Mouse anti-actin antibody (1:1.000)	Anti-mouse IgG antibody (1:5000)

4.8. PI3K/MAPK Dual Pathway Activation Assay

The Muse® PI3K/MAPK Dual Pathway activation kit purchased from Millipore (Billerica, MA, USA) was employed to examine both the PI3K and MAPK signaling pathways simultaneously using the Muse Cell Analyzer (Merck-Millipore®, Burlington, MA, USA). The protocol followed was performed as the manufacturer instructions. The samples (quadruplicates for three different populations of each cell line) were analyzed in the Muse™ Cell Analyzer.

4.9. Statistical Analysis

The results were expressed as mean ± standard error of the mean (SEM). Likewise, the study of the statistical difference between the data groups was carried out through the analysis of the two-way variance (two-way ANOVA). When interactions between the analyzed factors were present, the one-way analysis of the variance (one-way ANOVA) was used followed by the Tukey test. The differences were considered significant for p values < 0.05. Statistical analyses were performed using the IBM SPSS Statistics (IBM®, 22.0 version) software.

5. Conclusions

In conclusion, taken together, the results of this study suggest that uvaol exhibits a potential anti-proliferative effect through G_0/G_1 cell cycle arrest, apoptosis, by inhibition of AKT/PI3K signaling pathway, and decreased ROS levels on the HepG2 human hepatocarcinoma cell line.

Author Contributions: E.E.R.-P., A.P.-J. and J.P. have designed the research plan; G.C.B.-P., A.P.-J., I.G.-C., A.M.P.-P., F.J.R.-Z., E.S., R.C. and E.E.R.-P. have performed the different experiments; E.E.R.-P., J.P., A.P.-J. and J.A.L. have analyzed the data and G.C.B.-P., A.P.-J. and E.E.R.-P. have written the manuscript. All authors have read and agreed to the published version of the manuscript.

Funding: This research was funded by a research project (UJA2014/07/13) from the Research Plan, Action 7 of University of Jaén, Spain and by funds of the consolidated Researches Group BIO-157 and BIO-341, from the General Secretariat of Universities, Research and Technology of the Ministry of Economy, Innovation, Science and Employment.

Conflicts of Interest: The authors declare no conflict of interest. The funders had no role in the design of the study; in the collection, analyses, or interpretation of data; in the writing of the manuscript, or in the decision to publish the results.

References

1. Bray, F.; Ferlay, J.; Soerjomataram, I.; Siegel, R.L.; Torre, L.A.; Jemal, A. Global Cancer Statistics 2018: GLOBOCAN Estimates of incidence and mortality worldwide for 36 cancers in 185 countries. *CA Cancer J. Clin.* **2018**, *68*, 394–424. [CrossRef]
2. European Association for the Study of the Liver. EASL Clinical practice guidelines: Management of hepatocellular carcinoma. *J. Hepatol.* **2018**, *69*, 182–236. [CrossRef] [PubMed]
3. Shin, J.W.; Chung, Y.H. Molecular targeted therapy for hepatocellular carcinoma: Current and future. *World J. Gastroenterol.* **2013**, *19*, 6144–6155. [CrossRef] [PubMed]

4. Justman, Q.A. Looking beyond the stop sign: Cell-cycle checkpoints reconsidered. *Cell Syst.* **2017**, *5*, 438–440. [CrossRef] [PubMed]

5. Chen, J.D. The cell-cycle arrest and apoptotic functions of p53 in tumor initiation and progression. *Csh. Perspect Med.* **2016**, *6*, a026104. [CrossRef] [PubMed]

6. Villanueva, A.; Newell, P.; Chiang, D.Y.; Friedman, S.L.; Llovet, J.M. Genomics and signaling pathways in hepatocellular carcinoma. *Semin. Liver Dis.* **2007**, *27*, 55–76. [CrossRef]

7. Koutb, F.; Abdel-Rahman, S.; Hassona, E.; Haggag, A. Association of C-myc and p53 gene expression and polymorphisms with hepatitis c (HCV) chronic infection, cirrhosis and hepatocellular carcinoma (HCC) stages in Egypt. *Asian Pac. J. Cancer Prev.* **2017**, *18*, 2049–2057. [CrossRef]

8. Merle, P.; Trepo, C. Molecular mechanisms underlying hepatocellular carcinoma. *Viruses Basel* **2009**, *1*, 852–872. [CrossRef]

9. Zhang, X.R.; Chen, Y.L.; Cai, G.S.; Li, X.; Wang, D. Carnosic acid induces apoptosis of hepatocellular carcinoma cells via ROS-mediated mitochondrial pathway. *Chem. Biol. Interact.* **2017**, *277*, 91–100. [CrossRef]

10. Pilz, R.B.; Willis, R.C.; Boss, G.R. The influence of ribose 5-phosphate availability on purine synthesis of cultured human lymphoblasts and mitogen-stimulated lymphocytes. *J. Biol. Chem.* **1984**, *259*, 2927–2935.

11. Kan, B.; London, I.M.; Levin, D.H. Role of reversing factor in the inhibition of protein synthesis initiation by oxidized glutathione. *J. Biol. Chem.* **1988**, *263*, 15652–15656. [PubMed]

12. Lupiáñez, J.A.; Adroher, F.J.; Vargas, A.M.; Osuna, A. Differential behaviour of glucose 6-phosphate dehydrogenase in two morphological forms of *Trypanosoma cruzi*. *Int. J. Biochem.* **1987**, *19*, 1085–1089. [CrossRef] [PubMed]

13. Barroso, J.B.; García-Salguero, L.; Peragón, J.; de la Higuera, M.; Lupiáñez, J.A. The influence of dietary-protein on the kinetics of NADPH production systems in various tissues of rainbow-trout (*Oncorhynchus mykiss*). *Aquaculture* **1994**, *124*, 47–59. [CrossRef]

14. Barroso, J.B.; Peragón, J.; García-Salguero, L.; de la Higuera, M.; Lupiáñez, J.A. Carbohydrate deprivation reduces NADPH-production in fish liver but not in adipose tissue. *Int. J. Biochem. Cell Biol.* **2001**, *33*, 785–796. [CrossRef]

15. Peragón, J.; Barroso, J.B.; García-Salguero, L.; Aranda, F.; de la Higuera, M.; Lupiáñez, J.A. Selective changes in the protein-turnover rates and nature of growth induced in trout liver by long-term starvation followed by re-feeding. *Mol. Cell Biochem.* **1999**, *201*, 1–10. [CrossRef] [PubMed]

16. Peragón, J.; Barroso, J.B.; García-Salguero, L.; de la Higuera, M.; Lupiáñez, J.A. Dietary alterations in protein, carbohydrates and fat increase liver protein-turnover rate and decrease overall growth rate in the rainbow trout (*Oncorhynchus mykiss*). *Mol. Cell Biochem.* **2000**, *209*, 97–104. [CrossRef]

17. Rufino-Palomares, E.; Reyes-Zurita, F.J.; Fuentes-Almagro, C.A.; de la Higuera, M.; Lupiáñez, J.A.; Peragón, J. Proteomics in the liver of gilthead sea bream (*Sparus aurata*) to elucidate the cellular response induced by the intake of maslinic acid. *Proteomics* **2011**, *11*, 3312–3325. [CrossRef]

18. Rufino-Palomares, E.E.; Pérez-Jiménez, A.; Reyes-Zurita, F.J.; García-Salguero, L.; Mokhtari, K.; Herrera-Merchán, A.; Medina, P.P.; Peragón, J.; Lupiáñez, J.A. Anti-cancer and anti-angiogenic properties of various natural pentacyclic tri-terpenoids and some of their chemical derivatives. *Curr. Org. Chem.* **2015**, *19*, 919–947. [CrossRef]

19. Mokhtari, K.; Rufino-Palomares, E.E.; Pérez-Jiménez, A.; Reyes-Zurita, F.J.; Figuera, C.; García-Salguero, L.; Medina, P.P.; Peragón, J.; Lupiáñez, J.A. Maslinic acid, a triterpene from olive, affects the antioxidant and mitochondrial status of B16F10 melanoma cells grown under stressful conditions. *Evid. Based Compl. Alt.* **2015**, *2015*, 272457. [CrossRef]

20. Yuan, X.X.; Wang, B.Y.; Yang, L.; Zhang, Y.L. The role of ROS-induced autophagy in hepatocellular carcinoma. *Clin. Res. Hepatol. Gas* **2018**, *42*, 306–312. [CrossRef]

21. Brieger, K.; Schiavone, S.; Miller, F.J.; Krause, K. H, Reactive oxygen species: From health to disease. *Swiss Med. Wkly.* **2012**, *142*, w13659. [CrossRef]

22. Kulik, L.; El-Serag, H.B. Epidemiology and management of hepatocellular carcinoma. *Gastroenterology* **2019**, *156*, 477–491. [CrossRef] [PubMed]

23. Llovet, J.M.; Villanueva, A.; Lachenmayer, A.; Finn, R.S. Advances in targeted therapies for hepatocellular carcinoma in the genomic era. *Nat. Rev. Clin. Oncol.* **2015**, *12*, 408–424. [CrossRef] [PubMed]

24. Yuan, H.D.; Ma, Q.Q.; Ye, L.; Piao, G.C. The traditional medicine and modern medicine from natural products. *Molecules* **2016**, *21*, 559. [CrossRef] [PubMed]

25. Escrich, E.; Ramírez-Tortosa, M.C.; Sánchez-Rovira, P.; Colomer, R.; Solanas, M.; Gaforio, J.J. Olive oil in cancer prevention and progression. *Nutr. Rev.* **2006**, *64*, S40–S52. [CrossRef]

26. López-García, E.; Rodríguez-Artalejo, F.; Li, T.Y.; Fung, T.T.; Li, S.S.; Willett, W.C.; Rimm, E.B.; Hu, F.B. The mediterranean-style dietary pattern and mortality among men and women with cardiovascular disease. *Am. J. Clin. Nutr.* **2014**, *99*, 172–180. [CrossRef]

27. Yubero-Serrano, E.M.; López-Moreno, J.; Gómez-Delgado, F.; López-Miranda, J. Extra virgin olive oil: More than a healthy fat. *Eur. J. Clin. Nutr.* **2019**, *72*, 8–17. [CrossRef]

28. Preedy, V.R.; Watson, R.R. *Olives and Olive Oil in Health and Disease Prevention*; Academic Press: San Diego, CA, USA, 2010; p. 1520.

29. Sánchez-Quesada, C.; López-Biedma, A.; Gaforio, J.J. The differential localization of a methyl group confers a different anti-breast cancer activity to two triterpenes present in olives. *Food Funct.* **2015**, *6*, 249–256. [CrossRef]

30. Sánchez-Quesada, C.; López-Biedma, A.; Warleta, F.; Campos, M.; Beltrán, G.; Gaforio, J.J. Bioactive properties of the main triterpenes found in olives, virgin olive oil, and leaves of *Olea europaea*. *J. Agric. Food Chem.* **2013**, *61*, 12173–12182. [CrossRef]

31. Martín, R.; Ibeas, E.; Carvalho-Tavares, J.; Hernández, M.; Ruiz-Gutiérrez, V.; Nieto, M.L. Natural triterpenic diols promote apoptosis in astrocytoma cells through ROS-mediated mitochondrial depolarization and JNK activation. *PLoS ONE* **2009**, *4*, e5975. [CrossRef]

32. Allouche, Y.; Warleta, F.; Campos, M.; Sánchez-Quesada, C.; Uceda, M.; Beltrán, G.; Gaforio, J.J. Antioxidant, antiproliferative, and pro-apoptotic capacities of pentacyclic triterpenes found in the skin of olives on MCF-7 human breast cancer cells and their effects on DNA damage. *J. Agric. Food Chem.* **2011**, *59*, 121–130. [CrossRef]

33. Grandhi, M.S.; Kim, A.K.; Ronnekleiv-Kelly, S.M.; Kamel, I.R.; Ghasebeh, M.A.; Pawlik, T.M. Hepatocellular carcinoma: From diagnosis to treatment. *Surg. Oncol.* **2016**, *25*, 74–85. [CrossRef] [PubMed]

34. Peragón, J.; Rufino-Palomares, E.E.; Muñoz-Espada, I.; Reyes-Zurita, F.J.; Lupiáñez, J.A. A new HPLC-MS method for measuring maslinic acid and oleanolic acid in HT29 and HepG2 human cancer cells. *Int. J. Mol. Sci.* **2015**, *16*, 21681–21694. [CrossRef] [PubMed]

35. Wang, X.Y.; Zhang, W.; Gao, K.; Lu, Y.Y.; Tang, H.F.; Sun, X.L. Oleanane-type saponins from *Anemone taipaiensis* and their cytotoxic activities. *Fitoterapia* **2013**, *89*, 224–230. [CrossRef] [PubMed]

36. Zhu, Y.Y.; Huang, H.Y.; Wu, Y.L. Anticancer and apoptotic activities of oleanolic acid are mediated through cell cycle arrest and disruption of mitochondrial membrane potential in HepG2 human hepatocellular carcinoma cells. *Mol. Med. Rep.* **2015**, *12*, 5012–5018. [CrossRef] [PubMed]

37. Halliwell, B.; Gutteridge, J.M.C. *Free Radicals in Biology and Medicine*, 5th ed.; Oxford University Press: New York, NY, USA, 2015.

38. Bretones, G.; Delgado, M.D.; León, J. Myc and cell cycle control. *Biochim. Biophys. Acta* **2015**, *1849*, 506–516. [CrossRef] [PubMed]

39. Frenzel, A.; Grespi, F.; Chmelewskij, W.; Villunger, A. Bcl2 Family proteins in carcinogenesis and the treatment of cancer. *Apoptosis* **2009**, *14*, 584–596. [CrossRef]

40. Wang, J.J.; Han, H.; Chen, X.F.; Yi, Y.H.; Sun, H.X. Cytotoxic and apoptosis-inducing activity of triterpene glycosides from *Holothuria scabra* and *Cucumaria frondosa* against HepG2 cells. *Mar. Drugs* **2014**, *12*, 4274–4290. [CrossRef]

41. Parra, A.; Martín-Fonseca, S.; Rivas, F.; Reyes-Zurita, F.J.; Medina-O'Donnell, M.; Rufino-Palomares, E.E.; Martínez, A.; García-Granados, A.; Lupiáñez, J.A.; Albericio, F. Solid-phase library synthesis of bi-functional derivatives of oleanolic and maslinic acids and their cytotoxicity on three cancer cell lines. *Acs Comb. Sci.* **2014**, *16*, 428–447. [CrossRef]

42. Zhang, J.; Zhou, X.C.; Chang, H.L.; Huang, X.J.; Guo, X.; Du, X.H.; Tian, S.Y.; Wang, L.X.; Lyv, Y.H.; Yuan, P.; et al. Hsp60 exerts a tumor suppressor function by inducing cell differentiation and inhibiting invasion in hepatocellular carcinoma. *Oncotarget* **2016**, *7*, 68976–68989. [CrossRef]

43. Kauffmann-Zeh, A.; Rodríguez-Viciana, P.; Ulrich, E.; Gilbert, C.; Coffer, P.; Downward, J.; Evan, G. Suppression of c-Myc-induced apoptosis by Ras signalling through PI(3)K and PKB. *Nature* **1997**, *385*, 544–548. [CrossRef] [PubMed]

44. Yang, Y.; Yang, L.; You, Q.D.; Nie, F.F.; Gu, H.Y.; Zhao, L.; Wang, X.T.; Guo, Q.L. Differential Apoptotic Induction of gambogic acid, a novel anticancer natural product, on hepatoma cells and normal hepatocytes. *Cancer Lett.* **2007**, *256*, 259–266. [CrossRef] [PubMed]

45. Duronio, V.; Scheid, M.P.; Ettinger, S. Downstream signalling events regulated by phosphatidylinositol 3-kinase activity. *Cell Signal.* **1998**, *10*, 233–239. [CrossRef]

46. Vivanco, I.; Sawyers, C.L. The phosphatidylinositol 3-kinase-AKT pathway in human cancer. *Nat. Rev. Cancer* **2002**, *2*, 489–501. [CrossRef]

47. Moelling, K.; Schad, K.; Bosse, M.; Zimmermann, S.; Schweneker, M. Regulation of Raf-Akt cross-talk. *J. Biol. Chem.* **2002**, *277*, 31099–31106. [CrossRef] [PubMed]

48. Chuang, W.L.; Lin, P.Y.; Lin, H.C.; Chen, Y.L. The apoptotic effect of ursolic acid on SK-Hep-1 cells is regulated by the PI3K/Akt, p38 and JNK MAPK signaling pathways. *Molecules* **2016**, *21*, 460. [CrossRef] [PubMed]

49. Qin, H.; Du, X.Y.; Zhang, Y.; Wang, R. Platycodin D, a triterpenoid saponin from *Platycodon grandiflorum*, induces G2/M arrest and apoptosis in human hepatoma HepG2 cells by modulating the PI3K/Akt pathway. *Tumor Biol.* **2014**, *35*, 1267–1274. [CrossRef]

50. Tang, C.; Lu, Y.H.; Xie, J.H.; Wang, F.; Zou, J.N.; Yang, J.S.; Xing, Y.Y.; Xi, T. Downregulation of survivin and activation of caspase-3 through the PI3K/Akt pathway in ursolic acid-induced HepG2 cell apoptosis. *AntiCancer Drugs* **2009**, *20*, 249–258. [CrossRef]

51. Pérez-Jiménez, A.; Rufino-Palomares, E.E.; Fernández-Gallego, N.; Ortuño-Costela, M.C.; Reyes-Zurita, F.J.; Peragón, J.; García-Salguero, L.; Mokhtari, K.; Medina, P.P.; Lupiáñez, J.A. Target Molecules in 3T3-L1 adipocytes differentiation are regulated by maslinic acid, a natural triterpene from *Olea europaea*. *Phytomedicine* **2016**, *23*, 1301–1311. [CrossRef]

Sample Availability: Samples of the compound of uvaol are available from the authors.

Article

From a Medicinal Mushroom Blend a Direct Anticancer Effect on Triple-Negative Breast Cancer: A Preclinical Study on Lung Metastases

Elisa Roda [1,2,*,†], Fabrizio De Luca [1,†], Carlo Alessandro Locatelli [2], Daniela Ratto [1], Carmine Di Iorio [1], Elena Savino [3], Maria Grazia Bottone [1] and Paola Rossi [1,*]

[1] Department of Biology and Biotechnology "L. Spallanzani", University of Pavia, 27100 Pavia, Italy; fabrizio.deluca01@universitadipavia.it (F.D.L.); daniela.ratto01@universitadipavia.it (D.R.); carmine.diiorio01@universitadipavia.it (C.D.I.); mariagrazia.bottone@unipv.it (M.G.B.)

[2] Laboratory of Clinical & Experimental Toxicology, Pavia Poison Centre, National Toxicology Information Centre, Toxicology Unit, Istituti Clinici Scientifici Maugeri IRCCS, 27100 Pavia, Italy; carlo.locatelli@icsmaugeri.it

[3] Department of Earth and Environmental Science, University of Pavia, 27100 Pavia, Italy; elena.savino@unipv.it

* Correspondence: elisa.roda@unipv.it (E.R.); paola.rossi@unipv.it (P.R.); Tel.: +39-0382-5924-14 (E.R.); +39-0382-8960-76 (P.R.)

† These authors contributed equally to this work.

Academic Editors: José Antonio Lupiáñez, Amalia Pérez-Jiménez and Eva E. Rufino-Palomares
Received: 28 October 2020; Accepted: 16 November 2020; Published: 18 November 2020

Abstract: Bioactive metabolites isolated from medicinal mushrooms (MM) used as supportive treatment in conventional oncology have recently gained interest. Acting as anticancer agents, they interfere with tumor cells and microenvironment (TME), disturbing cancer development/progression. Nonetheless, their action mechanisms still need to be elucidated. Recently, using a 4T1 triple-negative mouse BC model, we demonstrated that supplementation with Micotherapy U-Care, a MM blend, produced a striking reduction of lung metastases density/number, paralleled by decreased inflammation and oxidative stress both in TME and metastases, together with QoL amelioration. We hypothesized that these effects could be due to either a direct anticancer effect and/or to a secondary/indirect impact of Micotherapy U-Care on systemic inflammation/immunomodulation. To address this question, we presently focused on apoptosis/proliferation, investigating specific molecules, i.e., PARP1, p53, BAX, Bcl2, and PCNA, whose critical role in BC is well recognized. We revealed that Micotherapy U-Care is effective to influence balance between cell death and proliferation, which appeared strictly interconnected and inversely related (p53/Bax vs. Bcl2/PARP1/PCNA expression trends). MM blend displayed a direct effect, with different efficacy extent on cancer cells and TME, forcing tumor cells to apoptosis. Yet again, this study supports the potential of MM extracts, as adjuvant supplement in the TNBC management.

Keywords: breast cancer; lung metastases; in vivo; apoptosis; complementary medicine; medicinal mushrooms

1. Introduction

Breast cancer (BC) is the most frequently diagnosed malignant neoplasm in women [1]. Due to the progress achieved both in early diagnosis and in novel therapeutic treatments, the death rate of BC is progressively decreasing, but BC still remains one of the leading causes of morbidity and mortality in females worldwide [2]. In particular, in Western countries BC metastases are the second cause of mortality among tumor patients [3,4].

Triple-negative breast cancer (TNBC) is characterized by the lack of expression of specific receptors, i.e., estrogens, progesterone, and epidermal growth factor 2, and represents about 15–20% of all BC diagnoses. Indeed, TNBC differs from other subgroups of BC for its increased growth and fast spreading, with reduced treatment possibilities (due to the absence of the three above reported receptors) and a worse outcome [5,6]. Actually, TNBC patients are extremely prone to metastasis and relapse [7] which mainly affect brain, liver, and, in particular, lungs [8]. Tumor progression to metastasis is a complex and many-sided process, affected by both intrinsic cellular mutational burden and several interactions between malignant and non-malignant cell types, and also constantly regulated by the various extrinsic microenvironmental niches [9].

The tumor microenvironment (TME), consisting of immune cells, fibroblasts, satellite cells, as well as blood and lymphatic vessels, plays a fundamental role in the tumor biology, i.e., the behavior of a bulk tumor, co-evolving in a delicate ecosystem with the tumor niche. This never-ending evolving process is governed by local and distant microenvironments, being also regulated by systemic inflammation. In fact, under cytokines, growth factors, and chemotactic stimuli induction, cancer cells recruit and transform stromal fibroblasts into malignant cells, which alter the extracellular matrix by secreting tumor-stimulating factors to facilitate tumor invasion and metastasis. Thus, based on the strict interaction between tumor cells and multicellular proinflammatory TME, cancer typical features also include the evasion of immune destruction, tumor promoting inflammation, and angiogenesis induction. Indeed, nowadays it is well-known that TME plays a crucial role in cancer progression, therapeutic response, and patient outcome [9–11]. Immunotherapy is a recent approach in cancer therapy, specifically addressed on TME [12]. Currently, the elite therapy for treating TNBC is the cytotoxic chemotherapy, characterized by several side effects [13,14]. A bulk of literature demonstrated that chemotherapy induces apoptosis and cancer is a pathology characterized by a dysregulation in cell cycle/death, in which the disruption of the apoptotic pathway leads to uncontrolled cell proliferation [15]. Anomalies in the apoptotic pathway are crucial for cancer genesis, evolution, and regression after treatment [16]. Indeed, the apoptosis rate is crucial in determining the fate between cancer progression and regression as well as in response to current available treatments, i.e., chemotherapy, radiotherapy, surgery, and hormonal therapies.

As a matter of fact, cells undergo apoptosis by changing the balance of proapoptotic and antiapoptotic genes [17]. The main actors involved in apoptosis pathway can be reduced to a few crucial proteins largely preserved through species [18]. In humans and mice, these key proteins belong to the Bcl-2 family [18], Poly (ADP-ribose) polymerase (PARP) [19], and p53 [20]. Bcl-2 family contains both inhibitors and promoters of apoptosis [18]. Acting as proapoptotic protein, Bax is an important tumor suppressor factor, and its reduced levels provide tumor cells with a selective survival advantage, contributing to their expansion [21]. Contrarily, acting as antiapoptotic protein, Bcl-2 plays an essential role as a cell survivor enhancer, and its increased expression level gives cancer cells with a selective survival gain [22]. Concerning PARPs, PARP1 is a nuclear protein which contributes in DNA single-strand break repair, and its dysregulation is involved in tumorigenesis phenomena to such a degree that PARP1 inhibitors have been authorized for the treatment of some types of BC, stimulating both apoptosis induction and synthetic lethality mechanism/DNA repair [23–27]. p53 protein is a nuclear protein known as tumor suppression factor, which plays a critical role deciding whether DNA would be repaired or the damaged cell will self-destruct, inducing programmed cell death [20,28]. Dysregulation of these mentioned proteins is a frequent feature of human malignant diseases and causal for therapy resistance. Tumor cells usually escape from apoptosis by downregulation of proapoptotic genes and/or hyperactivation of antiapoptotic genes. Therefore, apoptosis induction in cancer cells is considered an excellent approach for treating tumors [29].

It has also to be mentioned the PCNA pivotal role in cancer, owing to its function in cell proliferation. Since cancer is caused by and manifest through multiple mechanisms, many of which converging to deregulated proliferation at primary and metastatic sites, and PCNA is an indispensable factor for cell cycle control, DNA replication, DNA nucleotide excision repair, and chromatin assembly, PCNA inhibition is considered to be another viable anticancer strategy [30,31].

In the last decades, considerable attention has been focused on developing/identifying new compounds for TNBC treatment with absent/minimal side adverse effects. Interestingly, in recent years the importance of naturally derived compounds has been highlighted as a source of anticancer and proapoptotic drugs [32]. One of the most promising sources for drug discovery in integrative oncology are medicinal mushrooms (MM), which have an established story of use in traditional oriental medicine and as nutritionally functional foods. MM display antitumor, proapoptotic, onco-immunological, and immunomodulatory effects in vitro and improve the quality of life in cancer patients during conventional anticancer treatments [33,34]. Indeed, in the last years, the use of several MM has been approved as adjuvant supplements in antitumor therapy in different countries.

Different MM produce hundreds of bioactive compounds which are able to influence, often in a synergistic way, numerous cancer-related pathways, also modulating cellular targets typically involved in cell proliferation, survival, and angiogenesis [33,35].

Several studies showed that the use of MM extracts or their compounds, alone or combined with conventional anticancer treatments, is safe and beneficial [33,35].

The present investigation is strictly linked to a previous study employing a 4T1 triple-negative mouse BC model to explore the effects of an oral supplementation with "Micotherapy U-care" (M. U-care), a medicinal mushroom blend, consisting of a mixture of 20% extracts of mycelia and sporophores of five MM, i.e., *Agaricus blazei*, *Ophiocordyceps sinensis*, *Ganoderma lucidum*, *Grifola frondosa*, and *Lentinula edodes* [36]. Each of these MM have been shown displays anticancer and immunomodulatory effects, both in vitro and in vivo, and in preclinical and clinical studies [34,37].

Using a syngeneic tumor-bearing mouse model of TNBC, we demonstrated that M. U-Care supplementation, starting before 4T1 cells injection and lasting throughout the whole experimental time (about 3 months), elicited (i) an increase in the quality of life, (ii) a dramatic decrease of lung metastases density and nodules number, and (iii) a substantial decrease of inflammatory and oxidative stress pathways, characterized by a similar protein expression trend in both lung TME and metastases [36]. Based on these results, we hypothesized that the supplement effects could have been ascribable either to a direct M. U-care anticancer effect on lung cells and/or to secondary/indirect impacts of the MM blend on systemic inflammation and immunomodulation. To punctually address this question, in the present study, we focused on the programmed cell death pathway, namely, apoptosis, by investigating specific molecules whose critical role in human BC is well known. With this aim, using immunohistochemistry, we evaluated the expression, localization and changes of the following proteins, i.e., PARP1, p53, Bax, and Bcl2, performing a comparative assessment on lung metastases and pulmonary parenchyma, namely, TME. For the sake of clarity, we focused on the pathological outcomes in the murine pulmonary tissue, as the lung is the one of the distant organs recurrently implicated in typical metastatic pattern of primary TNBC [7]. The apoptotic cell death was also examined by in situ detection of DNA fragmentation using the terminal deoxynucleotidyl-transferase (TUNEL) assay. In addition, based on the notion that proliferative activity of cancer cells is also a crucial prognostic marker in the tumor diagnosis, we investigated the role of Proliferating Cell Nuclear Antigen (PCNA), being a pivotal protein directly related to the degrees of tumor malignancy and diagnosis [38,39].

2. Results

Figure 1 emphasizes the key outcomes of the present investigation and also summarizes some major results of our previous study (for details see the work in [36]).

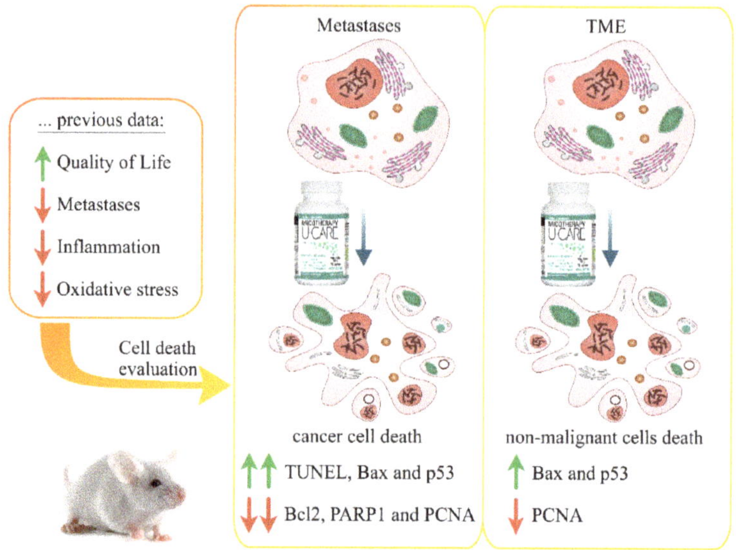

Figure 1. Graphic illustration underlining the main results.

2.1. Raw Materials, Extract Procedure, and Main Active Metabolites of Micotherapy U-Care Blend

The MM blend Micotherapy U-care was produced and supplied by A.V.D. Reform s.r.l. (Noceto, Parma, Italy) and it consists of a mixture of five fungal species, as reported in Table 1.

Table 1. Details on Micotherapy U-care supplement composition.

Medicinal Mushroom	Fungal Part Used in Micotherapy U-Care	% Contained in Micotherapy U-Care	ID Code
Agaricus blazei	Fruiting body	20%	7700
Ophiocordyceps sinensin	Fruiting body and mycelium	20%	Cm2
Ganoderma lucidum	Fruiting body	20%	Gač
Grifola frondosa	Fruiting body	20%	Gf3
Lentinula edodes	Fruiting body	20%	Le.ed.1

The specific MM species strains were established by genetic analyses, sequencing ITS regions and confirming the ID code. Each MM was ground, and total genomic DNA was extracted through the DNeasy mini plant kit (Qiagen NV, Venlo, Netherlands). The Internal Transcribed Spacer (ITS) regions of nuclear DNA were amplified by using PCR, applying two set of primers: ITS F 5′-AGAAAGTCGTAACAAGGTTTCCGTAG-3′, ITS R 5′-TTTTCCTCCGCTCATTGATATGCTT-3′, ITS-g F 5′-TCCGTAGGTGAACCTGCGG-3′, and ITS-g R 5′-TCCTCCGCTTATTGATATGC-3′. Next, the PCR products were purified, sequenced by Eurofins Genomics (Konstanz, Germany), and identified by using NCBI Nucleotide Blast software, version 2.9.0 (Table 1, ID code).

Next, the sporophores and/or mycelia were cultivated, harvested, and the fresh material was extracted for 3 h at 95 °C in distilled water with ethanol 10% (for 1 kg of raw material, 15 L of water-ethanol solution was used). After water-ethanol extraction, solid and liquid components were divided, and the fluid part was dehydrated until the humidity amount was smaller than 7%. Dry extracts were ground and blended to have the 20% of each selected mushroom in the MM blend Micotheraphy U-care (Table 1).

Finally, the polysaccharide content of Micotherapy U-care was determined by using a β-Glucan Assay Kit, and the polysaccharide content was more than 30%. Of this 30%, more than 15% of the polysaccharides were 1,3-1,6 β-glucans, the main active metabolites in Micotherapy U-care.

2.2. TUNEL Assay

TUNEL staining, as a typical marker of apoptotic events [40], revealed an extensive spreading in alveolar pneumocytes (both type I and II) and bronchiolar epithelial cells as well as in metastatic nodules (Figure 2).

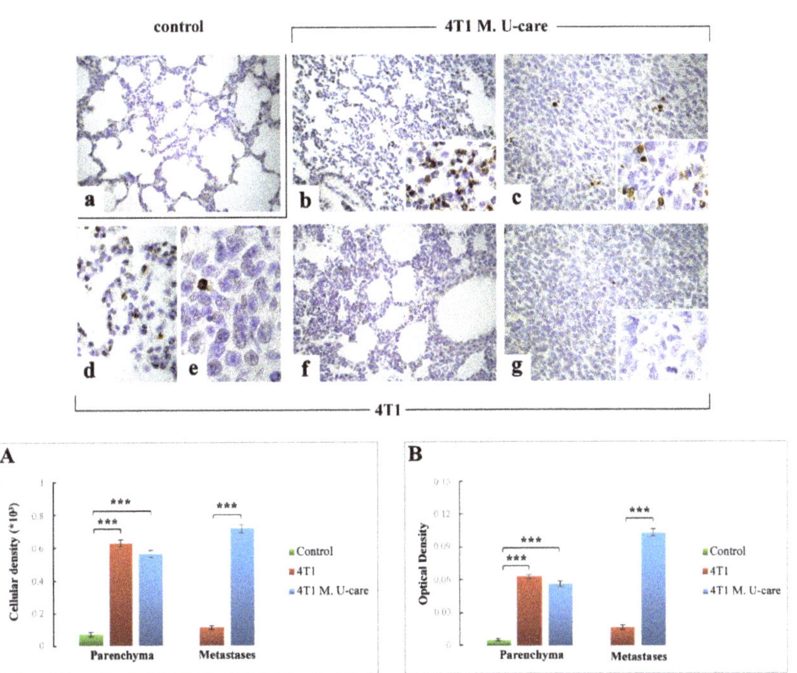

Figure 2. Terminal deoxynucleotidyl-transferase (TUNEL) immunostaining in healthy control (**a**), 4T1 M. U-care (**b,c**) and 4T1 (**d–g**) mice. Light microscopy magnification: 40× (**a–c,f,g**); 100× (**d,e**, insert in panels (**b,c,g**)). Panel (**A**) and (**B**): Histograms presenting immunopositive cell density and OD, respectively. *p* values calculated by unpaired Student's *t*-test: (***) < 0.001.

In particular, concerning the TME, TUNEL-immunoreactivity significantly increased in both 4T1 and 4T1 M. U-care mice (Figure 2b,d,f, respectively) compared to controls (Figure 2a).

The quantitative analysis showed an extremely significant increase of TUNEL-immunopositive cell density and OD comparing 4T1 animals to controls (0.63 ± 0.02 vs. 0.07 ± 0.01 and 0.06 ± 0.00 vs. 0.01 ± 0.00, for cell density and OD, respectively) (Figure 2, Panels (A,B)). In a similar manner, an extremely significant immunoreactivity enhancement was detected when comparing 4T1 M. U-care mice to controls (0.56 ± 0.02 vs. 0.07 ± 0.01 and 0.06 ± 0.00 vs. 0.01 ± 0.00, for cell density and OD, respectively). Diversely, a slight decrease of TUNEL-immunopositive cell density and OD was revealed in 4T1 M. U-care animals compared to 4T1 mice (0.56 ± 0.02 vs. 0.63 ± 0.02 and 0.06 ± 0.00 vs. 0.06 ± 0.00, for cell density and OD, respectively) (Figure 2, Panels (A,B)).

Regarding metastases, an extremely significant increase of TUNEL-immunopositive cell density and OD was detected in 4T1 M. U-care mice compared to 4T1 animals (Figure 2c,e,g, respectively):

0.72 ± 0.02 vs. 0.11 ± 0.01 and 0.10 ± 0.00 vs. 0.02 ± 0.00, for density and OD, respectively (Figure 2, Panels (A,B)). Notably, in 4T1 mice several mitoses were also observable (Figure 2, insert in Figure 2g).

2.3. PARP1, p53, Bax, Bcl2, and PCNA Immunohistochemical Assessment

The cellular expression, localization, and distribution of PARP1, p53, Bax, Bcl2, and PCNA, all involved in cell death and proliferation pathways, were explored.

The immunohistochemical evaluation of all these molecules revealed a widespread labeling in the metastatic nodules and/or in TME, at bronchiolar and alveolar level, evidencing a different efficacy of the MM blend, with the more marked effect on tumor cells.

2.3.1. PARP1

A very significant increase of PARP1-immunoreactive cell density was measured in the TME of 4T1 animals (Figure 3d,f) compared to controls (Figure 2a): 7.46 ± 0.83 vs. 1.85 ± 0.14, respectively (Figure 3, Panel (A)); similarly, a significant increase was detected when comparing 4T1 M. U-care mice (Figure 3b) to controls (6.14 ± 0.87 vs. 1.85 ± 0.14, respectively). Notably, any difference was determined evaluating PARP1-immunopositive cell density in 4T1 animals and 4T1 M. U-care mice (7.46 ± 0.83 vs. 6.14 ± 0.87, respectively) (Figure 3, Panel (A)).

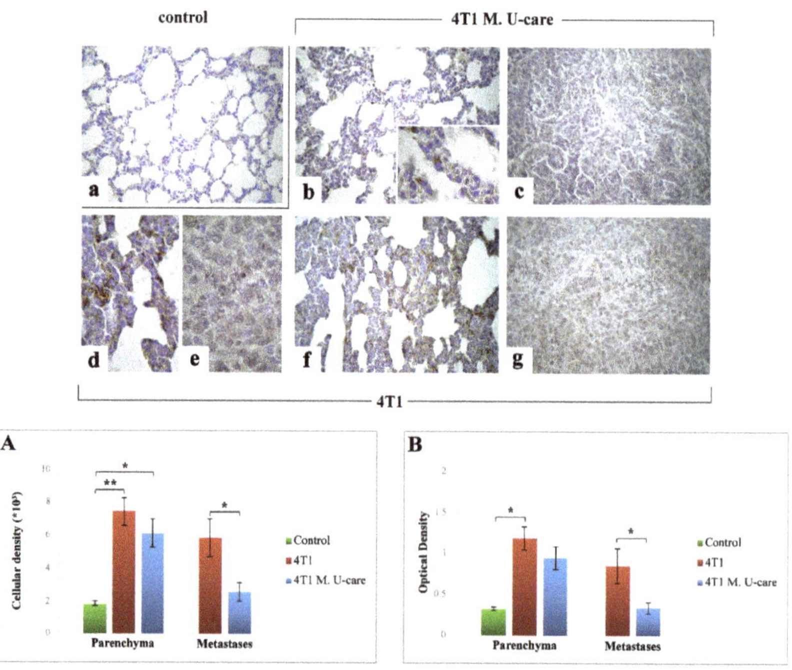

Figure 3. Immunostaining reaction for PARP1 in healthy control (**a**), 4T1 M. U-care (**b**,**c**) and 4T1 (**d–g**) mice. Light microscopy magnification: 40× (**a–c,f,g**); 100× (**d,e**, insert in panel (**b**)). Panels (**A**) and (**B**): Histograms displaying immunopositive cell density and OD, respectively. *p* values calculated by Unpaired Student's *t*-test: (*) < 0.05 and (**) < 0.01.

Likewise, a significant increase of PARP1-immunoreactive OD was measured in 4T1 animals compared to controls (1.19 ± 0.14 vs. 0.32 ± 0.02, respectively) (Figure 3, Panel (B)), while, differently, any difference was revealed evaluating PARP1-immunopositive OD in 4T1 animals compared to 4T1

M. U-care mice (1.19 ± 0.14 vs. 0.95 ± 0.14, respectively), or even between 4T1 M. U-care and controls (0.95 ± 0.14 vs. 0.32 ± 0.02, respectively) (Figure 3, Panel (B)).

Concerning the metastatic tissue, a significant increase of PARP1-immunopositive cell density and OD was determined in 4T1 mice compared to 4T1 M. U-care animals (5.85 ± 1.16 vs. 2.57 ± 0.57 and 0.85 ± 0.21 vs. 0.34 ± 0.07, for cell density and OD, respectively) (Figure 3, Panels (A,B)).

2.3.2. p53

Comparably to the above reported TUNEL immunostaining trend, p53 immunoreactivity was significantly increased in 4T1 M. U-care mice (Figure 4b,c) compared to both 4T1 animals (Figure 4d–g) and controls (Figure 4a). Specifically, a very significant increase of p53-immunopositive cell density and OD was determined in 4T1 M. U-care animals compared to 4T1 mice (6.42 ± 0.78 vs. 3.09 ± 0.65 and 1.11 ± 0.14 vs. 0.53 ± 0.12, for cell density and OD, respectively). A significant increase of p53-immunopositive cell density and OD was also observed when comparing 4T1 M. U-care animals to control (6.42 ± 0.78 vs. 2.93 ± 0.59 and 1.11 ± 0.14 vs. 0.49 ± 0.11, for cell density and OD, respectively). Differently, any significant difference was calculated when comparing 4T1 mice to controls (3.09 ± 0.65 vs. 2.93 ± 0.59 and 0.53 ± 0.12 vs. 0.49 ± 0.11, for cell density and OD, respectively) (Figure 4, Panel (A,B)).

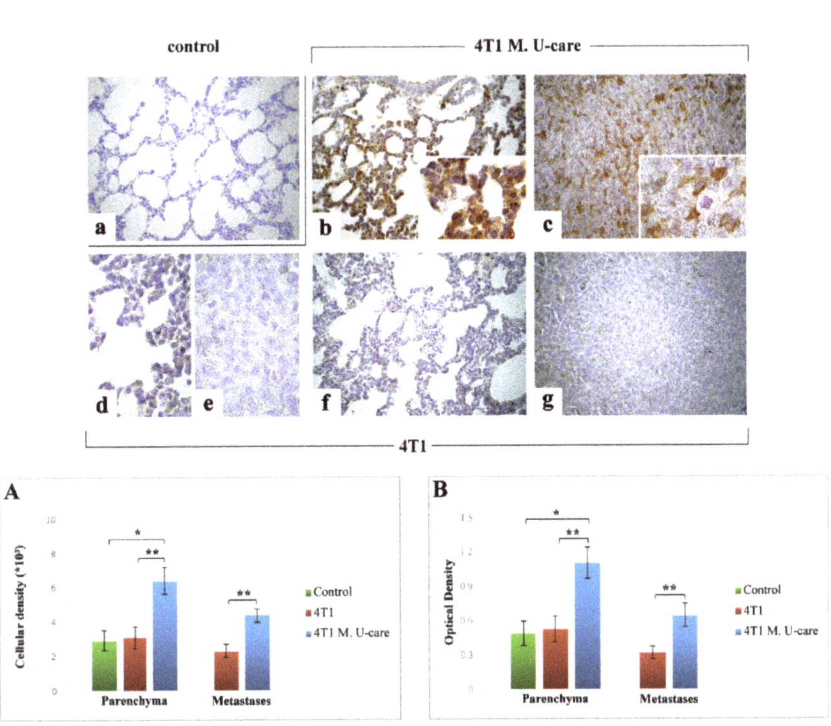

Figure 4. p53-immunostaining reaction in healthy control (**a**), 4T1 M. U-care (**b,c**) and 4T1 (**d–g**) mice. Light microscopy magnification: 40× (**a–c,f,g**); 100× (**d,e**, insert in panels (**b,c**)). Panels (**A**) and (**B**): Histograms showing immunopositive cell density and OD, respectively. *p* values calculated by Unpaired Student's *t*-test: (*) < 0.05 and (**) < 0.01.

With regard to metastatic tissue, a very significant increase of both p53-immunoreactive cell density and OD was determined in 4T1 M. U-care mice compared to 4T1 animals (4.38 ± 0.63 vs. 2.29 ± 0.39 and 0.64 ± 0.10 vs. 0.32 ± 0.06, for density and OD, respectively) (Figure 4, Panels (A,B)).

2.3.3. Bax

Similarly to p53, a significant increase in Bax immunopositivity was observed in the TME of 4T1 M. U-care mice at alveolar and stromal level, with several immunopositive endothelial cells in bronchiolar areas (Figure 5b), compared to both 4T1 mice (Figure 5d,f) and controls (Figure 5a). In the same manner, Bax resulted overexpressed in metastatic nodules 4T1 M. U-care mice (Figure 5c), compared to 4T1 animals (Figure 5e,g). Notably, a significant increase of Bax-immunoreactive cell density was measured in 4T1 M. U-care animals compared to 4T1 mice (4.75 ± 0.48 vs. 3.13 ± 0.40, respectively). Likewise, a significant increase was observed in 4T1 M. U-care mice compared to control (4.75 ± 0.48 vs. 2.47 ± 0.28, respectively), while any difference was determined when comparing 4T1 animals and controls (3.13 ± 0.4 and 2.47 ± 0.28, respectively) (Figure 5, Panel (A)). Moreover, a significant increase of Bax-immunostaining OD was evidenced in 4T1 M. U-care mice, compared to both 4T1 and controls (0.76 ± 0.09 vs. 0.48 ± 0.07 and 0.76 ± 0.09 vs. 0.37 ± 0.04, respectively). Any significant difference was measured when comparing 4T1 mice to controls (0.48 ± 0.07 vs. 0.37 ± 0.04, respectively) (Figure 5, Panel (B)).

Concerning the metastases, an extremely significant increase of both Bax-immunopositive cell density and OD was determined in 4T1 M. U-care mice compared to 4T1 animals (4.47 ± 0.40 vs. 1.59 ± 0.31, and 0.58 ± 0.07 vs. 0.24 ± 0.06, respectively) (Figure 5, Panels (A,B)).

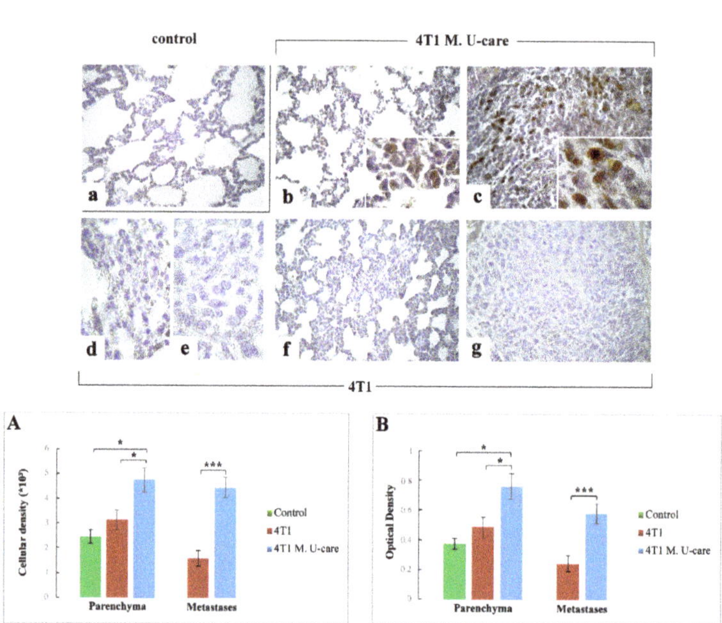

Figure 5. Immunohistochemical staining for Bax in healthy control (**a**), 4T1 M. U-care (**b,c**) and 4T1 (**d–g**) mice. Light microscopy magnification: 40× (**a–c,f,g**); 100× (**d,e**, insert in panels (**b,c**)). Panels (**A**) and (**B**): Histograms showing immunopositive cell density and OD, respectively. *p* values calculated by Unpaired Student's *t*-test: (*) < 0.05 and (***) < 0.001.

2.3.4. Bcl2

Concerning the TME, Bcl2-immunoreactivity appeared slightly enhanced in both 4T1 and 4T1 M. U-care mice (Figure 6b,d,f, respectively) compared to controls (Figure 6a). Notably, a slight non-significant increase of Bcl2-immunoreactive cell density and OD was measured both in 4T1 and 4T1 M. U-care animals compared to controls (4.88 ± 0.69 vs. 4.89 ± 0.68 vs. 1.74 ± 0.15 and 0.86 ± 0.12 vs. 0.89 ± 0.13 vs. 0.32 ± 0.03, for density and OD, respectively). Any difference was evidenced evaluating Bcl2-immunopositive cell density and OD in 4T1 animals and 4T1 M. U-care mice (4.88 ± 0.69 vs. 4.89 ± 0.68 and 0.86 ± 0.12 vs. 0.89 ± 0.13, for density and OD, respectively) (Figure 6, Panels (A,B)).

Regarding the metastatic nodules (Figure 6c,e,g), contrarily to Bax, a very significant decrease of Bcl2-immunopositive cell density and OD was detected in 4T1 M. U-care mice compared to 4T1 (1.77 ± 0.32 vs. 3.29 ± 0.34 and 0.24 ± 0.04 vs. 0.52 ± 0.06, for density and OD, respectively) (Figure 6, Panels (A,B)).

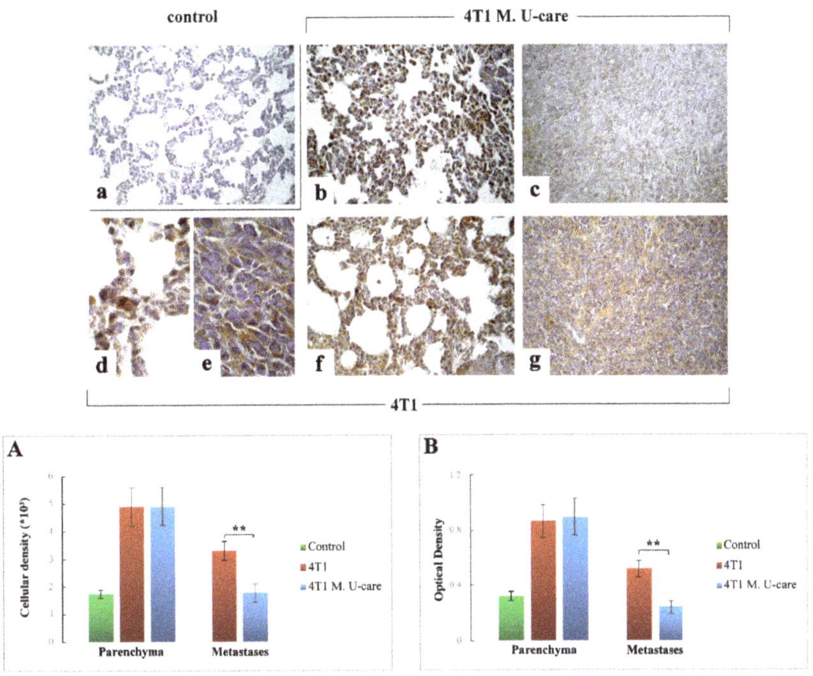

Figure 6. Bcl2-immunostaining reaction in healthy control (**a**), 4T1 M. U-care (**b,c**) and 4T1 (**d–g**) mice. Light microscopy magnification: 40× (**a–c,f,g**); 100× (**d,e**). Panels (**A**) and (**B**): Histograms showing immunopositive cell density and OD, respectively. *p* values calculated by Unpaired Student's *t*-test: (**) < 0.01.

2.3.5. PCNA

PCNA-immunopositivity was extremely enhanced in TME of 4T1 mice (Figure 7d,f) compared to both 4T1 M. U-care (Figure 7b) and controls (Figure 7a). In particular, the quantitative analysis highlighted an extremely significant increase in PCNA-immunoreactive cell density and OD when comparing 4T1 animals to controls (0.40 ± 0.03 vs. 0.06 ± 0.01 and 0.93 ± 0.01 vs. 0.01 ± 0.00, for cell density and OD, respectively). In a similar manner, an extremely significant enhancement was detected when comparing 4T1 mice to 4T1 M. U-care animals (0.40 ± 0.03 vs. 0.17 ± 0.02 and 0.93 ± 0.01 vs.

0.02 ± 0.00, for cell density and OD, respectively). No significant differences were revealed in 4T1 M. U-care animals compared to control (0.17 ± 0.02 vs. 0.06 ± 0.01 and 0.02 ± 0.00 vs. 0.01 ± 0.00, for cell density and OD, respectively) (Figure 7, Panels (A,B)).

Concerning the metastatic tissue, an extremely significant decrease of PCNA-immunopositive cell density and OD was detected in 4T1 M. U-care mice (Figure 7c) compared to 4T1 animals (Figure 7e,g): 4.74 ± 0.28 vs. 9.57 ± 0.44 and 0.64 ± 0.04 vs. 1.77 ± 0.08, for density and OD, respectively (Figure 7, Panels (A,B)).

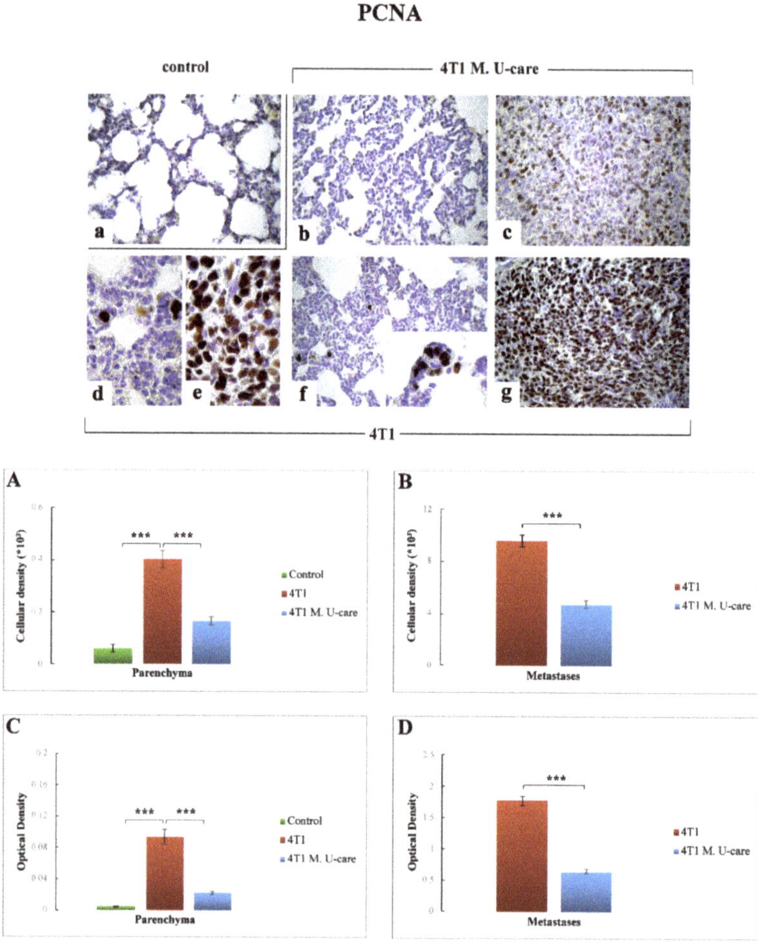

Figure 7. DAB-immunostaining reaction for PCNA in healthy control (**a**) 4T1 M. U-care (**b**,**c**), and 4T1 (**d–g**) mice. Light microscopy magnification: 40× (**a–c**,**f**,**g**); 100× (**d**,**e**, insert in panel (**f**)). Panels (**A**) and (**B**): Histograms showing immunopositive cell density. Panels (**C**) and (**D**): histograms exhibiting immunoreactive OD. *p* values calculated by Unpaired Student's *t*-test: (***) < 0.001.

3. Discussion

Tumor development and progression are influenced by rearrangement of TME components, e.g., immune cells, fibroblasts, satellite cells, blood, and lymphatic vessels. Tumor cells are known to manipulate the function of cellular and non-cellular components through a complex signaling network to gain tumorigenesis, tumor maintenance, and drug resistance (MDR), taking advantage of the non-malignant cells. Experimental and clinical data evidence that an in-depth analysis of the bidirectional communications and interactions between tumor cells and their surrounding dynamic TME is essential to identify the existing mechanisms of tumor expansion and invasion [41]. This complex network moreover involved apoptosis, also engaged to efficiently eliminate dysfunctional cells, plays an important role in both carcinogenesis and cancer treatment [42]. In particular, to guarantee its nourished growth, cancer is able to provide both endurance signals as well as mechanisms saving malignant cells from apoptosis. It is well known that an imbalance between cell proliferation and cell death typically characterizes cancer condition, in which several genetic aberrations may drive malignant cells to an uncontrolled progression and survival [43,44].

In our previous paper using the same preclinical model, i.e., 4T1 triple-negative mouse BC, we demonstrated that M. U-Care supplementation, starting 2 months before 4T1 injection and lasting throughout the whole experimental time (including tumor development and metastatization), produced a dramatic reduction of both lung metastases density and number. These effects were accompanied by a substantial decrease of inflammation and oxidative stress both in lung TME and metastases, together with a bettering of the QoL [36]. Therefore, we hypothesized that the supplement effects could have been ascribable either to a direct anticancer effect and/or to the secondary/indirect impacts of the MM blend on systemic inflammation and immunomodulation. Therefore, aiming at addressing this crucial question, in the present investigation, we focused on the programmed cell death, investigating specific molecules, i.e., PARP1, p53, BAX, and Bcl2, pivotally involved in apoptotic pathway, whose critical role in human BC is well known. Further, in parallel we investigated PCNA expression pathway, based on the great interest on the pivotal role of PCNA in cancer cells proliferation, also taking into consideration that PCNA modifications may determine both tumor progression as well as the outcome of anticancer treatment.

Firstly, it has to be highlighted that performing a comparative assessment on cancer tissue, i.e., lung metastases, and surrounding TME by TUNEL assay, we revealed the existence of a different expression trend in 4T1 animals compared to 4T1 M. U-care mice. In detail, in these latter animals, an extremely intense immunopositivity was observed in the metastases, indicating an evident activation of the apoptotic pathway in cancer cells after the oral supplementation with the MM blend. Interestingly, this effect was definitely weaker in 4T1 animals, in which several mitoses were clearly observed in the tumor nodules, thus demonstrating the occurrence of an already active proliferation process. Concerning the TME, any statistical difference was measured comparing 4T1 M. U-care animals and 4T1 mice. These data corroborate the action of the MM blend, who display a powerful action able to drive metastatic cells to apoptosis, whereas was less effective with regards to the surrounding TME.

Concerning all evaluated apoptotic and proliferation markers, the comparative assessment of TME and metastases revealed diverse trends of protein expression, with the metastatic nodules being the most affected.

With regards to PARP1 levels, crucially implicated in tumorigenesis phenomena [24,26,27], the significant reduction observed in metastatic nodules of 4T1 M. U-care mice compared to 4T1 animals seemed to demonstrate a valuable action of the adjuvant micotherapic supplementation, suggesting a direct influence of the blend on tumor cells.

Our putative idea of a direct MM blend action able to determine an imbalance between proliferation and apoptosis, driving to a significant increase in apoptotic events, was further supported by p53 and Bax high expression levels measured in 4T1 M. U-care mice, showing the most striking effect in the metastatic nodules. By contrast, concerning both Bax and p53, an almost complete lack of effect was determined in 4T1 animals, in which the protein expression levels were similar to that observed in controls. Acting as proapoptotic protein, Bax should essentially play as a tumor suppressor factor. Consequently, reductions in its expression levels would provide tumor cells with a selective survival advantage, contributing to their expansion and invasion [21]. In particular, the reduced Bax expression levels can be traced back to an alteration of the p53 function which, by itself, is able to alter the levels of this proapoptotic protein [21]. Certainly, TP53 mutations are the most common genetic alterations in breast cancer [28]. Notably, the increased immunopositivity for the wild type p53 isoform detected in 4T1 M. U-care mice could indicate a possible attempt to cell cycle arrest, in order to allow cellular repair processes and inhibit the proliferation of damaged cells.

Based on the notion that (i) several apoptotic stimuli induce cell death through a Bcl-2-regulated pathway and (ii) proteins belonging to Bcl2 family can play a dichotomous role, acting both as promoters as well as inhibitors of apoptotic events [21,45,46], we may suppose that the lower expression of Bcl-2 determined in metastatic tissue of 4T1 M. U-care mice could be related to a decrease in cancer cell proliferation, evidencing an alteration of the proliferation/apoptosis balance. This effect could also counteract the capacity of Bcl2 to trigger drug resistance at high expression [21,47]. Moreover, the overexpression of p53 in 4T1 M. U-Care mice may support a possible role of p53 in downregulating Bcl-2, feasibly explaining the apoptosis induction by wild type p53. Notably, it has to be underlined that, differently to all other evaluated markers, but similarly to TUNEL staining, we determined any difference in the Bcl-2 expression trend comparing metastases and TME in 4T1 M. U-care mice and 4T1 animals. Specifically, TME appeared to be unaffected by the mycotherapic oral supplementation, whereas the MM blend display a specific and selective effect restricted to the metastatic area.

In addition, concerning the proliferation marker PCNA, we evidenced a reduced expression in metastatic nodules in 4T1 M. U-care mice, thus highlighting an evident inhibitory effect of the MM blend on proliferation activity in tumor cells. Notably, a strong positive correlation between the expression of PCNA and COX2 in breast cancer has been recently highlighted [48], also in accordance with our previous findings [36].

Taken together the present data supported that the oral supplementation is effective to induce a peculiar swinging balance of cell death and proliferation, in which these two essential mechanisms appeared strictly interconnected and inversely related (see p53 and Bax vs. Bcl2 and PCNA expression trends). Notably, the MM blend bettered the cancer state with a direct effect, able to force tumor metastatic cells to an apoptotic fate.

In summary, all the present findings confirmed the protective role played by Micotherapy U-care blend in the lung metastases and surrounding TME. Our data further suggest that both an immunomodulatory anti-inflammatory systemic effect and a direct, selective anticancer effect act in a positive pleiotropic way. This double action mechanism, which (i) disturbs the TME signaling and (ii) targets the complex apoptotic pathway, could represent a promising approach for patient's treatment. In fact, in the search for novel therapeutic agents targeting tumor development and progression, MM-derived natural bioactive compounds, displaying multi-targeting potential, can overcome the disadvantages of monotherapy such as side effects and drug resistance. In particular, TNBC is the most aggressive malignant BC, difficult to treat due to its unresponsiveness to current clinical targeted treatments (e.g., hormonal therapy protocols or chemotherapeutics targeting HER2 protein receptors) and high rate of recurrence. In this scenario, the biological effects of combining M. U-care supplement with conventional therapies to target crucial cancer signaling pathways, i.e., proliferation and apoptosis, may interfere with cellular and molecular processes fueling TNBC growth, trying to create a new joint medical protocol able to hinder the typical TNBC metastatic pattern, i.e., frequent occurrence of distant metastases, mainly localized in lung, central nervous system, and bones, often associated with poor prognosis.

4. Materials and Methods

4.1. Cell Culture

The mice breast cancer cell line, 4T1, was acquired from American Type Culture Collection (ATCC) and maintained at 37 °C in a humidified atmosphere (95% air/5% CO_2) [36].

4.2. Animals and Experimental Plan

The detailed experimental design was previously describe in Roda et al. [36]. Briefly, 34 two-month-old wild type female BALB/c mice were obtained from Charles River Italia (Calco, Italy) and acclimatized for at least 3 weeks before the experiments.

All experiments were achieved in agreement with the European Council Directive 2010/63/EU and the Ethics Committee of Pavia University guidelines (Ministry of Health, License number 364/2018-PR). Therefore, all mice have been treated humanely, with due consideration for the reduction of pain and distress.

For execution of experiments and subsequent analyses, researchers were blinded to the designed group.

Sixteen (4T1 M. U-care mice) out of thirty-four mice were supplemented until sacrifice with a medicinal mushroom blend, namely, Micotherapy U-care (provided by A.V.D. Reform s.r.l., Noceto, Parma, Italy) consisting of a mixture of 20% extracts of sporophores and mycelia of five fungal species: *Agaricus blazei*, *Cordyceps sinensis*, *Ganoderma lucidum*, *Grifola frondosa*, and *Lentinula edodes* (see Table 1 and Results section). The mycotherapic blend was solubilized in water, selecting a dose of 4 mg supplement/mice per day to mimic the oral supplementation in humans (about 1.5 g/day). Otherwise, control ($n = 4$) and non-treated (4T1, $n = 14$) mice did not received any diet supplementation.

For the syngeneic tumor-bearing mice (4T1 and 4T1 M. U-care) generation see the work in [36].

The syngeneic tumor-bearing mice (4T1 and 4T1 M. U-care) were generated by injecting 1×10^6 of the 4T1 cells into the nape of the neck of the female Balb/C mice. Control animals were injected with phosphate-buffer saline (PBS). A survival rate of 100% was kept in all experimental groups, throughout the whole experimental time course.

Lung preparation was done by vascular perfusion of fixative [49]. Then, lungs were accurately removed, sectioned and then processed for immunohistochemistry.

4.3. Tissue Sampling and Immunohistochemistry

4.3.1. Lung Specimens Preparation

The lung specimens preparation was previously described in detail [36]. Briefly, the top and the bottom regions of the right lungs of mice from each experimental group were dissected. Tissue samples were obtained according to a stratified random sampling scheme, fixed and processed as previously described [36]. Eight micrometer thick sections were cut in transversal plane and placed on silane-coated slides.

4.3.2. TUNEL Staining

The reaction was performed using the terminal deoxynucleotidyl-transferase (TUNEL) assay (Oncogene Res. Prod., Boston, MA, USA). The lung sections were incubated for 5 min with 20 µg mL^{-1} proteinase-K solution at room temperature, followed by treatment with 3% H_2O_2 to quench endogenous peroxidase activity. After incubation with the TUNEL solution (90 min with TdT/biotinylated dNTP and 30 min with HRP-conjugate streptavidin) in a humidified chamber at 37 °C, the reaction was developed using a 0.1% DAB solution. After nuclear counterstaining employing Carazzi's Hematoxylin, the sections were dehydrated in ethanol, cleared in xylene, and finally mounted in Eukitt (Kindler, Freiburg, Germany).

As a negative control, the TdT incubation was omitted; no staining was observed in these conditions.

4.3.3. Immunohistochemistry: Apoptotic Pathway Assessment

Commercial antibodies were employed on murine lung specimens to investigate the expression of different specific apoptotic markers: (i) Poly (AD-ribose) polymerase 1(PARP1), (ii) p53, (iii) Bcl2 associated X-protein (Bax), (iv) B-cell lymphoma/leukemia protein (Bcl2), and (v) the proliferating cell nuclear antigen (PCNA). Table 2 shows both primary and secondary antibodies as well as respective dilutions used for immunohistochemical experiments.

Immunohistochemical procedures have been conducted exactly as previously described [36].

Table 2. Primary/secondary antibodies and respective dilution used for immunohistochemical experimental procedures.

	Antigen	Immunogen	Manufacturer, Species, Mono-Polyclonal, Cat./Lot. No., RRID	Dilution
Primary Antibodies	Anti-poly (ADP-ribose) polymerase (46D11)	Purified antibody raised against the residues surrounding Gly623 of human PARP-1	Cell Signaling Technology (Danvers, MA, USA), Rabbit monoclonal IgG, Cat# 9532, RRID:AB_659884	1:100
	Anti-p53 (Ab-5)	Purified antibody raised against the ~53 kDa wild type p53 protein of mouse origin	Sigma-Aldrich (St. Louis, MO, USA), Mouse monoclonal IgG2a, Cat# OP33-100UG, RRID:AB_564977	1:100
	Anti-Bcl-2-associated X protein (P-19)	Purified antibody Raised against a peptide mapping at the amino terminus of Bax of mouse origin	Santa Cruz Biotechnology (Santa Cruz, CA, USA), Rabbit polyclonal IgG, Cat# sc-526, RRID:AB_2064668	1:100
	Anti-B-Cell Leukemia/Lymphoma 2 protein (N-19)	Purified antibody raised against a peptide mapping at the N-terminus of Bcl-2 of human origin	Santa Cruz Biotechnology (Santa Cruz, CA, USA), Rabbit polyclonal IgG, Cat# sc-492, RRID:AB_2064290	1:100
	Anti-Proliferating Cell Nuclear Antigen (Ab-1)	Purified antibody raised against the ~37 kDa PCNA protein of mouse origin	Sigma-Aldrich (St. Louis, MO, USA), Mouse monoclonal IgG2a, Cat# NA03-200UG, RRID:AB_213111	2:1000
Secondary Antibodies	Biotinylated horse anti-mouse IgG	Gamma immunoglobulin	Vector Laboratories (Burlingame, CA, USA), Horse, Cat# PK-6102, RRID:AB_2336821	1:200
	Biotinylated goat anti-rabbit IgG	Gamma immunoglobulin	Vector Laboratories (Burlingame, CA, USA), Goat, lot# PK-6101, RRID: AB_2336820	1:200

4.3.4. Immunohistochemical Evaluations

To prevent differences due to small procedural changes, immunohistochemical reactions were performed simultaneously on samples from different experimental groups. The expression of each selected marker was examined in six slides (about 30 sections) per mouse. The shown micrographs display the most representative pulmonary conditions and modifications for each immunohistochemical reaction.

Immunohistochemical labeling extent evaluation was previously described in detail [36].

The optical density (OD), intended as immunohistochemical intensity, was assessed in 30 cells/section per six slides/mouse. OD was related to the immunopositive cell density. In addition, immunopositive cells density count was evaluated, intended as number of immunopositive cells/area in mm^2.

4.4. Statistics

Data were expressed as mean ± standard error of the mean (SEM). Regarding the TME, the statistical differences among the three experimental groups were calculated by using one-way ANOVA followed by Bonferroni's post hoc test. Otherwise, for metastases, the statistical differences between 4T1 and 4T1 M. U-care mice were evaluated by using unpaired Student's *t*-test.

The differences were considered statistically significant for $p < 0.05$ (*), $p < 0.01$ (**), and $p < 0.001$ (***). Statistical analyses were performed by using GraphPad Prism 7.0 (GraphPad Software Inc., La Jolla, CA, USA).

5. Conclusions

Overall, these results support the use of oral supplementation with Micotherapy U-care blend as a new effective strategy to be used in the field of integrative oncology to decrease adverse side effects caused by conventional cancer therapies. All obtained results corroborated the Micotherapy U-care protective role in metastases and TME, in which an immunomodulatory anti-inflammatory systemic action together with a direct, selective anticancer mechanism exerted a positive pleiotropic effect. The Micotherapy U-care preventive and protective effect could affect the TME signaling and, at the same time, target the multifaceted apoptotic pathway. Once again, the present investigation highlights the importance of translational research in the development of clinically relevant therapeutic strategies. In particular, the growing use of translational research "from bench to bedside" in cancer medicine, could allow to overcome challenges which everlastingly hinder medicinal advancements, yielding significant advances in cancer therapeutics and also improvements in the ability to predict clinical course of patient's disease based on individual tumor characteristics. In this view, medicinal mushrooms extracts, being natural sources of novel drugs, could be used as effective adjuvant therapy in the critical management of TNBC.

Author Contributions: Conceptualization, E.R., E.S. and P.R.; methodology, F.D.L., D.R. and C.D.I.; software, F.D.L. and D.R.; formal analysis, E.R., F.D.L. and P.R.; investigation, E.R., F.D.L. and C.D.I.; writing—original draft preparation, E.R. and P.R.; writing—review and editing, F.D.L. and D.R.; supervision, E.R., C.A.L., M.G.B. and P.R. All authors have read and agreed to the published version of the manuscript.

Funding: This research was supported by the Italian Ministry of Education, University and Research (MIUR): Dipartimenti di Eccellenza Program (2018–2022)—Dept. of Biology and Biotechnology "L. Spallanzani", University of Pavia.

Acknowledgments: We thank Andrei Gregori (Biotechnical Faculty, University of Ljubljana, Slovenia) for raw materials, extract procedure, and main active metabolites of MM blend Micotherapy U-care. We thank A.V.D. Reform s.r.l. (Noceto, Parma) for providing us the supplement "Micotherapy U-Care". We thank Rita Vaccarone, Department of Biology and Biotechnology, University of Pavia, for her excellent technical assistance.

Conflicts of Interest: The authors declare no conflict of interest.

References

1. Jemal, A.; Bray, F.; Center, M.M.; Ferlay, J.; Ward, E.; Forman, D. Global cancer statistics. *CA Cancer J. Clin.* **2011**, *61*, 69–90. [CrossRef] [PubMed]
2. Cardoso, F.; Harbeck, N.; Fallowfield, L.; Kyriakides, S.; Senkus, E.; ESMO Guidelines Working Group. Locally recurrent or metastatic breast cancer: ESMO Clinical Practice Guidelines for diagnosis, treatment and follow-up. *Ann. Oncol.* **2012**, *23*, vii11–vii19. [CrossRef] [PubMed]
3. Zeeshan, R.; Mutahir, Z. Cancer metastasis: Tricks of the trade. *Bosn. J. Basic Med. Sci.* **2017**, *17*, 172–182. [CrossRef] [PubMed]
4. Schito, L.; Rey, S. Hypoxic pathobiology of breast cancer metastasis. *Biochim. Biophys. Acta Rev. Cancer* **2017**, *1868*, 239–245. [CrossRef]
5. Yadav, B.S.; Chanana, P.; Jhamb, S. Biomarkers in triple negative breast cancer: A review. *World J. Clin. Oncol.* **2015**, *6*, 252. [CrossRef]
6. Yeo, S.K.; Guan, J.L. Breast Cancer: Multiple Subtypes within a Tumor? *Trends Cancer* **2017**, *3*, 753–760. [CrossRef]

7. Yao, Y.; Chu, Y.; Xu, B.; Hu, Q.; Song, Q. Risk factors for distant metastasis of patients with primary triple-negative breast cancer. *Biosci. Rep.* **2019**, *39*. [CrossRef]

8. Iriondo, O.; Liu, Y.; Lee, G.; Elhodaky, M.; Jimenez, C.; Li, L.; Lang, J.; Wang, P.; Yu, M. TAK1 mediates microenvironment-triggered autocrine signals and promotes triple-negative breast cancer lung metastasis. *Nat. Commun.* **2018**, *9*, 1994. [CrossRef]

9. Chitty, J.L.; Filipe, E.C.; Lucas, M.C.; Herrmann, D.; Cox, T.R.; Timpson, P. Recent advances in understanding the complexities of metastasis. *F1000Research* **2018**, *7*, 1169. [CrossRef]

10. Katsuta, E.; Rashid, O.M.; Takabe, K. Clinical relevance of tumor microenvironment: Immune cells, vessels, and mouse models. *Hum. Cell* **2020**, *33*, 930–937. [CrossRef]

11. Buhrmann, C.; Shayan, P.; Banik, K.; Kunnumakkara, A.B.; Kubatka, P.; Koklesova, L.; Shakibaei, M. Targeting NF-κB Signaling by Calebin A, a Compound of Turmeric, in Multicellular Tumor Microenvironment: Potential Role of Apoptosis Induction in CRC Cells. *Biomedicines* **2020**, *8*, 236. [CrossRef] [PubMed]

12. Cacho-Diaz, B.; García-Botello, D.R.; Wegman-Ostrosky, T.; Reyes-Soto, G.; Ortiz-Sánchez, E.; Herrera-Montalvo, L.A. Tumor microenvironment differences between primary tumor and brain metastases. *J. Transl. Med.* **2020**, *18*, 1. [CrossRef] [PubMed]

13. Nounou, M.I.; Elamrawy, F.; Ahmed, N.; Abdelraouf, K.; Goda, S.; Syed-Sha-Qhattal, H. Breast Cancer: Conventional Diagnosis and Treatment Modalities and Recent Patents and Technologies. *Breast Cancer* **2015**, *9*, 17–34. [CrossRef] [PubMed]

14. Zhao, Y.; Jing, Z.; Li, Y.; Mao, W. Berberine in combination with cisplatin suppresses breast cancer cell growth through induction of DNA breaks and caspase-3-dependent apoptosis. *Oncol. Rep.* **2016**, *36*, 567–572. [CrossRef] [PubMed]

15. Czabotar, P.E.; Lessene, G.; Strasser, A.; Adams, J.M. Control of apoptosis by the BCL-2 protein family: Implications for physiology and therapy. *Nat. Rev. Mol. Cell Biol.* **2014**, *15*, 49–63. [CrossRef]

16. Kadam, C.Y.; Abhang, S.A. Apoptosis Markers in Breast Cancer Therapy. *Adv. Clin. Chem.* **2016**, *74*, 143–193. [CrossRef]

17. Elmore, S. Apoptosis: A Review of Programmed Cell Death. *Toxicol. Pathol.* **2007**, *35*, 495–516. [CrossRef]

18. Parton, M.; Dowsett, M.; Smith, I. Studies of apoptosis in breast cancer. *BMJ* **2001**, *322*, 1528–1532. [CrossRef]

19. Hong, S.J.; Dawson, T.M.; Dawson, V.L. PARP and the Release of Apoptosis-Inducing Factor from Mitochondria; Madame Curie Bioscience Database. 2013. Available online: https://www.ncbi.nlm.nih.gov/books/NBK6179/ (accessed on 8 October 2020).

20. Aubrey, B.J.; Kelly, G.L.; Janic, A.; Herold, M.J.; Strasser, A. How does p53 induce apoptosis and how does this relate to p53-mediated tumour suppression? *Cell Death Differ.* **2018**, *25*, 104–113. [CrossRef]

21. Krajewski, S.; Thor, A.D.; Edgerton, S.M.; Moore, D.H.; Krajewska, M.; Reed, J.C. Analysis of Bax and Bcl-2 expression in p53-immunopositive breast cancers. *Clin. Cancer Res.* **1997**, *3*, 199–208.

22. Frenzel, A.; Grespi, F.; Chmelewskij, W.; Villunger, A. Bcl2 family proteins in carcinogenesis and the treatment of cancer. *Apoptosis* **2009**, *14*, 584–596. [CrossRef] [PubMed]

23. Slade, D. PARP and PARG inhibitors in cancer treatment. *Genes Dev.* **2020**, *34*, 360–394. [CrossRef] [PubMed]

24. Ossovskaya, V.; Koo, I.C.; Kaldjian, E.P.; Alvares, C.; Sherman, B.M. Upregulation of poly (ADP-Ribose) polymerase-1 (PARP1) in triple-negative breast cancer and other primary human tumor types. *Genes Cancer* **2010**, *1*, 812–821. [CrossRef] [PubMed]

25. Yu, S.W.; Andrabi, S.A.; Wang, H.; No, S.K.; Poirier, G.G.; Dawson, T.M.; Dawson, V.L. Apoptosis-inducing factor mediates poly(ADP-ribose) (PAR) polymer-induced cell death. *Proc. Nat. Acad. Sci. USA* **2006**, *103*, 18314–18319. [CrossRef]

26. Domagala, P.; Huzarski, T.; Lubinski, J.; Gugala, K.; Domagala, W. PARP-1 expression in breast cancer including BRCA1-associated, triple negative and basal-like tumors: Possible implications for PARP-1 inhibitor therapy. *Breast Cancer Res. Treat.* **2011**, *127*, 861–869. [CrossRef]

27. Schreiber, V.; Dantzer, F.; Amé, J.C.; De Murcia, G. Poly(ADP-ribose): Novel functions for an old molecule. *Nat. Rev. Mol. Cell Biol.* **2006**, *7*, 517–528. [CrossRef]

28. Bertheau, P.; Lehmann-Che, J.; Varna, M.; Dumay, A.; Poirot, B.; Porcher, R.; Turpin, E.; Plassa, L.F.; De Roquancourt, A.; Bourstyn, E.; et al. P53 in breast cancer subtypes and new insights into response to chemotherapy. *Breast* **2013**, *22*, S27–S29. [CrossRef]

29. Hanahan, D.; Weinberg, R.A. The hallmarks of cancer. *Cell* **2000**, *100*, 57–70. [CrossRef]

30. Wang, S.-C. PCNA: A silent housekeeper or a potential therapeutic target? *Trends Pharmacol. Sci.* **2014**, *35*, 178–186. [CrossRef]

31. Juríková, M.; Danihel, Ľ.; Polák, Š.; Varga, I. Ki67, PCNA, and MCM proteins: Markers of proliferation in the diagnosis of breast cancer. *Acta Histochem.* **2016**, *118*, 544–552. [CrossRef]

32. Varghese, E.; Samuel, S.M.; Sadiq, Z.; Kubatka, P.; Liskova, A.; Benacka, J.; Pazinka, P.; Kruzliak, P.; Büsselberg, D. Anti-Cancer Agents in Proliferation and Cell Death: The Calcium Connection. *Int. J. Mol. Sci.* **2019**, *20*, 3017. [CrossRef] [PubMed]

33. Rossi, P.; Difrancia, R.; Quagliariello, V.; Savino, E.; Tralongo, P.; Randazzo, C.L.; Berretta, M. B-glucans from Grifola frondosa and Ganoderma lucidum in breast cancer: An example of complementary and integrative medicine. *Oncotarget* **2018**, *9*, 24837–24856. [CrossRef] [PubMed]

34. Wasser, S.P. Medicinal mushrooms in human clinical studies. Part I. anticancer, oncoimmunological, and immunomodulatory activities: A review. *Int. J. Med. Mush.* **2017**, *19*, 279–317. [CrossRef]

35. Blagodatski, A.; Yatsunskaya, M.; Mikhailova, V.; Tiasto, V.; Kagansky, A.; Katanaev, V.L. Medicinal mushrooms as an attractive new source of natural compounds for future cancer therapy. *Oncotarget* **2018**, *9*, 29259–29274. [CrossRef] [PubMed]

36. Roda, E.; De Luca, F.; Di Iorio, C.; Ratto, D.; Siciliani, S.; Ferrari, B.; Cobelli, F.; Borsci, G.; Priori, E.C.; Chinosi, S.; et al. Novel medicinal mushroom blend as a promising supplement in integrative oncology: A multi-tiered study using 4t1 triple-negative mouse breast cancer model. *Int. J. Mol. Sci.* **2020**, *21*, 3479. [CrossRef]

37. Zmitrovich, I.V.; Belova, N.V.; Balandaykin, M.E.; Bondartseva, M.A.; Wasser, S.P. Cancer without Pharmacological Illusions and a Niche for Mycotherapy (Review). *Int. J. Med. Mush.* **2019**, *21*, 105–119. [CrossRef]

38. Bechtel, P.E.; Hickey, R.J.; Schnaper, L.; Sekowski, J.W.; Long, B.J.; Freund, R.; Liu, N.; Rodriguez-Valenzuela, C.; Malkas, L.H. A Unique Form of Proliferating Cell Nuclear Antigen Is Present in Malignant Breast Cells. *Cancer Res.* **1998**, *58*, 3264–3269.

39. Malkas, L.H.; Herbert, B.S.; Abdel-Aziz, W.; Dobrolecki, L.E.; Liu, Y.; Agarwal, B.; Hoelz, D.; Badve, S.; Schnaper, L.; Arnold, R.J.; et al. A cancer-associated PCNA expressed in breast cancer has implications as a potential biomarker. *Proc. Natl. Acad. Sci. USA* **2006**, *103*, 19472–19477. [CrossRef]

40. Coccini, T.; Manzo, L.; Roda, E. Safety Evaluation of Engineered Nanomaterials for Health Risk Assessment: An Experimental Tiered Testing Approach Using Pristine and Functionalized Carbon Nanotubes. *ISRN Toxicol.* **2013**, *2013*, 825427. [CrossRef]

41. Baghban, R.; Roshangar, L.; Jahanban-Esfahlan, R.; Seidi, K.; Ebrahimi-Kalan, A.; Jaymand, M.; Kolahian, S.; Javaheri, T.; Zare, P. Tumor microenvironment complexity and therapeutic implications at a glance. *Cell Commun. Signal.* **2020**, *18*, 59. [CrossRef]

42. Liu, Z.; Ding, Y.; Ye, N.; Wild, C.; Chen, H.; Zhou, J. Direct Activation of Bax Protein for Cancer Therapy. *Med. Res. Rev.* **2016**, *36*, 313–341. [CrossRef] [PubMed]

43. Wong, R.S. Apoptosis in cancer: From pathogenesis to treatment. *J. Exp. Clin. Cancer Res.* **2011**, *30*, 87. [CrossRef] [PubMed]

44. Yaacoub, K.; Pedeux, R.; Tarte, K.; Guillaudeux, T. Role of the tumor microenvironment in regulating apoptosis and cancer progression. *Cancer Lett.* **2016**, *378*, 150–159. [CrossRef] [PubMed]

45. Oltval, Z.N.; Milliman, C.L.; Korsmeyer, S.J. Bcl-2 heterodimerizes in vivo with a conserved homolog, Bax, that accelerates programed cell death. *Cell* **1993**, *74*, 609–619. [CrossRef]

46. Bodrug, S.E.; Aimé-Sempé, C.; Sato, T.; Krajewski, S.; Hanada, M.; Reed, J.C. Biochemical and functional comparisons of Mcl-1 and Bcl-2 proteins: Evidence for a novel mechanism of regulating Bcl-2 family protein function. *Cell Death Diff.* **1995**, *2*, 173–182.

47. Teixeira, C.; Reed, J.C.; Pratt, M.A. Estrogen promotes chemotherapeutic drug resistance by a mechanism involving Bcl-2 proto-oncogene expression in human breast cancer cells. *Cancer Res.* **1995**, *55*, 3902–3907.

48. Qiu, X.; Mei, J.; Yin, J.; Wang, H.; Wang, J.; Xie, M. Correlation analysis between expression of PCNA, Ki-67 and COX-2 and X-ray features in mammography in breast cancer. *Oncol. Lett.* **2017**, *14*, 2912–2918. [CrossRef]

49. Roda, E.; Bottone, M.; Biggiogera, M.; Milanesi, G.; Coccini, T. Pulmonary and hepatic effects after low dose exposure to nanosilver: Early and long-lasting histological and ultrastructural alterations in rat. *Toxicol. Rep.* **2019**, *6*, 1047–1060. [CrossRef]

Sample Availability: Samples of the compounds are available from the authors.

Publisher's Note: MDPI stays neutral with regard to jurisdictional claims in published maps and institutional affiliations.

Article

Pogostemon cablin Triggered ROS-Induced DNA Damage to Arrest Cell Cycle Progression and Induce Apoptosis on Human Hepatocellular Carcinoma In Vitro and In Vivo

Xiao-Fan Huang [1,2], Gwo-Tarng Sheu [1], Kai-Fu Chang [1,2], Ya-Chih Huang [1,2], Pei-Hsiu Hung [3,4,†] and Nu-Man Tsai [2,5,*,†]

[1] Institute of Medicine, Chung Shan Medical University, Taichung 40201, Taiwan; s9870509@gmail.com (X.-F.H.); gtsheu@csmu.edu.tw (G.-T.S.); kfchang1015@gmail.com (K.-F.C.); hwangjy319@gmail.com (Y.-C.H.)
[2] Department of Medical Laboratory and Biotechnology, Chung Shan Medical University, Taichung 40201, Taiwan
[3] Director of Traditional Chinese Medicine, Ditmanson Medical Foundation Chia-Yi Christian Hospital, Chiayi 60002, Taiwan; cych05086@gmail.com
[4] Department of BioIndustry Technology, Da-Yeh University, Changhua 51591, Taiwan
[5] Clinical Laboratory, Chung Shan Medical University Hospital, Taichung 40201, Taiwan
*   Correspondence: numan@csmu.edu.tw; Tel.: +886-4-2473-0022 (ext. 12411)
†   These authors contributed equally to this work.

Academic Editors: José Antonio Lupiáñez, Amalia Pérez-Jiménez, Eva E. Rufino-Palomares and Emerson F. Queiroz

Received: 1 October 2020; Accepted: 27 November 2020; Published: 30 November 2020

Abstract: The purpose of the study was to elucidate the anti-hepatoma effects and mechanisms of *Pogostemon cablin* essential oils (PPa extract) in vitro and in vivo. PPa extract exhibited an inhibitory effect on hepatocellular carcinoma (HCC) cells and was less cytotoxic to normal cells, especially normal liver cells, than it was to HCC cells, exerting a good selective index. Additionally, PPa extract inhibited HCC cell growth by blocking the cell cycle at the G_0/G_1 phase via p53 dependent or independent pathway to down regulated cell cycle regulators. Moreover, PPa extract induced the FAS-FASL-caspase-8 system to activate the extrinsic apoptosis pathway, and it increased the bax/bcl-2 ratio and reduced $\Delta\Psi$m to activate the intrinsic apoptosis pathway that might be due to lots of reactive oxygen species (ROS) production which was induced by PPa extract. In addition, PPa extract presented to the potential to act synergistically with sorafenib to effectively inhibit HCC cell proliferation through the Akt/mTOR pathway and reduce regrowth of HCC cells. In an animal model, PPa extract suppressed HCC tumor growth and prolonged lifespan by reducing the VEGF/VEGFR axis and inducing tumor cell apoptosis in vivo. Ultimately, PPa extract demonstrated nearly no or low system-wide, physiological, or pathological toxicity in vivo. In conclusion, PPa extract effectively inhibited HCC cell growth through inducing cell cycle arrest and activating apoptosis in vitro and in vivo. Furthermore, PPa extract exhibits less toxicity toward normal cells and organs than it does toward HCC cells, which might lead to fewer side effects in clinical applications. PPa extract may be developed into a clinical drug to suppress tumor growth or functional food to prevent HCC initiation or chemoprotection of HCC recurrence.

Keywords: hepatocellular carcinoma (HCC); *Pogostemon cablin* (PPa extract); cell cycle; apoptosis; synergism; chemoprevention

1. Introduction

Hepatocellular carcinoma (HCC) is the third leading cause of cancer death worldwide [1]. In addition, HCC has a poor prognosis because of chronic hepatitis, with cirrhosis leading to the deterioration of liver function. Moreover, intrahepatic metastasis result in highly recurrence [2]. Sorafenib is generally acknowledged as the standard of care to improve the overall survival (OS) of patients with advanced HCC. Though sorafenib improves the OS of patients with HCC, the clinical benefit is transient, and the toxicity as well as poor antitumor effects of sorafenib remain unsolved issues. With increasing advances in medicine, the combination of chemotherapy agents remains a promising therapeutic strategy for increasing the response rate of advanced HCC patients, for instance, regorafenib and bevacizumab and so on. Another type of agent that is attracting considerable interest is immune checkpoint inhibitors, such as anti-PD-1/PD-L1 (Nivolumab, Pembrolizumab) or CTLA-4 antibodies, and phase III studies of such inhibitors are currently under investigation [3]. Thus, there is obviously a need for effective therapeutic options for HCC patients. Furthermore, new strategies are needed not only to prevent the development or posttreatment recurrence of HCC but also to enhance survival or quality of life [4,5].

Herbal medicine is considered a great way to improve therapeutic efficacy and reduce toxic effects. In the past, many chemotherapeutic agents have been derived from natural products with effective therapeutic effects or low toxicity in treating various illnesses [6,7]. A large number of herbal products have been used worldwide to manage many kinds of liver diseases because of their safety, curative effects and minimal adverse effects. In addition, a number of studies have shown that medicinal herbs function via several mechanisms, such as suppressing carcinogenesis, inhibiting oxidative injury, and reducing inflammation, which protect the normal function of the liver [8]. Hence, the development of new pharmacologically effective chemotherapeutic agents from natural plants that can trigger cancer cell death would be a significant clinical benefit.

Pogostemon cablin has been orally and topically administered in Asia for centuries as a pharmaceutical product for curing exogenous fever, headache, hypotension, allergy, thirst, ache, dysentery, diarrhea, and inflammation [9]. Scientific studies have revealed that *Pogostemon cablin* has the following biological activities: antidepressant [10,11], antimicrobial [12,13], antiviral [14], anti-inflammatory [15], gastroprotective [16,17], antiaging [18], and antitumor activities [19,20]. Moreover, *Pogostemon cablin* has been found to inhibit colon cancer proliferation through the induction of cell cycle arrest at the G_0/G_1 phase. However, it is still unknown what the role of *Pogostemon cablin* is in hepatocellular carcinoma, and the molecular mechanisms behind its anti-hepatoma activity are also unclear.

Our group has demonstrated that *Pogostemon cablin* essential oils which abbreviated form of PPa extract in this study induced apoptosis in human hepatocellular carcinoma $HepG_2$ cells through oxidative stress-regulated mitochondrial dysfunction involving the p53/p21 and apoptotic pathways. Based on the findings of previous work, we investigated the role of apoptosis in the anticancer effect of PPa extract in $HepG_2$ cells in vitro and the underlying mechanisms of apoptosis-related signaling pathways. These findings demonstrated for the first time that PPa extract induced apoptosis by activating the caspase cascade, and we revealed the underlying antitumor effects of PPa extract plus sorafenib in vitro. These results could provide novel insights into the mechanisms underlying the anticancer effects of PPa on human hepatoma cells.

2. Results

2.1. PPa Extract Inhibited HCC Cell Growth

To address the effect of PPa extract on cell proliferation in human hepatoma cells, cells were treated with PPa extract at increasing doses over various periods of time. As shown in Figure 1A, treatments with increasing concentrations of PPa extract over increasing periods of time decreased the cell viability from 100% to 5%, showing that PPa extract inhibited HCC cell proliferation in a dose-dependent manner. The PPa extract showed a 50% inhibition at concentrations ranging from 7.34 ± 3.09 to 33.29 ± 2.72 µg/mL in hepatoma cells. Moreover, 5-FU, VP-16 and sorafenib are known for their inhibitory effects on hepatoma cells, and the IC$_{50}$ of those drugs ranged from 1.77 ± 4.31 to 18.79 ± 0.91 µg/mL in hepatoma cells (Table 1). To further determine the effects of PPa extract on the growth of normal cells which were in a non-proliferative state, the results showed that PPa extract possessed less inhibitory ability than normal cells (Figure 1B). The IC$_{50}$ of PPa extract on SVEC and MDCK cells was 69.68 ± 4.63 and 73.61 ± 0.16 µg/mL, respectively. Interestingly, the IC$_{50}$ of PPa extract in a normal liver cell type, BNL CL.2 cells, was 147.24 ± 7.71 µg/mL, indicating that PPa extract apparently exhibited a smaller inhibitory effect on normal cells. After that, to explore the selective index (SI), which is defined greater than 2 presenting good selectivity [21], Table 2 shows that the PPa extract exhibited better SI values (2.1–20.1) than sorafenib (1.4–6.1) and VP-16 (0.4–2.1). Consequently, PPa extract demonstrated inhibitory effects on hepatoma cells but exerted less cytotoxic effects on normal cells. In addition, the PPa extract exhibited a good selective index, suggesting that the PPa extract might possess fewer side effects than other agents.

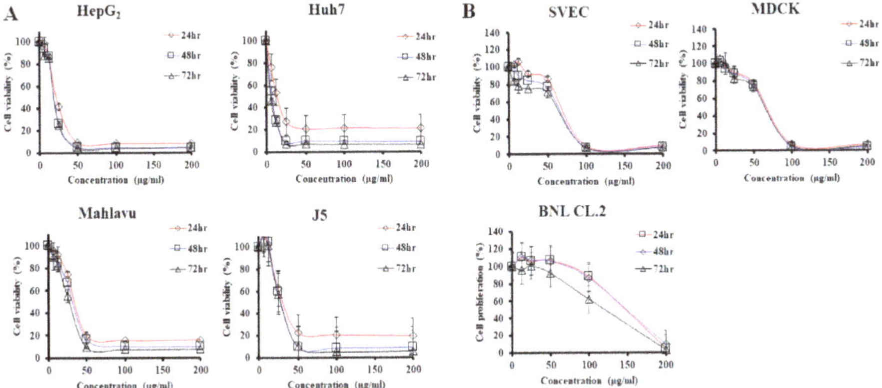

Figure 1. PPa extract inhibited HCC cell growth with less toxicity to normal cells. (**A**) Cell viability of HCC cells after PPa extract treatment (0–200 µg/mL), as assessed by MTT assay. (**B**) The viability of normal cells treated with PPa extract (0–200 µg/mL). The results are presented as the mean ± SD.

Table 1. The IC$_{50}$ of PPa extract and clinical drugs in HCC and normal cells.

Cell Line	Tumor Type	PPa Extract	SOR	VP-16	5-FU
Hepatocellular Carcinoma Cells					
HepG$_2$	Human HCC cell	20.09 ± 2.21 [a,b]	3.52 ± 1.97	4.73 ± 3.57	2.09 ± 1.63
Mahlavu	Human HCC cell	33.29 ± 2.72 [a,b]	6.07 ± 1.06	4.54 ± 2.17	14.97 ± 3.71
J5	Human HCC cell	29.87 ± 3.62 [a,b]	2.39 ± 1.91	3.01 ± 2.75	18.79 ± 0.91
Huh7	Human HCC cell	7.34 ± 3.09 [a,b]	1.77 ± 4.31	5.11 ± 2.08	4.88 ± 3.37
Normal cells					
SVEC	Mouse vascular endothelial cell	69.68 ± 4.63 [c]	8.55 ± 2.73	1.85 ± 0.49	1.69 ± 2.8
MDCK	Canine epithelial kidney cell	73.61 ± 0.16 [c]	6.95 ± 1.45	3.38 ± 0.43	8.74 ± 0.53
BNL CL.2	Mouse liver embryonic cell	147.24 ± 7.71 [c]	10.75 ± 5.17	6.43 ± 3.02	>20

Note: Values are the mean ± SD (µg/mL) at 48 hr. [a]: HCC cells were significantly different from normal cells in the PPa extract treatment group ($p < 0.05$). [b] and [c]: PPa extract treatment was significantly different from SOR treatment in both HCC and normal cells ($p < 0.05$). SOR: sorafenib. VP-16: Etoposide. 5-FU: 5- Fluorouracil.

Table 2. Comparison of Selectivity index (SI) on drugs.

Normal Cells	/Tumor Cells	PPa Extract	SOR	VP-16
	/HepG	3.5	2.4	0.4
SVEC	/Mahlavu	2.1	1.4	0.4
	/J5	2.3	3.6	0.6
	/Huh7	9.5	4.8	0.4
	/HepG$_2$	3.7	2.0	0.7
MDCK	/Mahlavu	2.2	1.1	0.8
	/J5	2.5	2.9	1.1
	/Huh7	10	3.9	0.7
	/HepG$_2$	7.4	3.1	1.4
BNL CL.2	/Mahlavu	4.4	1.8	1.4
	/J5	4.9	4.5	2.1
	/Huh7	20.1	6.1	1.3

Note: Selectivity index (SI) = IC_{50} of normal cells/IC_{50} of HCC cells. SI > 2: indicated that drugs have high selectivity for tumor cells; SI < 2: indicated that drugs have poor selectivity for tumor cells [21]. SOR: sorafenib. VP-16: Etoposide.

2.2. PPa extract Altered the Cell Cycle Distribution in HCC cells

PPa extract was next studied to determine if it altered the cell cycle distribution to exert its inhibitory effect on hepatoma cells. The results indicated that while PPa extract treatment of HepG$_2$ cells did increase the number of cells in G$_0$/G$_1$ phase (from 57% to 71%), it also decreased the number of cells in S and G$_2$/M phase (from 16% to 3%; from 26% to 17%) (Figure 2A). Similar results were observed in Mahlavu cells, and the cell population in the G$_0$/G$_1$ phase increased from 45% to 62%; however, it decreased the cell population of the S and G$_2$/M phase (from 29% to 15%; from 26% to 22%) (Figure 2B). The results showed that both cell lines induced cell cycle arrest at the G$_0$/G$_1$ phase. Hence, we aimed to gain insight into the changes in tumor suppressors and cell cycle regulators. After PPa extract treatment in HepG$_2$ cells, PPa extract induced the expression of p53 and *p*-p53 proteins, and it increased the expression of p21 protein, resulting in decreased levels of the following downstream proteins: PCNA, cdk4, cdk2, cyclin D1, cyclin A, and cyclin B1. Mahlavu cells with a p53 mutant were exposed to PPa extract treatment, and the results showed that the expression of p-p53 was not obviously changed; however, the expression of p21 was modest increased, and the expression of downstream proteins decreased, suggesting that PPa extract also induced a p53-independent pathway and subsequently upregulated the expression of p21 (Figure 2C). Moreover, PPa extract also decreased the expression of total Rb and p-Rb in both cell lines. Taken together, the data revealed that PPa extract induced cell cycle arrest at the G$_0$/G$_1$ phase through induction of p53-dependent and p53-independent pathways to increase the expression level of p21, leading to the decreased expression of cell cycle regulators.

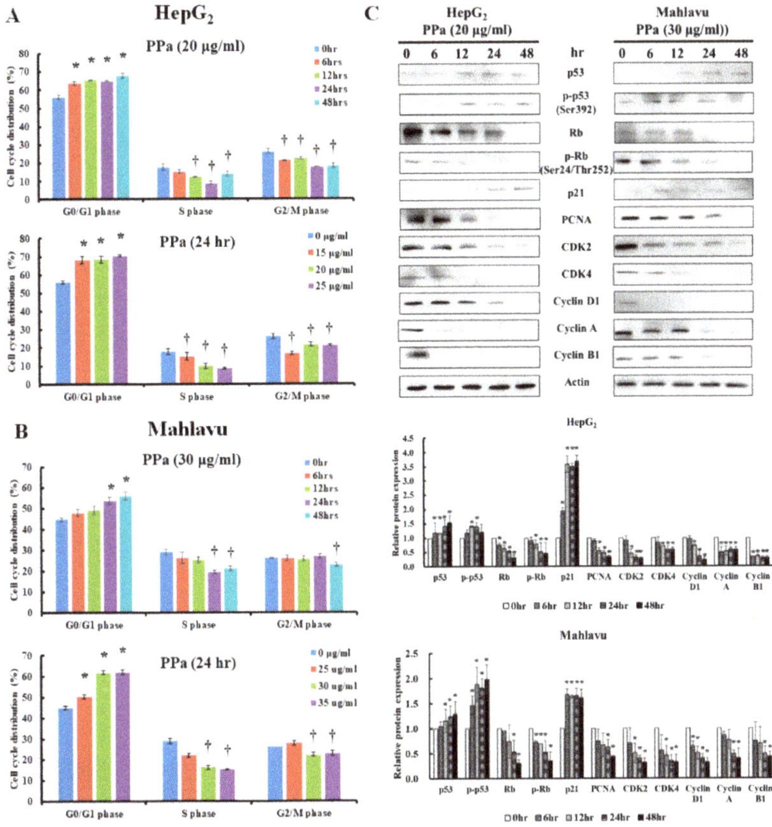

Figure 2. PPa extract blocked the cell cycle at G_0/G_1 phase in HCC cells. (**A,B**) HepG$_2$ and Mahlavu cells were treated with PPa extract for the indicated time intervals, and the cell cycle distribution was analyzed by flow cytometry. (**C**) After PPa extract treatment, cell lysates were collected and analyzed for cell cycle regulators by Western blotting with the indicated antibodies. *, †: Significant difference between the control group and experimental group, $p < 0.05$.

2.3. Ppa Extract Stimulated ROS Production and Imbalanced Mitochondrial Membrane Potential in HCC Cells

Former results revealed that p53/p21 pathway was activated to arrest cell cycle at G_0/G_1 phase. The high concentration of ROS production can damage DNA, causing activation of p53 (Ser392), which is phosphorylated by DNA damage signaling [22], and consequently increases p21 expression. Then, ultimately, cell cycle arrest will occur to repair damaged DNA for subsequent cell cycle proceed. Moreover, cells exposure to high dose of ROS may cause severe damage to DNA, proteins and lipids, and cells will arrest in all phases of the cell cycle and will undergo apoptosis. Consequently, we wondered whether PPa extract induced ROS production in PPa extract-treated HepG$_2$ and Mahlavu cells and the results found that PPa extract rapidly increased the ROS level within 3 h in both HCC cell lines (Figure 3). Further, ROS can also induce mitochondrial dysfunction resulting in mitochondrial membrane potential (MMPs, $\Delta\Psi$m) imbalance that can activate intrinsic apoptosis pathway. Further, $\Delta\Psi$m was evaluated, and the results showed significantly increased $\Delta\Psi$m (by 30–40%) within 12 h of treatment in HepG$_2$ and Mahlavu cells (Figure 4A). Moreover, after PPa extract treatment, JC-1 fluorescence was observed, and the results showed that green fluorescence was increased, suggesting that PPa extract induced mitochondrial membrane potential ($\Delta\Psi$m) loss (Figure 4B).

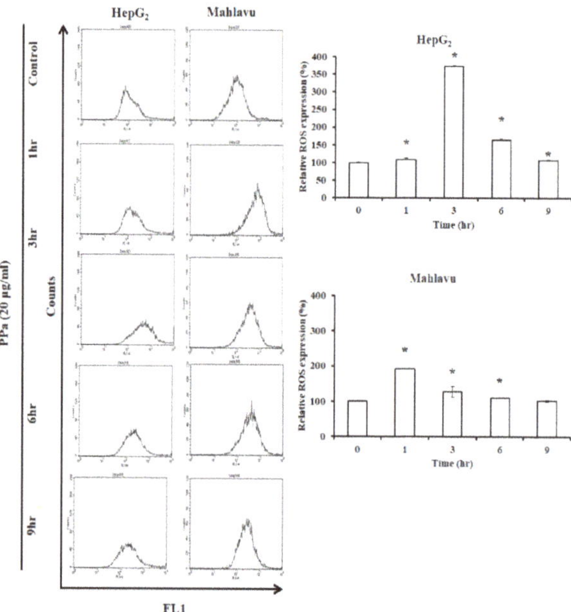

Figure 3. PPa extract induced ROS generation. HepG$_2$ and Mahlavu cells were treated with PPa extract (20 µg/mL) for time intervals and analyzed by flow cytometry at FL1-H to detect ROS production. *: Significant difference between the control group and experimental group; $p < 0.05$.

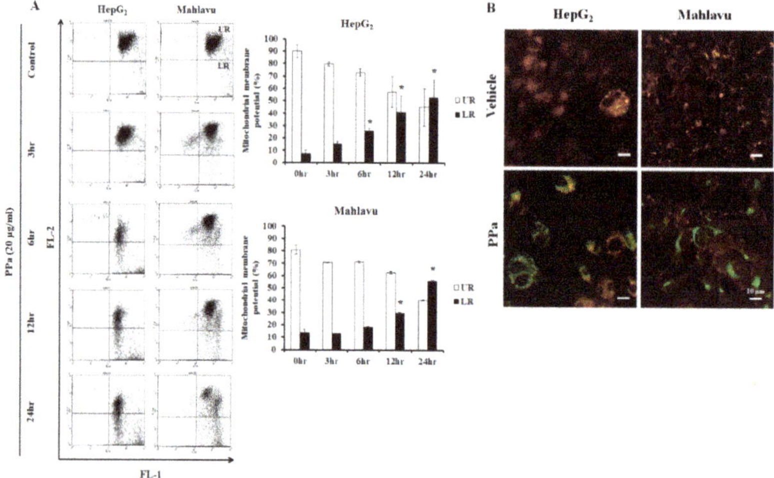

Figure 4. PPa extract stimulated mitochondrial membrane potential loss. (**A**) Mitochondrial membrane potentials (MMPs) were measured at FL2-H and FL1-H channels after PPa extract treatment in HCC cells. (**B**) After PPa extract treatment, JC-1 fluorescence was detected. *: Significant difference between the control group and experimental group; $p < 0.05$.

2.4. PPa Extract Induced Extrinsic and Intrinsic Apoptosis in HCC Cells

Subsequently, the percentage of PPa extract-treated HCC cells that died was detected by flow cytometry. HepG$_2$ and Mahlavu cells were exposed to serial doses PPa extract for 24 h, and the data showed that in both cell types PPa extract increased the number of cells in the sub-G_1 phase in a dose-dependent pattern (Figure 5A). Then, TUNEL assays were used to determine whether PPa extract induced apoptosis. As shown in Figure 5B, after PPa extract treatment, the two cell lines exhibited cell shrinkage under light field microscopy, and the number of TUNEL-positive cells increased, as indicated by anoikis, DNA fragmentation, chromatin condensation and apoptotic body formation. To further elucidate the PPa extract-induced apoptosis pathway in HCC cells, Western blotting was utilized. After exposure to PPa extract, the expression of FAS and FASL increased, the expression of procaspase-8 decreased, and cleaved caspase-8 increased, indicating that the extrinsic apoptotic pathway might be activated (Figure 5C). Furthermore, detecting Bax/Bcl2 ration was increased, resulting in the downregulation of procaspase-9 and cleaved caspase-9 increasing, suggesting that the intrinsic apoptosis pathway might be activated. Additionally, the expression of AIF increased after PPa extract treatment, suggesting that PPa extract might also induce the caspase-independent apoptosis pathway to cause cell death. And then, the expression of procaspase-3 was decreased and cleaved caspase-3 was increased, revealing that the caspase cascade might be involved. After that, to confirm that PPa extract activated the caspase cascade, HepG$_2$ and Mahlavu cells were pretreated with caspase-3, -8, or -9 inhibitors (1 µM) for 2 h. Then, the cells were treated with PPa extract (20 µg/mL) for 24 h and were analyzed by Western blot. The results revealed that PPa extract indeed induced extrinsic as well as intrinsic apoptosis pathway activation, which activated the caspase cascade (Figure 5D). These results validated that PPa extract activates the caspase cascade via extrinsic and intrinsic apoptosis pathways, leading to HCC cell death. Taken together, our data showed that PPa extract contributed to the production of ROS and triggered p53/p21 expression to affect mitochondrial membrane potential ($\Delta\Psi$m), resulting in the activation of the apoptosis pathway.

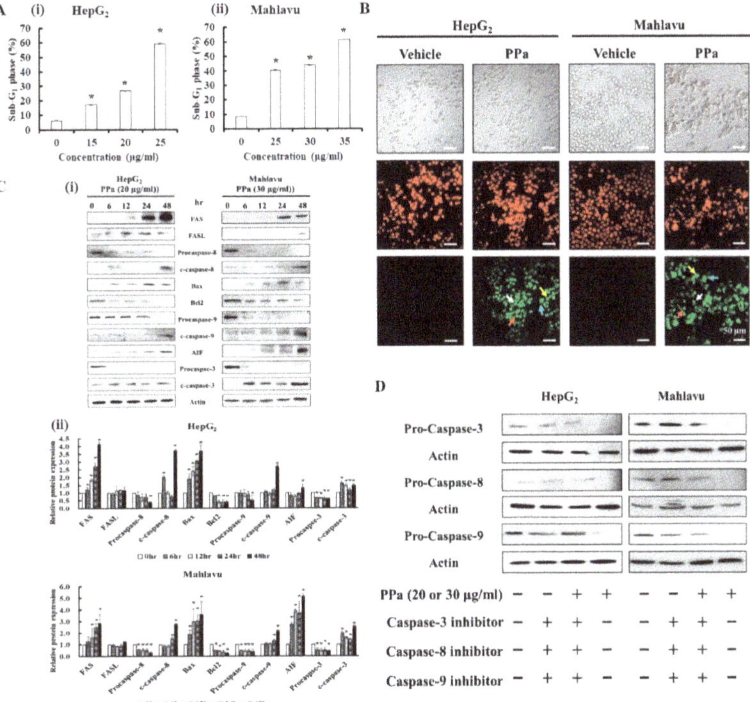

Figure 5. PPa extract induced HCC cell apoptosis by activating both the extrinsic and intrinsic apoptotic pathways. (**A**) After PPa treatment, the sub-G_1 phase of the cell population was analyzed by flow

cytometry; (i): HepG$_2$ cells; (ii): Mahlavu cells. (**B**) HepG$_2$ and Mahlavu cells were incubated with PPa extract for 24 h and analyzed by TUNEL assay. TUNEL positive (green); PI: propidium iodide (red); red arrow: chromatin condensation; yellow arrow: DNA fragments; blue arrow: anoikis; and white arrow: apoptotic bodies. (**C**) Western blots for pro-apoptotic and anti-apoptotic proteins in PPa extract-treated HCC cells; (i): indicated protein expressions; (ii): the quantitative data of protein expressions. (**D**) Western blots for caspase-3, -8 and -9 proteins in HepG$_2$ and Mahlavu cells which pretreated with caspase-3, -8 or -9 inhibitors (1 μM) for 2 h and then treated with PPa extract (20 or 30 μg/mL). *: Significant difference between the control group and experimental group; $p < 0.05$.

2.5. Synergistic Inhibitory Effects Induced by PPa Extract Plus Sorafenib in Hepatoma Cells

To examine synergism between PPa extract and sorafenib, HepG$_2$ and Mahlavu cells were treated in combination with indicated concentrations of drugs to calculate the combination index (CI). As shown in Figure 6, the results revealed that the combination of the PPa extract plus sorafenib exerted a synergistic effect in HepG$_2$ cells at 48 h, and significant synergy was observed in Mahlavu cells at 24 and 48 h. Both cell lines showed CI values of less than 1 at 48 h, and Mahlavu cells with the p53 mutant were more sensitive to PPa extract plus sorafenib. Next, we addressed the effects of the two drugs in combination on the induction of cell death by observing the sub-G$_1$ cell population. Indeed, HepG$_2$ and Mahlavu cells appeared to be sensitive to PPa extract plus sorafenib, resulting in a marked increase in the sub-G$_1$ cell population (Figure 7A). In addition, to assess the ability of PPa extract plus sorafenib to inhibit cell regrowth at day 4 and day 8. The results showed sorafenib no inhibitory effect at 0.2 μg/mL concentration on day 4 and PPa extract continuedly showed antiproliferation effects on both of cells. PPa extract combined with sorafenib showed inhibitory effects in both cell lines, indicating that the two-agent combination could prevent HCC cell regrowth (Figure 7B). Furthermore, to elucidate the antiproliferation mechanism of the two drugs, the AKT/mTOR, ERK and caspase cascades signaling pathway was examined. First, PPa extract used in combination with sorafenib reduced AKT, pAKT (Ser473), mTOR, p-mTOR (Ser2448), P70S6K, and p-P70S6K (Ser411) expression, suggesting that the combination might inhibit HCC cell growth by suppressing the AKT/mTOR signaling pathway. Second, ERK/pERK expression was also detected, and the results revealed that PPa extract plus sorafenib resulted in a more dramatic reduction in p-ERK (Tyr204) protein expression than what was observed in controls. These results revealed PPa extract plus sorafenib suppressed the expression of AKT/mTOR and ERK signaling in HepG$_2$ and Mahlavu cells. Then, to test the combination of PPa extract and sorafenib on induction of apoptosis, the results showed the combination of PPa extract and sorafenib was found to strongly reduce the protein expression of pro-caspase-8, -9, and -3. In converse, cleaved caspase-8, -9, and -3 protein expressions and Bax/Bcl2 ratio were significantly increasing., revealing that the combination of these two drugs enhanced the induction of cell apoptosis in HepG$_2$ and Mahlavu cells (Figure 7C). These results suggested that PPa extract in combination with sorafenib exhibited a synergistic effect that reduced HepG$_2$ and Mahlavu cell proliferation and regrowth via the induction of cell death and inhibition of the AKT/mTOR and ERK pathway.

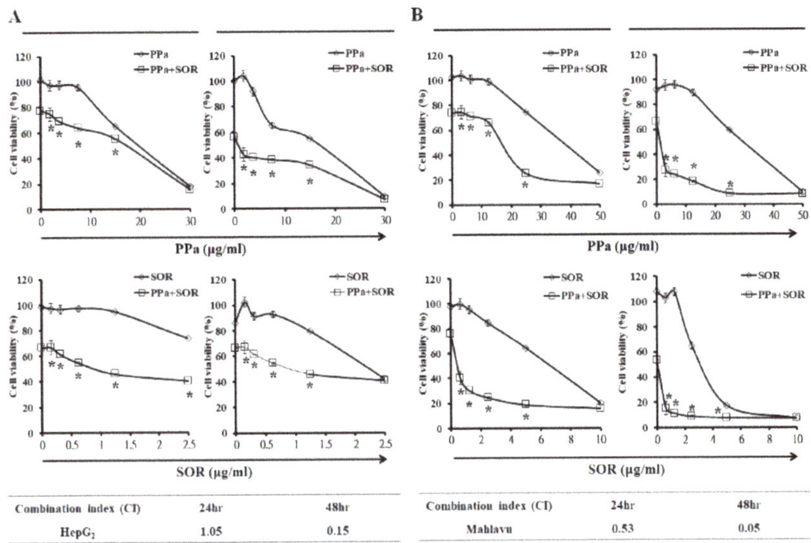

Figure 6. PPa extract synergized with sorafenib to enhance the inhibitory ability of PPa extract on HCC cell growth. (**A**), (**B**) HepG$_2$ and Mahlavu cells were treated with one drug or a combination of drugs and then were evaluated by the combination index. SOR: sorafenib. *: Significant difference between the control group and experimental group; $p < 0.05$.

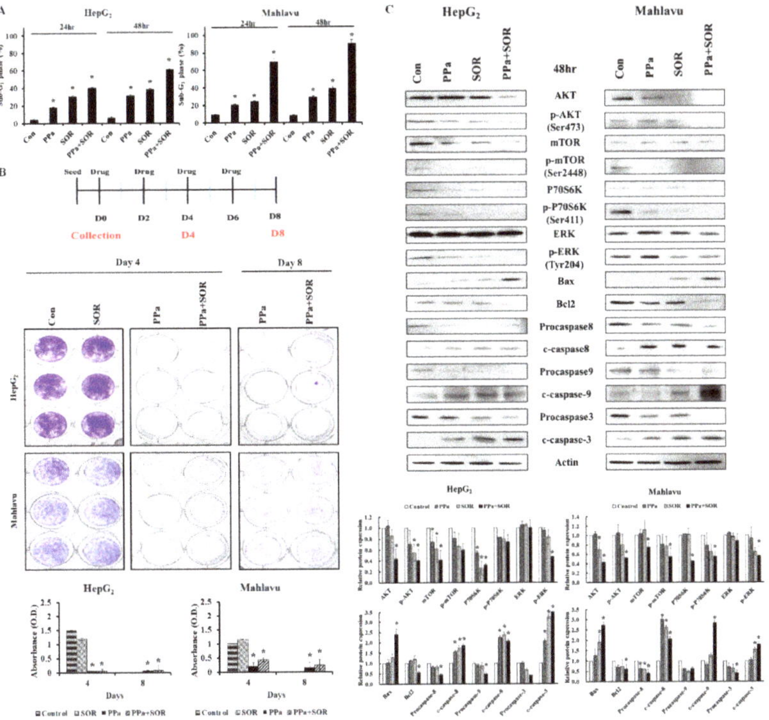

Figure 7. PPa extract plus sorafenib reinforced the suppression of cell proliferation and induction of cell apoptosis in HCC cells. (**A**) HepG$_2$ cells were treated with 2 μg/mL sorafenib plus 15 μg/mL PPa

extract, and Mahlavu cells were treated with 3 µg/mL sorafenib plus 25 µg/mL PPa extract for 24 and 48 h. After combined treatment, both cell lines were analyzed for the percentage of sub-G_1 phase by flow cytometry. (**B**) HepG$_2$ and Mahlavu cells were treated with PPa extract (15 µg/mL) and sorafenib (0.2 µg/mL) for 4 and 8 days. After treatment, cells were stained with crystal violet and absorbance was measured at 550 nm to calculate cell viability. (**C**) HepG$_2$ and Mahlavu cells were treated with combined treatment for 48 h. Cell extracts were prepared and analyzed by Western blotting with the indicated antibody. CON: control; SOR: sorafenib. *: Significant difference between the control group and experimental group; $p < 0.05$.

2.6. PPa Extract Suppressed Hepg$_2$ Tumor Growth and Exhibited Less Toxicity in HCC Xenograft Model

To further assess the inhibitory effect of PPa extract on growth was evaluated in HCC xenografts in nude mice. As shown in Figure 6, Balb/c nude mice bearing xenograft tumors were administered PPa extract (200 mg/kg, subcutaneous injection once every two days). Volumes of tumors and body weights of mice were measured every two days during the experimental period. The results revealed that PPa extract exerted greater antitumor effects than vehicle treatment (Figure 8A). In addition, we found that PPa extract prolonged the lifespan of mice by a range of 31 days to 51 days (Figure 8B). As shown in Figure 8C, we found no significant differences in body weight between vehicle- and PPa extract-treated mice. These results revealed that PPa extract exerted an antihepatoma capacity to suppress HCC tumor growth and extended survival time with no remarkability changes of body weight in vivo. As we had monitored the body weights throughout the study, PPa extract did not dramatically decrease the body weight, revealing that PPa extract might not cause severe systemic toxicity in vivo. Subsequently, we further assess the pathology of the following organs: heart, liver, spleen, lung, kidney, stomach, and intestine, after PPa extract treatment. Notably, the cell morphology of these organs did not appear obviously change and remained the integral structure of organs (Figure 9A). Further, no significant the blood and immune cell infiltration were observed after PPa extract administration. These results revealed that PPa extract might not cause severe organ damage to recruit immune cells and activate inflammation. Moreover, we found no significant differences in the WBC, RBC, and platelet counts (Figure 9B). Importantly, the values of AST and ALT showed no significant differences when comparing the control with PPa extract treatment groups, revealing that PPa extract might not further cause severe liver cell damage and toxicity, which lead to liver dysfunction. The data demonstrated that PPa extract presented a lower physiological and pathological toxicity in vivo.

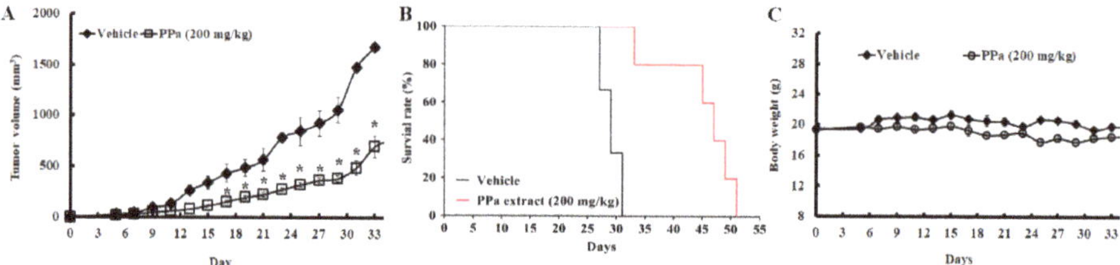

Figure 8. PPa extract suppressed HCC cell proliferation in the xenograft model. Balb/c nude mice were injected with HepG$_2$ cells on day 0 of the experiment, started treatment after five days and then were treated with PPa extract (200 mg/kg) every two days. When the tumor volume reached 1500 mm^3, the mice were sacrificed. (**A**) Tumor volume. (**B**) Survival rates. The data are expressed as the mean ± SEM. *: Significant difference between the control group and experimental group, $p < 0.05$. (**C**) Balb/c nude mice were injected with HepG$_2$ cells, which was followed by administration of PPa extract (200 mg/kg) and measurement of body weight.

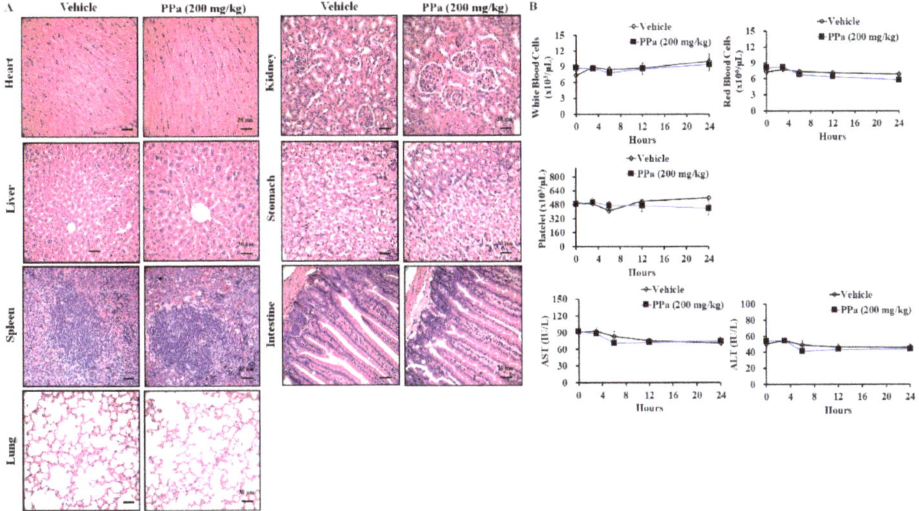

Figure 9. PPa extract displayed low pathological and physiological toxicity in vivo. (**A**) After sacrificing the animals, organs were collected and analyzed by HE staining. (**B**) After PPa treatment, blood was collected at the 0, 3, 6, 12, and 24 h for analysis of blood cells (white blood cell, red blood cell and platelet) and serum biochemistry (ALT, and AST).

2.7. PPa Extract Induced Apoptosis and Reduced Autocrine Proliferation in Xenograft Model

Next, we addressed the inhibitory effect of PPa extract in vivo with H&E and IHC staining. The results showed that PPa extract caused HCC tumor cell death due to the induction of ROS production and causing DNA damage in tumor cells, leading to an increase in 8-oxo-dG expression, which is the commonly used marker of oxidative stress-derived DNA damage [23,24]. (Figure 10A). Additionally, PPa extract increased the expression of cleaved caspase-3 and TUNEL positive (green) lead to apoptosis in vivo. These results indicated that PPa extract induced ROS production to damage tumor cells causing activation of apoptosis in vivo that was consistent with the funding in vitro. As shown in Figure 10B, PPa extract also suppressed the expression of PCNA, VEGF, VEGFR1 and VEGFR2, resulting in inhibition of HCC growth. In conclusion, PPa extract exhibited inhibitory effects on HCC through ROS production, induction of apoptosis, and suppression of autocrine proliferation. Ultimately, to elucidate the components of PPa extract, we utilized GC/MS analysis. The data revealed that patchouli alcohol (RT: 14.78; 32.12%), α-gurjunene (RT: 12.83; 21.67%) and α-guaiene (RT: 11.97; 17.98%) were three major components in the PPa extract and there were other components, including seychellene, α-patchoulene, caryophyllene, azulene, α-elemene, 2-butenal, caryophyllene oxide, globulol, α-humulene, longifolenaldehyde, longiborneol, and azulenone (Figure 11). Among these components, the content of patchouli alcohol was highest, and this indicated that patchouli alcohol might be the active ingredients to exert antihepatoma capacity.

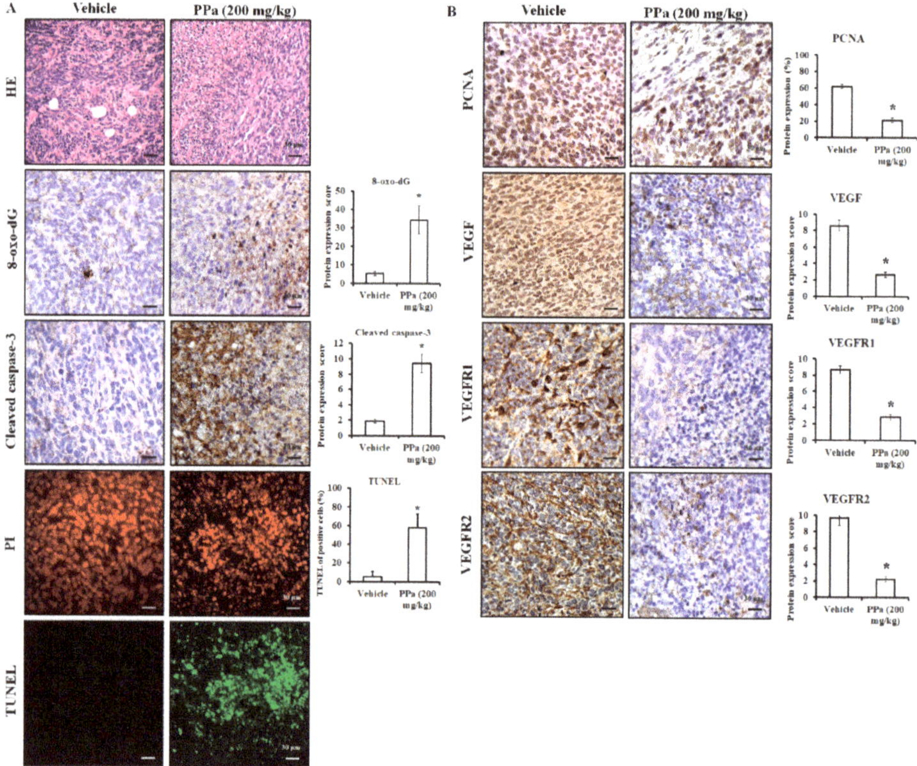

Figure 10. PPa extract affected ROS generation, cell apoptosis and proliferation. After the mice were sacrificed, the tumor mass was collected for HE and IHC analysis. (**A**) After PPa extract treatment, tumor cell damage, 8-oxo-dG, as well as cleaved caspase-3 were observed and TUNEL assay was performed to detect apoptosis in HCC tissue. PI: propidium iodide (red); TUNEL: green. (**B**) Autocrine proliferative proteins were examined by IHC staining. *: Significant difference between the control group and experimental group; $p < 0.05$.

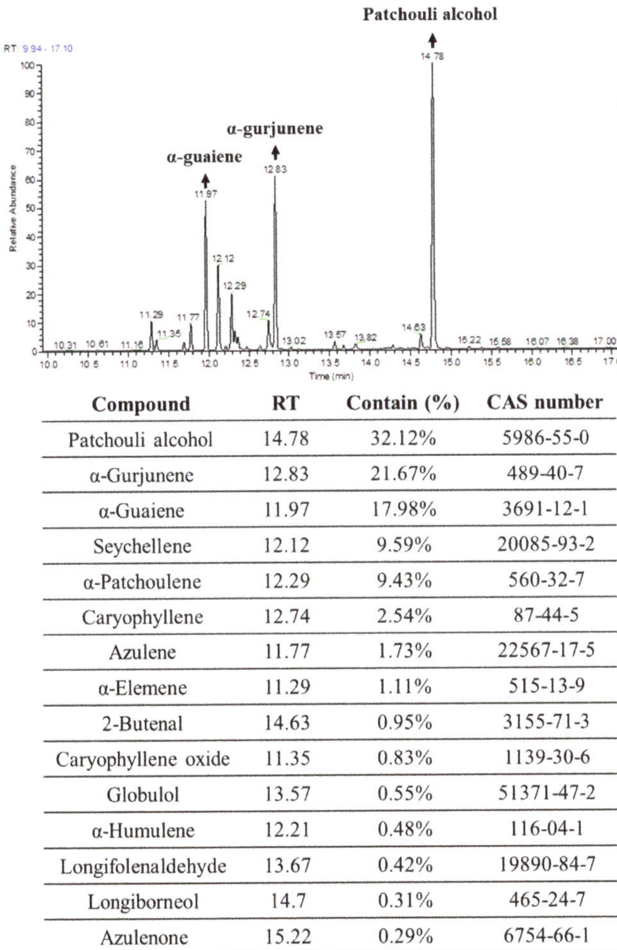

Compound	RT	Contain (%)	CAS number
Patchouli alcohol	14.78	32.12%	5986-55-0
α-Gurjunene	12.83	21.67%	489-40-7
α-Guaiene	11.97	17.98%	3691-12-1
Seychellene	12.12	9.59%	20085-93-2
α-Patchoulene	12.29	9.43%	560-32-7
Caryophyllene	12.74	2.54%	87-44-5
Azulene	11.77	1.73%	22567-17-5
α-Elemene	11.29	1.11%	515-13-9
2-Butenal	14.63	0.95%	3155-71-3
Caryophyllene oxide	11.35	0.83%	1139-30-6
Globulol	13.57	0.55%	51371-47-2
α-Humulene	12.21	0.48%	116-04-1
Longifolenaldehyde	13.67	0.42%	19890-84-7
Longiborneol	14.7	0.31%	465-24-7
Azulenone	15.22	0.29%	6754-66-1

Figure 11. GC/MS analysis of PPa extract. Gas chromatography-mass spectrometry (GC-MS) analyses were performed by the National Central Taiwan University Office of Research and Development's Center for Advanced Instrumentation (Hsinchu, Taiwan). Components were identified by comparing their mass spectra with those obtained from authentic samples or spectra of the Wiley/Nist libraries. RT: Retention time.

3. Discussion

Hepatocellular carcinoma (HCC) is the most frequent tumor and the third most common malignancy, and it causes high mortality worldwide; in addition, the incidence of HCC has been increasing. Although many treatment approaches have been used to treat advanced HCC, for now, only transarterial chemoembolization (TACE) and sorafenib have been shown to provide survival benefit [2]. As a result, better options for the prevention of HCC development might be a good approach. Here, we used *Pogostemon cablin*, a plant of the Lamiaceae family that is native to tropical regions of Asia, and studies have demonstrated many biofunction activities of *Pogostemon cablin*, including antidepressant [10,11], antimicrobial [12,13], antiviral [14], anti-inflammatory [15], gastroprotective [16,17], antiaging [18], and antitumor activities [19,20]. Among these, our previous study has demonstrated the anticancer activity of *Pogostemon cablin* extract on colon cancer in vitro and in vivo. In the study, *Pogostemon cablin* extract induces apoptosis and cell cycle arrest and presented no obvious pathological toxicity in vivo [20]. These results indicate that *Pogostemon cablin* extract is a potential anticancer agent for cancer treatment. As a result, our study explored the potential role of PPa extract in inhibiting human hepatocellular carcinoma. Here, we first demonstrated that PPa extract inhibited HCC cell proliferation, which was shown in the following cells: HepG$_2$ cells—a cell line with WT p53 derived from a patient from the United States [25]; Mahlavu cells—a cell line with mutated p53 with poor differentiation derived from a patient from Africa; Huh7 cells—a cell line with mutated p53 derived from a Japanese patient [26]; and J5 cells—a cell line derived from Taiwan patients. The p53 gene is the most commonly mutated tumor suppressor gene in various human cancers, and hepatocellular carcinoma is no exception [27,28]. Moreover, mutations in p53 are a poor prognostic indicator for survival, suggesting that patients with p53 mutations have a worse prognosis than those with WT p53 [29]. Therefore, PPa extract effectively repressed HCC cell growth and might provide an option for increasing patient benefit. Moreover, the IC$_{50}$ values of *Pogostemon cablin* extract were variety on different types of colon cancer cells and HCC cells, revealed that its' anticancer potential on these two cancers. On the other hand, HCC patients are common to have a different level of liver dysfunction that restricts the use of chemodrugs result in the limitation of therapeutic efficacy. Herein, present study had to assess inhibitory effect of PPa extract in normal cells including SVEC, MDCK and BNL CL.2 cells and the results revealed that the IC$_{50}$ values of tumor cells compared with normal cells were ranged from 2.1 to 20.1 in PPa extract treatment. These results demonstrated that PPa extract might have less cytotoxicity to normal cells, including epithelial cells, kidney cells and liver cells; moreover, the AST as well ALT values remained in normal range after PPa extract treatment. These results revealing that PPa extract might not induce severe side effects and adverse liver dysfunction.

Subsequently, we further examined the inhibitory effects of PPa extract on alteration of cell cycle distribution because of the cell cycle as an important mechanism for controlling cell growth. The results showed that PPa extract induced cell cycle arrest at the G_0/G_1 phase in both HepG$_2$ and Mahlavu cells in a dose- and time-dependent manner. After that, the regulation of cell cycle progression was explored. p53 as well as Rb is a tumor suppressor and plays a crucial role in governing the cell cycle and apoptosis [30]. The p21 is a downstream protein of p53 and a CKI that can block the cell cycle at the G_0/G_1 phase and G_2/M transitions by inhibiting cdk4,6/cyclin D and cdk2/cyclin E, respectively. Our results showed that PPa extract induced p53 and p-p53 protein expression to increase downstream proteins p21 expression and reduced the level of cdk2, ckd4, and cyclin D1, A, and B1, resulting in cell cycle arrest at the G_0/G_1 phase in WT p53 cells (HepG$_2$) and mutant p53 cells (Mahlavu). Moreover, p21 can bind to PCNA to inhibit DNA replication [31], and the data revealed that the expression of PCNA was decreased after PPa extract. The Rb promotes E2F-dependent gene expression to stimulate DNA replication and proceed G_1/S phase transition [32]. Our results showed that PPa extract reduced Rb and p-Rb expression, which decreased DNA replication to block the G1/S transition lead to cell cycle arrest in HepG$_2$ and Mahlavu cells.

Exogenous or endogenous ROS can activate p53 phosphorylation at Ser392 by directly damaging of nuclear and mitochondrial DNA [22] and stimulate transcription of proapoptotic genes that including intrinsic and extrinsic apoptosis pathway, such as Bax, Fas, and FasL [33,34]. Moreover, cytosolic p53 enhances mitochondrial membrane depolarization by causing rearrangement of Bax/Bak on outer mitochondrial membrane and subsequent release apoptotic factors, such as AIF which is involved in DNA fragmentation and activate caspase independent apoptosis. Then, apoptosome complex is formed to activated caspase-9 and effector caspases such as caspase-3 leading to intrinsic apoptosis [35,36]. ROS can directly activate extrinsic apoptosis pathway and recruits cleaved caspase-8 and -3 to trigger apoptosis [37]. As a result, we next to detect the ROS generation of PPa extract-induced in both HepG$_2$ and Mahlavu cells and the results indicated that PPa extract induced abundant ROS production contributing to mitochondrial membrane potential imbalance. Meanwhile, sub-G$_1$ phase was observed to increase after PPa extract treatment, and TUNEL positive data revealed that PPa extract induced cell apoptosis with classical cell death morphology, including anoikis, DNA fragments, chromatin condensation and apoptotic bodies. To gain insight into the mechanisms of the PPa extract-induced apoptosis pathway, extrinsic, intrinsic, and caspase-independent associated proteins were detected. Fas and FasL are recognized as major pathways involved in the cleavage of caspase-8, which induces extrinsic apoptosis. Moreover, some studies have revealed that HCC cells show resistance to apoptosis because of their suppression of FAS expression [38], and serum levels of soluble FASL in patients with hepatocellular carcinoma show potential as a clinical parameter to evaluate prognosis [39]. Our results showed that PPa extract enhanced FAS and FASL protein expression to induce HepG$_2$ and Mahlavu cell apoptosis via extrinsic apoptosis. On the other hand, p53 can directly activate several genes, including Bax, Bcl-2 and AIF. Increasing the Bax/Bcl-2 ratio and AIF expression participate in the regulation of apoptotic mitochondrial membrane permeabilization, resulting in the induction of caspase-9 cleavage and activation of the intrinsic apoptosis pathway. Therefore, PPa extract induced mitochondrial membrane potential loss via both the intrinsic apoptosis pathway and AIF regulation. AIF also plays an important role in inducing caspase-independent chromatin condensation and DNA fragmentation [40]. Consequently, our results found that PPa extract stimulated lots of ROS generation and activated p-p53 (Ser392) expression, which is correlated with DNA damage, leading to caspase dependent apoptosis (extrinsic and intrinsic) and caspase independent apoptosis pathway. Furthermore, these results showed that PPa extract exerted a similar anticancer activity on colon cancer cells and HCC [20].

Drug resistance and hepatotoxicity are important considerations during HCC treatment. Moreover, some clinical studies assessed the combination of standard therapies with traditional herbal medicine and observed significant survival benefits, such as a reduction in recurrence and prolonged survival time [41,42]. Hence, we found that PPa extract seemed to be a good adjuvant for sorafenib, which worked synergistically in both HepG$_2$ and Mahlavu cells by enhancement of inhibition cell viability, significantly. Moreover, previous statistical reports indicate that the HCC recurrence rate accounted for 20–25% per year after therapeutic procedure [43]. Our results demonstrated that PPa extract plus sorafenib also diminished the regrowth of HepG$_2$ and Mahlavu cells, suggesting that PPa extract combined with sorafenib might provide new approaches for posttreatment chemoprevention regimes. Furthermore, our former data showed that PPa extract stimulated abundant ROS production to repress HCC cell growth and contribute to cell death. Further, sorafenib is reported that it can lead to HCC cell death by inducing of ROS production in vitro and in vivo [44]. ROS can activate PI3K, inactivate PTEN that negatively regulates the synthesis of PIP3 and inhibits the activation of AKT [45]. Moreover, deregulation of the AKT/mTOR pathway has increasingly been implicated in HCC [46], and activation of AKT is thought to mediate the resistance of sorafenib [47]. Additionally, the ERK pathway may also be involved in chemodrugs-induced drug resistance in HCC [48]. Then, to examine the synergistic mechanisms of PPa extract plus sorafenib and the results revealed that the combinational treatment reinforced to reduce the AKT/mTOR pathway and level of ERK, indicated that PPa extract plus sorafenib might be induce more ROS production than drug alone to repress HCC cell growth and reduce the potential of sorafenib-resistant development on AKT/mTOR and ERK pathway. Furthermore, abundant ROS that generated by PPa extract plus sorafenib caused more cell death through induction of caspase dependent pathway. Therefore, the results showed a lower concentration of drugs was needed to induce more cell death and enhance the inhibitory ability by blocking the AKT/mTOR pathway. These results provided a new therapeutic approach for clinical utilization of PPa extract. Moreover, our previous data regarding the anticancer activity of *Pogostemon cablin* extract on colon cancer cells have also shown its capacity for combination with 5-FU [20], and these results might indicate that *Pogostemon cablin* extract probably would be good adjuvant for cancer treatment.

PPa extract induced ROS generation to affect p53/p21 expression, cell cycle-associated regulators, downstream death ligands/receptors and mitochondrial-mediated apoptosis pathways in vitro. Then, we had PPa extract administration in animal model and PPa extract effectively suppressed HCC tumor growth and prolonged the lifespan of mice. Subsequently, considering system toxicity and more specific toxicity, such as hepatoxicity or nephrotoxicity, is extremely important. Additionally, advanced hepatocellular carcinoma is commonly characterized by liver dysfunction attributable to the presence of chronic liver disease and cirrhosis. During PPa extract administration, we measured body weight every two days, and the differences in body weight between the groups appeared to be not statistically significant, suggesting that PPa extract might exhibit few systemic toxic effects in vivo. After two months of treatment, organ toxicity was evaluated, and no obvious pathological damage was observed, suggesting that PPa extract might not have no obviously accumulated toxicity to normal organs. Moreover, the numbers of WBCs, RBCs and platelets were not distinctly different from those in the vehicle group, suggesting that PPa extract might not be induce strong inflammation and blood lysis. In addition, ALT and AST were in normal ranges in both vehicle and PPa treatments, revealing that PPa extract presented little or low liver toxicity in vivo. Taken together, PPa extract might not be induce strong adverse effects in vivo and could be developed as an adjuvant or chemodrug on HCC treatment.

Subsequently, in HCC xenograft, PPa extract demonstrated that it stimulated ROS production, triggered level of cleaved caspase-3 and lead to HCC cell apoptosis that was consistent with the results we observed in vitro. On the other hand, many studies have revealed that HCC cells produce and secrete VEGF and express VEGFR to promote tumor proliferation, indicating the activity of VEGF/VEGFR autocrine and paracrine signaling pathways in HCC cells [49,50]. Moreover, VEGF/VEGFR signaling is positively correlated with tumor size, intrahepatic metastasis, vascular invasion, and TNM stage, which affect prognosis and survival time [51]. Hence, the VEGF/VEGFR signaling axis is an ideal target for treating HCC. After PPa extract administration, VEGF, VEGFR1 and VEGFR2 expression was reduced in HCC tumor tissue, suggesting that PPa extract effectively decreased the VEGF/VEGFR signaling axis to mitigate the autocrine and paracrine signaling pathways. Consequently, PPa extract induced p53-dependent or p53-independent signaling activation to trigger p21 protein expression and block cell cycle progression via downregulation of cell cycle regulators and reduction of DNA replication by inhibition of PCNA and Rb expression in vitro. In addition, PPa extract modulated the VEGF/VEGFR signaling axis to inhibit HCC tumorigenesis in vivo. Ultimately, the ingredient of PPa extract was analyzed by GC/MS and patchouli alcohol, α-gurjunene, and α-guaiene were three major components in PPa extract. However, our previous study showed that *Pogostemon cablin* extract from Republik Indonesia contained several compounds, including azulene, α-guaiene, patchouli alcohol, α-patchoulene, and γ-gurjunene [20]. We assumed that *Pogostemon cablin* from different origin might be the reason why it had different chemical composition. Among these, patchouli alcohol has been reported the anticancer activity on lung and colon cancer [52,53]. Patchouli alcohol presented anticancer effect by inhibiting of histone deacetylases (HDAC) activity and c-myc expression to activate p21 and downregulate cyclin D1 and cdk4, resulting in cell growth arrest and apoptosis on colon cancer [52]; it induced apoptosis and cell cycle arrest by blocking phosphorylation of EGFR pathways, activating JNK pathways and activating p53/p21 pathway to affect cyclin E and cdk2 complex in A549 cancer cells in vitro and in vivo [53]. Azulene have been found the antiproliferation activity on MCF7 breast cancer cells and DU145 prostate cancer cells [54]. As a result, we guessed that patchouli alcohol or azulene might be the major anticancer compound in PPa extract.

In conclusion, PPa extract seems to be a good herbal agent with higher safety margins and effectively suppresses hepatoma by reducing tumor cell growth and inducing tumor cell apoptosis. Moreover, PPa extract plus sorafenib yielded synergistic effects on the AKT/mTOR pathway, reducing cell proliferation and activating the caspase cascade. Thus, in vitro and in vivo data suggest that PPa extract might be an effective anti-hepatoma agent for HCC treatment (Figure 12).

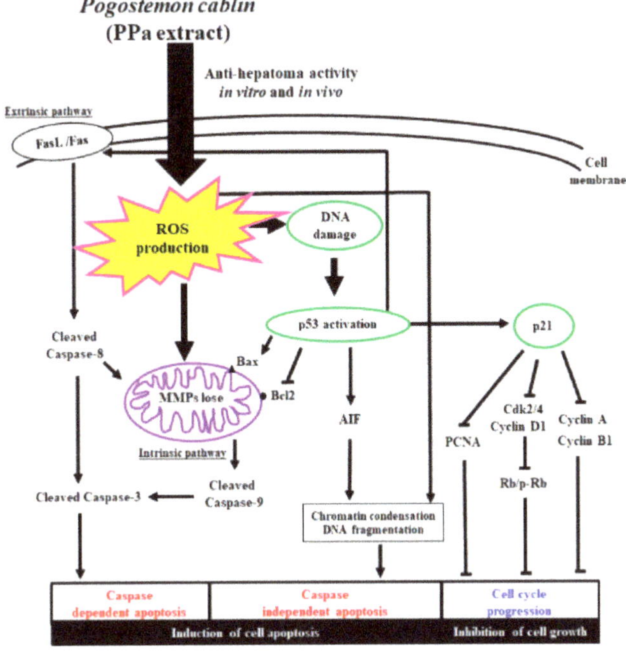

Figure 12. Overview of the mechanisms underlying the anti-hepatoma effects of PPa extract in vitro and in vivo.

4. Materials and Methods

4.1. Extration Essential Oils of Pogostemon Cablin (PPa Extract)

Pogostemon cablin plant had confirmation of identification by Professor Han-Ching Lin (Department & Graduate Institute of Pharmacology, National Defense Medical Center, Taiwan). The fresh leaves of *Pogostemon cablin*, which was origin from England (2.0 kg) were dried at temperature of 30 °C for 7 h/day for 3 days. After that, *Pogostemon cablin* essential oils was produced by using steam distillation. The dried leaves of *Pogostemon cablin* (500 g) was placed in a 2-L steam distillation steel apparatus unit with a flow rate of generated steam approximately 7.2 mL/min at 100 °C for 100 min and the yields were about 2.32% [20]. *Pogostemon cablin* extract (PPa extract) was commissioned to Phoenix (Red Bank, NJ, USA) for large scale extraction. After extraction, the PPa extract was sealed in a black glass bottle and stored at 4 °C. For long-term preservation, avoiding moisture and light was necessary. Before experiments conducting, the concentration of PPa extract was calculated as the equation: the weight of 20 μL PPa extract (g)/ (the weight of 180 μL DMSO + the weight of 20 μL PPa extract) (g) and the final concentration of DMSO in cells was less than 1%.

4.2. Cell Culture

HepG$_2$, Mahlavu, J5, Huh-7, SVEC, MDCK and BNL CL.2 cell lines were purchased from American Type Culture Collection (Manassas, VA, USA) or Bioresource Collection and Research Center (Hsinchu, Taiwan). The cells were cultured in DMEM or RPMI-1640 supplemented with 10% FBS (Gibco, Mexico), 1% sodium pyruvate, 1% HEPES and 1% penicillin/streptomycin at 37 °C in a humidified incubator supplemented with 5% carbon dioxide. All cell culture reagents were purchased from Gibco/Thermo Fisher Scientific (Waltham, MA, USA). HepG$_2$ and Mahlavu cells were analyzed with a Femtopath TP53 Exon8 Primer Set (HongJing Biotech., New Taipei City, Taiwan) to confirm their TP53 levels.

4.3. Cell Viability

HCC cells (5×10^3/100 µL), SVEC, MDCK (1×10^4/100 µL), and BNL CL.2 cells (4×10^4/100 µL) were seeded in a 96-well plate. After incubating overnight at 37 °C, the cells were treated with serially diluted drug concentrations for 24, 48, and 72 h. The cell viability was measured using an MTT assay [20].

4.4. Cell Cycle Analysis

Cells (2×10^6) were seeded in 10 cm dishes and were treated with indicated concentration of PPa extract for the indicated time points, and then incubated with PI stain (40 µg/mL) after harvesting the cells. Fluorescence was detected by a FACScan flow cytometer (FASCS Calibur, Franklin Lakes, NJ, USA) and the cell cycle distribution was analyzed by FlowJo 7.6.1 (Ashland, OR, USA) [20].

4.5. TUNEL Assay

A TUNEL in situ cell death detection kit (Roche Applied Science, ON, Canada) was used according to the manufacturer's instructions. Cells (2×10^5) were seeded in 6-well and were treated with PPa extract (20 or 30 µg/mL) for 24 h. After treatment, cells were harvested by 0.05% trypsin, fixed with 10% buffered formalin for 10 min, and smeared on the slides to incubate with 3% H_2O_2 for 5 min and 0.1% Triton X-100 for 1 min. Then, TUNEL reaction solution was added to incubate for 2 h and stained with 10 µg/mL propidium iodide (PI, red) for 10 min as counterstain. The average number of TUNEL-positive cells (green) was determined from ten fields under 400× magnification by microscope (ZEISS Axio Imager A2, Bremen, Germany).

4.6. Western Blot Analysis

Cells (2×10^6) were seeded in 10 cm dishes and treated with PPa extract (20 or 30 µg/mL) for 0, 6, 12, 24, and 48 h to detect cell proliferative, cell cycle related and apoptotic protein expressions. To verify whether PPa extract activated the caspase cascade, cells (2×10^6) were seeded in 10 cm dishes overnight and pretreated with caspase-3 inhibitor (CPI-370, Z-DEVD-FMK, 1 µM), capase-8 inhibitor (CPI-008, Z-LETD-FMK, 1 µM) or caspase-9 inhibitor (CPI-009, Z-LEHD-FMK 1 µM), which were purchased from G-Biosciences (Louis, MO, USA), for 1 h. After removing caspase inhibitors, PPa extract (20 or 30 µg/mL) were added and incubated for 24 h to detect indicated protein expressions. After getting end points of experiment, the cells were harvested and had cell lysate by RIPA buffer. After cell lysate colleting, protein concentration was quantified using the BCA Protein Assay Reagent (Thermo Fisher Scientific, Waltham, MA USA). Whole cell lysates were fractionated via 8–12% SDS-PAGE and transferred to PVDF membranes by electroblotting. The membranes were blocked with skim milk and were incubated at 4 °C overnight with each primary antibody. The membranes were then incubated with the appropriate secondary antibody and horseradish peroxidase, which were purchased from Santa Cruz Biotechnology (Dallas, TX, USA). Detection was carried out using a T-Pro LumiFast plus Chemiluminescence Detection Kit (T-Pro Biotechnology, New Taipei County, Taiwan). Bands were detected and photographed by a fluorescence imaging analyzer (GE LAS-4000, Little Chalfont, United Kingdom) and were quantified using ImageJ software (NIH, Bethesda, MD, USA). These experiments were independently achieved duplicated. Primary antibodies, including p53 (SC-6243), p-p53 (Ser392; SC-7997), Rb (SC-7905), p-Rb (Ser24/Thr252; SC-16671), p21 (SC-397), PCNA (SC-7907), CDK2 (SC-163), CDK4 (SC-260), cyclin A (SC-751), FAS (SC-715), FASL (SC-834), caspase-8 (SC-5263), caspase-9 (SC-7885), AIF (SC-9416), caspase-3 (SC-98785), AKT (SC-8312), p-AKT (Ser473; SC-7985-R), mTOR (SC-8319), p-mTOR (Ser2448; SC-101738), P70S6K (SC-8418), p-P70S6K (Ser411; SC-8416), ERK (SC-154), p-ERK (Tyr204; SC-7383), and Actin (SC-47778), were purchased from Santa Cruz Biotechnology (Dallas, TX, USA). The primary antibodies of bcl2 (IR94-392), bax (IR93-390), cyclin D1 (IR117-294) and cyclin B1 (IR116-289) were purchased from iReal Biotechnology Co., Ltd. (Hsinchu, Taiwan).

4.7. Detection of Reactive Oxygen Species (ROS)

HCC cells were subcultured in a 6-well plate at a density of 1×10^5 cells/mL and were treated with PPa extract (20 µg/mL) at different time points. After collecting cells, ROS generation was assessed using DCFH-DA staining according to the manufacturer's instructions. FACScan flow cytometry was performed and the data were analyzed by FlowJo 7.6.1.

4.8. Mitochondrial Membrane Potential (MMPs, $\Delta\Psi m$) Assay

Cells were subcultured in a 6-well plate at a density of 1×10^5 cells/mL and then were treated with PPa extract (20 µg/mL) for the indicated times. After the cells harvesting, cells stained with JC-1 probe (AAT Bioquest, Inc., Sunnyvale, CA, USA) according to the manufacturer's instructions. FACScan flow cytometer was performed and the data were analyzed by FlowJo 7.6.1. The $\Delta\Psi m$ fluorescence was observed and photographed under a microscope at a magnification of ×400 (ZEISS Axio Imager A2, Bremen, Germany).

4.9. Synergistic Effect of Ppa Extract Plus Sorafenib on Cell Proliferation and Regrowth

For combined treatment with PPa extract and sorafenib, MTT assay was performed and the data were converted to a readout of fraction of inhibition affected by the individual drug or the combination and analyzed using the combination index method [51,52]. Cells (5×10^3/100 µL) grown in 96-well overnight and treated with as the following for 24 and 48 h: PPa extract (0, 1.9, 3.8, 7.5, 15 and 30 µg/mL) plus sorafenib (2 µg/mL), or sorafenib (0, 0.16, 0.32, 0.64, 1.25 and 2.5 µg/mL) plus PPa extract (15 µg/mL) for HepG$_2$ cells; PPa extract (0, 3.2, 6.4, 12.5, 25 and 50 µg/mL) plus sorafenib (3 µg/mL), or sorafenib (0, 0.64, 1.25, 2.5, 5 and 10 µg/mL) plus PPa extract (25 µg/mL) for Mahlavu cells. After end points, the combination index (CI), which was calculated as the equation: CI = IC$_{50}$ of synergistic treatment I/IC$_{50}$ of PPa extract + IC$_{50}$ of synergistic treatment II/IC$_{50}$ of sorafenib. CI values significantly less than 1.0 indicated synergy. For analysis of sub-G$_1$ phase, cells (2×10^6) seeded at 10 cm dishes and treated with as following for 24 and 48 h: 2 µg/mL sorafenib plus 15 µg/mL PPa extract for HepG$_2$ cells; 3 µg/mL sorafenib plus 25 µg/mL PPa extract for Mahlavu cells. After that, both cell lines were analyzed for flow cytometry. For regrowth assay, cells (5×10^2) seeded at 96-well overnight and treated with PPa extract (15 µg/mL) and sorafenib (0.2 µg/mL) for 4 and 8 days. During treatments, fresh drugs were changed every two days and harvested the data at 4 and 8 days. After treatment, cells were stained with 0.1% crystal violet and absorbance was measured at 550 nm after dissolving in 0.5% acetic acid. For detection of indicated protein expressions, cells (2×10^6) seeded at 10 cm dishes. And then, HepG$_2$ cells were treated with 2 µg/mL sorafenib plus 15 µg/mL PPa extract and Mahlavu cells were treated with 3 µg/mL sorafenib plus 25 µg/mL PPa extract for 48 h. Cell lysates were collected for western blot.

4.10. Xenograft Animal Study

Balb/c nude mice (8–12 weeks, female) purchased from National Laboratory Animal Center (Taipei, Taiwan) were housed in a pathogen-free environment. All procedures of the liver cancer cell xenograft animal model were performed in the Laboratory Animal Center of Chung Shan Medical University (CSMU) following the Guide for the Care and Use of Laboratory Animals and were approved by the IACUC of CSMU (CSMU-IACUC-1662). To establish a subcutaneous liver cancer model in mice, HepG$_2$ cells (1×10^6 cells/100 µL/mouse) were injected into the right flank of mice. After 5 days, mice were randomly divided into two groups (n = 5 for each): vehicle treated with mineral oil or subcutaneously treated with PPa extract (200 mg/kg) on left flank of mice once every two days. All mice were observed until the tumor volume was greater than 1500 mm^3 (L × H×W mm^3), which was noted as the last survival day. The following analysis was performed by IHC and H&E staining. The slides were assessed by light microscopy at a magnification of ×400. For serum biochemical estimation, blood was collected from the tail vein at 0, 3, 6, 12, and 24 h for analysis of acute toxicity which is determining the short-term adverse effects of a drug when administered in a single dose, or in multiple doses during a period of 24 h [55], after PPa extract (200 mg/kg) treatment. Serum was separated via centrifugation at 3000 rpm for 10 min for examination of blood cells (white blood cell, red blood cell and platelet) and serum biochemistry (ALT, and AST).

4.11. H&E and Immunohistochemistry Staining

The colleting tissues were fixed in 10% buffered formalin, embedded in paraffin wax, cut into 4 µm thick sections. These sections were then stained with Mayer's hematoxylin and eosin Y solution and were observed under a light microscope after staining at ×200 magnification. Indicated protein detection was performed by immunohistochemical analysis. After antigen retrieval, the sections were treated with 10% BSA solution for blocking and 3% H_2O_2 for removing of the activity of endogenous enzymes. The primary antibodies, including 8-oxo-dG (bs-1278R, Bioss Antibodies Inc., Woburn, MA, USA.), caspase-3 (SC-98785), PCNA (SC-7907), VEGF (SC-152), VEGFR1 (SC-9029) and VEGFR2 (SC-6251), which were purchased from Santa Cruz Biotechnology (Dallas, TX, USA) were added and incubated at 4 °C overnight. Positive cells were detected with HRP-conjugated secondary antibody and visualized using 3,30-diaminobenzidine (DAB) staining. After immunostaining, sections were counterstained with hematoxylin. The experiment was randomly selected and counted ten ×200 field to achieve IHC score, which was evaluated by three experienced pathologists, independently, using quick score = intensity score × positive area score. The intensity scoring criteria: 0, any tumor cells with membrane staining at intensity of no staining; 1, weak staining; 2, moderate staining; 3, strong staining and 4, strongest staining. The percentage of positive area at the ×200 field are scored as 0 = 0%, 1 = 1–20%, 2 = 21–40%, 3 = 41–60%, 4 = 61–80% and 5 = 81–100%. Scoring of PCNA protein expressions were randomly counted the number of positive cells at ten fields under ×400 magnification and the results were presented as a percentage of the number of positive cells. The data were observed and photographed at ×400 magnification by microscope (ZEISS Axio Imager A2, Bremen, Germany).

4.12. Gas Chromatography-Mass Spectrometry Analysis

Gas chromatography-mass spectrometry (GC-MS) analyses were performed using an Agilent 7890CB gas chromatograph (AccuTOF-GCx, Jeol, MA, USA) with an Rxi-5MS capillary column (film thickness: 30 m × 0.25 mm × 0.25 µm) that was commissioned to the National Central Taiwan University Office of Research and Development's Center for Advanced Instrumentation (Hsinchu, Taiwan). The sample was diluted using hexane (1/500), the carrier gas was helium (1 mL/min), and the injector temperature was 300 °C with an injection flow rate of 1 mL/min. Components were identified by comparing their mass spectra with those obtained from authentic samples or spectra of the Wiley/Nist libraries.

4.13. Statistical Analysis

All data are presented as the mean ± SD. Statistical significance between groups was determined by Student's *t*-test. The evaluation of survival rate utilized Kaplan–Meier software. A p value < 0.05 was considered statistically significant.

Author Contributions: Conceptualization, N.-M.T.; methodology, X.-F.H., K.-F.C., and Y.-C.H.; software, X.-F.H., G.-T.S., K.-F.C., and Y.-C.H.; validation, G.-T.S., and Y.-C.H.; formal analysis, X.-F.H., G.-T.S., and K.-F.C.; investigation, X.-F.H.; resources, P.-H.H.; data curation, X.-F.H. and N.-M.T.; writing—original draft preparation, X.-F.H. and N.-M.T.; writing—review and editing, X.-F.H. and N.-M.T.; visualization, N.-M.T. and P.-H.H.; supervision, N.-M.T.; project administration, N.-M.T.; funding acquisition, N.-M.T. and P.-H.H. All authors have read and agreed to the published version of the manuscript.

Funding: The work was supported by MOST 105-2320-B-040-025 and MOST 109-2320-B-040-012 from the Ministry of Science and Technology, Taiwan and the grant number CSMU-CYC-101-01 was funded by Ditmanson Medical Foundation Chia-Yi Christian Hospital, Taiwan.

Acknowledgments: ZEISS Axio Imager A2 microscopy was performed in the Instrument Center of Chung Shan Medical University, which is supported by the National Science Council, the Ministry of Education, and Chung Shan Medical University.

Conflicts of Interest: The authors declare no conflict of interest.

References

1. Yu, W.-B.; Rao, A.; Vu, V.; Xu, L.; Rao, J.-Y.; Wu, J.-X. Management of centrally located hepatocellular carcinoma: Update 2016. *World J. Hepatol.* **2017**, *9*, 627–634. [CrossRef] [PubMed]

2. Stagos, D.; Amoutzias, G.D.; Matakos, A.; Spyrou, A.; Tsatsakis, A.M.; Kouretas, D. Chemoprevention of liver cancer by plant polyphenols. *Food Chem. Toxicol.* **2012**, *50*, 2155–2170. [CrossRef] [PubMed]

3. Kudo, M. Systemic Therapy for Hepatocellular Carcinoma: 2017 Update. *Oncology* **2017**, *93* (Suppl. 1), 135–146. [CrossRef] [PubMed]

4. Llovet, J.M.; Ricci, S.; Mazzaferro, V.; Hilgard, P.; Gane, E.; Blanc, J.F.; de Oliveira, A.C.; Santoro, A.; Raoul, J.L.; Forner, A.; et al. Sorafenib in advanced hepatocellular carcinoma. *N. Engl. J. Med.* **2008**, *359*, 378–390. [CrossRef]

5. Cheng, A.L.; Kang, Y.K.; Chen, Z.; Tsao, C.J.; Qin, S.; Kim, J.S.; Luo, R.; Feng, J.; Ye, S.; Yang, T.S.; et al. Efficacy and safety of sorafenib in patients in the Asia-Pacific region with advanced hepatocellular carcinoma: A phase III randomised, double-blind, placebo-controlled trial. *Lancet Oncol.* **2009**, *10*, 25–34. [CrossRef]

6. Tsai, N.M.; Lin, S.Z.; Lee, C.C.; Chen, S.P.; Su, H.C.; Chang, W.L.; Harn, H.J. The antitumor effects of Angelica sinensis on malignant brain tumors in vitro and in vivo. *Clin. Cancer. Res.* **2005**, *11*, 3475–3484. [CrossRef]

7. Tsai, N.M.; Chen, Y.L.; Lee, C.C.; Lin, P.C.; Cheng, Y.L.; Chang, W.L.; Lin, S.Z.; Harn, H.J. The natural compound n-butylidenephthalide derived from Angelica sinensis inhibits malignant brain tumor growth in vitro and in vivo. *J. Neurochem.* **2006**, *99*, 1251–1262. [CrossRef]

8. Abdel-Hamid, N.M.; Abass, S.A.; Mohamed, A.A.; Muneam Hamid, D. Herbal management of hepatocellular carcinoma through cutting the pathways of the common risk factors. *Biomed. Pharmacother.* **2018**, *107*, 1246–1258. [CrossRef]

9. Feng, X.X.; Yu, X.T.; Li, W.J.; Kong, S.Z.; Liu, Y.H.; Zhang, X.; Xian, Y.F.; Zhang, X.J.; Su, Z.R.; Lin, Z.X. Effects of topical application of patchouli alcohol on the UV-induced skin photoaging in mice. *Eur. J. Pharm. Sci.* **2014**, *63*, 113–123. [CrossRef]

10. Sangeet Pilkhwal Sah, C.S.M. Kanwaljit Chopra, Antidepressant effect of Valeriana wallichii patchouli alcohol chemotype in mice: Behavioural and biochemical evidence. *J. Ethnopharmacol.* **2011**, *135*, 197–200. [CrossRef]

11. Sah, S.P.; Mathela, C.S.; Chopra, K. Involvement of nitric oxide (NO) signalling pathway in the antidepressant activity of essential oil of Valeriana wallichii Patchouli alcohol chemotype. *Phytomedicine* **2011**, *18*, 1269–1275. [CrossRef] [PubMed]

12. Kocevski, D.; Du, M.; Kan, J.; Jing, C.; Lačanin, I.; Pavlović, H. Antifungal effect of Allium tuberosum, Cinnamomum cassia, and Pogostemon cablin essential oils and their components against population of Aspergillus species. *J. Food Sci.* **2013**, *78*, M731–M737. [CrossRef] [PubMed]

13. Farisa Banu, S.; Rubini, D.; Shanmugavelan, P.; Murugan, R.; Gowrishankar, S.; Karutha Pandian, S.; Nithyanand, P. Effects of patchouli and cinnamon essential oils on biofilm and hyphae formation by Candida species. *J. Mycol. Med.* **2018**, *28*, 332–339. [CrossRef] [PubMed]

14. Wu, X.L.; Ju, D.H.; Chen, J.; Yu, B.; Liu, K.L.; He, J.X.; Dai, C.Q.; Wu, S.; Chang, Z.; Wang, Y.P.; et al. Immunologic Mechanism of Patchouli Alcohol Anti-H1N1 Influenza Virus May through Regulation of the RLH Signal Pathway In Vitro. *Curr. Microbiol.* **2013**, *67*, 431–436. [CrossRef] [PubMed]

15. Xie, J.; Lin, Z.; Xian, Y.; Kong, S.; Lai, Z.; Ip, S.; Chen, H.; Guo, H.; Su, Z.; Yang, X.; et al. (-)-Patchouli alcohol protects against Helicobacter pylori urease-induced apoptosis, oxidative stress and inflammatory response in human gastric epithelial cells. *Int. Immunopharmacol.* **2016**, *35*, 43–52. [CrossRef] [PubMed]

16. Liang, J.; Dou, Y.; Wu, X.; Li, H.; Wu, J.; Huang, Q.; Luo, D.; Yi, T.; Liu, Y.; Su, Z.; et al. Prophylactic efficacy of patchoulene epoxide against ethanol-induced gastric ulcer in rats: Influence on oxidative stress, inflammation and apoptosis. *Chem. Biol. Interact.* **2018**, *283*, 30–37. [CrossRef] [PubMed]

17. Wu, J.Z.; Liu, Y.H.; Liang, J.L.; Huang, Q.H.; Dou, Y.X.; Nie, J.; Zhuo, J.Y.; Wu, X.; Chen, J.N.; Su, Z.R.; et al. Protective role of beta-patchoulene from Pogostemon cablin against indomethacin-induced gastric ulcer in rats: Involvement of anti-inflammation and angiogenesis. *Phytomedicine* **2018**, *39*, 111–118. [CrossRef] [PubMed]

18. Han, X.; Beaumont, C.; Stevens, N. Chemical composition analysis and in vitro biological activities of ten essential oils in human skin cells. *Biochimie Open* **2017**, *5*, 1–7. [CrossRef]

19. Tsai, C.C.; Chang, Y.H.; Chang, C.C.; Cheng, Y.M.; Ou, Y.C.; Chien, C.C.; Hsu, Y.C. Induction of Apoptosis in Endometrial Cancer (Ishikawa) Cells by Pogostemon cablin Aqueous Extract (PCAE). *Int. J. Mol. Sci.* **2015**, *16*, 12424–12435. [CrossRef]

20. Chien, J.-H.; Lee, S.-C.; Chang, K.-F.; Huang, X.-F.; Chen, Y.-T.; Tsai, N.-M. Extract of Pogostemon cablin Possesses Potent Anticancer Activity against Colorectal Cancer Cells In Vitro and In Vivo. *Evid. Based Complement. Alternat. Med.* **2020**, *2020*, 9758156. [CrossRef]

21. Singh, K.; Gangrade, A.; Jana, A.; Mandal, B.B.; Das, N. Design, Synthesis, Characterization, and Antiproliferative Activity of Organoplatinum Compounds Bearing a 1,2,3-Triazole Ring. *ACS Omega* **2019**, *4*, 835–841. [CrossRef]

22. Keller, D.M.; Zeng, X.; Wang, Y.; Zhang, Q.H.; Kapoor, M.; Shu, H.; Goodman, R.; Lozano, G.; Zhao, Y.; Lu, H. A DNA damage-induced p53 serine 392 kinase complex contains CK2, hSpt16, and SSRP1. *Mol. Cell* **2001**, *7*, 283–292. [CrossRef]

23. Sova, H.; Jukkola-Vuorinen, A.; Puistola, U.; Kauppila, S.; Karihtala, P. 8-Hydroxydeoxyguanosine: A new potential independent prognostic factor in breast cancer. *Br. J. Cancer* **2010**, *102*, 1018–1023. [CrossRef] [PubMed]

24. Lee, S.F.; Pervaiz, S. Assessment of oxidative stress-induced DNA damage by immunoflourescent analysis of 8-oxodG. *Methods Cell Biol.* **2011**, *103*, 99–113. [CrossRef] [PubMed]

25. Costantini, S.; Di Bernardo, G.; Cammarota, M.; Castello, G.; Colonna, G. Gene expression signature of human HepG2 cell line. *Gene* **2013**, *518*, 335–345. [CrossRef]

26. Lin, Y.; Ong, L.K.; Chan, S.H. Differential in situ hybridization for determination of mutational specific expression of the p53 gene in human hepatoma cell lines. *Pathology* **1995**, *27*, 191–196. [CrossRef] [PubMed]

27. Hussain, S.P.; Schwank, J.; Staib, F.; Wang, X.W.; Harris, C.C. TP53 mutations and hepatocellular carcinoma: Insights into the etiology and pathogenesis of liver cancer. *Oncogene* **2007**, *26*, 2166–2176. [CrossRef]

28. Oda, T.; Tsuda, H.; Scarpa, A.; Sakamoto, M.; Hirohashi, S. p53 gene mutation spectrum in hepatocellular carcinoma. *Cancer Res.* **1992**, *52*, 6358–6364.

29. Honda, K.; Sbisa, E.; Tullo, A.; Papeo, P.A.; Saccone, C.; Poole, S.; Pignatelli, M.; Mitry, R.R.; Ding, S.; Isla, A.; et al. p53 mutation is a poor prognostic indicator for survival in patients with hepatocellular carcinoma undergoing surgical tumour ablation. *Br. J. Cancer* **1998**, *77*, 776–782. [CrossRef]

30. Tornesello, M.L.; Buonaguro, L.; Tatangelo, F.; Botti, G.; Izzo, F.; Buonaguro, F.M. Mutations in TP53, CTNNB1 and PIK3CA genes in hepatocellular carcinoma associated with hepatitis B and hepatitis C virus infections. *Genomics* **2013**, *102*, 74–83. [CrossRef]

31. Karimian, A.; Ahmadi, Y.; Yousefi, B. Multiple functions of p21 in cell cycle, apoptosis and transcriptional regulation after DNA damage. *DNA Repair* **2016**, *42*, 63–71. [CrossRef] [PubMed]

32. Sherr, C.J.; McCormick, F. The RB and p53 pathways in cancer. *Cancer Cell* **2002**, *2*, 103–112. [CrossRef]

33. Abbas, T.; Dutta, A. p21 in cancer: Intricate networks and multiple activities. *Nat. Rev. Cancer* **2009**, *9*, 400–414. [CrossRef] [PubMed]

34. Liu, B.; Chen, Y.; St Clair, D.K. ROS and p53: A versatile partnership. *Free Radic. Biol. Med.* **2008**, *44*, 1529–1535. [CrossRef] [PubMed]

35. Redza-Dutordoir, M.; Averill-Bates, D.A. Activation of apoptosis signalling pathways by reactive oxygen species. *Biochim. Biophys. Acta* **2016**, *1863*, 2977–2992. [CrossRef] [PubMed]

36. Simon, H.U.; Haj-Yehia, A.; Levi-Schaffer, F. Role of reactive oxygen species (ROS) in apoptosis induction. *Apoptosis* **2000**, *5*, 415–418. [CrossRef] [PubMed]

37. Lee, S.H.; Shin, M.S.; Lee, H.S.; Bae, J.H.; Lee, H.K.; Kim, H.S.; Kim, S.Y.; Jang, J.J.; Joo, M.; Kang, Y.K.; et al. Expression of Fas and Fas-related molecules in human hepatocellular carcinoma. *Hum. Pathol.* **2001**, *32*, 250–256. [CrossRef] [PubMed]

38. Jodo, S.; Kobayashi, S.; Nakajima, Y.; Matsunaga, T.; Nakayama, N.; Ogura, N.; Kayagaki, N.; Okumura, K.; Koike, T. Elevated serum levels of soluble Fas/APO-1 (CD95) in patients with hepatocellular carcinoma. *Clin. Exp. Immunol.* **1998**, *112*, 166–171. [CrossRef] [PubMed]

39. Cande, C.; Cohen, I.; Daugas, E.; Ravagnan, L.; Larochette, N.; Zamzami, N.; Kroemer, G. Apoptosis-inducing factor (AIF): A novel caspase-independent death effector released from mitochondria. *Biochimie* **2002**, *84*, 215–222. [CrossRef]

40. Man, Y.N.; Liu, X.H.; Wu, X.Z. Chinese medicine herbal treatment based on syndrome differentiation improves the overall survival of patients with unresectable hepatocellular carcinoma. *Chin. J. Integr. Med.* **2015**, *21*, 49–57. [CrossRef]

41. Shu, X.; McCulloch, M.; Xiao, H.; Broffman, M.; Gao, J. Chinese herbal medicine and chemotherapy in the treatment of hepatocellular carcinoma: A meta-analysis of randomized controlled trials. *Integr. Cancer Ther.* **2005**, *4*, 219–229. [CrossRef] [PubMed]

42. Moriwaki, H. Prevention of liver cancer: Basic and clinical aspects. *Exp. Mol. Med.* **2002**, *34*, 319–325. [CrossRef] [PubMed]

43. Coriat, R.; Nicco, C.; Chéreau, C.; Mir, O.; Alexandre, J.; Ropert, S.; Weill, B.; Chaussade, S.; Goldwasser, F.; Batteux, F. Sorafenib-induced hepatocellular carcinoma cell death depends on reactive oxygen species production in vitro and in vivo. *Mol. Cancer Ther.* **2012**, *11*, 2284–2293. [CrossRef] [PubMed]

44. Zhang, J.; Wang, X.; Vikash, V.; Ye, Q.; Wu, D.; Liu, Y.; Dong, W. ROS and ROS-Mediated Cellular Signaling. *Oxid. Med. Cell. Longev.* **2016**, *2016*, 4350965. [CrossRef]

45. Zhou, Q.; Lui, V.W.; Yeo, W. Targeting the PI3K/Akt/mTOR pathway in hepatocellular carcinoma. *Future Oncol.* **2011**, *7*, 1149–1167. [CrossRef]

46. Zhai, B.; Hu, F.; Jiang, X.; Xu, J.; Zhao, D.; Liu, B.; Pan, S.; Dong, X.; Tan, G.; Wei, Z.; et al. Inhibition of Akt reverses the acquired resistance to sorafenib by switching protective autophagy to autophagic cell death in hepatocellular carcinoma. *Mol. Cancer Ther.* **2014**, *13*, 1589–1598. [CrossRef]

47. Zhang, Z.; Zhou, X.; Shen, H.; Wang, D.; Wang, Y. Phosphorylated ERK is a potential predictor of sensitivity to sorafenib when treating hepatocellular carcinoma: Evidence from an in vitro study. *BMC Med.* **2009**, *7*, 41. [CrossRef]

48. Tseng, P.L.; Tai, M.H.; Huang, C.C.; Wang, C.C.; Lin, J.W.; Hung, C.H.; Chen, C.H.; Wang, J.H.; Lu, S.N.; Lee, C.M.; et al. Overexpression of VEGF is associated with positive p53 immunostaining in hepatocellular carcinoma (HCC) and adverse outcome of HCC patients. *J. Surg. Oncol.* **2008**, *98*, 349–357. [CrossRef]

49. Zhang, L.; Wang, J.N.; Tang, J.M.; Kong, X.; Yang, J.Y.; Zheng, F.; Guo, L.Y.; Huang, Y.Z.; Zhang, L.; Tian, L.; et al. VEGF is essential for the growth and migration of human hepatocellular carcinoma cells. *Mol. Biol. Rep.* **2012**, *39*, 5085–5093. [CrossRef]

50. Mise, M.; Arii, S.; Higashituji, H.; Furutani, M.; Niwano, M.; Harada, T.; Ishigami, S.; Toda, Y.; Nakayama, H.; Fukumoto, M.; et al. Clinical significance of vascular endothelial growth factor and basic fibroblast growth factor gene expression in liver tumor. *Hepatology* **1996**, *23*, 455–464. [CrossRef]

51. Zhan, P.; Qian, Q.; Yu, L.K. Prognostic significance of vascular endothelial growth factor expression in hepatocellular carcinoma tissue: A meta-analysis. *Hepatobiliary Surg. Nutr.* **2013**, *2*, 148–155. [CrossRef] [PubMed]

52. Jeong, J.B.; Choi, J.; Lou, Z.; Jiang, X.; Lee, S.H. Patchouli alcohol, an essential oil of Pogostemon cablin, exhibits anti-tumorigenic activity in human colorectal cancer cells. *Int. Immunopharmacol.* **2013**, *16*, 184–190. [CrossRef] [PubMed]

53. Lu, X.; Yang, L.; Lu, C.; Xu, Z.; Qiu, H.; Wu, J.; Wang, J.; Tong, J.; Zhu, Y.; Shen, J. Molecular Role of EGFR-MAPK Pathway in Patchouli Alcohol-Induced Apoptosis and Cell Cycle Arrest on A549 Cells In Vitro and In Vivo. *Biomed. Res. Int.* **2016**, *2016*, 4567580. [CrossRef] [PubMed]

54. Ayaz, F.; Yuzer, A.; Ince, T.; Ince, M. Anti-Cancer and Anti-Inflammatory Activities of Bromo- and Cyano-Substituted Azulene Derivatives. *Inflammation* **2020**, *43*, 1009–1018. [CrossRef] [PubMed]

55. Colerangle, J.B. Chapter 25: Preclinical Development of Nononcogenic Drugs (Small and Large Molecules). In *A Comprehensive Guide to Toxicology in Nonclinical Drug Development*, 2nd ed.; Faqi, A.S., Ed.; Academic Press: Boston, MA, USA, 2017; pp. 659–683.

Sample Availability: Samples of the essential oils of *Pogostemon Cablin* are available from the authors.

Publisher's Note: MDPI stays neutral with regard to jurisdictional claims in published maps and institutional affiliations.

Article

Implication of Lactucopicrin in Autophagy, Cell Cycle Arrest and Oxidative Stress to Inhibit U87Mg Glioblastoma Cell Growth

Rossella Rotondo [1], Maria Antonietta Oliva [2], Sabrina Staffieri [2], Salvatore Castaldo [2], Felice Giangaspero [2,3] and Antonietta Arcella [2,*

[1]   Department of Molecular Medicine and Medical Biotechnologies, University of Naples Federico II, 80131 Naples, Italy; rossellaross1988@gmail.com
[2]   I.R.C.C.S Neuromed, Via Atinense, 18, 86077 Pozzilli IS, Italy; mariaantonietta.oliva@neuromed.it (M.A.O.); sabrina.staffieri@neuromed.it (S.S.); castaldosal90@gmail.com (S.C.); felice.giangaspero@uniroma1.it (F.G.)
[3]   Department of Radiologic, Oncologic and Anatomo Pathological Sciences, University of Rome La Sapienza, 00185 Rome, Italy
*   Correspondence: arcella@neuromed.it; Tel.: +39-0865-915220

Academic Editor: José Antonio Lupiáñez
Received: 19 November 2020; Accepted: 8 December 2020; Published: 10 December 2020

Abstract: In this study, we propose lactucopicrin (LCTP), a natural sesquiterpene lactone from Lactucavirosa, as a molecule able to control the growth of glioblastoma continuous cell line U87Mg. The IC50 of U87Mg against LCTP revealed a strong cytotoxic effect. Daily administration of LCTP showed a dose and time-dependent reduction of GBM cell growth and viability, also confirmed by inhibition of clonogenic potential and mobility of U87Mg cells. LCTP activated autophagy in U87Mg cells and decreased the phosphorylation of proliferative signals pAKT and pERK. LCTP also induced the cell cycle arrest in G2/M phase, confirmed by decrease of CDK2 protein and increase of p53 and p21. LCTP stimulated apoptosis as evidenced by reduction of procaspase 6 and the increase of the cleaved/full-length PARP ratio. The pre-treatment of U87Mg cells with ROS scavenger N-acetylcysteine (NAC), which reversed its cytotoxic effect, showed the involvement of LCTP in oxidative stress. Finally, LCTP strongly enhanced the sensitivity of U87Mg cells to canonical therapy Temozolomide (TMZ) and synergized with this drug. Altogether, the growth inhibition of U87Mg GBM cells induced by LCTP is the result of several synergic mechanisms, which makes LCTP a promising adjuvant therapy for this complex pathology.

Keywords: glioblastoma (GBM); lactucopicrin (LCTP); temozolomide (TMZ); autophagy; oxidative stress; NF-κB; p62/SQSM1

1. Introduction

Glioblastoma (GBM) is one of the most lethal brain tumors in adults, with survival rates which have remained substantially unchanged for 30 years. Common histological features of GBM comprise marked mitotic activity, high angiogenesis, cellular heterogeneity, necrosis, and pronounced proliferative rates [1]. Furthermore, the presence of cancer stem cells, able to proliferate and generate glial neoplastic cells [2,3], contributes to the unfavorable prognosis of GBM patients, whose median survival is approximately 14 months [4]. Gross-total resection of tumor tissue, followed by adjuvant chemo- and radiotherapy, remains the standard of care [5]. Numerous efforts have been made to identify the molecular pathways and potential "druggable" targets involved in gliomagenesis [6].

Temozolomide (TMZ) is a first-line chemotherapy that has significantly improved the prognosis of GBM patients [7]. However, development of TMZ resistance in some GBM patients during the therapeutic program limits the treatment of this brain tumor. The main mechanism of GBM resistance involves O6-methylguanine-DNA methyltransferase (MGMT) enzyme repair. MGMT acts by removing the methyl group attached to the O6 position in guanine from DNA of cancer cells, favoring a resistant phenotype [8]. However, other mechanisms contribute significantly to TMZ resistance [9]. Several molecules have been developed to overcome TMZ resistance, such as O6-benzyl-guanine which inhibits MGMT, tyrosine kinase inhibitors which modulate the epidermal growth factor receptor (EGFR) commonly overexpressed in GBMs, and nutlin-3 which inhibits murine double minute 2 (Mdm2), that in turn leads to restoration of p53 activity.

Due to the failure of classical chemotherapies and targeted drugs, research efforts are focusing on natural compounds that can overcome BBB, inhibit tumor growth and are able to promote the activity involving multiple pathways [10]. In fact, since multiple pathways are associated with TMZ resistance in GBM [11], the treatment of GMB with a selective target drug that blocks the activation of a single pathway could result in a compensatory mechanism that leads to the restoration of signals, conferring a TMZ resistance [12]. Interestingly, multiple natural compounds have already shown antitumor and apoptotic effects in TMZ resistant GBM cell lines and also displayed synergistic affects with TMZ [10,13,14]. Sesquiterpene lactones are a large and different group of biologically active plant compounds with anti-inflammatory and anti-tumor activity. Sesquiterpene lactones are secondary metabolites of the Asteraceae family, which exhibits great structural diversity and a broad range of biological activities [15,16]. As a member of the sesquiterpene superfamily, they are composed by 15-carbon skeletons, consisting of colorless, crisp and dry lipophilic compounds made up of three isopropyl units [17,18]. The biological activities of sesquiterpene lactones are due to the presence of α-methylene-γ-lactone, which reacts with nucleophilic structure of multiple targets through Michael's reaction, such as the thiol group of cysteinyl residues [17]. Therefore, they can exert their effects through the alkylation of transcription factors and various enzymes, interfering with several molecular pathways [19]. In the last few years, extensive research has confirmed the anticancer activity of different compounds within the sesquiterpene lactones and much effort has been undertaken to clarify the molecular mechanism and chemotherapeutic potential of some of these compounds in GMB treatment [20–22]. Very recently, Zhang et al. reported that lactucopicrin (or intybin), a secondary metabolite of lactucarium, derived from the plant *Lactuca virosa* (wild lettuce) and found in some related plants, such as Cichorium intybus, exhibits its anticancer activity in SKMEL-5 human skin cancer cells [23] and growth inhibition of Saos-2 osteosarcoma cells [24].

For a long time, lactucarium, a milky and bitter juice secreted by stem secretion of these plants, once dried, has been used as an opium substitute for its analgesic and sedative properties. Therefore, considering the ability of lactucopicrin to act on central nervous system, it can be speculated that this molecule can cross the BBB acting as multi-target molecule, potentially interfering with GMB cell growth. Therefore, we studied the effect of this natural substance to exploit the novel vulnerabilities of human glioblastoma cell line U87Mg.

2. Results

2.1. Dose-Response and Time-Course of LCTP Effects on Glioblastoma U87Mg Cells

The cytotoxic effects of LCTP on glioblastoma were investigated using continuous glioblastoma cell line U87Mg and estimating the IC50 of LCTP at 24, 48, and 72 h. As it is evident in Figure 1B, the IC50 of LCTP against U87Mg cells decreased from 12.5 ± 1.1 µM (5.1 ± 0.5 µg/mL) at 24 h to 3.6 ± 1.1 µM (1.5 ± 0.5 µg/mL) at 72 h. Considering the strong cytotoxicity of this molecule, LCTP was daily administered at concentrations of 7.5 and 10 µM at various time intervals (24, 48 and 72 h). We found that LCTP exhibited a statistically significant time- and dose-dependent growth inhibition of U87Mg, as measured by direct cell count (Figure 1D), which resulted in a growth rate inhibition that increased from 60% and 82% at 48 h, respectively, with LCTP 7.5 and 10 µM to 85% and 94% of inhibition at 72 h of treatment. Cell viability assay and morphological changes were also reported in Figure 1C,E. According to the growth rate inhibition, LCTP inhibited the viability of U87Mg in a time and dose-dependent manner, with about 50% of reduction at 72 h of treatment with LCTP 10 μM. LCTP induced dramatic morphological changes of U87Mg, visible at microscopically observation as rounded-shaped with a loss of filaments and cell shrinkage (Figure 1C).

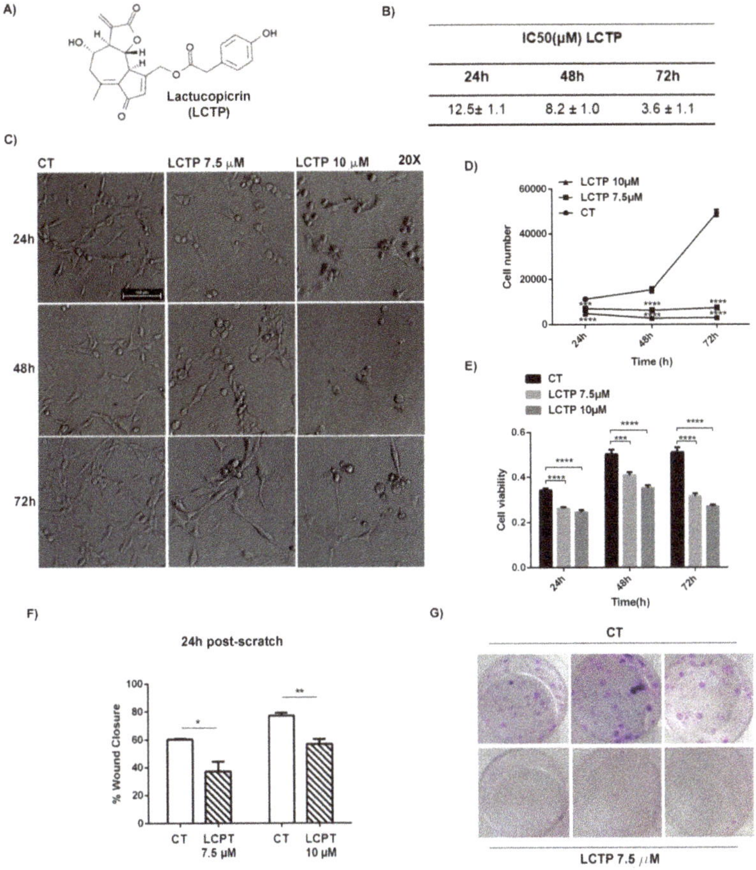

Figure 1. (**A**) Chemical structure of LCTP. (**B**) IC50-values of U87Mg cells after 24, 48 and 72 h of incubation with LCTP. (**C**) Morphological changes of U87Mg glioblastoma cells treated every 24 h with LCTP 7.5 and 10 μM for 24, 48 and 72 h. Magnification 20×. (**D**) Time and dose-dependent growth inhibition of LCTP-treated U87Mg cells. (**E**) Cytotoxic effect of various concentrations of LCTP (7.5 and 10 μM) at 24, 48 and 72 h post-treatment on U87Mg cell line. (**F**) Quantification of the cell-free area in wound healing assay at T0 and 24 h post-scratch in control and LCTP-treated U87Mg cells. (**G**) Colony-forming assay of U87Mg treated 24 h with LCTP 7.5. μM. For all the experiments, values are the means ± SEM of 3 individual determinations. Unpaired *t*-test, *p*-value < 0.05. According to GraphPad Prism, * *p*-value 0.01 to 0.05 (significant), ** *p*-value 0.001 to 0.01 (very significant), *** *p*-value 0.0001 to 0.001 (Extremely significant), **** *p*-value < 0.0001 (Extremely significant).

2.2. LCTP Interferes with Clonogenic Survival and Motility Capacity of GBM Cells

To further investigate the growth inhibition effects of LCTP, a colony assay of U87Mg treated with LCTP and control vehicle (DMSO) was performed. LCTP dramatically reduced the colony forming of U87Mg; at the lowest concentration used (7.5 μM), this natural molecule completely suppressed clonogenic growth (Figure 1G). LCTP also affected the cell motility of U87Mg, as it was demonstrated by the wound healing assay. The plot in Figure 1F showed that the percentage of wound closure significantly decreased with the increase of LCTP concentrations at 24 h post-scratch (Supplementary Figure S1).

2.3. Rapid Autophagy Response of U87Mg to LCTP Treatment Potentially Remodels the Cytoskeleton Proteins

The molecular mechanisms underlying LCTP-induced cytotoxicity were investigated exposing U87Mg cells to short-term treatment with LCTP 10 μM. In the Western blot analysis, in Figure 2A, the strong reduction of autophagic substrate p62/SQSM1 appears evident, already being visible after 30 min of induction followed by the increased of cleaved LC3BII, clearly indicating the activation of autophagy in GBM cells treated with LCTP, with respect to control cells. Autophagy activation may be sustained by a deep dephosphorylation of ^{Ser473}pAKT and pERK1/2 proteins in a LCTP- short-term treatment of U87Mg cells (Figure 2B). Interestingly, LCTP induced a strong cytoskeleton rearrangement in U87Mg cells, as it can be appreciated by immunofluorescence staining for intermediated filament Vimentin and the α-tubulin subunit of microtubules (Figure 2C).

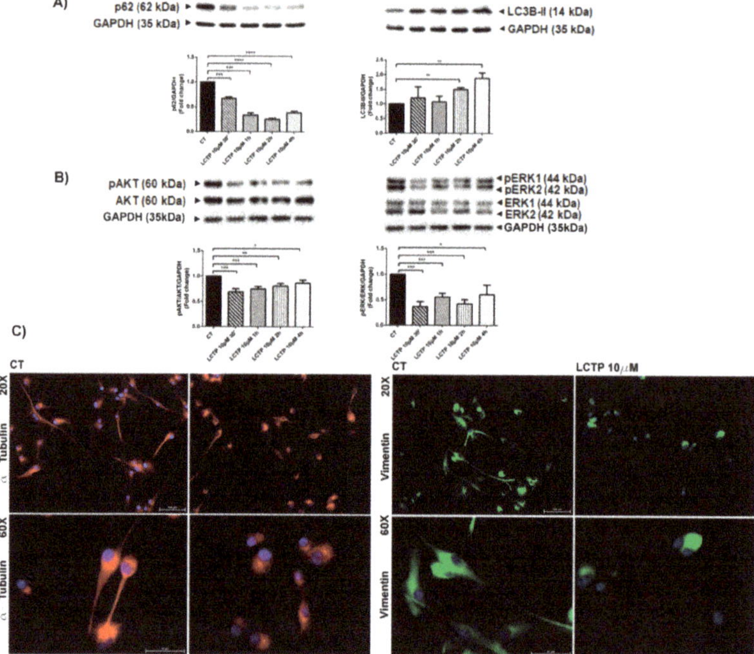

Figure 2. (**A**) Western blot analysis of autophagy-associated proteins p62 and LC3BII in short-termtreated-U87Mg cells with LCTP 10 μM. Densitometric analysis of protein levels represent the means ± SEM of 3 individual determinations. Data were normalized to GAPDH and are expressed as fold change over control-treated cells. * Unpaired *t*-test, *p*-value < 0.05. (**B**) Western blot analysis of proliferative signals pERK/ERK/GAPDH and pAKT/AKT/GAPDH in short-termtreated-U87Mg cells with LCTP 10 μM. Densitometric analysis of protein levels represent the means ± SEM of 3 individual determinations and are expressed as fold change over control-treated cells, normalized to GAPDH. * Unpaired *t*-test, *p*-value < 0.05. According to GraphPad Prism, * *p*-value 0.01 to 0.05 (significant), ** *p*-value 0.001 to 0.01 (very significant), *** *p*-value 0.0001 to 0.001 (Extremely significant), **** *p*-value < 0.0001 (Extremely significant). (**C**) Immunofluorescence analysis of cytoskeleton proteins α-tubulin and Vimentin in U87Mgcells treated with LCTP 10 μM and vehicle control (DMSO 0.08%) for 20 min. Magnification 20× and 60×.

2.4. Cell Cycle Arrest in G2/M Phase and Apoptosis Induction by LCTP in U87Mg Cells

To clarify the mechanism that regulates GBM cell growth inhibition, the long-term effects of LCTP on cell cycle distribution were investigated by flow cytometry. The analysis revealed that the cell cycle arrest in G2/M phase in LCTP-induced U87Mg cells was already visible at 24 h post-treatment (Figure 3A). The cell cycle distribution showed that the arrest in the G2/M phase—and concomitant decrease of cell percentage at the G0/G1 phase—is time-dependent (Figure 3B). In order to clarify the molecular mechanism underlying the growth inhibition of the U87Mg cells, the expression levels of key proteins which regulate the cell cycle were determined by Western blot analysis. As shown in Figure 3C, the long-term treatment of U87Mg cells with LCTP 7.5 μM induced the activation of CDK inhibitor p21, which was already up-regulated after 24 h of treatment. The tumor suppressor p53 was strongly activated at 72 h post-treatment in accordance, with the pronounced accumulation of cell population in the G2/M phase at the same time point (Figure 3B,C). The up-regulation of p21 and p53, after 72 h, is correlated to CDK2 inhibition in LCTP-treated cells, confirming the cytostatic activity of LCTP in U87Mg cells (Figure 3D). In the same conditions, to further investigate whether the anti-proliferative effect of LCTP was accompanied by the induction of cell death, Western blot analysis of apoptotic proteins was performed. In U87Mg-treated cells at 48 and 72 h post-treatment, procaspase 6 decreased, while cleaved/full-length ratio of Poly(ADP)ribose polymerase (PARP) significantly increased, clearly indicating the activation of programmed cell death by LCTP (Figure 3E,F). Loss of repair activity of cleaved PARP was also confirmed by the appearance of DNA fragments, as shown on 2% agarose gel electrophoresis at 72 h of LCTP-treatment (Figure 3G).

2.5. Involvement of Oxidative Stress in LCTP-Mediated Cytotoxicity in U87Mg Cells

The involvement of oxidative stress in LCTP-induced cytotoxicity is due to the presence of highly reactive α-methylene-γ-lactone group. This group is able to react with nucleophilic structure of multiple targets through Michael's reaction, such as the thiol group of cysteinyl residues [17]. The oxidative stress LCTP-induced cytotoxicity was demonstrated by pre-incubating U87Mg with N-acetylcysteine. Pre-treatment with ROS scavenger NAC reverted the LCTP effects on U87Mg cell viability and preserved cell morphology (Figure 4A,B). These results suggest that LCTP induces a redox imbalance that potentially mediates its anti-proliferative activity. ROS can influence tumor cell malignancy via the redox-regulated transcription factor NF-κB [25]. Therefore, Western blot analysis of short-term LCTP-treated GBM cells was performed to assess the expression levels of NF-κB p65 subunit. The down-regulation of NF-κB p65 expression upon LCTP treatment was already visible after 30 min of incubation with the drug and persisted throughout all of the time tested (1 h, 2 h, and 4 h) (Figure 4C). The expression of NF-κb p65 at long-term treatment was also tested (Supplementary Figure S2).

2.6. LCTP Enhances the Sensitivity of U87Mg to Canonical Therapy Temozolomide

To assess whether the effects of LCTP can influence the response to TMZ, U87Mg cells were pre-treated with LCTP 7.5 μM and 10 μM for 24 h and IC50 for TMZ at 24 h, which was determined to be compared with the standard IC50. As shown in Figure 5A,B, the IC50 of TMZ at 24 h was about 18 and 32 times higher compared to IC50 for TMZ at 24 h when U87Mg cells were pre-treated with LCTP (7.5 μM and 10 μM). It is worth noting that the lower IC50 of U87Mg to LCTP if compared with TMZ at 24 and 48 h, indicating a strong effect of LCTP in a short-term period, which could explain the rapid activation of autophagic pathway. At 72 h, the IC50 for TMZ and LCTP of U87Mg cells became comparable (Figures 1B and 5A).

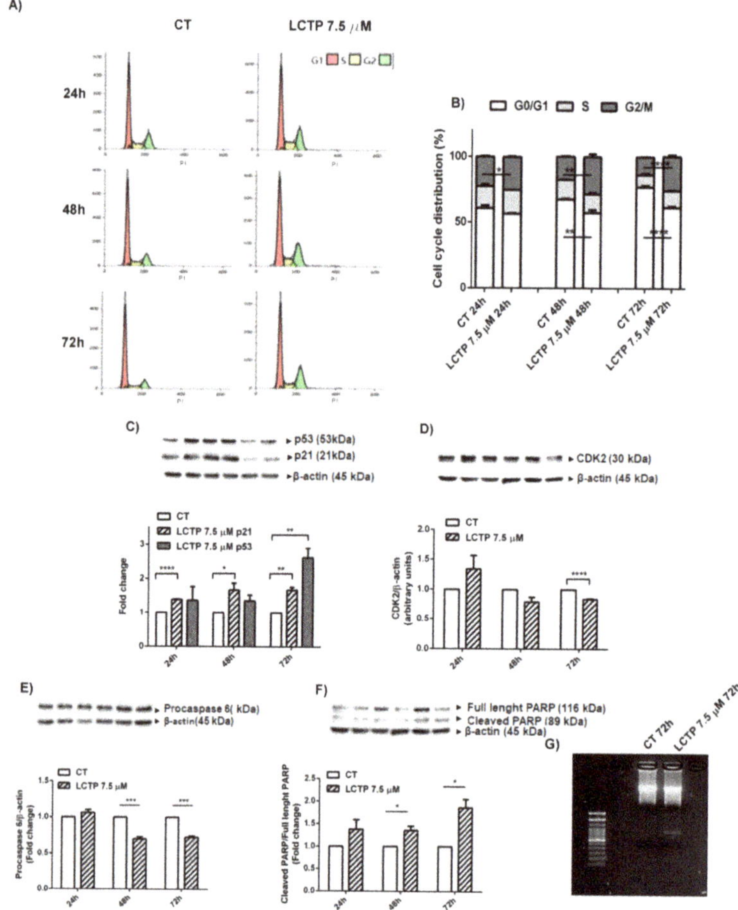

Figure 3. (**A**) Representative flow cytometry analysis of cell cycle arrest in G2/M and (**B**) plot of cell cycle distribution of propidium iodide (PI) staining in U87Mg, vehicle (left panel) and LCTP (7.5 µM) treated cells (right panel) for 24, 48 and 72 h. Values are the means ± SEM of 3 individual determinations. * Unpaired *t*-test, *p*-value < 0.05. (**C**) Western blot analysis of p53 and p21 in long-term LCTP–treated U87Mg cells. Data were normalized to β-actin and are expressed as fold change over control-treated cells of 3 individual determinations. * Unpaired *t*-test, *p*-value < 0.05. (**D**) Western blot analysis of CDK2: the figure shows CDK2 analysis of U87Mg cells treated with LCTP 7.5 µM at different time (24, 48, and 72 h). Data were normalized to β-actin and are expressed as fold change over control-treated cells of 3 individual determinations. * Unpaired *t*-test, *p*-value < 0.05. (**E**) Western blot and densitometric analysis of procaspase 6 of LCTP-treated U87Mg cells at different time (24, 48 and 72 h). Values are the means ± SEM of 3 individual determinations and protein expression levels, normalized to β-actin, are expressed as fold change over control-treated cells. * Unpaired *t*-test, *p*-value < 0.05. (**F**) Western blot analysis of PARP in LCTP-treated U87Mg cells at different time (24, 48 and 72 h). Values as ratio of cleaved PARP to full-length PARP represent the means ± SEM of 3 individual determinations and are expressed as fold change over control-treated cells. * Unpaired *t*-test, *p*-value < 0.05. According to GraphPad Prism, * *p*-value 0.01 to 0.05 (significant), ** *p*-value 0.001 to 0.01 (very significant), *** *p*-value 0.0001 to 0.001 (Extremely significant), **** *p*-value < 0.0001 (Extremely significant). (**G**) DNA laddering of long-term treatment of U87Mg cells with LCTP 7.5 with respect to untreated cells.

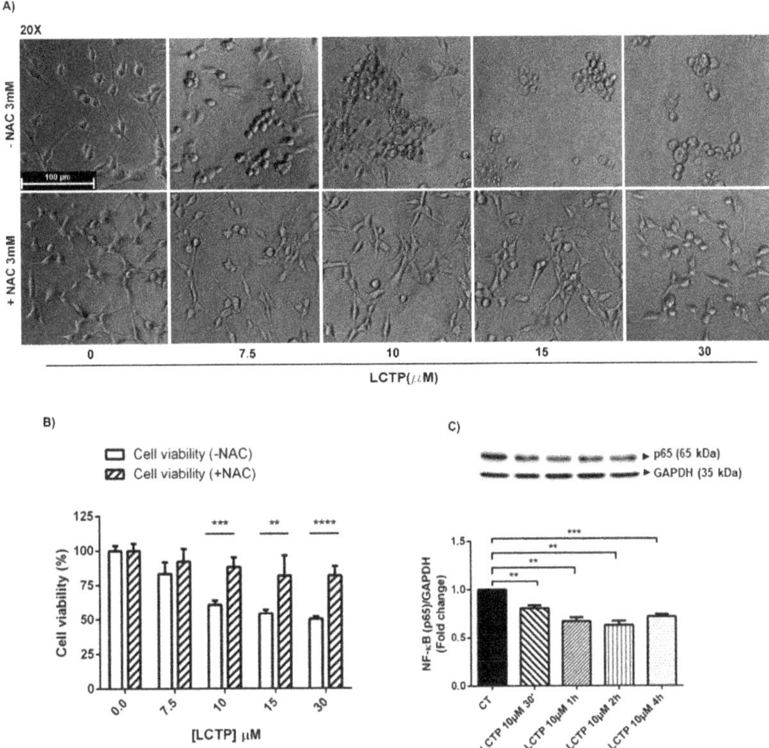

Figure 4. (**A**) Effect of pre-treatment with ROS scavenger NAC 3 mM for 4 h on morphological changes and (**B**) cell viability (%) induced by different concentrations of LCTP (7.5, 10, 15 and 30 µM) on U87Mg cells at 24 h from treatment. Magnification 20×. (**C**) Western blot analysis of NF-κB p65 subunit of short-term treated U87Mg cells with LCTP 7.5 µM (30 min, 1 h, 2 h and 4 h). Data were normalized to GAPDH and are expressed as fold change over control-treated cells of 3 individual determinations. According to GraphPad Prism, ** *p*-value 0.001 to 0.01 (very significant), *** *p*-value 0.0001 to 0.001 (Extremely significant), **** *p*-value < 0.0001 (Extremely significant).

2.7. Synergist Effect of LCTP and TMZ Affects the Cell Growth and Viability

In order to propose LCTP as an adjuvant therapy for GBM, in combination with conventional chemotherapy TMZ, U87Mg cells were treated every 24 h with LCTP 7.5 µM and 10 µM in combination with TMZ 1 µM for 24, 48. and 72 h. The effects of the combined therapy were assessed on cell growth and viability. Concomitant treatment with the two drugs significantly increased the effect of TMZ alone, inhibiting the replicative potential of U87Mg cells already from 24 h post-treatment (Figure 5C,D).

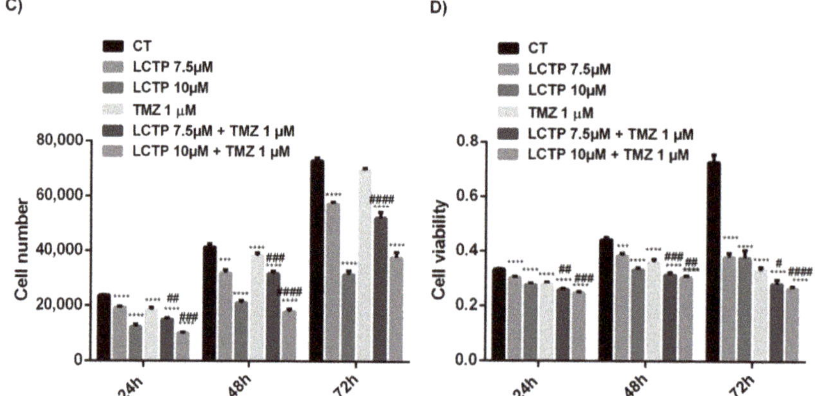

Figure 5. (**A**) IC50-values of U87Mg cells after 24, 48 and 72 h of incubation with TMZ. (**B**) IC50-values of TMZ at 24 h of U87Mg cells pre-treated with LCTP 7.5 and 10 µM for 24 h. (**C**) Synergistic effect of LCTP and conventional chemotherapy TMZ ongrowth of U87Mg cells. Values are the means ± SEM of 3 individual determinations. # *p*-value compared to cells treated with TMZ alone. (**D**) Synergistic effect of LCTP and conventional chemotherapy TMZ oncell viability of U87Mg cells. Values are the means ± SEM of 3 individual determinations. # *p*-value < 0.05 compared to cells treated with TMZ alone. According to GraphPad Prism, # *p*-value 0.01 to 0.05 (significant), ## *p*-value 0.001 to 0.01 (very significant), *** or ### *p*-value 0.0001 to 0.001 (Extremely significant), **** or #### *p*-value < 0.0001 (Extremely significant).

3. Discussion

The current oncology protocol for glioblastoma patients provides, after surgery, treatment with chemotherapeutic agents, such as TMZ, associated with radiotherapy (Stupp protocol) [5]. Over the past 10 years, however, therapeutic agents have not significantly increased the median survival rate of patients with glioblastoma. The 5-year survival rate for patients with the same disease, after treatment including surgical resection, radiotherapy, and chemotherapy, remains less than 9.8% [26].

Therapeutic approaches used against glioblastoma are associated with the development of resistance and with important side effects, which can very often represent a real obstacle for the patient in completing chemotherapy, and this can lead to therapeutic failure. For this reason, the current research is moving towards using natural substances as adjuvants to a traditional therapy that can, on the one hand, have an antitumor action (synergistic with temozolomide) and, on the other, soothe the side effects. A large number of clinical studies have demonstrated the benefits derived from the use of herbal medicines in combination with conventional therapies on the survival, immune modulation, and quality of life of cancer patients [27]. Here, we examined the anti-proliferative effects of a natural compound, Lactucopicrin, commonly found in Lactuca virosa, on human glioblastoma cells U87Mg. The effects of LCTP on the proliferation of GBM continuous cell line U87Mg were evaluated by setting up growth curves. To study the effect of LCTP on the growth rate of human glioma cells U87Mg, we applied the drug at chosen concentration (7.5 and 10 µM) to the growing medium once daily for three days, starting one day after plating. These applications had already reduced the growth in culture after only 24 h post-treatment. A time- and dose-dependent growth rate inhibition was observed in

U87Mg upon LCTP treatment. This trend was also confirmed by results obtained for cell viability assay. In accordance with results from cell count and MTT assay, the growth inhibitory effect of LCTP on U87Mg cell was also confirmed by clonogenic assay. In fact, at the lower concentration chosen (7.5 μM), LCTP completely suppressed the potential clonogenicity of U87Mg cells. Migration is a peculiar aspect which favors the aggressiveness and poor prognosis glioblastoma [28]; in this regard, our results evidence a dose-dependent, anti-migratory effect of LCTP on U87Mg cells, since the ability of cells to close the scratch decreased with LCTP concentrations.

The half-maximal inhibitory concentration of LCTP against U87Mg was determined at different time points, revealing a strong cytotoxic effect of LCTP. The IC50 for LCTP in the micromolar range is in line with the previously reported data in human SKMEL-5 [23]. Interestingly, compared with conventional chemotherapy TMZ, the IC50 of LCTP at 24 and 48 h resulted in being more than 10 times lower, becoming comparable only at 72 h of treatment.

In order to evaluate the potential synergism of TMZ with LCTP, the preliminary ability of this natural molecule to arrest the cell proliferation of U87Mg cells was investigated by comparing the IC50-values for TMZ before and after treatment with LCTP. LCTP strongly increased the sensitivity of U87Mg to TMZ. Finally, to test the synergism with TMZ (canonical drug for GBM) and to propose LCTP as adjuvant therapy, the GBM cell growth and viability were analyzed upon combined treatment which demonstrated a synergistic effect. We considered that it would be interesting to investigate the mechanisms involved in inhibiting cell growth after LCTP administration. Since members of the sesquiterpene lactone family have been reported to induce S and G2/M cell cycle arrest and cell death inducing apoptosis through PARP cleavage [20,23,29], the ability of LCTP to induce the cell cycle arrest in GBM U87Mg cells was evaluated by flow cytometry analysis. The results demonstrated a block of the cell cycle in the G2/M phase, especially 72 h after treatment with LCTP. The cell cycle arrest was supported by the increased expression of the CDK inhibitor p21 and oncosuppressor p53, while CDK2 protein decreased. The cell cycle blockage is not to be considered as the only mechanism that determines the growth inhibition of U87Mg cells after treatment with LCTP. In this order, the molecular activation of the apoptotic and autophagic responses, as well as the key pathways of proliferation U87Mg continuous cell line, were assessed by Western blot analysis of GBM cell extracts after treatments. Long-term exposure to LCTP 7.5 μM revealed an increase cleaved/full-length PARP ratio, confirming an involvement of the apoptotic pathway which is already visible by procaspase 6 decrease at 48 and 72 h of induction with LCTP.

The pro-apoptotic and cytostatic effects of LCTP may be the results of a rapid activation of autophagic pathway. The autophagy adaptor p62/SQSM1 has been reported to act as an oncogene in glioma and has been proposed as a novel therapeutic target for this pathology [30]. Moreover, a decreased expression of p62/SQSM1 induced a significant decrease of ERK phosphorylation [31]. According to previous literature, in our study, Western blot analysis revealed a strong and rapid reduction of p62/SQSM1 accompanied by the increase of cleaved LC3BII and a remarkable dephosphorylation of ERK1/2 kinases in LCTP–treated cells. The presence of α-methylene-γ-lactone promotes the nucleophilic attack of sesquiterpene lactones to multiple targets through Michael's reaction, such as the redox regulator NF-κB [17]. The constitutive activation of this transcription factor has been shown to stimulate the growth and survival of GBM [32]. Several natural sesquiterpene lactones act as NF-kB inhibitors [33], such as parthenolide and artemisinin, which enhance the sensitivity of cancer cells to chemotherapy [34,35]. In this regard, the effect of LCTP on the NF-kB p65 subunit was investigated by Western blot analysis, that revealed a significant reduction of p65 expression levels already appreciable at 30 min from induction. According to Jing et al., the autophagic degradation of p62/SQSM1 can lead the inhibition of NF-kB pathway and thus inhibit the proliferation of U87Mg glioblastoma cells [36]. The high nucleophilic reactivity of LCTP against cysteine-reactive electrophiles and the involvement of oxidative stress in LCTP-induced cytotoxicity was investigated pre-incubating U87Mg cells with NAC, a cysteine precursor acting as ROS scavenger. NAC completely reversed the cytotoxic effects of LCTP and protected cells from morphological changes.

AKT represents a central point of the RTK/PTEN/PI3K pathway. High levels of AKT and phospho-AKT have been detected in the majority of GBM tumor samples and cell lines, where it supports uncontrolled glioma cells growth, apoptotic blockage, and tumor invasion, thus representing an attractive pathway for GBM targeting therapy [37]. Western blot analysis displayed, in cells treated with LCTP for 30 min, 1 h, 2 h, and 4 h, a decrease in AKT phosphorylation, implying a strong reduction of the proliferation signaling pathway.

Finally, our study evidenced that treatment of U87Mg cells with LCTP induced—already in the first 30 min of treatment—a profound reshape of the glioblastoma cells' cytoskeleton, leading to the development of autophagic structures [38,39]. Immunofluorescence analysis revealed a profound change in α-tubulin distribution, that appeared clearly distributed in the cytoskeletal structures of the microtubules in the control glioblastoma cells, while α-tubulin is very concentrated in the cytoplasm after treatment with LCTP. On this side, LCTP treatment also remodeled the intermediate filament Vimentin, which appeared normally distributed in the cytoplasmic long stress fibers in control cells while accumulating near the nucleolemma, conferring a cell rounded-shaped morphology in LCTP-treated cells.

Concluding, the LCTP can be considered an excellent adjuvant substance for the treatment of glioblastoma, as it is involved in a series of mechanisms that control the growth of glioblastoma cells: autophagy, cell cycle arrest, and oxidative stress. Furthermore, the reduction of transcription factor NF-κB p65—a critical regulator of immune and inflammatory responses—showed a potential role of LCTP into reducing chronic inflammation. This natural substance—which potentially can pass through the BBB—used in combination with the canonical chemotherapy Temodal, opens the door to a multimodal therapy which could prove to be the most effective weapon against complex diseases, such as glioblastoma.

4. Materials and Methods

4.1. Cell Culture

The continuous human glioblastoma cell line U87Mgwas obtained from Sigma Aldrich Collection (LGCPromochem, Teddington, UK). U87Mg cells were growth in Dulbecco's Modified Eagle's Medium supplemented with 10% fetal bovine serum, 2 mmol/L-glutamine, 100IU/mL penicillin, 100 μg streptomycin at 37 °C, 5% CO_2, and 95% of humidity. For in vitro treatment, the pure molecule LCTP (Extrasynthase, Genay, France) and Temozolomide (Sigma Aldrich, St. Louis, MO, USA) were used.

4.2. Estimation of Half—Maximal Inhibitory Concentration (IC50) of LCTP and TMZ in U87Mg Cells

To estimate the IC50-values of LCTP and TMZ at 24, 48, and 72 h, U87Mg cells were plated in 96-well plates at density of 5×10^3 cells/well. The IC50-values for the selected drugs were determined by using, respectively, the concentrations of 7.5, 10, 15, and 30 μM for LCTP and 10, 50, 100, 150, and 200 μM for TMZ. The IC50-values for LCTP and TMZ in U87Mg cells were calculated using GraphPad Prism (GraphPad Software Inc., San Diego, CA, USA).

4.3. Proliferation Assay

The in vitro response of U87Mg cells to LCTP were evaluated by plating human GBM cell line in 48-well plates at 1×10^4 cells/well in DMEM supplemented with 10% FBS, incubating them at 37 °C in an atmosphere containing 5% CO_2. The following day, the cells were treated every 24 h with LCTP at concentrations of 7.5 μM and 10 μM for 24, 48, and 72 h. DMSO 0.08% was used as vehicle control. At the selected time point, cell count was performed using a Burker chamber.

4.4. Cell Viability Assay

U87Mg cells were seeded at density of 5×10^3 cells in 96-well plates. The cultures were treated every 24 h with concentrations of LCTP 7.5 μM and 10 μM for 24, 48, and 72 h, followed by MTT

(3-(4,5-dimethylthiazol-2-yl)-2,5-diphenyltetrazolium bromide) (Sigma–Aldrich) assay. DMSO 0.08% was used as vehicle control. Briefly, 5 mg/mL MTT was added in 100 µL of cultured cells in DMEM medium. Formazan crystals were dissolved in isopropanol/HCl 0.4% and the absorbance measured at 570 nm with a plate reading spectrophotometer.

LCTP-induced oxidative stress in U87Mg cells was investigated pre-treating starved cells with N-acetylcysteine (NAC) 3 mM for 4 h at 37 °C in DMEM with FBS 10%. After 4 h, the medium was replaced and GBM cells treated with different concentrations of LCTP (7.5, 10, 15, and 30 µM) for 24 h. DMSO 0.2% was used as vehicle control.

4.5. Microscopic Observation of Cell Morphology

U87Mg cells were seeded in 96-well plates and incubated with LCTP 7.5 µM every 24 h in DMEM with FBS 10% for 24, 48, and 72 h in the presence or absence of N-acetylcysteine 3 mM. After treatment, morphological changes of U87Mg were observed and imaged by a phase contrast microscope (Evos, Life technologies, Carlsbad, CA, USA).

4.6. Combined Treatment of TMZ and LCTP in Human GBM Cell Line

The in vitro response of U87Mg cells to combined LCTP and TMZ treatment was assessed seeding GBM cells in 48-well plates (10,000 cells/well) and treating the cells every 24 h with TMZ 1 µM alone or in combination with LCTP 7.5 µM and 10 µM for 24, 48, and 72 h. At the end of each treatment, cells were counted using a Burker chamber. The combined cytotoxic effects of LCTP and TMZ were evaluated by MTT assay as above described: U87Mg cells were plated at density of 5×10^3 cells in 96-well plates and treated with the same combination of drugs (LCTP 7.5 µM and 10 µM in combination with TMZ 1 µM) for 24, 48, and 72 h.

4.7. Estimation of IC50 of TMZ after Pre-Treatment with LCTP

To test the ability of LCTP to enhance the sensitivity of U87Mg cells to TMZ, cells were plated in 96-well plates at density of 5×10^3 cells/well and pre-treated for 24 h with LCTP 7.5 µM and 10 µM, before the estimation of IC50-values for TMZ. Cells were incubated with concentrations of TMZ 10, 50, 100, 150, and 200 µM and IC50-values for TMZ at 24 h determined by using GraphPad Prism (GraphPad Software Inc., San Diego, CA, USA).

4.8. Wound Healing Assay

To evaluate cell motility, GBM cells were seeded into 6-well culture plates. When the cells reached 90% confluence, a scratch was gently made through the cell monolayer by sterile 100 µL pipette tips, and the detached cells were washed away. The cell migration was observed and imaged under an Evos FL microscope (Life Technologies) for each condition (LCTP 7.5 µM and 10 µM) and timepoint (T0 and T24 h).

4.9. Clonogenic Assay

The clonogenic assay was performed seeding 10^3 U87Mg cells/well in triplicate in 6-well plates for 48 h. Cells were treated with LCTP 7.5 µM and control vehicle (0.08% DMSO) for 24 h and the medium was replaced every 3 days for 14 days. The colonies were fixed with 4% paraformaldehyde solution for 5 min, washed with PBS, and stained with crystal violet 0.05% for 30 min.

4.10. Western Blot Analysis of LCTP-Treated U87Mg Cells

Protein extraction from LCTP-treated U87MG cells was performed with Triton X-100 lysis buffer (Tris-HCL 10 mM, EDTA 1 mM, NaCl 150 mM, Triton X-100 1%, NaF 1 mM, $Na_4P_2O_7$ 1 mM, Na_3VO_4 1 mM, and protease inhibitors 1×). Protein lysates (15 µg) were resolved on a 12.5% SDS-PAGE transferred to PVDF membranes by electroblotting. The membranes were incubated for 1 h at room temperature in 5% not-fat dry milk or bovine serum albumin (BSA) diluted in 1× Tris-buffered saline containing Tween-20 (TBST) and then incubated overnight at 4 °C with primary selected antibodies. For protein normalization, each membrane was then incubated with mouse monoclonal anti–β-actin (1:10,000, Santa Cruz Biotechnology, Santa Cruz, CA, USA) or anti-GAPDH (1:1000, Santa Cruz Biotechnology). The membranes were incubated with the specific HRP-conjugated secondary antibodies (Calbiochem). The protein bands were detected by chemiluminescence using ECL Western blotting (Amersham) and the digital signals were quantified by densitometric analysis using Image Lab Software (Bio-Rad Laboratories, Hercules, CA, USA). For cell cycle proteins CDK2, p21, and p53 analysis, the cells were plated at a density of 5×10^5 cells in60 mm plates in DMEM without FBS for 48 h. After re-adding 10% FBS, inductions were performed every 24 h with LCTP7.5 μM and the cells collected after 24, 48, and 72 h of treatment. Anti-CDK2, anti-p21 antibody (1:1000, Cell Signaling) and anti-p53 (1:1000, Roche) were used.

Changes in the phosphorylation status of ERK1/2 and AKT, in the expression level of autophagy-associated proteins p62 and LC3B and NF-κB-p65 subunit were evaluated by treating U87Mg cells with LCTP 10 μM for 30 min, 1 h, 2 h, and 4 h. The membranes were incubated overnight with anti-pERK1/2 and anti-pAKT antibody (1:1000, Cell Signaling) in 2.5% bovine serum albumin (BSA) in TBST. The membranes were stripped to be re-probed with non-phosphorylated forms of anti-ERK1/2 and anti-AKT (1:1000, Cell Signaling). For autophagic proteins, anti-p62 (1:1000, Cell Signaling) in 2.5% milk in TBST and LC3B (1:1000, Cell Signaling) in BSA 2.5% in TBST were used. For NF-κB –p65 (1:1000, Santa Cruz) subunit in milk 2.5% in TBST were used.

4.11. Western Blot of Apoptosis-Associated Proteins

LCTP-induced apoptosis in U87Mg cells was assessed by Western blot analysis of caspases 6 and Poly (ADP-ribose) polimesare (PARP). The samples were prepared as previously reported for cell cycle proteins analysis and the PVDF membranes were incubated with caspase 6 (Cell Signaling, 1:1000), and PARP (Cell Signaling 1.100) in 2.5% milk in TBST overnight at 4 °C. The incubation with secondary antibodies and detection of proteins were performed as above described.

4.12. Immunofluorescence

U87Mg cells (5×10^3) were plated in 8-well chamber slides in DMEM with 2% serum for 48 h. Cells were treated with LCTP 10 μM in DMEM with 10% FBS for 20 min and the morphological change induced by LCTP treatment assessed by immunofluorescence staining for cytoskeleton proteins, α-tubulin, and Vimentin. In detail, at the end of treatment, U87Mg cells were fixed in 4% formalin (Diapath) for 20 min and permeabilized with 0.1% Triton (Invitrogen, Carlsbad, CA, USA) for 30 min. After blocking with 10% specific serum, the cells were incubated with anti-Vimentin (prediluted; Roche diagnostic) and anti-α-tubulin (1:200, Abcam, Cambridge, MA, USA) overnight at 4 °C. After washing with 0.025% PBS-Tween-20, cells were incubated with secondary antibody anti-mouse fluorescein (1:100; Vector, Stuttgart, Germany) in 2% serum for 1 h at room temperature. The slides were counterstained with DAPI Mounting Medium (Vectashield) for nuclei detection and analyzed with a fluorescence microscope at 20× and 40× magnification.

4.13. Cell Cycle Analysis by Flow Cytometry

U87MG human GBM cells were plated (25×10^4) in DMEM with 2% serum for 48 h and treated with LCTP 7.5 μM every 24 h in DMEM with FBS 10% for 24, 48, and 72 h. After treatment, cells were trypsinized, washed in sample buffer (glucose 0.1% in HBSS), fixed in 70% ethanol, and stored at 4 °C overnight until the day of analysis. Before analysis, Propidium iodide (50 μg/mL) was added for 30 min at room temperature. Flow cytometry analysis of the cell cycle was performed with a Gallios instrument (Beckman Coulter, Brea, CA, USA).

4.14. DNA Laddering

U87Mg cells were plated (4.5×10^4) in DMEM with 2% FBS for 48 h and induction performed every 24 h with LCTP: 7.5 μM and DMSO 0.08% as vehicle control for 72 h. At the end of treatment, the cells were collected, and the pellets were washed with phosphate buffer saline. DNA was extracted with QIAamp DNA Blood Mini kit (Qiagen, Hilden, Germany) and the DNA fragments were separated by 2% agarose gel electrophoresis and visualized by an SYBR Safe DNA Gel Stain (Invitrogen).

4.15. Statistical Analysis

Experiments were performed in triplicate and data were expressed as mean ± SEM. Statistical significance was determined by Student's *t*-test, considering a *p*-value of < 0.05 statistically significant.

Supplementary Materials: The following are available online. Figure S1(A) Representative images of wound healing assay of U87Mg cells treated with vehicle (DMSO 0.08%) and LCTP 7.5 and 10 μM at T0 and 24 h post-scratch (T24). Figure S2 Western blot analysis of NF-κB p65 subunit of long-term treated U87Mg cells with LCTP 7.5 μM for 24, 48, and 72 h. Data were normalized to β-actin and are expressed as fold change over control-treated cells of 3 individual determinations. * Unpaired *t*-test, *p*-value < 0.05.

Author Contributions: R.R. performed the main experiments; M.A.O. contributed to data analysis; S.S. setting up growth curves with T.M.Z., S.C. performed M.T.T. assays, F.G. contributed to a critical revision of the manuscript; A.A. drafted this study and wrote the paper. All authors have read and agreed to the published version of the manuscript.

Funding: This research was funded by Italian Ministry of Health with Ricerca Corrente and the APC was funded by I.R.C.C.S I.N.M. Neuromed.

Acknowledgments: We are grateful to Antonio Feliciello of University of Naples Federico II and Massimo Sanchez of Istituto Superiore di Sanità (ISS).

Conflicts of Interest: The authors declare no conflict of interest.

References

1. Kleihues, P.; Louis, D.N.; Scheithauer, B.W.; Rorke, L.B.; Reifenberger, G.; Burger, P.C.; Cavenee, W.K. The WHO classification of tumors of the nervous system. *J. Neuropathol. Exp. Neurol.* **2002**, *61*, 215–225. [CrossRef] [PubMed]

2. Ciceroni, C.; Bonelli, M.; Mastrantoni, E.; Niccolini, C.; Laurenza, M.; LaRocca, L.M.; Pallini, R.; Traficante, A.; Spinsanti, P.; Ricci-Vitiani, L.; et al. Type-3 metabotropic glutamate receptors regulate chemoresistance in glioma stem cells, and their levels are inversely related to survival in patients with malignant gliomas. *Cell Death Differ.* **2013**, *20*, 396–407. [CrossRef] [PubMed]

3. De Almeida Sassi, F.; Lunardi Brunetto, A.; Schwartsmann, G.; Roesler, R.; Abujamra, A.L. Glioma Revisited: From Neurogenesis and Cancer Stem Cells to the Epigenetic Regulation of the Niche. *J. Oncol.* **2012**, *2012*, 537861. [CrossRef]

4. Ohgaki, H.; Kleihues, P. Epidemiology and etiology of gliomas. *Acta Neuropathol.* **2005**, *109*, 93–108. [CrossRef] [PubMed]

5. Stupp, R.; Weber, D.C. The Role of Radio- and Chemotherapy in Glioblastoma. *Oncol. Res. Treat.* **2005**, *28*, 315–317. [CrossRef]

6. Hambardzumyan, D.; Squatrito, M.; Carbajal, E.; Holland, E.C. Glioma Formation, Cancer Stem Cells, and Akt Signaling. *Stem Cell Rev. Rep.* **2008**, *4*, 203–210. [CrossRef] [PubMed]

7. Friedman, H.S.; Kerby, T.; Calvert, H. Temozolomide and treatment of malignant glioma. *Clin. Cancer Res.* **2000**, *6*, 2585–2597.

8. Park, C.K.; Kim, J.E.; Kim, J.Y.; Song, S.W.; Kim, J.W.; Choi, S.H.; Kim, T.M.; Lee, S.H.; Kim, I.H.; Park, S. The Changes in MGMT Promoter Methylation Status in Initial and Recurrent Glioblastomas. *Transl. Oncol.* **2012**, *5*, 393–397. [CrossRef]

9. Messaoudi, K.; Clavreul, A.; Lagarce, F. Toward an effective strategy in glioblastoma treatment. Part I: Resistance mechanisms and strategies to overcome resistance of glioblastoma to temozolomide. *Drug Discov. Today* **2015**, *20*, 899–905. [CrossRef]

10. Vengoji, R.; Macha, M.A.; Batra, S.K.; Shonka, N. Natural products: A hope for glioblastoma patients. *Oncotarget* **2018**, *9*, 22194–22219. [CrossRef]

11. Yi, G.Z.; Xiang, W.; Feng, W.Y.; Chen, Z.Y.; Li, Y.M.; Deng, S.Z.; Guo, M.L.; Zhao, L.; Sun, X.G.; He, M.Y.; et al. Identification of Key Candidate Proteins and Pathways Associated with Temozolomide Resistance in Glioblastoma Based on Subcellular Proteomics and Bioinformatical Analysis. *Biomed. Res. Int.* **2018**, *2018*, 5238760. [CrossRef] [PubMed]

12. Sestito, S.; Runfola, M.; Tonelli, M.; Chiellini, G.; Rapposelli, S. New Multitarget Approaches in the War Against Glioblastoma: A Mini-Perspective. *Front. Pharmacol.* **2018**, *9*, 874. [CrossRef] [PubMed]

13. Arcella, A.; Oliva, M.A.; Sanchez, M.; Staffieri, S.; Esposito, V.; Giangaspero, F.; Cantore, G. Effects of hispolon on glioblastoma cell growth. *Environ. Toxicol.* **2017**, *32*, 2113–2123. [CrossRef] [PubMed]

14. Arcella, A.; Oliva, M.A.; Staffieri, S.; Sanchez, M.; Madonna, M.; Riozzi, B.; Esposito, V.; Giangaspero, F.; Frati, L. Effects of aloe emodin on U87MG glioblastoma cell growth: In vitro and in vivo study. *Environ. Toxicol.* **2018**, *33*, 1160–1167. [CrossRef]

15. Kreuger, M.R.; Grootjans, S.; Biavatti, M.W.; Vandenabeele, P.; D'Herde, K. Sesquiterpene lactones as drugs with multiple targets in cancer treatment: Focus on parthenolide. *Anticancer Drugs* **2012**, *23*, 883–896.

16. Amorim, M.H.; Gil da Costa, R.M.; Lopes, C.; Bastos, M.M. Sesquiterpene lactones: Adverse health effects and toxicity mechanisms. *Crit. Rev. Toxicol.* **2013**, *43*, 559–579. [CrossRef]

17. Chadwick, M.; Trewin, H.; Gawthrop, F.; Wagstaff, C. Sesquiterpenoids lactones: Benefits to plants and people. *Int. J. Mol. Sci.* **2013**, *14*, 12780–12805. [CrossRef]

18. Ghantous, A.; Gali-Muhtasib, H.; Vuorela, H.; Saliba, N.A.; Darwiche, N. What made sesquiterpene lactones reach cancer clinical trials? *Drug Discov. Today* **2010**, *15*, 668–678. [CrossRef]

19. Babaei, G.; Aliarab, A.; Abroon, S.; Rasmi, Y.; Aziz, S.G. Application of sesquiterpene lactone: A new promising way for cancer therapy based on anticancer activity. *Biomed. Pharmacother.* **2018**, *106*, 239–246. [CrossRef]

20. Wang, J.; Yu, Z.; Wang, C.; Tian, X.; Huo, X.; Wang, Y.; Sun, C.; Feng, L.; Ma, J.; Zhang, B.; et al. Dehydrocostus lactone, a natural sesquiterpene lactone, suppresses the biological characteristics of glioma, through inhibition of the NF-kappaB/COX-2 signaling pathway by targeting IKKbeta. *Am. J. Cancer Res.* **2017**, *7*, 1270–1284.

21. Wang, X.; Yu, Z.; Wang, C.; Cheng, W.; Tian, X.; Huo, X.; Wang, Y.; Sun, C.; Feng, L.; Xing, J.; et al. Alantolactone, a natural sesquiterpene lactone, has potent antitumor activity against glioblastoma by targeting IKKbeta kinase activity and interrupting NF-kappaB/COX-2-mediated signaling cascades. *J. Exp. Clin. Cancer Res.* **2017**, *36*, 93. [CrossRef] [PubMed]

22. Youn, U.J.; Miklóssy, G.; Chai, X.; Wongwiwatthananukit, S.; Toyama, O.; Songsak, T.; Turkson, J.; Chang, L.C. Bioactive sesquiterpene lactones and other compounds isolated from Vernonia cinerea. *Fitoterapia* **2014**, *93*, 194–200. [CrossRef] [PubMed]

23. Zhang, X.; Lan, D.; Ning, S.; Ruan, L. Anticancer action of lactucopicrin in SKMEL-5 human skin cancer cells is mediated via apoptosis induction, G2/M cell cycle arrest and downregulation of m=TOR/PI3K/AKT signalling pathway. *J. BUON* **2018**, *23*, 224–228. [PubMed]

24. Meng, Q.; Tang, B.; Qiu, B. Growth inhibition of Saos-2 osteosarcoma cells by lactucopicrin is mediated via inhibition of cell migration and invasion, sub-G1 cell cycle disruption, apoptosis induction and Raf signalling pathway. *J. BUON* **2019**, *24*, 2136–2140.

25. Wierzbicki, M.; Sawosz, E.; Strojny, B.; Jaworski, S.; Grodzik, M.; Chwalibog, A. NF-kappaB-related decrease of glioma angiogenic potential by graphite nanoparticles and graphene oxide nanoplatelets. *Sci. Rep.* **2018**, *8*, 14733. [CrossRef]

26. Stupp, R.; Hegi, M.E.; Mason, W.P.; van den Bent, M.J.; Taphoorn, M.J.; Janzer, R.C.; Ludwin, S.K.; Allgeier, A.; Fisher, B.; Belanger, K.; et al. Effects of radiotherapy with concomitant and adjuvant temozolomide versus radiotherapy alone on survival in glioblastoma in a randomised phase III study: 5-year analysis of the EORTC-NCIC trial. *Lancet Oncol.* **2009**, *10*, 459–466. [CrossRef]

27. Yin, S.Y.; Wei, W.C.; Jian, F.Y.; Yang, N.S. Therapeutic applications of herbal medicines for cancer patients. *Evid. Based Complement. Altern. Med.* **2013**, *2013*, 302426. [CrossRef]

28. Liu, C.A.; Chang, C.Y.; Hsueh, K.W.; Su, H.L.; Chiou, T.W.; Lin, S.Z.; Harn, H.J. Migration/Invasion of Malignant Gliomas and Implications for Therapeutic Treatment. *Int. J. Mol. Sci.* **2018**, *19*, 1115. [CrossRef]

29. Li, C.; Han, X. Anthecotulide Sesquiterpene Lactone Exhibits Selective Anticancer Effects in Human Malignant Melanoma Cells by Activating Apoptotic and Autophagic Pathways, S-Phase Cell Cycle Arrest, Caspase Activation, and Inhibition of NF-kappaBSignalling Pathway. *Med. Sci. Monit.* **2019**, *25*, 2852–2858. [CrossRef]

30. Deng, D.; Luo, K.; Liu, H.; Nie, X.; Xue, L.; Wang, R.; Xu, Y.; Cui, J.; Shao, N.; Zhi, F. p62 acts as an oncogene and is targeted by miR-124-3p in glioma. *Cancer Cell Int.* **2019**, *19*, 280. [CrossRef]

31. Xu, H.; Sun, L.; Zheng, Y.; Yu, S.; Ou-Yang, J.; Han, H.; Dai, X.; Yu, X.; Li, M.; Lan, Q. GBP3 promotes glioma cell proliferation via SQSTM1/p62-ERK1/2 axis. *Biochem. Biophys. Res. Commun.* **2018**, *495*, 446–453. [CrossRef] [PubMed]

32. Friedmann-Morvinski, D.; Narasimamurthy, R.; Xia, Y.; Myskiw, C.; Soda, Y.; Verma, I.M. Targeting NF-kappaB in glioblastoma: A therapeutic approach. *Sci. Adv.* **2016**, *2*, e1501292. [CrossRef] [PubMed]

33. Hehner, S.P.; Heinrich, M.; Bork, P.M.; Vogt, M.; Ratter, F.; Lehmann, V.; Schulze-Osthoff, K.; Droge, W.; Schmitz, M.L. Sesquiterpene lactones specifically inhibit activation of NF-kappa B by preventing the degradation of I kappa B-alpha and I kappa B-beta. *J. Biol. Chem.* **1998**, *273*, 1288–1297. [CrossRef] [PubMed]

34. Pajak, B.; Gajkowska, B.; Orzechowski, A. Molecular basis of parthenolide-dependent proapoptotic activity in cancer cells. *Folia Histochem. Cytobiol.* **2008**, *46*, 129–135. [CrossRef] [PubMed]

35. Efferth, T. Willmar Schwabe Award 2006: Antiplasmodial and antitumor activity of artemisinin—From bench to bedside. *Planta Med.* **2007**, *73*, 299–309. [CrossRef]

36. Su, J.; Liu, F.; Xia, M.; Xu, Y.; Li, X.; Kang, J.; Li, Y.; Sun, L. p62 participates in the inhibition of NF-kappaB signaling and apoptosis induced by sulfasalazine in human glioma U251 cells. *Oncol. Rep.* **2015**, *34*, 235–243. [CrossRef]

37. McDowell, K.A.; Riggins, G.J.; Gallia, G.L. Targeting the AKT pathway in glioblastoma. *Curr. Pharm. Des.* **2011**, *17*, 2411–2420. [CrossRef]

38. Izdebska, M.; Halas-Wisniewska, M.; Zielinska, W.; Klimaszewska-Wisniewska, A.; Grzanka, D.; Gagat, M. Lidocaine induces protective autophagy in rat C6 glioma cell line. *Int. J. Oncol.* **2018**, *54*, 1099–1111. [CrossRef]

39. Li, S.S.; Xu, L.Z.; Zhou, W.; Yao, S.; Wang, C.L.; Xia, J.L.; Wang, H.F.; Kamran, M.; Xue, X.Y.; Dong, L.; et al. p62/SQSTM1 interacts with vimentin to enhance breast cancer metastasis. *Carcinogenesis* **2017**, *38*, 1092–1103. [CrossRef]

Article

Macrocybin, a Natural Mushroom Triglyceride, Reduces Tumor Growth In Vitro and In Vivo through Caveolin-Mediated Interference with the Actin Cytoskeleton

Marcos Vilariño [1], Josune García-Sanmartín [1], Laura Ochoa-Callejero [1], Alberto López-Rodríguez [2], Jaime Blanco-Urgoiti [2] and Alfredo Martínez [1,*]

[1] Oncology Area, Center for Biomedical Research of La Rioja (CIBIR), 26006 Logroño, Spain; mvilarino@riojasalud.es (M.V.); jgarcias@riojasalud.es (J.G.-S.); locallejero@riojasalud.es (L.O.-C.)
[2] CsFlowchem, Campus Universidad San Pablo CEU, Boadilla del Monte, 28668 Madrid, Spain; alberto.lopez@csflowchem.com (A.L.-R.); jaime.blanco@csflowchem.com (J.B.-U.)
* Correspondence: amartinezr@riojasalud.es; Tel.: +34-941278775

Academic Editor: Thomas J. Schmidt
Received: 6 November 2020; Accepted: 17 December 2020; Published: 18 December 2020

Abstract: Mushrooms have been used for millennia as cancer remedies. Our goal was to screen several mushroom species from the rainforests of Costa Rica, looking for new antitumor molecules. Mushroom extracts were screened using two human cell lines: A549 (lung adenocarcinoma) and NL20 (immortalized normal lung epithelium). Extracts able to kill tumor cells while preserving non-tumor cells were considered "anticancer". The mushroom with better properties was *Macrocybe titans*. Positive extracts were fractionated further and tested for biological activity on the cell lines. The chemical structure of the active compound was partially elucidated through nuclear magnetic resonance, mass spectrometry, and other ancillary techniques. Chemical analysis showed that the active molecule was a triglyceride containing oleic acid, palmitic acid, and a more complex fatty acid with two double bonds. The synthesis of all possible triglycerides and biological testing identified the natural compound, which was named Macrocybin. A xenograft study showed that Macrocybin significantly reduces A549 tumor growth. In addition, Macrocybin treatment resulted in the upregulation of Caveolin-1 expression and the disassembly of the actin cytoskeleton in tumor cells (but not in normal cells). In conclusion, we have shown that Macrocybin constitutes a new biologically active compound that may be taken into consideration for cancer treatment.

Keywords: natural product; therapeutic triglyceride; xenograft study; Caveolin-1; actin cytoskeleton

1. Introduction

The discovery of new anticancer drugs can be accomplished through different approaches, including the screening of natural products, and the testing of synthetic compound libraries, computer-assisted design, and machine learning, among others [1]. Living organisms constitute an almost unlimited source of compounds with potential biological activity and, so far, have provided the majority of currently approved therapies for the treatment of cancer [2,3]. A few examples include taxol, which was obtained from the bark of the Pacific Yew tree [4,5], vinblastine and related alkaloids from *Vinca* [6], camptothecin from the bark of a Chinese tree [7], or trabectedin and other drugs which were purified from marine organisms [8]. In addition, many natural products are the basis for further chemical development and the rational design of synthetic drugs [9].

The mechanisms of action by which natural products reduce cancer cell growth are diverse. In the case of taxanes, direct interaction with tubulin results in cytoskeleton hyperpolymerization and mitosis arrest [10]; vinblastine has the reverse effect, depolymerizing the cytoskeleton, but achieves the same final outcome, stopping cell division [10]; others, such as some mushroom extracts, induce immunomodulation [11,12], or affect other cancer hallmarks [13].

Mushrooms have been used for millennia in Eastern traditional medicine as cancer remedies [13,14] and recent scientific publications have corroborated the presence of anticancer molecules in mushroom extracts. For instance, some polysaccharides from *Ganoderma* possess anticancer activity through immunomodulatory, anti-proliferative, pro-apoptotic, anti-metastatic, and anti-angiogenic effects [15]; lentinan, a β-glucan from *Lentinus* is licensed as a drug for treating gastric cancer where it activates the complement system [16]; other β-glucans from *Ganoderma* and *Grifola* [17] also show great potential as modulators of the immune system; lectins from *Clitocybe* provide distinct carbohydrate-binding specificities, showing immunostimulatory and adhesion-/phagocytosis-promoting properties [18]; furthermore, triterpenoids from *Inonotus* have been shown to induce apoptosis in lung cancer cell lines [19]. In addition, the immense number of mushroom species around the world [20], and their sessile nature, suggests that many fungal natural products may still be waiting for discovery.

Our goal in this study was to screen a number of scarcely known mushroom species collected in the rain forest of Costa Rica looking for new antitumor molecules. The most potent extract was found in specimens of *Macrocybe titans*, a local mushroom of very large proportions which usually grows in the anthills of leaf cutter ants [21]. The molecule responsible for the anticancer activity was isolated, characterized, and synthesized, and its "anticancer" properties were demonstrated both in vitro and in vivo.

2. Results

Several mushroom species were collected in the rainforests of Costa Rica, classified by expert botanists, and brought to the laboratory for extract preparation. Each specimen was subjected to several extraction procedures, including water- and ethanol-based techniques. All extracts were freeze-dried and sent to the Center for Biomedical Research of La Rioja (CIBIR) for anticancer activity testing in cell lines. Several extracts from different species showed some "anticancer" properties, defined as the ability to destroy cancer cells while being innocuous to nontumoral cells (Figure 1).

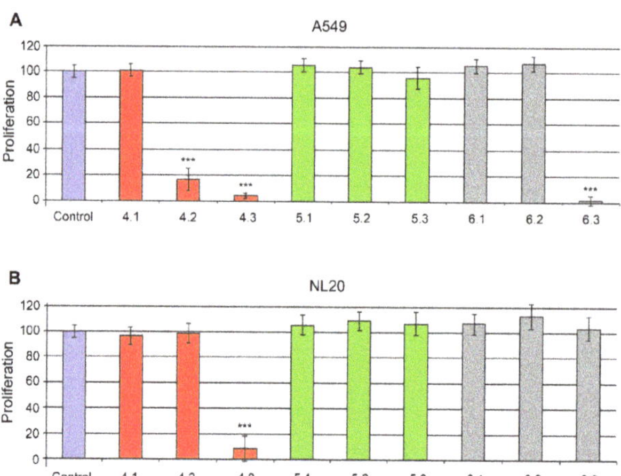

Figure 1. Toxicity of different mushroom extracts on cell lines A549 (**A**) and NL20 (**B**). All extracts were added at 0.3 mg/mL in a growth medium containing 1% FBS and incubated for 5 days. Comparing extract effects in both cell lines, three behaviors can be identified: (i) no effect on either cell line (extract 4.1, for example), (ii) toxic for both cell lines (extract 4.3), and (iii) "antitumoral" effect (toxic for A549 and nontoxic for NL20, extracts 4.2 and 6.3). Bars represent the mean ± SD of 8 independent measures. ***: $p < 0.001$ vs. control.

The most promising sample was the precipitated phase of a 95% ethanol extract from the mushroom *Macrocybe titans*. Specifically, we began with 3.5 Kg of fresh mushroom specimens. These were dried to obtain 367.2 g of dry mushroom powder, which was extracted with two washes, 30 min each, in 500 mL 95% ethanol at room temperature, with ultrasounds. The ethanol extracts were kept at −20 °C for 24 h and, then, the soluble phase was separated from the precipitate. This species was chosen for further analysis.

The initial extract was fractionated following different techniques and the resulting fractions were tested again in the cell lines to follow the anticancer activity. The finally successful strategy consisted of an initial alkaloidal separation, followed by column chromatography with polymeric resin (HP20-SS) using different mobile phases. The best ratio of selectivity was obtained with an IPA:CH2Cl2 8:2 mobile phase at pH 12.0 (Figure 2).

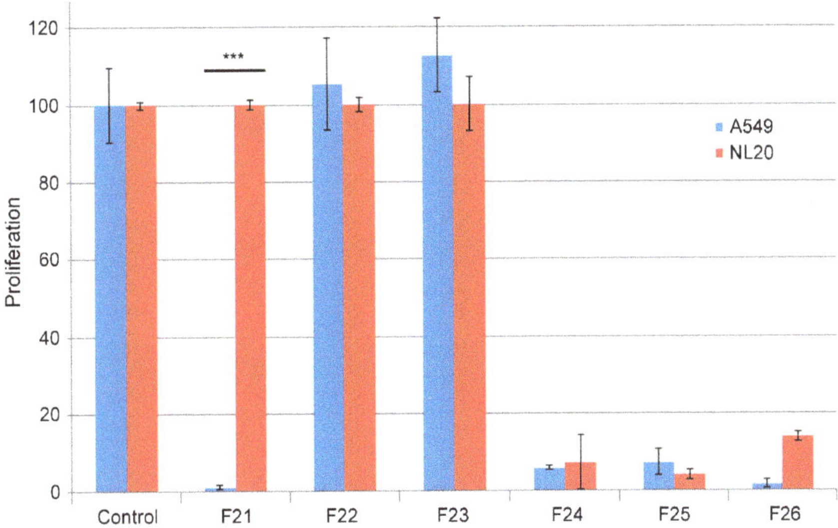

Figure 2. Toxicity of several fractions obtained by chromatography of the initial *M. titans* extract on cell lines A549 (blue bars) and NL20 (red bars). All extracts were added at 0.3 mg/mL in a growth medium containing 1% FBS and incubated for 5 days. Fraction 21 (F21) presents the desired characteristics of destroying the cancer cells while preserving the nontumoral cells. Bars represent the mean ± SD of 8 independent measures. ***: $p < 0.001$ between cell lines.

Once the molecule responsible for the differential growth inhibitory activity was pure enough, the final fraction was subjected to nuclear magnetic resonance (NMR), mass spectrometry, and 2D-COSY (Supplementary Materials, Natural extract characterization) to determine its chemical structure.

Data analysis identified our molecule as a triglyceride containing palmitic acid, oleic acid, and a more complex fatty acid of 18–20 carbons containing two double bonds. Since the final structure was impossible to determine, we decided to synthesize all compatible triglycerides (Table 1) and test them for biological activity with the cell lines. The presence of the two double bonds provides the molecule with optical activity. Some molecules were synthesized as racemic mixtures. For other molecules, both enantiomers were synthesized and tested. From the 10 triglycerides that were synthesized (Table 1) only one exerted differential growth inhibitory activity, specifically the R enantiomer of TG (C18:2, 9z,12z; C16:0; C18:1, 9z) (Supplementary Materials, Growth modulatory activity of synthetic triglycerides). A search of the Chemical Abstract database came out negative, indicating that this is a new structure. This triglyceride was named Macrocybin because of its origin from the *Macrocybe titans* mushroom.

Table 1. Triglycerides synthesized for this study and the chemical structure of Macrocybin. The structure of the other triglycerides is shown in Supplementary Materials (Triglyceride synthesis).

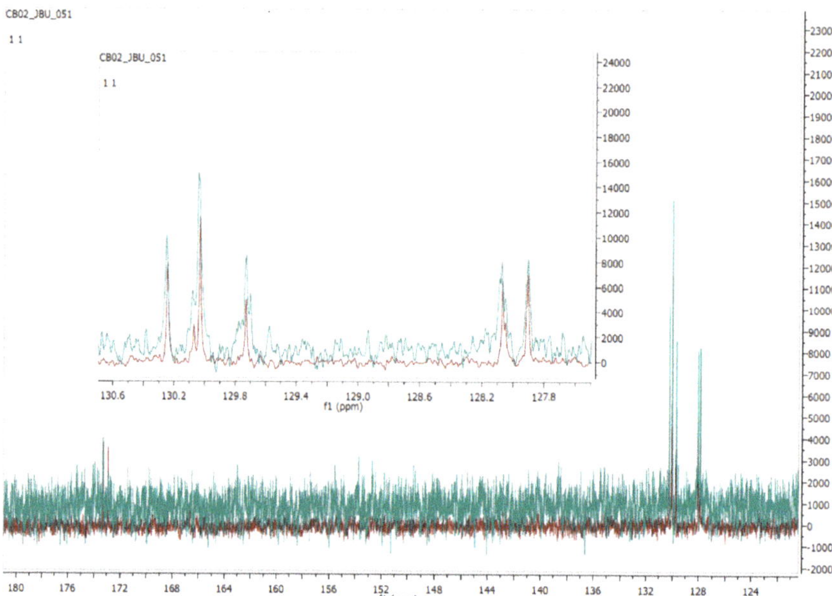

Internal Code	Chemical Formula
TG1	TG (C16:0; C18:1, 9z; C20:2, 11z,14z)
TG2	TG (C16:0; C20:2, 11z,14z; C18:1, 9z)
TG3	TG (C16:0; C18:1, 9z; C18:1, 9z)
TG4	TG (C18:0; C18:1, 9z; C18:2, 9z,12z)
TG5	2S-TG (C20:2, 11z,14z; C18:1, 9z; C16:0)
TG6	2R-TG (C20:2, 11z,14z; C18:1, 9z; C16:0)
TG7	2S-TG (C20:2, 11z,14z; C16:0; C18:1, 9z)
TG8	2R-TG (C20:2, 11z,14z; C16:0; C18:1, 9z)
TG9	2S-TG (C18:2, 9z,12z; C16:0; C18:1, 9z)
TG10 (Macrocybin)	2R-TG (C18:2, 9z,12z; C16:0; C18:1, 9z)

To determine the similarities between the synthetic molecule and the mushroom fraction, their NMR spectra were compared (Figure 3) and they were very similar.

Figure 3. Comparison of ^{13}C-NMR spectra of olefinic carbons of the synthetic triglyceride TG10 (red) and the natural extract (green). Spectra comparing the aliphatic carbons are shown in Supplementary Materials.

Toxicity assays for TG10 demonstrated a wide ratio of selectivity between A549 and NL20, with IC_{50} values of 13.4 and 50.1 µg/mL, respectively (Figure 4), thus confirming the in vitro anticancer activity of the synthetic triglyceride, Macrocybin.

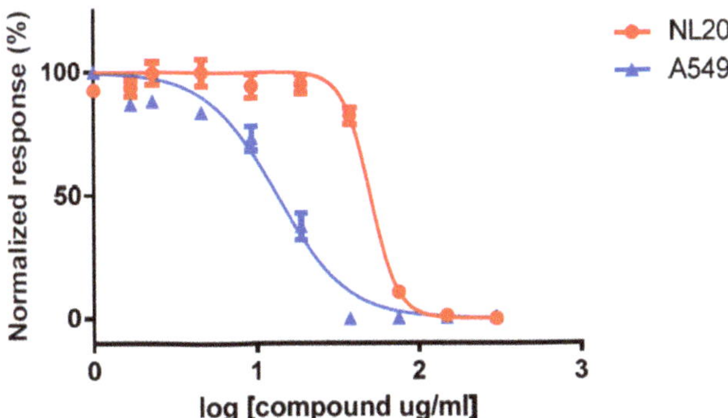

Figure 4. Concentration-dependent toxicity of synthetic Macrocybin on cell lines A549 (blue triangles) and NL20 (red circles). The synthetic triglyceride is more toxic for tumor cells than for nontumoral cells. Error bars represent the SD of 8 independent measures.

To determine whether this new molecule had antitumor properties in vivo, a xenograft study was performed, using A549 as the tumor-initiating cell. Tumors of mice receiving vehicle injections grew progressively until they reached the maximum volume allowed for humane reasons and the mice had to be sacrificed. On the other hand, tumors injected with Macrocybin grew somewhat more slowly ($p < 0.05$ after 40 days of treatment), indicating a therapeutic function for the triglyceride (Figure 5).

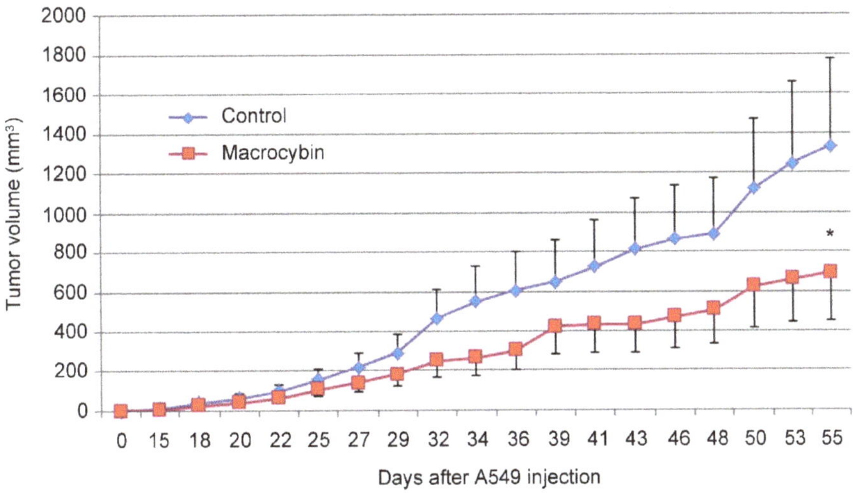

Figure 5. Xenograft experiment. Evolution of tumor volume in control (blue diamonds) and Macrocybin treated (red squares) A549 tumors grown in the flank of experimental mice. Cell injection was performed on day 0, and intratumoral injection began on day 15. There was a significant difference between treatments (ANOVA, $p < 0.05$). Each point represents the mean ± SD for 10 mice. *: $p < 0.05$ between treatments. Raw data values are shown in the Supplementary Materials.

To investigate the potential mechanism of action driving this antitumor activity, A549 and NL20 cells were stained with cytoskeleton-labeling moieties after exposure to Macrocybin. No differences were found in the tubulin cytoskeleton (results not shown) but the actin cytoskeleton, as stained with phallacidin, was dismantled in A549 tumor cells by the triglyceride, whereas it was unaffected in NL20 cells. The actin molecules of A549 translocated to the cell membrane forming filopodia (Figure 6).

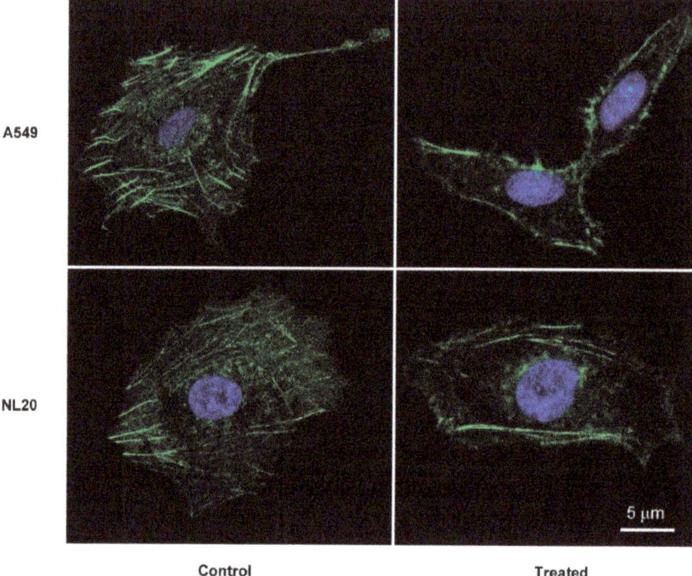

Figure 6. Representative confocal microscopy images of A549 and NL20 cells treated, or not (control), with 37 µg/mL Macrocybin for 24 h. The actin cytoskeleton was stained with Bodipy-phallacidin (green) and the nuclei with DAPI (blue). Following treatment, A549 cells lose their stress fibers and actin migrates to the cell membrane to produce small filopodia. Scale bar = 5 µm.

Macrocybin is a complex lipid and, as such, could be internalized into the cell through a number of lipid transport mechanisms [22]. To investigate whether a preferential transport into tumoral cells takes place, we analyzed Macrocybin-induced changes in lipid transport molecules (Table 2) through qRT-PCR. A549 and NL20 cells were seeded in 6-well plates and Macrocybin (or PBS as the vehicle) was added at 37 µg/mL for 6 h. The only significant differences were found for Caveolin-1 whose expression increased significantly ($p < 0.05$) in Macrocybin-treated A549 cells over the untreated controls. In contrast, Macrocybin did not affect Caveolin-1 expression in NL20 cells (Figure 7).

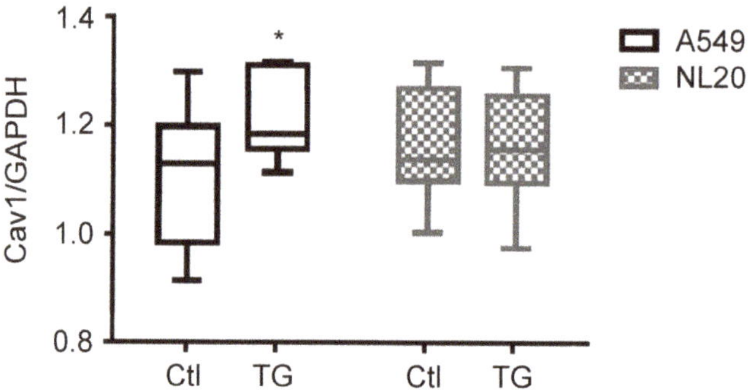

Figure 7. Gene expression for A549 and NL20 cells that were treated with 37 µg/mL Macrocybin (or control) for 6 h. Several lipid membrane transport proteins were analyzed by qRT-PCR. The only changes were observed in the expression of Caveolin-1 (Cav1) on treated A549 cells. Caveolin expression values were relativized to the housekeeping gene GAPDH. *: $p < 0.05$ vs. control (CTL). This is a representative example of 3 independent experiments. Raw data values are provided in the Supplementary Materials.

3. Discussion

We have shown that the Costa Rican mushroom *Macrocybe titans* has "anticancer" properties. The molecule responsible for this activity was isolated to purity and identified as a complex triglyceride with formula 2R-TG (C18:2, 9z,12z; C16:0; C18:1, 9z). This molecule was named Macrocybin and was synthesized in the laboratory. The synthetic molecule retained the "anticancer" properties both in vitro and in a xenograft model in vivo. The mechanism of action of the new molecule involves Caveolin-1 overexpression and actin cytoskeleton disorganization in the cancer cells.

The mushroom *M. titans* is a species that inhabits the tropical and subtropical regions of America and has been found between Florida and Argentina [23]. Other species of the same genus are found in tropical regions around the world, and include *M. gigantea* (India), *M. crassa* (Sri Lanka), *M. lobayensis* (Ghana), and *M. spectabilis* (Mauritius), among others [21]. *Macrocybe titans* arguably produces the largest carpophores in the world of fungi, with some specimens reaching 100 cm in diameter and weighing up to 20 Kg [21]. Interestingly, the specimens found in Costa Rica appear predominantly in the vicinity of the nests of gardening ants (*Atta cephalotes*) suggesting a potential symbiotic relationship between these two species [21]. All members of the *Macrocybe* genus are considered edible [24]. Although these mushrooms have not been reported as medicinal remedies in traditional pharmacopeia, recent studies have found that some polysaccharides of *M. titans* inhibit melanoma cell migration [25] and that the fruiting body of *M. gigantea* contains antimicrobial compounds [26]. These discoveries underscore the need for preserving biodiversity as a source for novel drugs and drug precursors [27].

The idea of a triglyceride acting as a therapy against cancer may seem rather counterintuitive. High levels of triglycerides in the blood constitute a clear risk for cancer initiation and progression since they provide a rich source of energy for developing tumors [28] and may increase metastatic potential [29]. Nevertheless, there are some specific types of cancer, such as breast [30] or prostate [28], where high levels of circulating triglycerides correlate with a better prognosis. In addition, a few studies have identified specific lipids with therapeutic functions. For instance, conjugated linoleic acid has been described as a natural anticarcinogenic compound [31].

Our studies testing the anticancer efficacy of different synthetic candidates indicate a high level of specificity in the anticancer actions of Macrocybin. Even the S enantiomer of Macrocybin was devoid of physiological activity. This indicates that the anticancer activity is dependent on a very specific interaction of Macrocybin with biological components (receptors and/or membrane transporters) of the tumor cells, rather than a bulk role, such as energy provider (a role that generic blood triglycerides may play).

Macrocybin was able to reduce tumor growth in a xenograft model of human lung cancer. As a triglyceride, Macrocybin has a modest solubility in water-based vehicles. Perhaps future formulations using micelles [32] or nanoparticles [33] may increase Macrocybin biodisponibility and antitumor efficacy. Furthermore, combinations of Macrocybin with other chemotherapeutic or immunotherapeutic drugs may represent promising avenues to reduce tumor burden [34].

Our results show that Macrocybin differentially affects the actin cytoskeleton of tumor vs. nontumor cells and that this mechanism is mediated by changes in the expression of Caveolin-1. Caveolin-1 has been shown as a lipid transporter in a variety of cell types [35–37]. Changes in the expression of this protein have been related to lung cancer prognosis [38], although some studies suggest that this relationship may be context-dependent [39]. This protein may constitute the entry point for Macrocybin into the cell and the fact that its expression is affected in tumor cells but not in nontumor cells may partially explain the preferential toxicity of the triglyceride for cancer cells.

Disruption of the actin cytoskeleton is a common mechanism of action for many natural products with antitumor properties. For instance, proteoglucans extracted from *Grifola* mushrooms affect actin cytoskeleton rearrangements, thus decreasing breast cancer cell motility [40]. Cucurbitacin I disrupts actin filaments in A549 cells, resulting in a reduction in lung cancer cell growth [41]. Enterolactone, a flaxseed-derived lignan, alters Focal Adhesion Kinase-Sarcoma Protein (FAK-Src) signaling and disorganizes the actin cytoskeleton, suppressing the migration and invasion of lung cancer cells [42]. Another example is narciclasine, an isocarbostyril alkaloid isolated from *Amaryllidaceae* plants, which impairs actin cytoskeleton organization and induces apoptosis in brain cancers [43]. All these studies suggest that interfering with the actin cytoskeleton of tumor cells provides a useful approach to induce tumor cell death and to prevent metastasis by reducing tumor cell motility.

Given the ubiquity among cancer cells of Caveolin-1 and the actin cytoskeleton, Macrocybin may constitute a common therapeutic agent for a variety of cancers. Many further investigations will be necessary to evaluate, in more detail, the usefulness of Macrocybin as an antitumoral drug or lead structure. These studies must include, among others, studies on the systemic applicability of Macrocybin. Future studies would also have to investigate whether other tumors are also susceptible to this new antitumor compound.

4. Materials and Methods

4.1. Mushroom Collection and Extract Preparation

Several mushroom specimens belonging to different species were collected in the rainforests of Costa Rica by expert personnel of the Instituto Nacional de Biodiversidad de Costa Rica (INBio, Santo Domingo, Costa Rica), under a specific permit issued by the Costa Rican Government (Oficina Técnica de la Comisión Nacional de la Gestión de la Biodiversidad, CONAGEBIO, San José, Costa Rica). Mushroom fragments were subjected to different extraction procedures, including crude extracts, extraction in hot water (80 °C) for several 30 min incubations, or in 95% ethanol at 50 °C for several 30 min incubations, with different exposures to ultrasound treatments. Final fractions, including the pellet and supernatant of each extraction, were freeze-dried and sent to CIBIR for screening. Voucher samples (FRS 01) of the fungal collection are kept at INBio (Santo Domingo, Costa Rica).

4.2. Anticancer Screening Strategy (Toxicity Assays)

Two human cell lines were used to screen the "anticancer" properties of the mushroom extracts: lung adenocarcinoma A549, and non-tumoral immortalized bronchiolar epithelial cell line NL20 (ATCC). Both cell lines were exposed in parallel to different concentrations of particular extracts for 5 days, and cytotoxicity was measured by the MTS method (Cell Titer, Promega, Madison, WI, USA), as reported in [44]. To identify a sample as having "anticancer" properties, it had to destroy cancer cells while preserving non-tumoral cells (or at least showing a wide therapeutic window between both lines). This strategy was used iteratively to guide extract purification until the final molecule could be identified.

Following ATCC's protocols, A549 cells were maintained in Ham's F12 medium containing 10% fetal bovine serum (FBS) and NL20 cells in complete medium (Ham's F12 with 2 mM L-glutamine, 1.5 g/L sodium bicarbonate, 2.7 g/L D-glucose, 0.1 mM NEAA, 0.005 mg/mL insulin, 10 ng/mL EGF, 0.001 mg/mL transferrin, 500 ng/mL hydrocortisone, and 4% FBS).

Before starting the project, optimization of the toxicity assay was performed, testing different numbers of cells per well, days of incubation with the extracts, and FBS contents. We wanted all cells to receive the same number of extracts in the same medium. We, therefore, chose the NL20 medium, which is the most complex and restrictive. FBS concentration was reduced to 1% to allow for a 5-day incubation period. Specifically, A549 and NL20 were seeded in 96-well plates at different densities (2000 and 10,000 cells/well, respectively) in a complete NL20 medium containing 1% FBS, in a final volume of 50 μL/well, and incubated at 37 °C in a humidified atmosphere, containing 5% CO_2. The next day, extracts were added at the indicated concentrations, in another 50 μL/well, in the same medium, and incubated for 5 days. At the end of this period, 15 μL of Cell Titer were added per well and, after an additional incubation of 4 h, color intensity was assessed in a plate reader (POLARstar Omega, BMG Labtech, Ortenberg, Germany) at 490 nm.

To determine IC_{50} values, treatments were diluted at 0.6 mg/mL in test medium and 8 serial double dilutions were prepared. These solutions were added to cells and incubated as above. For each concentration point, 8 independent repeats were performed. Graphs and IC$_{50}$ values were obtained with GraphPad Prism 8.3.0 software.

Cell lines were authenticated by STR profiling (IDEXX BioAnalytics, Westbroock, ME, USA).

4.3. Fractionation Strategies

The mushroom with the highest anticancer activity was chosen for further analysis. The initial extract was subjected to different fractionation protocols, which included different chromatographic techniques. Each time new fractions were separated, they were analyzed for their "anticancer" properties through cell line screening (as above) and the positive fractions were subjected to further fractionation until a pure enough compound, that could be identified through analytical chemistry methods, was obtained.

Many combinations were tested, but the one that resulted in final successful purification included alkaloidal separation, followed by column chromatography with polymeric resin HP20-SS, using different mobile phases.

4.4. Elucidation of the Compound's Chemical Structure

The final extract was subjected to analytical techniques, including ^{1}H-nuclear magnetic resonance (NMR), ^{13}C-NMR, DEPT-135-NMR, mass spectroscopy (Q-TOF), and bidimensional experiments heteronuclear multiplke-quantum correlation (HMQC) and correlation spectroscopy (COSY). Interpretation of the data indicated that the new compound was a triglyceride and the component fatty acids were identified to a certain degree of confidence, but the exact location of the fatty acids in the triglyceride, and the position of the two double bonds in the more complex fatty acid, could not be completely determined. Therefore, we synthesized the most probable molecules (Table 1) and tested them with our cell line screening strategy. Given the presence of two double bonds in one of the fatty acids, the triglyceride demonstrated optical activity. Therefore, both enantiomers were synthesized (Table 1) and tested.

4.5. Triglyceride Synthesis

The molecules in Table 1 were synthesized following standard techniques. The details are shown in the Supplementary Materials (Triglyceride synthesis).

4.6. Anticancer Activity in a Xenograft Model

Once the active synthetic molecule (Macrocybin) was available, it was tested in vivo following standard protocols [45]. Briefly, 10×10^6 A549 cells were injected in the flank of twenty 8-week-old non-obese diabetic (NOD) *scid* gamma mice (NSG, Stock No. 005557, The Jackson Laboratory). Animals were randomly divided into 2 experimental groups and when tumors became palpable, they were injected intratumorally with 100 µL of the vehicle (PBS) in the control group ($n = 10$), or with 0.1 mg/mL Macrocybin in 100 µL of PBS in the treatment group ($n = 10$), three times a week. Just before injection, tumor volume was estimated with a caliper by measuring the maximum length and width of the tumor and by applying the following formula: Volume = (width)^2 × length/2 [46]. When tumor volume reached 2000 mm^3, mice were sacrificed for ethical consideration.

All procedures involving animals were carried out in accordance with the European Communities Council Directive (2010/63/UE), and Spanish legislation (RD53/2013) on animal experiments, and with approval from CIBIR's committee on the ethical use of animals (Órgano Encargado del Bienestar Animal del Centro de Investigación Biomédica de La Rioja, OEBA-CIBIR).

4.7. Cytoskeleton Staining and Confocal Microscopy

To test whether Macrocybin, like other natural products, acts through cytoskeleton interactions, A549 and NL20 cells were seeded in 8-well chamber slides (Lab-Tek II), treated with different concentrations of Macrocybin (or the vehicle) for 24 h, washed, fixed with 10% formalin, and permeabilized for 10 min with 0.1% Triton X-100. For cytoskeleton imaging, cells were exposed to 1:1000 mouse anti-tubulin antibody (T6074, Sigma-Aldrich, Sant Louis, MI, USA) overnight at 4 °C, washed with PBS, incubated with a mixture of 1:200 Bodipy-phallacidin (Molecular Probes) and 1:400 goat-anti mouse IgG labeled with Alexa Fluor 633 (A-21052, Invitrogen, Waltham, MA, USA), and washed again. Nuclear staining was achieved with DAPI (ProLong Gold Antifade Mountant, Invitrogen). Slides were observed with a confocal microscope (TCS SP5, Leica, Badalona, Spain).

4.8. Gene Expression

To identify the cellular pathways potentially involved in Macrocybin action, mRNA was extracted from treated and untreated A549 and NL20 cells using the RNeasy MiniKit (Qiagen, Hilden, Germany). Total RNA (1 µg) was reverse transcribed using the SuperScriptR III Reverse Transcriptase Kit (Thermo Fisher Scientific, Waltham, MA, USA), and quantitative real-time PCR was performed as described [47]. Specific primers are shown in Table 2. GAPDH was used as a housekeeping gene.

Table 2. Primers used for qRT-PCR in this study. Annealing temperature was 60 °C for all primers. GAPDH was used as a housekeeping gene.

Target Gene	Forward Primer	Reverse Primer	Amplicon Size
CAV1	GCGACCCTAAACACCTCAAC	CAGCAAGCGGTAAAACCAGT	149
hGOT2_F (=FABPpm)	TGGTGCCTACCGGGATGATA	GGCAGAAAGACATCTCGGCT	153
hSLC27A1_F (=FATP1)	GCCAAATCGGGGAGTTCTAC	TTGAAACCACAGGAGCCGA	85
hSLC27A4_F (=FATP4)	CAAGACCATCAGGCGCGATA	CCGAACGGTAGAGGCAAACA	118
GAPDH	AAATCCCATCACCATCTTCC	GACTCCACGACGTACTCAGC	81

4.9. Statistical Analysis

All datasets were tested for normalcy and homoscedasticity. Normally distributed data were evaluated by Student's t test or by ANOVA followed by the Dunnet's post hoc test while data not following a normal distribution were analyzed with the Kruskal–Wallis test followed by the Mann–Whitney U test. All data were analyzed with GraphPad Prism 8.3.0 software and were considered statistically significant when $p < 0.05$.

5. Conclusions

In conclusion, we have shown that the natural mushroom product, Macrocybin, is a new, biologically active compound that reduces tumor growth by disassembling the actin cytoskeleton, providing a potential new strategy to fight cancer.

Supplementary Materials: The following are available online: Natural extract characterization, triglyceride synthesis, growth modulatory activity of synthetic triglycerides, spectral comparison of aliphatic components, and actual values for xenograft and qRT-PCR studies.

Author Contributions: Conceptualization, M.V., and A.M.; methodology, J.G.-S., L.O.-C., A.L.-R., and J.B.-U.; formal analysis, J.G.-S., L.O.-C., A.L.-R., and J.B.-U.; investigation, M.V., J.G.-S., L.O.-C., A.L.-R., and J.B.-U.; data curation, J.G.-S., L.O.-C., J.B.-U., and A.M.; writing—original draft preparation, A.M.; writing—review and editing, A.M.; supervision, A.M. All authors have read and agreed to the published version of the manuscript.

Funding: This research was funded by Fundación Rioja Salud (FRS), la Agencia de Desarrollo Económico de La Rioja (ADER), project 2017-I-IDD-00067, and Fondos FEDER.

Acknowledgments: The authors gratefully acknowledge the work of Kattia Rosales and her colleagues collecting specimens at the Instituto Nacional de Biodiversidad de Costa Rica (INBio).

Conflicts of Interest: The authors declare no conflict of interest. The funders had no role in the design of the study; in the collection, analyses, or interpretation of data; in the writing of the manuscript, or in the decision to publish the results.

References

1. Hu, Y.; Zhao, T.; Zhang, N.; Zhang, Y.; Cheng, L. A Review of Recent Advances and Research on Drug Target Identification Methods. *Curr. Drug Metab.* **2019**, *20*, 209–216. [CrossRef] [PubMed]
2. Rates, S. Plants as source of drugs. *Toxicon* **2001**, *39*, 603–613. [CrossRef]
3. Newman, D.J.; Cragg, G.M. Natural Products as Sources of New Drugs over the Last 25 Years. *J. Nat. Prod.* **2007**, *70*, 461–477. [CrossRef] [PubMed]
4. Guéritte-Voegelein, F.; Guenard, D.; Potier, P. Taxol and Derivatives: A Biogenetic Hypothesis. *J. Nat. Prod.* **1987**, *50*, 9–18. [CrossRef]
5. Barbuti, A.M.; Chen, Z.-S. Paclitaxel Through the Ages of Anticancer Therapy: Exploring Its Role in Chemoresistance and Radiation Therapy. *Cancers* **2015**, *7*, 2360–2371. [CrossRef]
6. Martino, E.; Casamassima, G.; Castiglione, S.; Cellupica, E.; Pantalone, S.; Papagni, F.; Rui, M.; Siciliano, A.M.; Collina, S. Vinca alkaloids and analogues as anti-cancer agents: Looking back, peering ahead. *Bioorg. Med. Chem. Lett.* **2018**, *28*, 2816–2826. [CrossRef]
7. Wall, M.E.; Wani, M.C.; Cook, C.E.; Palmer, K.H.; McPhail, A.T.; Sim, G.A. Plant Antitumor Agents. I. The Isolation and Structure of Camptothecin, a Novel Alkaloidal Leukemia and Tumor Inhibitor from Camptotheca acuminata1,2. *J. Am. Chem. Soc.* **2005**, *88*, 3888–3890. [CrossRef]
8. Jimenez, P.; Wilke, D.V.; Branco, P.C.; Bauermeister, A.; Rezende-Teixeira, P.; Susana, G.; Costa-Lotufo, L.V. Enriching cancer pharmacology with drugs of marine origin. *Br. J. Pharmacol.* **2020**, *177*, 3–27. [CrossRef]
9. Hamburger, M.; Hostettmann, K. 7. Bioactivity in plants: The link between phytochemistry and medicine. *Phytochemistry* **1991**, *30*, 3864–3874. [CrossRef]
10. Matson, D.R.; Stukenberg, P.T. Spindle Poisons and Cell Fate: A Tale of Two Pathways. *Mol. Interv.* **2011**, *11*, 141–150. [CrossRef]
11. Guggenheim, A.G.; Wright, K.M.; Zwickey, H.L. Immune Modulation From Five Major Mushrooms: Application to Integrative Oncology. *Integr. Med.* **2014**, *13*, 32–44.
12. Benson, K.F.; Stamets, P.; Davis, R.; Nally, R.; Taylor, A.; Slater, S.; Jensen, G.S. The mycelium of the Trametes versicolor (Turkey tail) mushroom and its fermented substrate each show potent and complementary immune activating properties in vitro. *BMC Complement. Altern. Med.* **2019**, *19*, 1–14. [CrossRef] [PubMed]
13. Blagodatski, A.; Yatsunskaya, M.; Mikhailova, V.; Tiasto, V.; Kagansky, A.; Katanaev, V.L. Medicinal mushrooms as an attractive new source of natural compounds for future cancer therapy. *Oncotarget* **2018**, *9*, 29259–29274. [CrossRef] [PubMed]
14. Sullivan, R.; Smith, J.E.; Rowan, N.J. Medicinal Mushrooms and Cancer Therapy: Translating a traditional practice into Western medicine. *Perspect. Biol. Med.* **2006**, *49*, 159–170. [CrossRef] [PubMed]
15. Sohretoglu, D. Ganoderma lucidum Polysaccharides as An Anti-cancer Agent. *Anti-Cancer Agents Med. Chem.* **2018**, *18*, 667–674. [CrossRef] [PubMed]
16. Ina, K.; Kataoka, T.; Ando, T. The Use of Lentinan for Treating Gastric Cancer. *Anti-Cancer Agents Med. Chem.* **2013**, *13*, 681–688. [CrossRef]
17. Rossi, P.; Difrancia, R.; Quagliariello, V.; Savino, E.; Tralongo, P.; Randazzo, C.L.; Berretta, M. B-glucans from Grifola frondosa and Ganoderma lucidum in breast cancer: An example of complementary and integrative medicine. *Oncotarget* **2018**, *9*, 24837–24856. [CrossRef]
18. Pohleven, J.; Kos, J.; Sabotic, J. Medicinal Properties of the Genus Clitocybe and of Lectins from the Clouded Funnel Cap Mushroom, C. nebularis (Agaricomycetes): A Review. *Int. J. Med. Mushrooms* **2016**, *18*, 965–975. [CrossRef]
19. Baek, J.; Roh, H.-S.; Baek, K.; Lee, S.; Lee, S.; Song, S.-S.; Kim, K.H. Bioactivity-based analysis and chemical characterization of cytotoxic constituents from Chaga mushroom (Inonotus obliquus) that induce apoptosis in human lung adenocarcinoma cells. *J. Ethnopharmacol.* **2018**, *224*, 63–75. [CrossRef]
20. Mueller, G.M.; Schmit, J.P. Fungal biodiversity: What do we know? What can we predict? *Biodivers. Conserv.* **2007**, *16*, 1–5. [CrossRef]
21. Pegler, D.N.; Lodge, D.J.; Nakasone, K.K. The Pantropical Genus Macrocybe Gen. nov. *Mycologia* **1998**, *90*, 494. [CrossRef]
22. Lev, S. Non-vesicular lipid transport by lipid-transfer proteins and beyond. *Nat. Rev. Mol. Cell Biol.* **2010**, *11*, 739–750. [CrossRef] [PubMed]

23. Ramirez, N.A.; Niveiro, N.; Michlig, A.; Popoff, O. First record of Macrocybe titans (Tricholomataceae, Basidiomycota) in Argentina. *Check List.* **2017**, *13*, 153–158. [CrossRef]

24. Razaq, A.; Nawaz, R.; Khalid, A.N. An Asian edible mushroom, Macrocybe gigantea: Its distribution and ITS-rDNA based phylogeny. *Mycosphere* **2016**, *7*, 525–530. [CrossRef]

25. Milhorini, S.D.S.; Smiderle, F.R.; Biscaia, S.M.P.; Rosado, F.R.; Trindade, E.S.; Iacomini, M. Fucogalactan from the giant mushroom Macrocybe titans inhibits melanoma cells migration. *Carbohydr. Polym.* **2018**, *190*, 50–56. [CrossRef]

26. Gaur, T.; Rao, P.B. Analysis of Antibacterial Activity and Bioactive Compounds of the Giant Mushroom, Macrocybe gigantea (Agaricomycetes), from India. *Int. J. Med. Mushrooms* **2017**, *19*, 1083–1092. [CrossRef]

27. Sen, T.; Samanta, S.K. Medicinal plants, human health and biodiversity: A broad review. *Adv. Biochem. Eng. Biotechnol.* **2015**, *147*, 59–110.

28. Ulmer, H.; Borena, W.; Rapp, K.; Klenk, J.; Strasak, A.; Diem, G.; Concin, H.; Nagel, G. Serum triglyceride concentrations and cancer risk in a large cohort study in Austria. *Br. J. Cancer* **2009**, *101*, 1202–1206. [CrossRef]

29. Shen, Y.; Wang, C.; Ren, Y.; Ye, J. A comprehensive look at the role of hyperlipidemia in promoting colorectal cancer liver metastasis. *J. Cancer* **2018**, *9*, 2981–2986. [CrossRef]

30. Ni, H.; Liu, H.; Gao, R. Serum Lipids and Breast Cancer Risk: A Meta-Analysis of Prospective Cohort Studies. *PLoS ONE* **2015**, *10*, e0142669. [CrossRef]

31. Cannella, C.; Giusti, A. Conjugated linoleic acid: A natural anticarcinogenic substance from animal food. *Ital. J. Food Sci.* **2000**, *12*, 123–127.

32. Hanafy, N.A.N.; El-Kemary, M.; Leporatti, S. Micelles structure development as a strategy to improve smart cancer therapy. *Cancers* **2018**, *10*, 238. [CrossRef] [PubMed]

33. Nicolas, J.; Couvreur, P. Polymer nanoparticles for the delivery of anticancer drug. *Med. Sci.* **2017**, *33*, 11–17.

34. Joshi, S.; Durden, D.L. Combinatorial Approach to Improve Cancer Immunotherapy: Rational Drug Design Strategy to Simultaneously Hit Multiple Targets to Kill Tumor Cells and to Activate the Immune System. *J. Oncol.* **2019**, *2019*, 1–18. [CrossRef] [PubMed]

35. Meshulam, T.; Simard, J.R.; Wharton, J.; Hamilton, J.A.; Pilch, P.F. Role of Caveolin-1 and Cholesterol in Transmembrane Fatty Acid Movement. *Biochemistry* **2006**, *45*, 2882–2893. [CrossRef] [PubMed]

36. Siddiqi, S.; Sheth, A.; Patel, F.; Barnes, M.; Mansbach, C.M. Intestinal caveolin-1 is important for dietary fatty acid absorption. *Biochim. Biophys. Acta* **2013**, *1831*, 1311–1321. [CrossRef] [PubMed]

37. Gerbod-Giannone, M.-C.; Dallet, L.; Naudin, G.; Sahin, A.; Decossas, M.; Poussard, S.; Lambert, O. Involvement of caveolin-1 and CD36 in native LDL endocytosis by endothelial cells. *Biochim. Biophys. Acta* **2019**, *1863*, 830–838. [CrossRef]

38. Chen, D.; Shen, C.; Du, H.; Zhou, Y.; Che, G.-W. Duplex value of caveolin-1 in non-small cell lung cancer: A meta analysis. *Fam. Cancer* **2014**, *13*, 449–457. [CrossRef]

39. Shi, Y.-B.; Li, J.; Lai, X.-N.; Jiang, R.; Zhao, R.-C.; Xiong, L.-X. Multifaceted Roles of Caveolin-1 in Lung Cancer: A New Investigation Focused on Tumor Occurrence, Development and Therapy. *Cancers* **2020**, *12*, 291. [CrossRef]

40. Alonso, E.N.; Ferronato, M.J.; Fermento, M.E.; Gandini, N.A.; Romero, A.L.; Guevara, J.A.; Facchinetti, M.M.; Curino, A.C. Antitumoral and antimetastatic activity of Maitake D-Fraction in triple-negative breast cancer cells. *Oncotarget* **2018**, *9*, 23396–23412. [CrossRef]

41. Guo, H.; Kuang, S.; Song, Q.-L.; Liu, M.; Sun, X.-X.; Yu, Q. Cucurbitacin I inhibits STAT3, but enhances STAT1 signaling in human cancer cells in vitro through disrupting actin filaments. *Acta Pharmacol. Sin.* **2018**, *39*, 425–437. [CrossRef] [PubMed]

42. Chikara, S.; Lindsey, K.; Borowicz, P.; Christofidou-Solomidou, M.; Reindl, K. Enterolactone alters FAK-Src signaling and suppresses migration and invasion of lung cancer cell lines. *BMC Complement. Altern. Med.* **2017**, *17*, 1–12. [CrossRef] [PubMed]

43. Van Goietsenoven, G.; Mathieu, V.; Lefranc, F.; Kornienko, A.; Evidente, A.; Kiss, R. Narciclasine as well as other Amaryllidaceae Isocarbostyrils are Promising GTP-ase Targeting Agents against Brain Cancers. *Med. Res. Rev.* **2013**, *33*, 439–455. [CrossRef] [PubMed]

44. Iwai, N.; Martinez, A.; Miller, M.-J.; Vos, M.; Mulshine, J.L.; Treston, A.M. Autocrine growth loops dependent on peptidyl α-amidating enzyme as targets for novel tumor cell growth inhibitors. *Lung Cancer* **1999**, *23*, 209–222. [CrossRef]

45. Zitvogel, L.; Pitt, J.M.; Daillère, L.Z.J.M.P.R.; Smyth, M.J.; Kroemer, G. Mouse models in oncoimmunology. *Nat. Rev. Cancer* **2016**, *16*, 759–773. [CrossRef]

46. Monga, S.P.S.; Wadleigh, R.; Sharma, A.; Adib, H.; Strader, D.; Singh, G.; Harmon, J.W.; Berlin, M.; Monga, D.K.; Mishra, L. Intratumoral Therapy of Cisplatin/Epinephrine Injectable Gel for Palliation in Patients With Obstructive Esophageal Cancer. *Am. J. Clin. Oncol.* **2000**, *23*, 386–392. [CrossRef]

47. Ochoa-Callejero, L.; Garcia-Sanmartin, J.; Martínez-Herrero, S.; Rubio-Mediavilla, S.; Narro-Íñiguez, J.; Martinez, A. Small molecules related to adrenomedullin reduce tumor burden in a mouse model of colitis-associated colon cancer. *Sci. Rep.* **2017**, *7*, 17488. [CrossRef]

Sample Availability: Samples of Macrocybin are available from the authors, while supplies last.

Publisher's Note: MDPI stays neutral with regard to jurisdictional claims in published maps and institutional affiliations.

Article

Glycoconjugation of Betulin Derivatives Using Copper-Catalyzed 1,3-Dipolar Azido-Alkyne Cycloaddition Reaction and a Preliminary Assay of Cytotoxicity of the Obtained Compounds

Mirosława Grymel [1,2,*], Gabriela Pastuch-Gawołek [1,2], Anna Lalik [2,3], Mateusz Zawojak [1], Seweryn Boczek [1], Monika Krawczyk [1,2] and Karol Erfurt [4]

[1] Department of Organic Chemistry, Bioorganic Chemistry and Biotechnology, Silesian University of Technology, B. Krzywoustego 4, 44-100 Gliwice, Poland; gabriela.pastuch@polsl.pl (G.P.-G.); mateuszzawojak@wp.pl (M.Z.); seweryn.boczek16@gmail.com (S.B.); monika.krawczyk@polsl.pl (M.K.)

[2] Biotechnology Center, Silesian University of Technology, B. Krzywoustego 8, 44-100 Gliwice, Poland; anna.lalik@polsl.pl

[3] Department of Systems Biology and Engineering, Silesian University of Technology, Akademicka 16, 44-100 Gliwice, Poland

[4] Department of Chemical Organic Technology and Petrochemistry, Silesian University of Technology, B. Krzywoustego 4, 44-100 Gliwice, Poland; karol.erfurt@polsl.pl

* Correspondence: miroslawa.grymel@polsl.pl; Tel.: +48-032-237-1873

Academic Editors: José Antonio Lupiáñez, Amalia Pérez-Jiménez and Eva E. Rufino-Palomares

Received: 26 November 2020; Accepted: 16 December 2020; Published: 18 December 2020

Abstract: Pentacyclic lupane-type triterpenoids, such as betulin and its synthetic derivatives, display a broad spectrum of biological activity. However, one of the major drawbacks of these compounds as potential therapeutic agents is their high hydrophobicity and low bioavailability. On the other hand, the presence of easily transformable functional groups in the parent structure makes betulin have a high synthetic potential and the ability to form different derivatives. In this context, research on the synthesis of new betulin derivatives as conjugates of naturally occurring triterpenoid with a monosaccharide via a linker containing a heteroaromatic 1,2,3-triazole ring was presented. It has been shown that copper-catalyzed 1,3-dipolar azide-alkyne cycloaddition reaction (CuAAC) provides an easy and effective way to synthesize new molecular hybrids based on natural products. The chemical structures of the obtained betulin glycoconjugates were confirmed by spectroscopic analysis. Cytotoxicity of the obtained compounds was evaluated on a human breast adenocarcinoma cell line (MCF-7) and colorectal carcinoma cell line (HCT 116). The obtained results show that despite the fact that the obtained betulin glycoconjugates do not show interesting antitumor activity, the idea of adding a sugar unit to the betulin backbone may, after some modifications, turn out to be correct and allow for the targeted transport of betulin glycoconjugates into the tumor cells.

Keywords: betulin glycoconjugates; click chemistry; 1,3-dipolar cycloaddition; anticancer activity

1. Introduction

For several decades, natural products (*NPs*) have been widely researched in terms of searching for potential therapeutic agents. One important class of natural plant products is pentacyclic lupane-type triterpenoids, among them betulin (*BN*). In recent years, due to a strong biological activity and low toxicity, the great interest of scientists has been focused on its naturally occurring bioactive skeleton, as evidenced by numerous new semi-synthetic derivatives reported every year. Betulin (*BN*, 3-lup-20(29)-ene-3,28-diol) is cheap, easily accessible from natural resources, and can be readily extracted from the bark of several species of trees, especially from the white birch (*Betula pubescens*) [1,2]. *BN* possesses a broad spectrum of biological activity confirmed by its anticancer [3–13], antibacterial [14,15], anti-HIV [14,16], anti-inflammatory [17–19], antiviral [20,21], antimalarial activity [17,19], and hepatoprotective properties [9,22,23].

One of the major drawbacks of betulin and analogs as potential therapeutic agents is its poor hydrophilicity, poor solubility in aqueous media like blood serum and polar solvents used for bioassays, and low bioavailability. However, due to the presence of easily transformable functional groups (including C3OH, C28OH) in the parent structure, betulin has a high synthetic potential and the ability to form numerous semisynthetic derivatives. Thus, the parent structure of betulin, being a biologically attractive scaffold with a high safety profile, makes it possible to carry out a variety of structural modifications to improve its pharmacokinetic properties. In some cases [5,6,9,12,24–26], the transformations gave compounds with a higher bioavailability and higher biological activity than the parent compound (Figure 1).

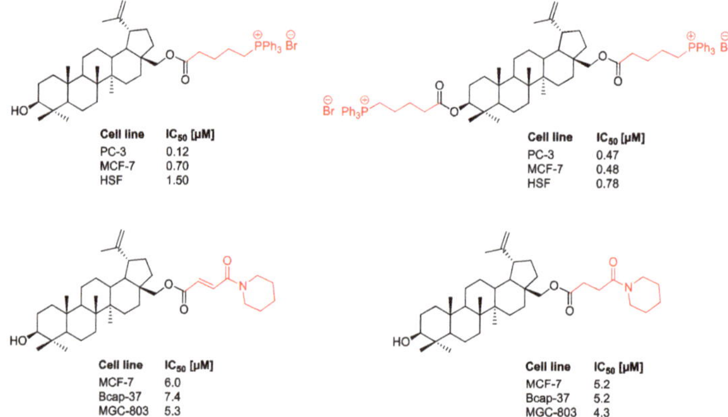

Figure 1. Antitumor activities of betulin derivatives against human breast carcinoma (MCF-7 and Bcap-37), human gastric carcinoma (MGC-803), human prostate carcinoma (PC-3), and human skin fibroblast (HSF) cell lines [12,27].

One of the strategies to improve the pharmacokinetic properties of betulin is its link to the 1,2,3-triazole ring, which is an important five-membered heterocyclic scaffold due to the extensive biological activities. This framework can be obtained through the copper-catalyzed azide-alkyne cycloaddition (CuAAC). In addition, the 1,2,3-triazole ring can be successfully used for conjugation in mild conditions and with high yields, with two or more molecules of interest. Due to the stability of triazoles in typical physiological conditions, they act as ideal linkers and could improve the hydrophilic property of the parent skeleton of NPs via the ability to create hydrogen bonding. The 1,2,3-triazole moiety mimics the naturally occurring amide bonds, and at the same time, it is more stable [28–31]. This methodology allows the functionalization of *BN* at position C3, C28, and C30, with different substrate (Scheme 1). Although the triazole derivatives of betulin have been widely researched, no structures have been found so far that would demonstrate cytotoxicity at a level that would allow them to be used as drugs [32].

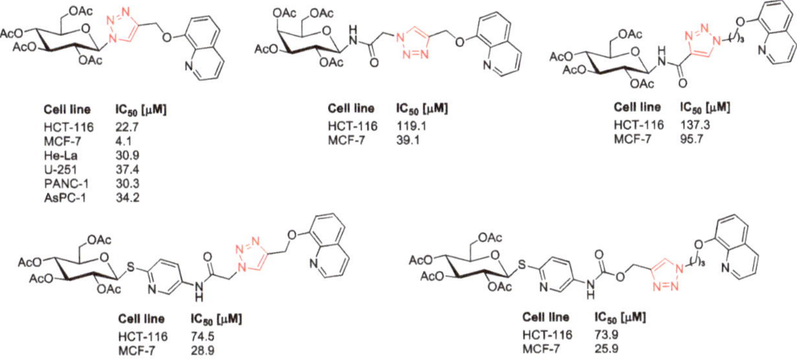

Scheme 1. Functionalization of betulin at the C3, C28, and C30 position via CuAAC [33–35].

The motivation for our research was the successes in the synthesis of molecular hybrids presented in Figure 2, obtained with the use of a natural biologically active compound, where the introduction of a triazole ring and a sugar unit into the parent structure of quinoline resulted in the improvement of pharmacological properties of the obtained glycoconjugates [36–39].

Cell line	IC$_{50}$ [µM]
HCT-116	22.7
MCF-7	4.1
He-La	30.9
U-251	37.4
PANC-1	30.3
AsPC-1	34.2

Cell line	IC$_{50}$ [µM]
HCT-116	119.1
MCF-7	39.1

Cell line	IC$_{50}$ [µM]
HCT-116	137.3
MCF-7	95.7

Cell line	IC$_{50}$ [µM]
HCT-116	74.5
MCF-7	28.9

Cell line	IC$_{50}$ [µM]
HCT-116	73.9
MCF-7	25.9

Figure 2. Selected quinoline glycoconjugates with anticancer activity.

The presence of a sugar unit in the hybrid molecules improves their pharmacokinetic properties, including solubility and intermembrane transport, as well as the selectivity in targeting drugs for a specific purpose. Sugar moieties are known to influence the pharmacokinetic properties of the respective compounds, such as absorption, distribution, and metabolism [40]. Since it is generally accepted that glycosides are more water soluble than the respective aglycones, glycosidation of triterpenes should increase hydrosolubility and improve the pharmacological properties. Furthermore, it is known that cancerous cells need a more significant sugar contribution than normal cells [41]. It is related to the specific metabolism of glucose by cancer cells as well as with their often-increased proliferation compared to healthy cells. This phenomenon is known as the Warburg effect and arises from mitochondrial metabolic changes. It consists in the fact that cancer cells produce their energy through glycolysis followed by lactic acid fermentation, characteristic of hypoxic conditions, and its level is much higher (more than 100 times) than in healthy cells, for which the main source of energy is mitochondrial oxidative phosphorylation [42]. An increased glycolysis process in cancer cells is associated with overexpression of GLUT transporters (special transmembrane proteins that facilitate concentration-dependent glucose uptake inside the cell) [43]. Consequently, this difference could be exploited to support the absorption of the therapeutic agent by the tumoral site, which allows for controlled drug delivery.

Our research group decided to focus on the synthesis of the new betulin glycoconjugates via click chemistry reaction. We designed semi-synthetic betulin derivatives as a combination of naturally occurring triterpenoid (BN) with proven broad biological activity, with a monosaccharide through a linker containing a heteroaromatic 1,2,3-triazole ring. The choice of the sugar unit for glycoconjugation was determined by the willingness to increase the selectivity of potential drugs by using the overexpression of GLUT transporters in some types of cancers. These transporters also recognize other sugars, such as galactose, mannose, and glucosamine [44]. It was decided to attach either D-glucose or D-galactose units due to the common occurrence of these sugars in nature. These sugars were supposed to perform the function of a drug carrier across the cell membrane by adapting to the structure of the GLUT transporter [43]. The configuration at the anomeric center of the sugar also seems important from the point of view of matching with the transporter. The results of research on the structural requirements for binding variously substituted sugars to the sugar transporters indicate that the β-configuration is preferred for binding to the GLUT transporters [45,46]. In order to obtain a library of compounds intended for the initial assessment of biological cytotoxicity, we tested two strategies for conjugating the sugar unit with the parent skeleton of betulin, both based on the 1,3-dipolar azido-alkyne cycloaddition reaction (Scheme 2).

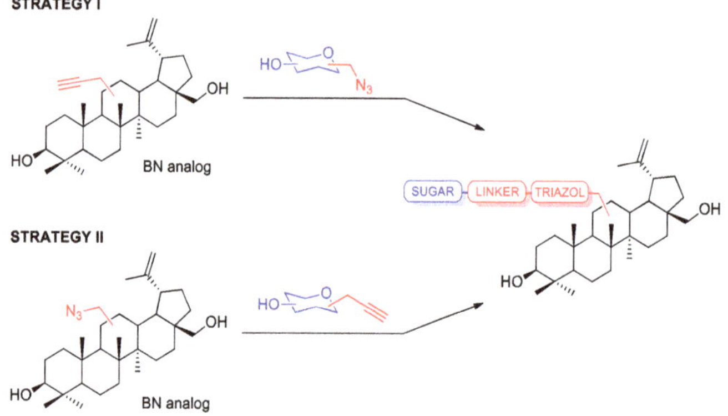

Scheme 2. Strategies for the synthesis of betulin glycoconjugates.

Strategy I involves the introduction of a propargyl moiety into the betulin structure at position C28 (monosubstituted betulin analogs) or C3 and C28 (disubstituted betulin analogs), followed by the synthesis of glycoconjugates containing the 1,2,3-triazole ring, resulting from combining propargyl betulin derivatives and appropriate sugar derivatives containing an azide moiety. *Strategy II*, on the other hand, consists of adding an appropriate linker (O(CO)CH$_2$N$_3$) to the betulin backbone one at position C28 or at two positions, C3 and C28, and the synthesis of glycoconjugates via click chemistry with the use of betulin azides and propargyl sugar derivatives. The linker that connects the betulin scaffold to the sugar unit is an important element that may influence the potential biological activity of the discussed betulin glycoconjugates. In this study, we report the synthesis of several glycoconjugates, having a linker containing a 1,2,3-triazole moiety and an additional ether (*O*-glycosidic bond), ester, or amide moiety (Figure 3). The triazole ring seems to be an ideal linker component as it improves water solubility, thus allowing in vivo administration; it is analogous to an amide function for its electronic properties but simultaneously is resistant to hydrolysis, it is sufficiently stable in biological systems, and finally, it is rigid enough, which allows avoidance of internal interaction between the two linked moieties. On the other hand, the presence of a glycosidic, amide, or ester linkage gives a chance for the hydrolysis of the glycoconjugate into its biologically active components after the compound enters the cell under the action of hydrolytic enzymes. The cytotoxicity of the obtained compounds was evaluated on a human breast adenocarcinoma cell line (MCF-7) and colorectal carcinoma cell line (HCT 116).

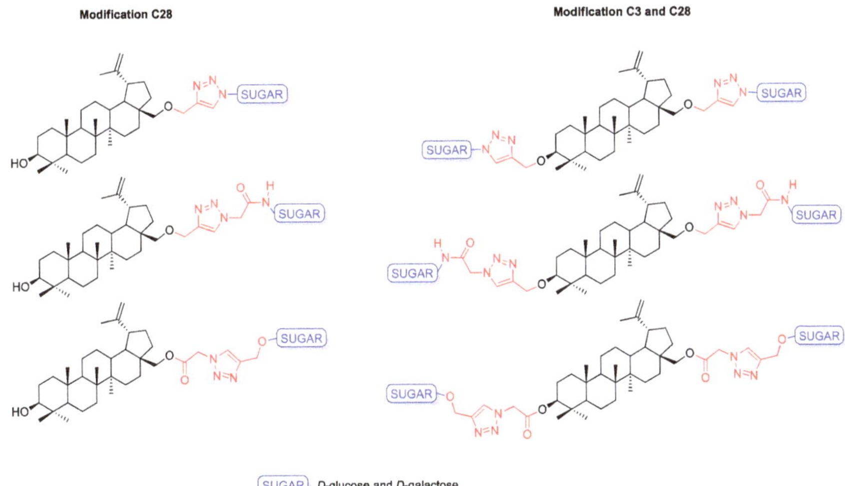

Figure 3. General structure of the betulin glycoconjugates.

2. Results and Discussion

2.1. Synthesis of Betulin Analogs

The introduction of a propargyl moiety into the betulin molecule was based on the procedure described by Khan et al. for betulinic acid [47]. It was assumed that both mono- and dipropargyl derivatives of betulin could be formed due to the presence in the substrate of the primary OH group at the C28 position and the secondary OH group at the C3 position. The propargylated betulin was prepared as shown in Scheme 3. Crystalline betulin was reacted with propargyl bromide in alkaline medium (NaH) in tetrahydrofuran. In the search for the optimal procedure, a number of modifications were made, by selecting the proportions of reagents, reaction time, and temperature. It was found that it is advantageous to use the reagents in a molar ratio (BN/NaH/HC≡CCH$_2$Br, 1:4:3.2) at ambient temperature for 24 h.

Scheme 3. Synthesis of propargyl derivatives of betulin **2** and **3**. Reagents and Conditions: (**i**) propargyl bromide, NaH, THF, r.t., 24 h.

The synthesis of 28-*O*-chloroacetylbetulin **4** and 3,28-*O*,*O*′-di(chloroacetyl)betulin **5** was performed by modifying the procedure described by Komissarova et al. in 2017 [48]. Chloroacetic chloride was added dropwise to the betulin solution in THF, while triethylamine (Et$_3$N) was replaced with *N*,*N*-diisopropylethylamine (DIPEA) and 4-(dimethylamino)pyridine (DMAP). We found that the application of twice the excess of chloroacetic chloride to BN results in the formation of only 3,28-*O*,*O*′-di(chloroacetyl)betulin **5** with a high yield (95%, Scheme 4). On the other hand, when the reagents were used in a molar ratio (Cl(CO)CH$_2$Cl/DIPEA/DMAP, 1.2:1.5:0.1), a modification of the betulin skeleton was observed at both the C28 and C3 positions. The maximum yield with which a monosubstituted product **4** can be obtained is 34%. Replacement of the chlorine anion in betulin analogs **4** and **5** by azide moiety was performed at an elevated temperature (90 °C) using NaN$_3$ and DMF as a solvent. 28-*O*-Azidoacetylbetulin **6** and 3,28-*O*,*O*′-di(azidoacetyl)betulin **7** were purified by column chromatography to give pure compounds with 64% and 72% yields, respectively (Scheme 4, Table 1).

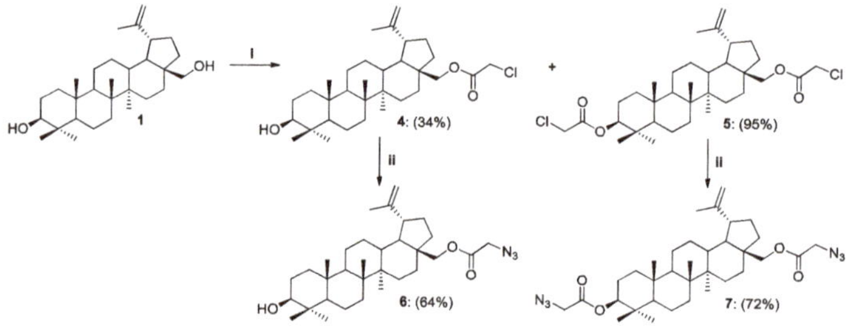

Scheme 4. Synthesis of betulin derivatives **4–7**. Reagents and Conditions: (i) Cl(CO)CH$_2$Cl, DIPEA, DMAP, THF, r.t., 24 h; (ii) NaN$_3$, DMF, 90 °C, 3 h.

Table 1. Synthesis of betulin derivatives **2–7**.

Entry	Substrate				Product			
	No.	Solvent	Temp., °C	Time, h	No.	R (C3)	R^1 (C28)	Yield, %
1	1	THF	r.t.	24	2	OH	OCH2C≡CH	54
					3	OCH2C≡CH	OCH2C≡CH	31
2	1	THF	r.t.	24	4	OH	O(CO)CH2Cl	34
3	1	THF	r.t.	24	5	O(CO)CH2Cl	O(CO)CH2Cl	95
4	4	DMF	90	3	6	OH	O(CO)CH2N3	64
5	5	DMF	90	3	7	O(CO)CH2N3	O(CO)CH2N3	72

2.2. Synthesis of Sugar Derivatives

Sugar derivatives substituted at the anomeric position were the second necessary structural element for the synthesis of betulin glycoconjugates. The synthetic route to the corresponding derivatives of D-glucose and D-galactose **9–14** is shown in Scheme 5. As mentioned before, the choice of such sugar moieties is dictated by the frequency of their natural occurrence as well as their importance for cell metabolism. All sugar derivatives were prepared according to known procedures involving the acetylation of free sugars **8a** or **8b** and conversion of per-*O*-acetylated derivatives **9a** or **9b** into the corresponding glycosyl bromides **11a** or **11b** [49,50]. The glycosyl bromides were used for further reactions leading to obtain 2,3,4,6-tetra-*O*-acetyl-β-glycosyl azides **12a** and **12b** [36]. D-Glucose derivative **10a**, in which the alkynyl moiety was introduced by the formation of an *O*-glycosidic linkage, was prepared by reacting per-*O*-acetylated D-glucose **9a** with propargyl alcohol in the presence of BF$_3$·Et$_2$O as a Lewis acid catalyst [51]. Sugar derivatives **13a** and **13b** containing an amide moiety at the sugar anomeric position were obtained in a two-step procedure. First, glycosyl azides **12a** and **12b** were converted to the corresponding 1-aminosugars through a hydrogenation reaction in a Parr apparatus using palladium hydroxide deposited on activated charcoal and then such obtained intermediates were reacted with chloroacetyl chloride in the presence of TEA, which neutralized the formed hydrogen chloride. Finally, the terminal chlorine atom in compounds **13a** and **13b** was exchanged with an azide group, which made it possible to obtain sugar derivatives **14a** and **14b** as a result of the reaction with sodium azide in anhydrous DMF [37].

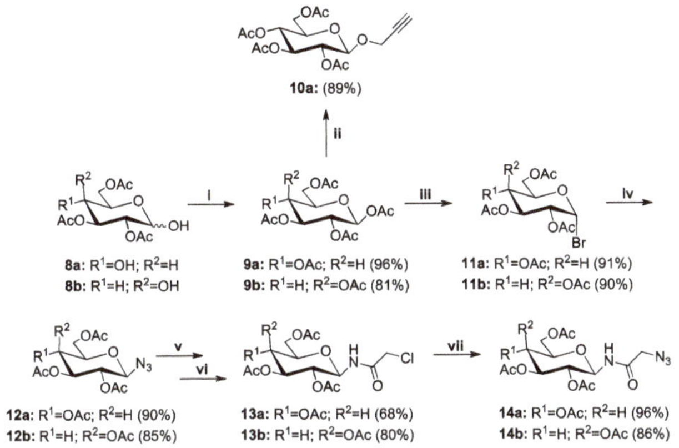

10a: (89%)

8a: R^1=OH; R^2=H
8b: R^1=H; R^2=OH

9a: R^1=OAc; R^2=H (96%)
9b: R^1=H; R^2=OAc (81%)

11a: R^1=OAc; R^2=H (91%)
11b: R^1=H; R^2=OAc (90%)

12a: R^1=OAc; R^2=H (90%)
12b: R^1=H; R^2=OAc (85%)

13a: R^1=OAc; R^2=H (68%)
13b: R^1=H; R^2=OAc (80%)

14a: R^1=OAc; R^2=H (96%)
14b: R^1=H; R^2=OAc (86%)

Scheme 5. Synthesis of sugar derivatives **9–14**. Reagents and Conditions: (**i**) CH_3COONa, Ac_2O, b.p., 1 h; (**ii**) propargyl alcohol, $BF_3 \cdot Et_2O$, DCM, 0 °C→r.t, 2 h; (**iii**) CH_3COOH, 33% HBr/AcOH, r.t., 1 h; (**iv**) NaN_3, TBASH, $CHCl_3$/$NaHCO_3$, r.t., 2 h; (**v**) 20% $Pd(OH)_2$/C, THF:EtOH (2:1, v/v), H_2, 1.5 bar, 2 h; (**vi**) chloroacetyl chloride, TEA, DCM, r.t., 1 h; (**vii**) NaN_3, DMF, r.t., 24 h.

2.3. Modifications of the Natural Betulin Skeleton via Click Chemistry

The copper-catalyzed azide-alkyne cycloaddition reaction (CuAAC) is widely used to modify natural bioactive compounds, and its main advantage is the ability to synthesize new molecular hybrids with high yields under mild conditions. In the course of our research, we developed a methodology for the synthesis of new previously unknown betulin derivatives, consisting of joining the triterpenoid skeleton with a monosaccharide through an appropriate linker containing a heteroaromatic 1,2,3-triazole ring. Glycoconjugation of betulin derivatives was based on the click chemistry reaction (CuAAC) using propargylbetulin derivatives **2**, **3** and protected D-glucose **12a**, **14a** or D-galactose **12b**, **14b** derivatives containing an azide group (Strategy I, Schemes 6 and 7) or propargyl O-glycoside derivative of *per*-O-acetylated-D-glucose **10a** with appropriate betulin azides **6**, **7** (Strategy II, Scheme 8). Building blocks designed in this way make it possible to obtain a whole range of betulin glycoconjugates, containing in the linker structure different combinations of an O-glycosidic linkage, an ester bond, an O-methylene 1,2,3-triazole linker, and an amide moiety.

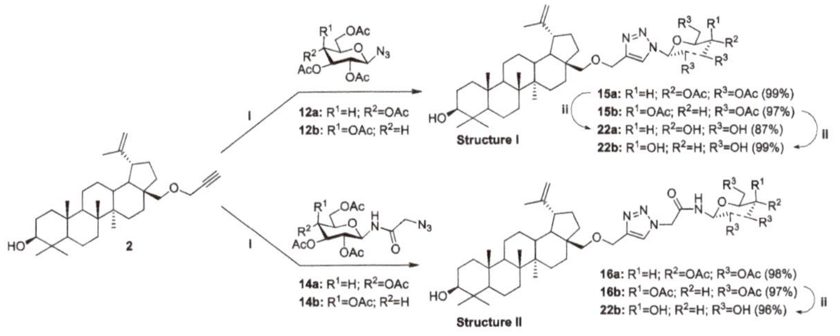

12a: R^1=H; R^2=OAc
12b: R^1=OAc; R^2=H

Structure I

15a: R^1=H; R^2=OAc; R^3=OAc (99%)
15b: R^1=OAc; R^2=H; R^3=OAc (97%)
22a: R^1=H; R^2=OH; R^3=OH (87%)
22b: R^1=OH; R^2=H; R^3=OH (99%)

14a: R^1=H; R^2=OAc
14b: R^1=OAc; R^2=H

Structure II

16a: R^1=H; R^2=OAc; R^3=OAc (98%)
16b: R^1=OAc; R^2=H; R^3=OAc (97%)
22b: R^1=OH; R^2=H; R^3=OH (96%)

Scheme 6. Synthesis of glycoconjugates type I and II according to Strategy I. Reagents and Conditions: (**i**) $CuSO_4 \cdot 5H_2O$, NaAsc, *i*-PrOH/H_2O (1:1, v/v), r.t., 24 h; (**ii**) 1. NaOMe, MeOH, r.t., 0.5 h; 2. *Amberlyst*-15.

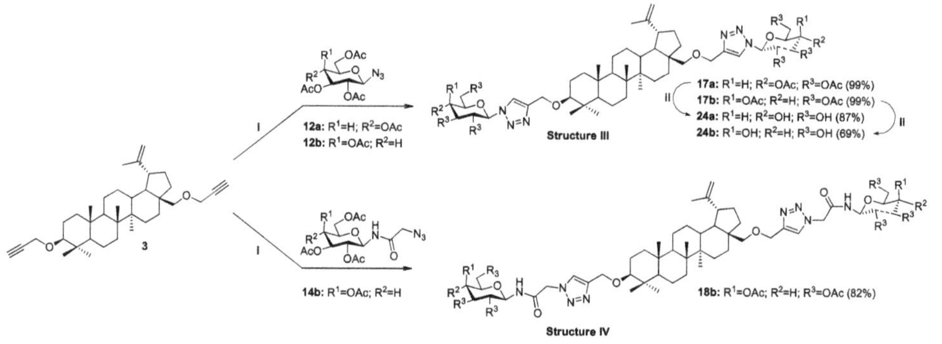

Scheme 7. Synthesis of glycoconjugates type **III** and **IV** according to Strategy I. Reagents and Conditions: (**i**) $CuSO_4 \cdot 5H_2O$, NaAsc, i-PrOH/H_2O (1:1, *v/v*), r.t., 24 h; (**ii**) 1. NaOMe, MeOH, r.t., 0.5 h: 2. *Amberlyst-15*.

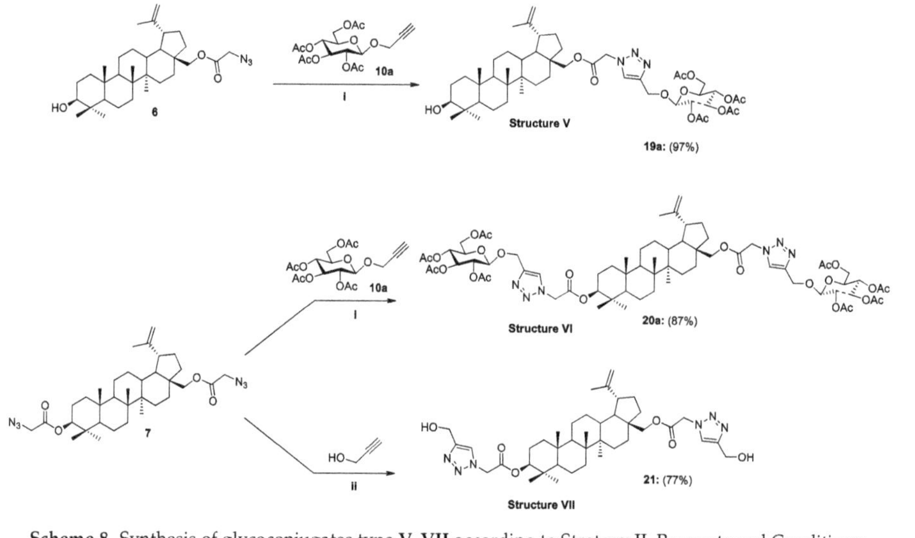

Scheme 8. Synthesis of glycoconjugates type **V–VII** according to Strategy II. Reagents and Conditions: (**i**) $CuSO_4 \cdot 5H_2O$, NaAsc, *i*-PrOH/THF/H_2O, r.t., 24 h, mol. ratio 6/10a/$CuSO_4 \cdot 5H_2O$/NaAsc, 1:1:0.5:1.1; 7/10a/$CuSO_4 \cdot 5H_2O$/NaAsc, 1:2:1.1:0.6; (**ii**) propargyl alcohol, $CuSO_4 \cdot 5H_2O$, NaAsc, *i*-PrOH/THF/H_2O (1.6:1:1.1, *v/v/v*), r.t., 24 h, mol ratio 7/C_3H_4O /$CuSO_4 \cdot 5H_2O$/NaAsc, 1:3:1.2:2.2.

The general procedure of glycoconjugation of betulin derivatives involves mixing of 28-*O*-propargylbetulin **2** or 3,28-*O*,*O'*-di(propargyl)betulin **3** with the appropriate monosugar derivative (D-glucose or D-galactose) in the water-alcohol solvent system (*i*-PrOH/H_2O). In the case of 28-*O*-azidoacetylbetulin **6** and 3,28-*O*,*O'*-di(azidoacetyl)betulin **7**, due to the too low solubility of the substrates in the *i*-PrOH/H_2O solvent system, it was necessary to use an additional solvent (THF). In each case, the reaction was carried out under an argon atmosphere at room temperature for 24 h, using $CuSO_4 \cdot 5H_2O$ as a catalyst and sodium ascorbate (NaAsc) as an agent reducing Cu(II) to Cu(I).

As a result of the described reactions, five new monosubstituted betulin derivatives (**15a**, **15b**, **16a**, **16b**, and **19a**) modified at position C28, as well as four new disubstituted betulin derivatives (**17a**, **17b**, **18b**, and **20a**), modified at positions C3 and C28 were isolated by column chromatography in very good yields (97–99% and 82–99%, respectively, Table 2).

Table 2. Summary synthesis of betulin glycoconjugates **15–24**.

Entry	Substrate				Product		
	Analog BN	No.	Sugar	R^1, R^2, R^3	No.	Linker	Yield, %
STRATEGY I							
1	2	12a	Glc	OAc	15a	28-OCH2Triaz	99
2	2	12b	Gal	OAc	15b	28-OCH2Triaz	97
3	2	14a	Glc	OAc	16a	28-OCH2TriazCH2(CO)NH	98
4	2	14b	Gal	OAc	16b	28-OCH2TriazCH2(CO)NH	97
5	3	12a	Glc	OAc	17a	3,28-di-OCH2Triaz	99
6	3	12b	Gal	OAc	17b	3,28-di-OCH2Triaz	99
7	3	14b	Gal	OAc	18b	3,28-di-OCH2TriazCH2(CO)NH	82
STRATEGY II							
8	6	10a	Glc	OAc	19a	28-O(CO)CH2TriazCH2O	97
9	7	10a	Glc	OAc	20a	3,28-di-O(CO)CH2TriazCH2O	87
10	7	–	–	–	21	3,28-di-O(CO)CH2TriazOH	77
DEPROTECTION							
11	15a	–	Glc	OH	22a	28-OCH2Triaz	87
12	15b	–	Gal	OH	22b	28-OCH2Triaz	99
13	16b	–	Gal	OH	23b	28-OCH2TriazCH2(CO)NH	96
14	17a	–	Glc	OH	24a	3,28-di-OCH2Triaz	87
15	17b	–	Gal	OH	24b	3,28-di-OCH2Triaz	69

It has been considered that the use of an acyl protected sugar moiety may not be sufficient to achieve the desired hydrophilicity of the novel BN analogs. In order to compare the properties of glycoconjugates and assess the effect of acyl group presence, for the five per-*O*-acetylated glycoconjugates BN (**15a, 15b, 16a, 16b, 17b**), the deprotection of the sugar unit was performed by applying a standard Zemplén procedure [52], under mild conditions using 1 M methanolic solution of sodium methoxide in methanol. The reaction was carried out at room temperature for 120 min. The final step was to neutralize the reaction mixture with the use of *Amberlyst-15* ion exchange resin, after which the mixture was filtered to give betulin glycoconjugates (**22a, 22b, 23b, 24a**) in high yield (87-99%), except compound **24b** (69%), as shown in Table 2.

Our proposed concept of adding a linker to the scaffold of betulin is based on introducing a chloroacetyl moiety into its structure and then converting the obtained analogs into azidoacetyl derivatives. In the case of the analog **21**, in the final stage, the azide moiety was coupled with propargyl alcohol using the click chemistry reaction. As a result, a 3,28-disubstituted betulin **21** modified with a linker containing a 1,2,3-triazole ring substituted in the C4 position with a hydroxymethyl group was obtained in a yield of 77% (Scheme 8). The motivation to obtain a 3,28-*O*,*O'*-di(2-(4-(hydroxymethyl-*1H*-1,2,3-triazol-1-yl)acetyl)betulin **21** was the desire to check whether the addition of a sugar unit is necessary to improve the biological properties of BN or whether the introduction of the linker containing the 1,2,3-triazole ring alone would be sufficient. The structures of all synthesized compounds were confirmed by nuclear magnetic resonance (^{1}H-, ^{13}C-NMR, gHSQC), infrared spectroscopy (FT-IR), and high-resolution mass spectrometry (HRMS).

2.4. Cytotoxicity Studies

The obtained betulin glycoconjugates were screened to determine their cytotoxicity. The research was conducted on two cell lines: HCT 116 (colorectal carcinoma cell line) and MCF-7 (human breast adenocarcinoma cell line). In these lines, overexpression of the glucose and galactose transporters was observed [53–55]. The aim of the research was to initially estimate the influence of the modifications of the parent betulin backbone by adding a sugar unit/units, as well as to examine the influence of the structure of linker connecting betulin with sugar on the cytotoxicity of the tested compounds.

In the first stage, betulin glycoconjugates in which the hydroxyl groups of the sugar unit were protected with acetyl groups were selected for research. The choice of the type of hydroxyl group protection was dictated by the sensitivity of acetyl groups to the action of hydrolytic enzymes and the fact that these groups provide glycoconjugates with increased hydrophobicity, which should facilitate passive transport into the cell. The proliferation of tumor cells (HCT 116, MCF-7) treated with tested compounds (**15a, 15b, 16a, 17a, 17b, 19a**) at 50- and 25-µM concentrations were determined after 24 and 48 h of incubation with the Cell Counting Kit-8 (CCK-8) based on the water-soluble tetrazole salt of WST-8. The CCK-8 kit, selected for the determination of viability and cytotoxicity, is less toxic and has higher sensitivity than other tetrazole salt-based tests, such as MTT or MTS [56]. The effect of these compounds was compared with the effect of betulin **1** doses, as shown in Figure 4.

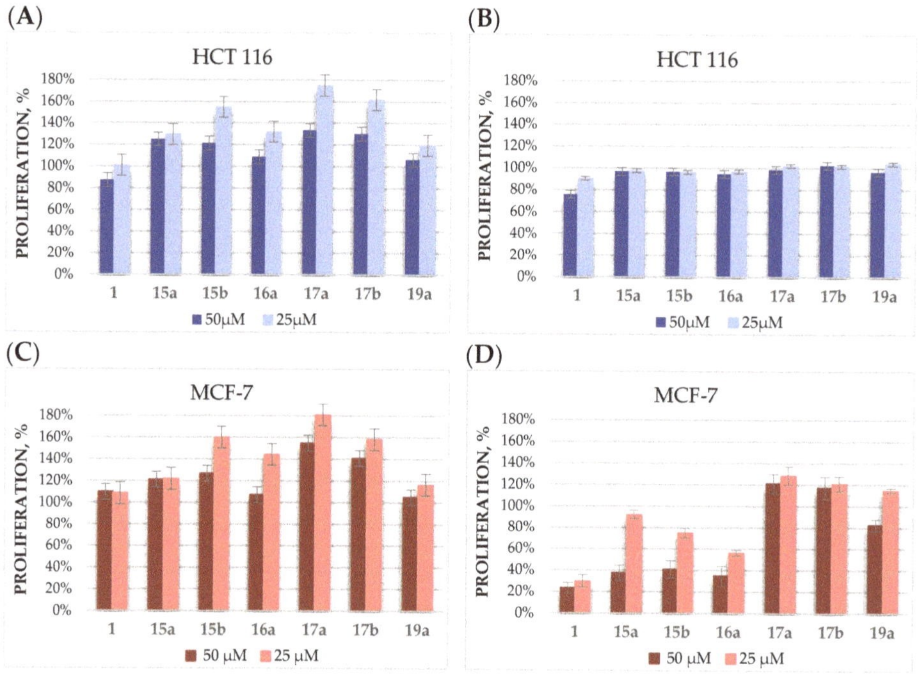

Figure 4. The dependence of HCT 116 and MCF-7 cell proliferation on the concentration of the betulin glycoconjugates, after 24 (**A,C**) or 48 h (**B,D**) of incubation.

Unfortunately, in the case of the HCT 116 cell line, it was found that the designed betulin glycoconjugates, both at 24 and 48 h, did not inhibit the proliferation of tumor cells. Additionally, it was observed that in some cases, the tested compounds even stimulate tumor cells to grow faster. It is especially visible after 24 h for analogs **15b, 17a,** and **17b** (Figure 4A). Our hypothesis explaining this fact is based on the assumption that these derivatives are able to enter cancer cells, where they are degraded by hydrolytic enzymes with the release of sugar molecules, which is an additional source of energy allowing the proliferation growth. In the case of the MCF-7 cell line, after 24 h of incubation with the tested glycoconjugates, a similar effect was observed as for the HCT 116 cell line. However, after 48 h, a concentration-dependent cytotoxic effect of the glycoconjugates **15a, 15b,** and **16a** was observed (Figure 4D). These glycoconjugates, given to cells at the 50-µM concentration, showed cytotoxicity comparable to that of betulin. In glycoconjugates **15a, 15b,** and **16a**, the betulin skeleton was modified at position C28 by adding a sugar unit via an *O*-methylene-1,2,3-triazole linker or the same linker extended with an *N*-methyleneamide group. Presumably, the protection of the sugar hydroxyl groups improved the passive transport of the glycoconjugates inside the cell, where they released betulin under the action of hydrolytic enzymes. In the case of the disubstituted analogs BN **17a** and **17b**, modified at both the C3 and C28 positions, such a beneficial effect was not observed, which may be related to the prolonged effect of the release of as many as two glucose molecules (per each unit of betulin), which are the source of energy for increased proliferation. Quite unexpectedly, for the monosubstituted glycoconjugate **19a**, no significant cytotoxic activity was observed on tumor cells. Perhaps, the ester bond connecting the betulin backbone with the sugar unit is too labile and some part of the compound undergoes hydrolysis even before it penetrates the cell.

In order to check whether the introduction of the sugar unit increases the affinity of the obtained molecules for GLUT transporters, the acetyl protections in the sugar fragment of the obtained glycoconjugates were removed, and then cytotoxicity tests of the obtained compounds were carried out on the same cell lines (HCT 116, MCF-7). The cytotoxicity of the disubstituted BN derivative **21**, in which a linker contains only the 1,2,3-triazole system, without a sugar unit, was also tested. This study was designed to check the influence of the presence of the 1,2,3-triazole ring on the activity of the betulin derivative.

When the research was carried out on the HCT 116 cell line, both after 24 h and after 48 h, no significant effect of inhibition of tumor cell proliferation by the tested glycoconjugates was observed (**22a**, **22b**, **23b**, **24a**, **24b**). Interestingly, derivative **21** after 48 h shows slightly higher cytotoxicity compared to the activity of the parent BN backbone (Figure 5B). In the case of the MCF-7 cell line, after 24 h of incubation, the effect caused by the glycoconjugates (**22a**, **22b**, **23b**, **24a**) is only slightly better than that observed with BN. However, already after 48 h of incubation, the significantly higher cytotoxic activity of BN was observed compared to glycoconjugates. Surprisingly, as was in the case with the HCT 116 cell line, BN derivative **21** without a sugar unit, but only with an attached linker with the 1,2,3-triazole system, proved to be the most active. After 48 h of incubation, only 20% cell proliferation for this compound was observed as shown in Figure 5D.

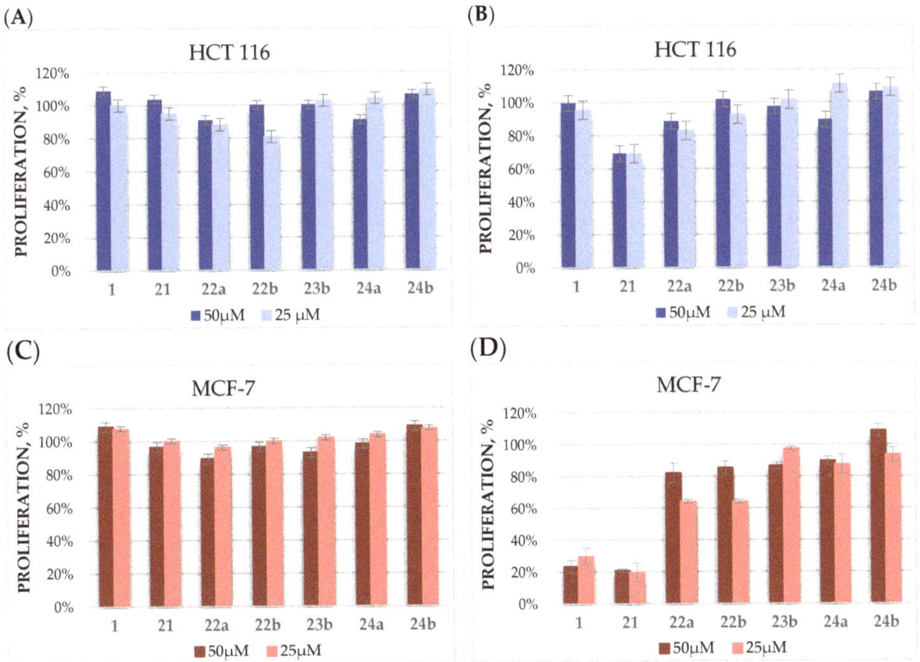

Figure 5. The dependence of HCT 116 and MCF-7 cell proliferation on the concentration of the betulin glycoconjugates, after 24 (**A,C**) or 48 h (**B,D**) of incubation.

In the course of further studies, for betulin **1** and its derivatives **15a, 15b, 16a,** and **21**, the ability to inhibit the proliferation of MCF-7 cells in a wider spectrum of concentrations was carried out in order to determine their IC_{50} value. The same compounds were also tested against NHDF cell line (normal human dermal fibroblast-neonatal cells), to assess the safety of their uses. Assays were performed in the concentration range of 50–6.25 μM for both tested cell lines. Unfortunately, it turned out that glycoconjugates **15a, 15b,** and **16a** are equally cytotoxic in relation to healthy cells and to the tested tumor lines (Table 3). Additionally, the most active of betulin derivative **21**, even at the lowest tested dose, is extremely cytotoxic and allows the proliferation of both cell lines to be kept at only 20%. The cytotoxicity of compound **21** significantly exceeds that of betulin, which may suggest that further modification of this compound is necessary to improve its selectivity.

Table 3. Summary of cytotoxicity of BN and the betulin glycoconjugates.

Compound	Activity IC_{50} [μM] [a,b]	
	MCF-7	**NHDF**
1	12.40 ± 0.14	11.60 ± 0.22
15a	65.28 ± 0.68	60.62 ± 0. 64
15b	47.70 ± 0.72	47.61 ± 0.86
16a	29.77 ± 0.52	29.82 ± 0.45

[a] Cytotoxicity was evaluated using the CCK-8 assay; [b] Incubation time 48 h.

2.5. Cytotoxic Activity Predicted Lipophilicity

In medicinal chemistry, the aqueous solubility and lipophilicity of a drug are important molecular parameters determining the absorption and the bioavailability. The lipophilicity is indicated by the logarithm of a partition coefficient (logP), which reflects the concentration ratio of the drug at equilibrium partitioning between octanol and water phases [57,58].

Betulin **1**, which is the subject of our research, is characterized by a highly limited solubility in polar solvents. This property may result from the complex structure, the presence of five fused aliphatic rings, and the isopropenyl moiety. The only polar fragments of the molecule are the two hydroxyl groups at C3 and C28. In this study, in order to determine the effect of various substituents on the value of the logP coefficient for several betulin derivatives, we used the prediction base *MolInspirations* to evaluate in silico the parameter (Table 4).

As we expected, adding a sugar moiety decreases the lipophilicity (<logP) of betulin analogs. Moreover, in the case of triazole derivatives of betulin modified with a sugar unit, the difference was revealed when the structures containing a per-*O*-acetylated sugar unit **15–20** and an unprotected sugar unit **22–24** were compared. The linker type is also crucial. For compounds bearing a substituent of the OCH_2 TriazGlc/Gal type at C3 and C28 position, the logP values are higher than for compounds bearing substituents with an ester moiety in the same position. On the other hand, the lowest values of the coefficient have derivatives in which an amide group is present in the linker structure. Only betulin glycoconjugates **23** and **24** have logP values smaller than 5 (Figure 6).

Table 4. LogP of BN and betulin analogs **2–24**.

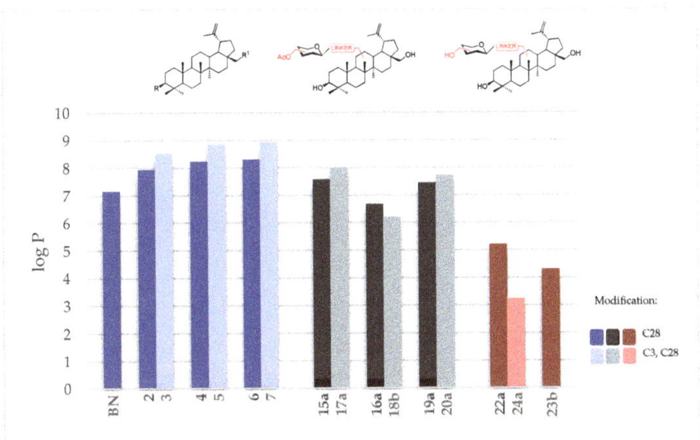

No.	R (C3)	R^1 (C28)	logP
1	OH	OH	7.16
2	OH	OCH2C≡CH	7.94
3	OCH2C≡CH	OCH2C≡CH	8.54
4	OH	O(CO)CH2Cl	8.23
5	O(CO)CH2Cl	O(CO)CH2Cl	8.86
6	OH	O(CO)CH2N3	8.31
7	O(CO)CH2N3	O(CO)CH2N3	8.94
15a	OH	OCH2TriazGlc(OAc)	7.59
16a	OH	OCH2TriazCH2(CO)NHGlc(OAc)	6.69
17a	OCH2TriazGlc(OAc)	OCH2TriazGlc(OAc)	8.01
18b	OCH2TriazCH2(CO)NHGal(OAc)	OCH2TriazCH2(CO)NHGal(OAc)	6.22
19a	OH	O(CO)CH2TriazCH2OGlc(OAc)	7.45
20a	O(CO)CH2TriazCH2OGlc(OAc)	O(CO)CH2TriazCH2OGlc(OAc)	7.74
21	O(CO)CH2TriazOH	O(CO)CH2TriazOH	6.38
22a	OH	OCH2TriazGlc(OH)	5.21
23b	OH	OCH2TriazCH2(CO)NHGal(OH)	4.31
24a	OCH2TriazGlc(OH)	OCH2TriazGlc(OH)	3.25

Figure 6. LogP of BN and betulin analogs **2–24**.

3. Materials and Methods

3.1. General Information

NMR spectra were recorded on a Varian spectrometer at operating frequencies of 600 or 400 and 150 or 100 MHz, respectively, using TMS as the resonance shift standard and NMR solvents (CDCl$_3$, CD$_3$OD or DMSO-d_6), which was purchased from ACROS Organics (Geel, Belgium). All chemical shifts (δ) are reported in ppm and coupling constants (J) in Hz. The following abbreviations were used to explain the observed multiplicities: s: singlet, d: doublet, dd: doublet of doublets, ddd: doublet of doublet of doublets, t: triplet, dd~t: doublet of doublets resembling a triplet (with similar values of coupling constants), m: multiplet, br: broad. IR-spectra were measured on an FTIR spectrophotometer

(ATR method). High-resolution mass spectrometry (HRMS) analyses were performed on a Waters LCT Premier XE system using the electrospray-ionization (ESI) technique. Optical rotations were measured with a JASCO P-2000 polarimeter using a sodium lamp (589.3 nm) at room temperature. Melting points were determined using a Boethius M HMK hot-stage apparatus. Reactions were monitored by TLC on precoated plates of silica gel 60 F254 (Merck Millipore). TLC plates were inspected under UV light (λ = 254 nm) or charring after spraying with 10% solution of sulfuric acid in ethanol. Crude products were purified using column chromatography performed on Silica Gel 60 (70–230 mesh, Fluka).

Cell viability was measured using a Cell Counting Kit-8 (Bimake) according to the manufacturer's protocol. The absorbance on the CKK-8 assay was measured spectrophotometrically at the 450-nm wavelength using a plate reader (Epoch, BioTek, USA).

28-*O*-Chloroacetylbetulin **4** and 3,28-*O*,*O*'-di(2-chloroacetyl)betulin **5** [48], 1,2,3,4,6-penta-*O*-acetyl-β-D-glucopyranose **9a**, 1,2,3,4,6-penta-*O*-acetyl-β-D-galactopyranose **9b** [53], propargyl 2,3,4,6-tetra-*O*-acetyl-β-D-glucopyranoside **10a** [55], 2,3,4,6-tetra-*O*-acetyl-α-D-glucopyranosyl bromide **11a**, 2,3,4,6-tetra-*O*-acetyl-α-D-galactopyranosyl bromide **11b** [54], 2,3,4,6-tetra-*O*-acetyl-β-D-glucopyranosyl azide **12a**, 2,3,4,6-tetra-*O*-acetyl-β-D-galactopyranosyl azide **12b** [36], 2,3,4,6-tetra-*O*-acetyl-*N*-(β-D-glucopyranosyl)chloroacetamide **13a**, 2,3,4,6-tetra-*O*-acetyl-*N*-(β-D-galacto-pyranosyl)chloroacetamide **13b**, 2,3,4,6-tetra-*O*-acetyl-*N*-(β-D-glucopyranosyl)azidoacetamide **14a**, and 2,3,4,6-tetra-*O*-acetyl-*N*-(β-D-galactopyranosyl)azidoacetamide **14b** [37] were prepared according to the respective published procedures. All used chemicals were purchased from Sigma-Aldrich, Fluka, Avantor, and ACROS Organics and were used without purification.

3.2. Chemistry

3.2.1. Procedure for the Synthesis of Betulin Analogs

Procedure for the Synthesis of 28-*O*-propargylbetulin **2** and 3,28-*O*,*O*'-di(propargyl)betulin **3**

Betulin **1** (0.226 mmol, 100.0 mg) and sodium hydride (NaH, 0.904 mmol, 21.7 mg) were dissolved in dry THF (1.3 mL) followed by addition of propargyl bromide (0.723 mmol, 0.062 mL). The heterogenous reaction mixture was left on the magnetic stirrer for 24 h at room temperature. The reaction progress was monitored on TLC in an eluents system DCM:AcOEt (10:1). Then, NH$_4$Cl (0.411 mmol, 22.0 mg, 5 mL H$_2$O) was added, extracted with Et$_2$O (5 × 4 mL), brine (5 mL) added, and dried over MgSO$_4$. After, this was followed by filtration and evaporation in vacuo. The crude products were purified using column chromatography (DCM:AcOEt, gradient: 50:1 to 10:1).

28-O-Propargylbetulin **2** was obtained as a white solid (58.8 mg, 54% yield); m.p.: 81—85 °C; $[\alpha]^{25}_D$ = +11.3 (c 0.5, CHCl$_3$); HRMS (ESI$^+$): calcd for C$_{33}$H$_{53}$O$_2$ ([M + H]$^+$): *m/z* 481.4046; found: *m/z* 481.4050; ^{1}H NMR (400 MHz, CDCl$_3$): δ_H 4.68 (s, br, 1H, H-29a), 4.58 (s, br, 1H, H-29b), 4.18 (dd, 1H, J_1 2.4 Hz, J_2 16.0 Hz, OCH$_a$C≡), 4.11 (dd, 1H, J_1 2.4 Hz, J_2 16.0 Hz, OCH$_b$C≡), 3.70 (d, 1H, *J* 9.2 Hz, H-28a), 3.21-3.15 (m, 2H, H-28b, H-3), 2.45-2.38 (m, 1H, H-19), 2.42 (dd~t, 1H, *J* 2.4 Hz, ≡CH), 2.1-0.60 (m, 25H, CH, CH$_2$), 1.68 (s, 3H, H-30), 1.05, 0.97 br, 0.83, 0.76 (all s, 3H each, H-23—H-27) ppm; ^{13}C NMR (150 MHz, CDCl$_3$) δ_C 150.64, 109.56, 80.51, 78.97, 74.01, 68.27, 58.61, 55.28, 50.42, 48.91, 47.99, 42.66, 40.98, 38.86, 38.69, 37.17, 34.69, 34.22, 29.90, 29.79, 27.99, 27.39, 27.14, 25.22, 20.87, 19.09, 18.29, 16.11, 15.98, 15.37, 14.83 ppm (Supplementary Materials, Table S1); IR (ATR) ν: 3350-3200, 2940, 2871, 1640, 1457, 1092, 882 cm^{-1}.

3,28-O,O'-Di(propargyl)betulin **3** was obtained as a yellow resin (36.5 mg, 31% yield); m.p.: resin; $[\alpha]^{25}_D$ = +31.9 (c 0.5, CHCl$_3$); HRMS (ESI$^+$): calcd for C$_{36}$H$_{55}$O$_2$ ([M + H]$^+$): *m/z* 519.4202; found: *m/z* 519.4202; ^{1}H NMR (600 MHz, CDCl$_3$): δ_H 4.68 (d, 1H, *J* 2.4 Hz, H-29a), 4.58 (dd, 1H, J_1 1.8 Hz, J_2 2.4 Hz, H-29b), 4.21-4.12 (m, 4H, 2x CH$_2$C≡), 3.70 (dd, 1H, J_1 1.2 Hz, J_2 8.4 Hz, H-28a), 3.16 (d, 1H, *J* 9.0 Hz, H-28b), 3.00 (dd, 1H, J_1 4.2 Hz, J_2 11.4 Hz, H-3), 2.44-2.39 (m, 1H, H-19), 2.41 i 2.36 (all dd~t, 1H each, *J* 2.4 Hz, ≡CH), 2.10-0.60 (m, 24H, CH, CH$_2$), 1.68 (s, 3H, H-30), 1.04, 0.98, 0.97, 0.83, 0.76 (all s, 3H each, H-23—H-27) ppm; ^{13}C NMR (150 MHz, CDCl$_3$) δ_C 150.64, 109.56, 85.86, 80.96, 80.50, 74.01, 73.41, 68.25, 58.62, 56.42, 55.87, 50.41, 48.92, 48.00, 47.02, 42.65, 41.03, 38.55, 37.12, 34.69, 34.24, 29.91, 29.71, 27.99,

27.12, 25.24, 20.91, 18.25, 19.11, 16.10, 16.00, 15.99, 14.79 ppm (Supplementary Materials, Table S1); IR (ATR) ν: 3314, 3261, 2952, 1610, 1453, 1377, 1085, 1068, 881 cm^{-1}.

Procedure for the Synthesis of 28-*O*-(2-azidoacetyl)betulin **6**

28-*O*-(2-Chloroacetyl)betulin **4** (0.193 mmol, 100.0 mg) was dissolved in DMF (1.93 mL) and sodium azide (0.963 mmol, 62.6 mg) was added. The heterogenous reaction mixture was stirred at 90 °C for 3 h. Then, H_2O (1 mL) and saturated $NaHCO_3$ solution (4 mL) was added. It was extracted with DCM (5 × 4 mL) and dried over $MgSO_4$. After that, this was followed by filtration and evaporation in vacuo. The crude product was purified using column chromatography (DCM:MeOH, gradient: 100:1 to 10:1).

28-O-(2-Azidoacetyl)betulin **6** was obtained as a white solid (65.0 mg, 64% yield); ^{1}H NMR (600 MHz, CDCl$_3$): δ$_H$ 4.70 (d, 1H, *J* 1.8 Hz, H-29a), 4.60 (d, 1H, *J* 1.2 Hz, H-29b), 4.41 (dd, 1H, *J$_1$* 1.2 Hz, *J$_2$* 10.8 Hz, H-28a), 3.99 (d, 1H, *J* 11.4 Hz, H-28b), 3.89 (s, 2H, CH$_2$N$_3$), 3.18 (dd, 1H, *J$_1$* 4.8 Hz, *J$_2$* 11.4 Hz, H-3), 2.44 (td, 1H, *J$_1$* 5.8 Hz, *J$_2$* 10.9 Hz, H-19), 2.10–0.60 (m, 25H, CH, CH$_2$), 1.68 (s, 3H, H-30), 1.04, 0.98, 0.97, 0.83, 0.76 (all s, 3H each, H-23–H-27) ppm; ^{13}C NMR (150 MHz, CDCl$_3$) δ$_C$ 168.65, 149.76, 109.92, 78.84, 64.31, 55.21, 50.41, 50.26, 48.72, 47.57, 46.33, 42.62, 40.79, 38.77, 38.62, 37.06, 34.39, 34.09, 29.59, 27.89, 27.30, 26.92, 25.09, 20.67, 19.04, 18.19, 16.00, 15.94, 15.27, 14.69 ppm (Supplementary Materials, Table S1).

Procedure for the Synthesis of 3,28-*O*,*O'*-di(2-azidoacetyl)betulin **7**

3,28-*O*,*O'*-Di(2-chloroacetyl)betulin **5** (0.168 mmol, 100.0 mg) was dissolved in 0.71 mL DMF and sodium azide (0.897 mmol, 58.3 mg) was added. The heterogenous reaction mixture was stirred at 90 °C for 3 h. Then, H_2O and saturated $NaHCO_3$ solution (4 mL) was added. It was extracted with DCM (4 × 4 mL) and dried over $MgSO_4$. After that, this was followed by filtration and evaporation in vacuo. The crude product was purified using column chromatography (DCM:MeOH, gradient: 100:1 to 10:1).

3,28-O,O'-Di(2-azidoacetyl)betulin **7** was obtained as a white solid (73.6 mg, 72% yield); ^{1}H NMR (600 MHz, CDCl$_3$): δ$_H$ 4.70 (d, 1H, *J* 1.2 Hz, H-29a), 4.61–4.58 (m, 2H, H-29b, H-3), 4.42 (d, 1H, *J* 11.4 Hz, H-28a), 3.98 (d, 1H, *J* 10.8 Hz, H-28b), 3.89 and 3.85 (all s, 2H each, 2x CH$_2$N$_3$), 2.44 (td, 1H, *J$_1$* 5.7 Hz, *J$_2$* 10.8 Hz, H-19), 2.10–0.60 (m, 24H, CH, CH$_2$), 1.69 (s, 3H, H-30), 1.04, 0.98, 0.87, 0.866, 0.862 (all s, 3H each, H-23–H-27) ppm; ^{13}C NMR (150 MHz, CDCl$_3$) δ$_C$ 168.75, 168.15, 149.84, 110.07, 83.21, 64.40, 55.36, 50.67, 50.52, 50.25, 48.80, 47.69, 46.43, 42.67, 40.91, 38.34, 37.89, 37.06, 34.49, 34.08, 29.67, 27.94, 27.00, 25.06, 23.68, 20.74, 19.05, 18.15, 16.54, 16.15, 16.04, 14.75 ppm (Supplementary Materials, Table S1); IR (ATR) ν: 2942, 2867, 1728, 1243, 1181, 1017, 979, 754 cm^{-1}.

3.2.2. General Procedure for the Synthesis of Betulin Glycoconjugates **15–16** according to Strategy I (Modification C28)

To a well-stirred 28-*O*-propargylbetulin **2** (0.208 mmol, 100.0 mg) and 2,3,4,6-tetra-*O*-acetyl-β-D-galactopyranosyl or 2,3,4,6-tetra-*O*-acetyl-β-D-glucopyranosyl azide **12a–b** (0.208 mmol, 77.7 mg) or **14a–b** (0.208 mmol, 89.5 mg) in *i*-PrOH (2 mL) were added CuSO$_4$·5H$_2$O (0.125 mmol, 31.2 mg) dissolved in H_2O (1 mL) and sodium ascorbate (0.229 mmol, 45.3 mg) dissolved in H_2O (1 mL) succesively. The reaction mixture was stirred under argon atmosphere for 24 h at room temperature. After completion of the reaction (was monitored by TLC), the reaction mixture was extracted with DCM (**15a–b**: 5 × 5 mL) or AcOEt (**16a–b**: 5 × 5 mL), then the mixture was dried over $MgSO_4$ and was concentrated in vacuo. The crude residue was separated/purified using column chromatography (**15a–b**: DCM:AcOEt, gradient: 20:1 to 1:1; **16a–b**: DCM:MeOH, gradient: 50:1 to 10:1).

Product 15a (28-OCH$_2$TriazGlcBN) was obtained as a white solid (175.1 mg, 99% yield); m.p.: 108–109 °C; [α]^{25}$_D$ = −21.8 (c 0.5, CHCl$_3$); HRMS (ESI$^+$): calcd for C$_{47}$H$_{72}$N$_3$O$_{11}$ ([M + H]$^+$): *m/z* 854.5167; found: *m/z* 854.5165; ^{1}H NMR (600 MHz, CDCl$_3$): δ$_H$ 7.75 (s, br, 1H, H-5$_{triaz.}$), 5.89 (d, 1H, *J*

9.0 Hz, H-1$_{Glc}$), 5.46 (dd~t, 1H, J_1 9.1 Hz, J_2 9.5 Hz, H-3$_{Glc}$), 5.42 (dd~t, 1H, J_1 9.0 Hz, J_2 9.5 Hz, H-2$_{Glc}$), 5.25 (dd~t, 1H, J_1 9.1 Hz, J_2 10.2 Hz, H-4$_{Glc}$), 4.67 (d, 1H, J 13.2 Hz, OCH$_a$), 4.66 (s, br, 1H, H-29a), 4.60 (d, 1H, J 12.6 Hz, OCH$_b$), 4.55 (dd, 1H, J_1 1.2 Hz, J_2 2.4 Hz, H-29b), 4.30 (dd, 1H, J_1 5.0 Hz, J_2 12.6 Hz, H-6a$_{Glc}$), 4.14 (dd, 1H, J_1 2.1 Hz, J_2 12.6 Hz, H-6b$_{Glc}$), 4.00 (ddd, 1H, J_1 2.1 Hz, J_2 5.0 Hz, J_3 10.2 Hz, H-5$_{Glc}$), 3.63 (d, 1H, J 8.4 Hz, H-28a), 3.19–3.17 (m, 2H, H-28a, H-3), 2.38 (td, 1H, J_1 5.8 Hz, J_2 11.1 Hz, H-19), 2.10–0.60 (m, 25H, CH, CH$_2$), 2.08, 2.07, 2.03, 1.88 (all s, 3H each, 4x CH$_3$CO), 1.66 (s, 3H, H-30), 1.02, 0.97, 0.96, 0.83, 0.76 (all s, 3H each, H-23–H-27) ppm; ^{13}C NMR (150 MHz, CDCl$_3$) δ_C 170.45, 169.90, 169.32, 168.87, 150.57, 146.62, 120.58, 109.60, 85.73, 78.95, 75.12, 72.73, 70.32, 68.94, 67.72, 64.95, 61.58, 55.31, 50.42, 48.85, 47.88, 47.26, 42.68, 40.92, 38.88, 38.73, 37.51, 37.17, 34.77, 34.26, 29.91, 29.85, 28.00, 27.42, 25.22, 20.85, 20.66, 20.53, 20.51, 20.20, 19.06, 18.32, 16.11, 16.03, 15.37, 14.79 ppm (Supplementary Materials, Table S2); IR (ATR) ν: 3400–3500, 2941, 1751, 1650, 1455, 1367, 1216, 1038, 731 cm^{-1}.

Product **15b** *(28-OCH$_2$TriazGalBN)* was obtained as a white solid (201.7 mg, 97% yield); m.p.: 117–118 °C; [α]^{25}$_D$ = −9.4 (c 0.5, CHCl$_3$); HRMS (ESI$^+$): calcd for C$_{47}$H$_{71}$N$_3$O$_{11}$Na ([M + Na]$^+$): *m/z* 876.4986; found: *m/z* 876.4989; ^{1}H NMR (600 MHz, CDCl$_3$): δ_H 7.80 (s, br, 1H, H-5$_{triaz.}$), 5.84 (d, 1H, J 9.3 Hz, H-1$_{Gal}$), 5.58 (dd, 1H, J_1 9.3 Hz, J_2 10.2 Hz, H-2$_{Gal}$), 5.55 (dd, 1H, J_1 0.8 Hz, J_2 3.4 Hz, H-4$_{Gal}$), 5.25 (dd, 1H, J_1 3.4 Hz, J_2 10.2 Hz, H-3$_{Gal}$), 4.69–4.57 (m, 4H, OCH$_2$, H-29a, H-29b), 4.24–4.18 (m, 2H, H-6a$_{Gal}$, H-5$_{Gal}$), 4.13 (d, 1H, J_1 6.3 Hz, J_2 10.9 Hz, H-6b$_{Gal}$), 3.63 (d, 1H, J 8.2 Hz, H-28a), 3.21–3.17 (m, 2H, H-28a, H-3), 2.40 (td, 1H, J_1 6.0 Hz, J_2 11.1 Hz, H-19), 2.30–0.60 (m, 25H, CH, CH$_2$), 2.22, 2.04, 2.01, 1.90 (all s, 3H each, 4x CH$_3$CO), 1.66 (s, 3H, H-30), 1.02, 0.97, 0.96, 0.83, 0.76 (all s, 3H each, H-23–H-27). ppm; ^{13}C NMR (150 MHz, CDCl$_3$) δ_C 170.28, 169.95, 169.80, 169.04, 150.61, 146.46, 120.76, 109.60, 86.27, 78.96, 74.02, 70.89, 68.94, 67.85, 66.87, 64.91, 61.15, 55.30, 50.41, 48.84, 47.87, 47.26, 42.69, 40.93, 38.88, 38.72, 37.50, 37.17, 34.79, 34.25, 29.95, 29.86, 28.00, 27.42, 27.23, 25.22, 20.84, 20.66, 20.62, 20.49, 20.30, 19.06, 18.33, 16.11, 16.00, 15.37, 14.80 ppm (Supplementary Materials, Table S2); IR (ATR) ν: 3600–3200, 2930, 2868, 1749, 1650, 1456, 1370, 1214, 1042, 731 cm^{-1}.

Product **16a** *(28-OCH$_2$TriazCH$_2$(CO)NHGlcBN)* was obtained as a white solid (190.0 mg, 99% yield); m.p.: 127-129 °C; [α]^{25}$_D$ = +11.4 (c 0.5, CHCl$_3$); HRMS (ESI$^+$): calcd for C$_{49}$H$_{74}$N$_4$O$_{12}$Na ([M + Na]$^+$): *m/z* 933.5201; found: *m/z* 933.5202; ^{1}H NMR (600 MHz, CDCl$_3$): δ_H 7.66 (s, br, 1H, H-5$_{triaz.}$), 6.99 (d, 1H, J 8.8 Hz, NH), 5.29 (dd~t, 1H, J 9.5 Hz, H-1$_{Glc}$), 5.21 (dd, 1H, J_1 8.9 Hz, J_2 9.4 Hz, H-3$_{Glc}$^{a}), 5.12 (d, 1H, J 16.6 Hz, OCH$_a$^{b}), 5.05 (dd, 1H, J_1 9.4 Hz, J_2 10.1 Hz, H-4$_{Glc}$^{a}), 5.04 (d, 1H, J 16.5 Hz, OCH$_b$^{b}), 4.90 (dd~t, 1H, J 9.6 Hz, H-2$_{Glc}$^{a}), 4.67 (s, br, 3H, H-29a, CH$_2$CO^b), 4.57 (dd, 1H, J_1 1.4 Hz, J_2 2.1 Hz, H-29b), 4.28 (dd, 1H, J_1 4.4 Hz, J_2 12.5 Hz, H-6a$_{Glc}$), 4.09 (dd, 1H, J_1 2.2 Hz, J_2 12.5 Hz, H-6b$_{Glc}$), 3.81 (ddd, 1H, J_1 2.2 Hz, J_2 4.4 Hz, J_3 10.1 Hz, H-5$_{Glc}$), 3.63 (d, 1H, J 9.1 Hz, H-28a), 3.22-2.18 (m, 2H, H-28b, H-3), 2.39 (td, 1H, J_1 5.8 Hz, J_2 10.8 Hz, H-19), 2.15–0.60 (m, 25H, CH, CH$_2$), 2.08, 2.03, 2.02, 2.01 (all s, 3H each, 4x CH$_3$CO), 1.67 (s, 3H, H-30), 0.97, 0.96, 0.956, 0.82, 0.77 (all s, 3H each, H-23–H-27) ppm; ^{13}C NMR (150 MHz, CDCl$_3$) δ_C 170.86, 170.57, 169.84, 169.49, 165.68, 150.53, 146.71, 123.76, 109.64, 78.99, 78.41, 73.80, 72.51, 70.32, 68.95, 68.01, 65.14, 61.56, 55.28, 52.59, 50.38, 48.83, 47.88, 47.23, 42.64, 40.88, 38.86 38.70, 37.49, 37.15, 34.73, 34.23, 29.86, 29.69, 28.01, 27.40, 27.16, 25.18, 20.83, 20.73, 20.56, 20.54, 19.08, 18.33, 16.12, 15.93, 15.41, 14.79 ppm (Supplementary Materials, Table S2); IR (ATR) ν: 3200–3400, 2937, 1749, 1650, 1450, 1367, 1223, 1037, 993, 907, 882, 753 cm^{-1}. ^{a,b} Reverse signal assignment is possible.

Product **16b** *(28-OCH$_2$TriazCH$_2$(CO)NHGalBN)* was obtained as a white solid (184.0 mg, 97% yield); m.p.: 117-120 °C; [α]^{25}$_D$ = +14.3 (c 0.5, CHCl$_3$); HRMS (ESI$^+$): calcd for C$_{49}$H$_{74}$N$_4$O$_{12}$Na ([M + Na]$^+$): *m/z* 933.5201; found: *m/z* 933.5202; ^{1}H NMR (600 MHz, CDCl$_3$): δ_H 7.66 (s, br, 1H, H-5$_{triaz.}$), 6.93 (d, 1H, J 8.9 Hz, NH), 5.43 (dd, 1H, J_1 0.9 Hz, J_2 3.3 Hz, H-4$_{Gal}$), 5.19 (dd, 1H, J_1 8.9 Hz, J_2 9.2 Hz, H-1$_{Gal}$), 5.12 and 4.99 (qAB, 2H, J 3.3 Hz, CH$_2$CO), 5.11 (dd, 1H, J_1 3.3 Hz, J_2 10.3 Hz, H-3$_{Gal}$), 5.06 (dd, 1H, J_1 9.2 Hz, J_2 10.3 Hz, H-2$_{Gal}$), 4.68 (s, 2H, OCH$_2$), 4.67 and 4.57 (s, br, 1H each, H-29a, H-29b), 4.11 (dd, 1H, J_1 6.9 Hz, J_2 11.3 Hz, H-6a$_{Gal}$), 4.07 (dd, 1H, J_1 6.2 Hz, J_2 11.23 Hz, H-6b$_{Gal}$), 4.02 (m, 1H, H-5$_{Gal}$), 3.63 (d, 1H, J 8.9 Hz, H-28a), 3.22–3.18 (m, 2H, H-28b, H-3), 2.39 (td, 1H, J_1 5.9 Hz, J_2 10.8 Hz, H-19), 2.30–0.60 (m, 25H, CH, CH$_2$), 2.15, 2.04, 2.03, 1.98 (all s, 3H each, 4x CH$_3$CO), 1.67 (s, 3H, H-30), 0.97, 0.96, 0.957, 0.82, 0.77 (all s, 3H each, H-23–H-27) ppm; ^{13}C NMR (150 MHz, CDCl$_3$) δ_C 171.00, 170.37,

169.98, 169.73, 165.67, 150.53, 146.60, 123.86, 109.63, 78.98, 78.61, 72.56, 70.71, 68.98, 68.06, 67.08, 65.13, 61.08, 55.29, 52.55, 50.39, 48.84, 47.89, 47.24, 42.65, 40.89, 38.87, 38.71, 37.50, 37.16, 34.74, 34.24, 29.88, 29.86, 28.01, 27.39, 27.17, 25.19, 20.84, 20.67, 20.64, 20.60, 20.50, 19.08, 18.33, 16.12, 15.94, 15.40, 14.79 ppm (Supplementary Materials, Table S2); IR (ATR) ν: 3100–2800, 1749, 1650, 1368, 1222, 1084, 1046, 752 cm^{-1}.

3.2.3. General Procedure for the Synthesis of Betulin Glycoconjugates 17–18 according to Strategy I (Modification C3 and C28)

To a well-stirred 3,28-*O,O'*-di(2-propargyl)betulin **3** (0.193 mmol, 100.0 mg) and per-*O*-acetylated glucopyranosyl or galactopyranosyl azides **12a–b** (0.385 mmol, 143.9 mg) or **14a–b** (0.385 mmol, 165.9 mg) in *i*-PrOH (6.6 mL), CuSO$_4$·5H$_2$O (0.231 mmol, 57.8 mg) dissolved in H$_2$O (3.3 mL) and sodium ascorbate (0.424 mmol, 84.0 mg) dissolved in H$_2$O (3.3 mL) were added successively. The reaction mixture was stirred under argon atmosphere for 24 h at room temperature. After completion of the reaction (was monitored by TLC), the reaction mixture was extracted with (**17a–b**: DCM 5 × 5 mL; **18b**: AcOEt 5 × 5 mL), then the mixture was dried over MgSO$_4$ and was concentrated in vacuo. The crude residue was separated/purified using column chromatography (**17a–b**: DCM:AcOEt, gradient: 15:1 to 1:1; **18b**: DCM:MeOH, gradient: 40:1 to 10:1).

*Product **17a** (3,28-di(OCH$_2$TriazGlc)BN)* was obtained as a white solid (242.0 mg, 99% yield); m.p.: 126–127 °C; $[\alpha]^{25}_D = -6.1$ (c 0.5, CHCl$_3$); HRMS (ESI$^+$): calcd for C$_{64}$H$_{93}$N$_6$O$_{20}$ ([M + H]$^+$): *m/z* 1265.6445; found: *m/z* 1265.6426; ^{1}H NMR (600 MHz, CDCl$_3$): δ_H 7.75 and 7.74 (all s, 1H each, 2x H-5$_{triaz.}$), 5.90 and 5.89 (all d, 1H each, J_1 9.0 Hz and J_2 9.4 Hz, 2x H-1$_{Glc}$), 5.50-5.40 (m, 4H, 2x H-2$_{Glc}$, 2x H-3$_{Glc}$)), 5.24 and 5.25 19 (all dd, 1H each, J_1 9.4 Hz, J_2 10.1 Hz and J_1 9.2 Hz, J_2 10.1 Hz, 2x H-4$_{Glc}$), 4.78–4.54 (m, 3H, OCH$_2$, H-29b), 4.32–4.28 (m, 2H, 2x H-6a$_{Glc}$), 4.36-4.13 (m, 2H, 2x H-6b$_{Glc}$), 4.03-4.00 (m, 2H, 2x H-5$_{Glc}$), 3.62 (d, 1H, J 8.4 Hz, H-28a), 3.18 (d, 1H, J 9.0 Hz, H-28b), 2.95 (dd, 1H, J_1 4.2 Hz, J_2 12.0 Hz, H-3), 2.38 (td, 1H, J_1 5.7 Hz, J_2 11.0 Hz, H-19), 2.20-0.60 (m, 24H, CH, CH$_2$), 2.08, 2.07, 2.069, 2.031, 2.029, 1.88, 1.87 (all s, 3H each, 8x CH$_3$CO), 1.66 (s, 3H, H-30), 1.01, 0.95, 0.89, 0.84, 0.78 (all s, 3H each, H-23–H-27) ppm; ^{13}C NMR (150 MHz, CDCl$_3$) δ_C 170.46, 170.44, 169.91, 169.89, 169.32, 168.86, 168.79, 150.57, 147.26, 146.59, 120.58, 120.50, 109.61, 86.65, 85.69, 85.62, 75.08 75.04, 72.81 72.71, 70.31 70.18, 68.92, 67.71 67.70, 64.93, 62.88, 61.57, 55.68, 50.36, 48.82, 47.84, 47.25, 42.65, 40.94, 38.81, 38.52, 37.48, 37.11, 34.76, 34.24, 29.89, 29.83, 27.97, 27.18, 25.20, 22.83, 20.85, 20.66, 20.53, 20.51, 20.17, 19.05, 18.24, 16.25, 16.12, 16.02, 14.73 ppm (Supplementary Materials, Table S2); IR (ATR) ν: 2942, 1750, 1457, 1368, 1216, 1091, 1055, 1036, 913, 732 cm^{-1}.

*Product **17b** (3,28-di(OCH$_2$TriazGal)BN)* was obtained as a white solid (242.0 mg, 99% yield); m.p.: 122-123 °C; $[\alpha]^{25}_D = +4.9$ (c 0.5, CHCl$_3$); HRMS (ESI$^+$): calcd for C$_{64}$H$_{92}$N$_6$O$_{20}$Na ([M + Na]$^+$): *m/z* 1287.6264; found: *m/z* 1287.6206; ^{1}H NMR (600 MHz, CDCl$_3$): δ_H 7.80 and 7.79 (all s, br, 1H each, 2x H-5$_{triaz.}$), 5.86 and 5.856 (all d, 1H each, J 9.4 Hz, 2x H-1$_{Gal}$), 5.61 and 5.58 (all dd, 1H each, J_1 9.4 Hz, J_2 10.2 Hz, 2x H-2$_{Gal}$), 5.56–5.54 (m, 2H, 2x H-4$_{Gal}$), 5.25 and 5.26 (all dd, 1H each, J_1 5.4 Hz, J_2 10.2 Hz, 2x H-3$_{Gal}$), 4.78–4.56 (m, 6H, 2x OCH$_2$, H-29a, H-29b), 4.26–4.23 (m, 2H, 2x H-5$_{Gal}$), 4.22–4.10 (m, 4H, 2x H-6a$_{Gal}$, 2x H-6b$_{Gal}$), 3.62 (d, 1H, J 8.8 Hz, H-28a), 3.20 (d, 1H, J 9.1 Hz, H-28b), 2.97 (dd, 1H, J_1 4.3 Hz, J_2 11.7 Hz, H-3), 2.40 (td, 1H, J_1 5.7 Hz, J_2 11.1 Hz, H-19), 2.23, 2.22, 2.044, 2.04, 2.013, 2.01, 1.90, 1.88 (all s, 3H each, 8x CH$_3$CO), 2.30–0.60 (m, 24H, CH, CH$_2$), 1.67 (s, 3H, H-30), 1.01, 0.95, 0.90, 0.84, 0.78 (all s, 3H each, H-23–H-27) ppm; ^{13}C NMR (150 MHz, CDCl$_3$) δ_C 170.31, 170.28, 169.95, 169.79, 169.03, 168.93, 150.59, 147.06, 146.42, 120.77, 120.66, 109.62, 86.63, 86.22, 86.17, 73.98, 70.94, 70.86, 68.91, 67.86, 67.74, 66.91, 64.88, 61.18, 55.68, 52.88, 50.36, 48.83, 47.86, 47.25, 42.66, 40.95, 38.82, 38.53, 37.48, 37.12, 34.78, 34.24, 29.94, 29.84, 27.96, 27.20, 22.87, 25.20, 20.86, 20.66, 20.63, 20.49, 20.29, 20.27, 19.05, 18.26, 16.26, 16.13, 16.00, 14.74 ppm (Supplementary Materials, Table S2); IR (ATR) ν: 2942, 1748, 1457, 1368, 1213, 1089, 1043, 921, 730 cm^{-1}.

*Product **18b** (3,28-di(OCH$_2$TriazCH$_2$(CO)NHGal)BN)* was obtained as a white solid (216.9 mg, 82% yield); m.p.: 142–145 °C; $[\alpha]^{25}_D = +4.9$ (c 0.5, CHCl$_3$); HRMS (ESI$^+$): calcd for C$_{68}$H$_{99}$N$_8$O$_{22}$ ([M + H]$^+$): *m/z* 1379.6900; found: *m/z* 1379.6533; ^{1}H NMR (600 MHz, CDCl$_3$): δ_H 7.66 and 7.64 (all s, 1H each, 2x H-5$_{triaz.}$), 6.86 and 6.85 (all d, 1H each, *J* 8.8 Hz, 2x NH), 5.45-5.40 (m, 2H, 2x H-4$_{Gal}$), 5.22-5.18 (m, 2H, 2x H-1$_{Gal}$), 5.14–4.98 (m, 8H, 2x H-2$_{Gal}$, 2x H-3$_{Gal}$, 2x CH$_2$), 4.84–4.57 (m, 6H, H-29a, H-29b, 2x CH$_2$), 4.14-4.07 (m, 4H, 2x H-6a$_{Gal}$, 2x H-6b$_{Gal}$), 4.06–4.01 (m, 2H, 2x H-5$_{Gal}$), 3.63 (d, 1H, *J* 8.9 Hz, H-28a), 3.22 (d, 1H, *J* 9.0 Hz, H-28b), 2.97 (dd, 1H, J_1 4.2 Hz, J_2 11.7 Hz, H-3), 2.39 (td, 1H, J_1 5.9 Hz, J_2 10.6 Hz, H-19), 2.30–0.60 (m, 24H, CH, CH$_2$), 2.15, 2.14, 2.04, 2.03, 2.02, 1.98, 1.977, 1.96 (all s, 3H each, 8x CH$_3$CO), 1.67 (s, 3H, H-30), 0.96, 0.95, 0.92, 0.83, 0.77 (all s, 3H each, H-23–H-27) ppm; ^{13}C NMR (150 MHz, CDCl$_3$) δ_C 171.03, 170.92, 170.34, 169.98, 169.71, 165.68, 165.55, 150.54, 147.49, 146.77, 123.75, 123.61, 109.65, 86.89, 78.75, 78.72, 72.59, 72.57, 70.67, 70.64, 69.01, 68.06, 68.00, 67.06, 65.21, 63.20, 61.11, 61.00, 55.69, 52.71, 52.65, 50.36, 48.85, 47.90, 47.25, 42.65, 40.93, 38.85, 38.52, 37.50, 37.14, 34.75, 34.24, 29.88, 29.69, 28.10, 27.17, 25.20, 22.86, 20.87, 20.67, 20.62, 20.60, 20.59, 20.50, 19.09, 18.26, 16.30, 16.15, 15.96, 14.76, ppm (Supplementary Materials, Table S2); IR (ATR) ν: 2950, 1748, 1620, 1510, 1450, 1368, 1218, 1084, 1045, 970, 890, 720 cm^{-1}.

3.2.4. Procedure for the Synthesis of Betulin Glycoconjugates **19a** according to Strategy II (Modification C28)

To a well-stirred 28-*O*-(2-azidoacetyl)betulin **6** (0.190 mmol, 100.0 mg) and *per-O*-acetylated glucopyranosyl propargyl **10a** (0.190 mmol, 73.5 mg) in *i*-PrOH (2 mL) and THF (1 mL), CuSO$_4$·5H$_2$O (0.114 mmol, 22.6 mg) dissolved in H$_2$O (1 mL) and sodium ascorbate (0.209 mmol, 52.2 mg) dissolved in H$_2$O (1 mL) were added successively. The reaction mixture was stirred under argon atmosphere for 24 h at room temperature. After completion of the reaction (was monitored by TLC), the reaction mixture was extracted with DCM (5 × 5 mL), then the mixture was dried over MgSO$_4$ and was concentrated in vacuo. The crude residue was separated/purified using column chromatography (DCM:MeOH, gradient: 50:1 to 10:1).

*Product **19a** (28-O(CO)CH$_2$TriazCH$_2$OGlcBN)* was obtained as a white solid (168.3 mg, 97% yield); m.p.: 99–101 °C; [α]^{25}$_D$ = −15.6 (c 0.5, CHCl$_3$); HRMS (ESI+): calcd for C$_{49}$H$_{73}$N$_3$O$_{13}$Na ([M + Na]$^+$): m/z 934.5041; found: m/z 934.5045; ^{1}H NMR (600 MHz, CDCl$_3$): δ_H 7.69 (s, br, 1H, H-5$_{triaz.}$), 5.22-5.17 (m, 2H, H-1$_{Glc}$, H-3$_{Glc}$), 5.10 (dd~t, 1H, J_1 9.6 Hz, J_2 9.8 Hz, H-2$_{Glc}$), 5.02 (ddt, 1H, J_1 8.0 Hz, J_2 9.5 Hz, H-4Glc), 4.97 and 4.86 (qAB, 2H, CH$_2$), 4.68 (s, br, 2H, CH$_2$), 4.68 and 4.60 (all s, br, 1H each, H-29a, H-29b), 4.43 (d, 1H, *J* 10.9 Hz, H-28a), 4.26 (dd, 1H, J_1 4.7 Hz, J_2 12.3 Hz, H-6a$_{Glc}$), 4.17 (dd, 1H, J_1 2.1 Hz, J_2 12.3 Hz, H-6b$_{Glc}$), 3.97 (d, 1H, *J* 11.0 Hz, H-28b), 3.73 (ddd, 1H, J_1 2.1 Hz, J_2 4.7 Hz, J_3 9.5 Hz, H-5$_{Glc}$), 3.18 (dd, 1H, J_1 4.7 Hz, J_2 11.5 Hz, H-3), 2.41 (td, 1H, J_1 5.7 Hz, J_2 10.7 Hz, H-19), 2.20-0.60 (m, 25H, CH, CH$_2$), 2.10, 2.03, 2.004, 2.00 (all s, 3H each, 4x CH$_3$CO), 1.68 (s, 3H, H-30), 1.01, 0.98, 0.97, 0.82, 0.76 (all s, 3H each, H-23–H-27) ppm; ^{13}C NMR (150 MHz, CDCl$_3$) δ_C 170.65, 170.18, 169.45, 169.42, 166.55, 149.73, 124.28, 110.10, 78.94, 72.79, 71.94, 71.21, 68.36, 61.80, 50.86, 64.96, 62.93, 55.29, 50.85, 50.33, 48.80, 47.64, 46.45, 42.71, 40.88, 38.87, 38.71, 37.68, 37.15, 34.39, 34.18, 29.60, 29.44, 27.99, 27.01, 27.39, 25.16, 20.78, 20.74, 20.67, 20.60, 19.11, 18.28, 16.10, 16.01, 15.36, 14.78 ppm (Supplementary Materials, Table S2); IR (ATR) ν: 3400–3200, 2942, 1747, 1640, 1456, 1366, 1218, 1037, 982, 883, 769 cm^{-1}.

3.2.5. Procedure for the Synthesis of Betulin Glycoconjugates **20a** according to Strategy II (Modification C3 and C28)

To a well-stirred 3,28-*O,O'*-di(2-azidoacetyl)betulin **7** (0.164 mmol, 100.0 mg) and per-*O*-acetylated glucopyranosyl propargyl **10a** (0.329 mmol, 126.9 mg) in *i*-PrOH (4.2 mL) and THF (1.8 mL), CuSO$_4$·5H$_2$O (0.181 mmol, 45.1 mg) dissolved in H$_2$O (2.1 mL) and sodium ascorbate (0.099 mmol, 19.5 mg) dissolved in H$_2$O (2.1 mL) were added successively. The reaction mixture was stirred under argon atmosphere for 24 h at room temperature. After completion of the reaction (was monitored by TLC), the reaction mixture was extracted with DCM (5 × 5 mL), then the mixture was dried over MgSO4 and was concentrated in vacuo. The crude residue was separated/purified using column chromatography (DCM:MeOH, gradient: 30:1 to 10:1).

*Product **20a** (3,28-di(O(CO)CH$_2$TriazCH$_2$OGlc)BN)* was obtained as a white solid (198.0 mg, 87% yield); m.p.: 109–112 °C; [α]^{25}$_D$ = −19.0 (c 0.5, CHCl$_3$); HRMS (ESI+): calcd for C$_{68}$H$_{97}$N$_6$O$_{24}$ ([M +

H]$^+$): m/z 1381.6554; found: m/z 1381.6218; ^{1}H NMR (600 MHz, CDCl$_3$): δ_H 7.68 and 7.66 (all s, 1H each, 2x H-5$_{triaz.}$), 5.23-5.13 (m, 4H, 2x (H-1$_{Glc}$ H-3$_{Glc}$)), 5.12–5.06 (m, 2H, 2x H-2$_{Glc}$), 5.05-4.99 (m, 2H, 2x H-4$_{Glc}$), 4.99-4.83 (m, 4H, 2x CH$_2$), 4.69–4.66 (m, 5H, H-29a, 2x CH$_2$), 4.60 (s, br, 1H, H-29b), 4.43 (d, 1H, J 10.6 Hz, H-28a), 4.26 and 4.27 (all dd, 1H each, J_1 1.0 Hz, J_2 12.4 Hz, 2x H-6a$_{Glc}$), 4.19–4.14 (m, 2H, 2x H-6b$_{Glc}$), 3.96 (d, 1H, J 11.0 Hz, H-28b), 3.75–3.70 (m, 2H, 2x H-5$_{Glc}$), 4.56 (dd, 1H, J_1 5.2 Hz, J_2 10.1 Hz, H-3), 2.41 (td, 1H, J_1 5.7 Hz, J_2 10.7 Hz, H-19), 2.20–0.60 (m, 24H, CH, CH$_2$), 2.093, 2.09, 2.028, 2.025, 2.03, 2.00, 1.998, 1.994 (all s, 3H each, 8x CH$_3$CO), 1.68 (s, 3H, H-30), 1.01, 0.98, 0.97, 0.84, 0.77 (all s, 3H each, H-23–H-27) ppm; ^{13}C NMR (150 MHz, CDCl$_3$) δ_C 170.64, 170.16, 169.43, 166.56, 165.90, 160.42, 149.68, 144.65, 144.55, 124.25, 124.24, 110.16, 83.86, 68.37, 72.80, 71.94, 71.22, 68.37, 62.94, 61.81, 61.80, 55.30, 51.05, 50.83, 50.21, 48.77, 47.65, 46.45, 42.73, 40.89, 38.27, 37.90, 37.65, 37.03, 34.38, 34.05, 29.57, 29.43, 28.01, 27.00, 25.08, 23.61, 21.96, 20.77, 20.66, 20.65, 20.60, 19.10, 18.11, 16.39, 16.12, 16.00, 14.75 ppm (Supplementary Materials, Table S2); IR (ATR) v: 2930, 1744, 1620, 1367, 1216, 1037, 979, 880, 752 cm^{-1}.

3.2.6. Procedure for the Synthesis of 3,28-*O,O′*-di(2-(4-(hydroxymethyl-1H-1,2,3-triazol-1-yl)acetyl)betulin **21**

To a well-stirred 3,28-*O,O′*-di(2-azidoacetyl)betulin **7** (0.663 mmol, 403.6 mg) in *i*-PrOH (7.6 mL) and THF (4.8 mL), propargyl alcohol (1.989 mmol, 111.5 mg, 0.116 mL) and CuSO$_4$·5H$_2$O (0.796 mmol, 198.6 mg) dissolved in H$_2$O (2.7 mL) and sodium ascorbate (1.458 mmol, 288.9 mg) dissolved in H$_2$O (2.7 mL) were added successively. The reaction mixture was stirred under argon atmosphere for 24 h at room temperature. After completion of the reaction (was monitored by TLC), the reaction mixture was extracted with DCM (9 × 30 mL), then mixture was dried over MgSO$_4$ and was concentrated in vacuo. The crude residue was separated/purified using column chromatography (DCM:MeOH, gradient: 20:1 to 5:1).

Product **21** *(3,28-di(O(CO)CH$_2$TriazCH$_2$OH)BN)* was obtained as a white solid (367.2 mg, 77% yield); m.p.: 105–108 °C; [α]$^{25}_D$ = +4.7 (c 0.5, CHCl$_3$); HRMS (ESI$^+$): calcd for C$_{40}$H$_{61}$N$_6$O$_6$ ([M + H]$^+$): *m/z* 721.4653; found: *m/z* 721.4650; ^{1}H NMR (600 MHz, CDCl$_3$): δ_H 7.68 and 7.66 (all s, 1H each, 2x H-5$_{triaz.}$), 5.19 (d, 2H, J 1.0 Hz, O(CO)CH$_2$), 5.15 (d, 2H, J 0.6 Hz, O(CO)CH$_2$), 4.82 (br s, 4H, 2x C<u>H</u>$_2$OH), 4.69 (d, 1H, J 1.6 Hz, H-29a), 4.60 (s, 1H, H-29b), 4.57–4.54 (m, 1H, H-3), 4.42 (d, 1H, J 10.9 Hz, H-28a), 3.96 (d, 1H, J 10.9 Hz, H-28b), 2.41 (m, 3H, H-19, 2x CH$_2$O<u>H</u>), 2.10–0.60 (m, 24H, CH, CH$_2$), 1.67 (s, 3H, H-30), 1.00, 0.96, 0.83, 0.828, 0.75 (all s, 3H each, H-23–H-27) ppm; ^{13}C NMR (150 MHz, CDCl$_3$) δ_C 165.97, 165.61, 149.96, 149.66, 148.09, 123.08 110.12, 83.85, 64.92, 56.50; 55.28, 55.23, 51.08; 50.86, 50.20, 48.75, 47.65, 46.43, 42.71, 40.88, 38.26, 37.88, 37.64, 37.02, 34.37, 34.04, 29.56, 29.43, 28.00, 26.98, 25.07, 23.58, 20.73, 19.09, 18.10, 16.36, 16.11, 16.00, 14.75 ppm (Supplementary Materials, Table S3); IR (ATR) v: 3500–300, 2927, 2871, 1743, 1456, 1376, 1266, 1219, 1040, 1006, 977, 882, 805 cm^{-1}.

3.2.7. Procedure for the Deprotection of Betulin Glycoconjugates **15–17**

Glycoconjugates **15–17** (1.0 mmol) were dissolved in MeOH (**15a–b**: 16 mL, **16a–b**: 35 mL, **17a–b**: 30 mL). Then, 1M solution of MeONa in MeOH (**15a–b**: 0.2 mmol, 0.2 mL, **16a–b**: 0.4 mmol, 0.4 mL, **17a–b**: 0.15 mmol, 0.15 mL) was added. Reaction was carried out for 120 min at room temperature. The reaction progress was monitored on TLC. After the reaction was complete, the mixture was neutralized with *Amberlyst-15*, filtered, and the filtrate was evaporated in vacuo. The crude products were crystallized from methanol.

Product **22a** *(28-OCH$_2$TriazGlcBN)* was obtained as a white solid (69.9 mg, 87% yield); m.p.: 174–175 °C; [α]$^{25}_D$ = +4.1 (c 0.5, MeOH); HRMS (ESI$^+$): calcd for C$_{39}$H$_{63}$N$_3$O$_7$Na ([M + Na]$^+$): *m/z* 708.4564; found: *m/z* 708.4555; ^{1}H NMR (400 MHz, MeOD): δ_H 8.19 (s, br, 1H, H-5$_{triaz.}$), 5.62 (d, 1H, J 9.2 Hz, H-1$_{Glc}$), 4.68–4.55 (m, 4H, H-29a, H-29b, OCH$_2$), 3.91–3.86 (m, 2H, H-6a$_{Glc}$, H-5$_{Glc}$), 3.73 (dd, 1H, J_1 5.2 Hz, J_2 12.1 Hz, H-6b$_{Glc}$), 3.65 (d, 1H, J 8.9 Hz, H-28a), 3.61–3.49 (m, 3H, H-4$_{Glc}$, H-3$_{Glc}$, H-2$_{Glc}$), 3.23 (d, 1H, J 9.0 Hz, H-28b), 3.12 (dd, 1H, J_1 4.9 Hz, J_2 11.2 Hz, H-3), 2.43 (td, 1H, J_1 5.2 Hz, J_2 10.9 Hz, H-19), 2.05-0.60 (m, 25H, CH, CH$_2$), 1.67 (s, 3H, H-30), 1.00, 0.98, 0.95, 0.86, 0.76 (all s, 3H each, H-23–H-27) ppm; ^{13}C NMR (150 MHz, MeOD) δ_C 151.86, 146.36, 124.41, 110.25, 89.70, 81.18, 79.75, 78.60, 74.10, 71.00, 69.88, 65.31, 62.47, 56.90, 51.94, 50.24, 43.82, 42.21, 40.14, 38.99, 38.35, 35.83, 35.53, 31.05, 30.75, 28.68, 28.39, 28.12, 26.71, 22.06, 19.52, 18.40, 16.76, 16.68, 16.15, 15.33 ppm (Supplementary Materials, Table S3); IR (ATR) ν: 3600-3100, 2941, 2871, 1620, 1450, 1390, 1100, 1043, 1013, 880 cm^{-1}.

Product **22b** *(28-OCH$_2$TriazGalBN)* was obtained as a white solid (79.5 mg, 99% yield); m.p.: 218–220 °C; $[\alpha]^{25}_D$ = +9.1 (c 0.5, MeOH); HRMS (ESI $^+$): calcd for C$_{39}$H$_{63}$N$_3$O$_7$Na ([M + Na]$^+$): *m/z* 708.4564; found: *m/z* 708.4553; ^{1}H NMR (400 MHz, MeOD): δ_H 8.21 (s, br, 1H, H-5$_{triaz.}$), 5.57 (d, 1H, *J* 9.2 Hz, H-1$_{Glc}$), 4.68-4.56 (m, 4H, OCH$_2$, H-29a, H-29b), 4.15 (dd~t, 1H, *J* 9.3 Hz, H-2$_{Gal}$), 4.00 (dd, 1H, *J$_1$* 0.9 Hz, *J$_2$* 3.3 Hz, H-4$_{Gal}$), 3.85–3.69 (m, 3H, 2H-6$_{Gal}$, H-5$_{Gal}$), 3.70 (dd, 1H, *J$_1$* 3.3 Hz, *J$_2$* 9.5 Hz, H-3$_{Gal}$), 3.64 (d, 1H, *J* 8.6 Hz, H-28a), 3.23 (d, 1H, *J* 8.8 Hz, H-28b), 3.12 (dd, 1H, *J$_1$* 5.0 Hz, *J$_2$* 11.1 Hz, H-3), 2.43 (td, 1H, *J$_1$* 5.9 Hz, *J$_2$* 10.9 Hz, H-19), 2.01-0.60 (m, 25H, CH, CH$_2$), 1.67 (s, 3H, H-30), 0.99, 0.98, 0.95, 0.86, 0.76 (all s, 3H each, H-23–H-27) ppm; ^{13}C NMR (150 MHz, MeOD) δ_C 151.87, 146.54, 123.87, 110.28, 90.28, 79.90, 79.71, 75.40, 71.50, 70.32, 69.67, 65.33, 62.30, 56.86, 51.90, 50.17, 43.79, 42.15, 39.99, 40.09, 38.95, 38.32, 35.82, 35.46, 30.99, 30.79, 28.65, 28.34, 28.09, 26.64, 22.02, 19.41, 19.50, 16.76, 16.62, 16.16, 15.28 ppm (Supplementary Materials, Table S3); IR (ATR) ν: 3600–3100, 2932, 2869, 1630, 1456, 1374, 1093, 1044, 880 cm^{-1}.

Product **23b** *(28-OCH$_2$TriazCH$_2$(CO)NHGalBN)* was obtained as a white solid (78.3 mg, 96% yield); m.p.: 187-190 °C; $[\alpha]^{25}_D$ = +6.9 (c 0.5, CHCl$_3$); HRMS (ESI $^+$): calcd for C$_{41}$H$_{67}$N$_4$O$_8$ ([M + H]$^+$): *m/z* 743.4959; found: *m/z* 743.4962; ^{1}H NMR (400 MHz, MeOD): 7.96 (s, br, 1H, H-5$_{triaz.}$), 7.46 (s, 1H, NH), 5.22 (q, 2H, *J* 16.4 Hz, CH$_2$N), 4.90 (d, 1H, *J* 8.8 Hz, H-1$_{Gal}$), 4.68 and 4.57 (s, br, 1H, H-29a, H-29b), 4.65 (s, 2H, OCH$_2$), 3.94 (d, 1H, *J* 2.8 Hz, H-4$_{Gal}$), 3.83–3.54 (m, 5H, H-28a, H-2$_{Gal}$, H-3$_{Gal}$, H-5$_{Gal}$, H-6a$_{Gal}$, H-6b$_{Gal}$), 3.39 (s, 2H, CH$_2$CO), 3.24 (d, 1H, *J* 9.2 Hz, H-28b), 3.18–3.14 (m, 1H, H-3), 2.39 (td, 1H, *J$_1$* 5.9 Hz, *J$_2$* 10.8 Hz, H-19), 2.05–0.60 (m, 25H, CH, CH$_2$), 1.68 (s, 3H, H-30), 0.99, 0.98, 0.96, 0.84, 0.76 (all s, 3H each, H-23–H-27) ppm; ^{13}C NMR (150 MHz, MeOD) δ_C 168.57, 151.91, 146.42, 126.86, 110.24, 81.77, 79.75, 78.48, 75.82, 71.51, 70.49, 69.49, 65.15, 56.90, 51.92, 50.23, 43.80, 42.18, 40.12, 40.01, 38.98, 38.35, 35.86, 35.52, 31.04, 30.78, 28.67, 28.36, 28.12, 26.67, 22.05, 19.53, 19.44, 16.77, 16.59, 16.17, 15.30 ppm (Supplementary Materials, Table S3); IR (ATR) ν: 3600–3000, 2942, 1698, 1374, 1228, 1083, 1045, 879 cm^{-1}.

Product **24a** *(3,28-di(OCH$_2$TriazGlc)BN)* was obtained as a white solid (63.9 mg, 87% yield); m.p.: 176–178 °C; $[\alpha]^{25}_D$ = +16.4 (c 0.5, MeOH); HRMS (ESI $^+$): calcd for C$_{48}$H$_{76}$N$_6$O$_{12}$Na ([M + Na]$^+$): *m/z* 951.5419; found: *m/z* 951.5441; ^{1}H NMR (600 MHz, MeOD): δ_H 8.18 i 8.14 (all s, br, 1H each, 2x H-5$_{triaz.}$), 5.61 and 5.59 (all d, 1H each, *J* 8.7 Hz, 2x H-1$_{Glc}$), 4.74–4.50 (m, 6H, H-29a, H-29b, 2x OCH$_2$), 3.90–3.87 (m, 4H, 2x H-6a$_{Glc}$, 2x H-5$_{Glc}$), 3.74–3.48 (m, 9H, H-28a, 2x H-2$_{Glc}$, 2x H-3$_{Glc}$, 2x H-4$_{Glc}$, 2x H-6b$_{Glc}$), 3.23 (d, 1H, *J* 9.0 Hz, H-28b), 2.98 (dd, 1H, *J$_1$* 4.2 Hz, *J$_2$* 11.4 Hz, H-3), 2.43 (td, 1H, *J$_1$* 5.9 Hz, *J$_2$* 11.0 Hz, H-19), 2.00–0.65 (m, 25H, CH, CH$_2$), 1.68 (s, 3H, H-30), 0.99, 0.98, 0.92, 0.87, 0.75 (all s, 3H each, H-23–H-27) ppm; ^{13}C NMR (150 MHz, MeOD) δ_C 151.89, 146.99, 146.42, 124.33, 124.18, 110.24, 88.07, 89.65, 81.20, 78.64, 74.11, 71.00, 69.82, 65.36, 63.50, 62.49, 57.23, 51.93, 50.25, 43.85, 42.25, 39.85, 39.96, 39.00, 38.37, 35.84, 35.52, 31.07, 30.76, 28.67, 28.41, 28.40, 26.71, 22.10, 19.43, 18.40, 16.82, 16.79, 16.70, 15.32 ppm (Supplementary Materials, Table S3); IR (ATR) ν: 3600–3100, 2936, 1620, 1460, 1374, 1093, 1042, 896 cm^{-1}.

Product **24b** *(3,28-di(OCH$_2$TriazGal)BN)* was obtained as a white solid (50.7 mg, 69% yield); m.p.: 178–179 °C; $[\alpha]^{25}_D$ = +22.4 (c 0.5, MeOH); HRMS (ESI$^+$): calcd for C$_{48}$H$_{77}$N$_6$O$_{12}$ ([M + H]$^+$): *m/z* 929.5599; found: *m/z* 929.5621; ^{1}H NMR (600 MHz, MeOD): δ_H 8.21 and 8.18 (all s, br, 1H each, 2x H-5$_{triaz.}$), 5.57 and 5.56 (all d, 1H each, *J* 9.0 Hz, 2x H-1$_{Gal}$), 4.74-4.51 (m, 6H, 2x OCH$_2$, H-29a, H-29b), 4.15 and 4.14 (all dd~t, 2H each, *J* 9.4 Hz, 2x H-2$_{Gal}$), 3.99 (m, 2H, 2x H-4$_{Gal}$), 3.85-3.69 (m, 8H, 2x H-3$_{Gal}$, 2x H-5$_{Gal}$, 2x H-6a$_{Gal}$, 2x H-6b$_{Gal}$), 3.64 (d, 1H, *J* 9.4 Hz, H-28a), 3.23 (d, 1H, *J* 9.0 Hz, H-28b), 2.98 (dd, 1H, *J$_1$* 4.3 Hz, *J$_2$* 11.8 Hz, H-3), 2.43 (td, 1H, *J$_1$* 5.9 Hz, *J$_2$* 10.7 Hz, H-19), 2.00–0.70 (m, 25H, CH, CH$_2$), 1.68 (s, 3H, H-30), 0.99, 0.98, 0.92, 0.87, 0.76 (all s, 3H each, H-23–H-27)ppm; ^{13}C NMR (150 MHz, MeOD) δ_C 151.89, 147.19, 146.56, 123.89, 123.64, 110.27, 90.27, 88.04, 80.04, 79.02, 75.44, 71.56, 71.55, 69.77, 65.38; 63.56, 62.50, 62.34, 57.22, 51.91, 50.22, 43.83, 42.22, 39.83, 39.97, 39.96, 38.34, 35.84, 35.48,

31.05, 30.78, 28.66, 28.37, 28.42, 26.68, 22.08, 19.42, 18.41, 16.83, 16.80, 16.66, 15.30 ppm (Supplementary Materials, Table S3); IR (ATR) ν: 3600–3100, 2925, 2869, 1650, 1456, 1374, 1219, 1090, 1045, 879 cm^{-1}.

3.3. Biological Assays

3.3.1. Cell Lines

HCT 116 and MCF-7 cells were purchased from the American Type Culture Collection. NHDF cells were purchased from Lonza. All cells were cultured under standard conditions at 37 °C in a humidified atmosphere at 5% CO_2 in DMEM/F12 medium (PAA) supplemented with 10% FBS (EURx).

3.3.2. CCK-8 Assay

A Cell Counting Kit-8 (CCK-8) from Bimake was used to assess cell viability. Briefly, cells were seeded in 96-well plates, with three duplicate wells in each group. Cells were treated with betulin or its derivatives and incubated for 24 or 48 h. Then, CCK-8 solution was added to each well and the plate was incubated for 2 h at 37 °C. Subsequently, the absorbance at 450 nm was measured using a microplate reader (Epoch, BioTek). Cell viability rate was calculated using *CalcuSyn software* (version 2.0, Biosoft, Cambridge, UK).

4. Conclusions

In conclusion, we designed and synthesized a library of novel glycoconjugates of natural pentacyclic triterpenoid (BN) by employing click chemistry. The CuAAC reactions are extensively employed in the conjugate synthesis of various organic compounds, owing to their versatility, high chemoselectivity, and mild conditions. In this study, we successfully prepared new mono- and disubstituted betulin derivatives containing a sugar unit attached via a linker inclusive of the 1,2,3-triazole ring at the C3 and C28 position of the parent skeleton of BN. We confirmed that the click chemistry approach in a simple, easy and inexpensive way leads to a wide range of products with high yield, purity, and selectivity.

The methodology developed by us extends the synthetic possibilities of modifying the betulin skeleton. Importantly, it enables the simple preparation of both mono-(modification C28) and disubstituted betulin derivatives (modification C3 and C28) with high yields. The construction of the linker connecting betulin to the sugar unit depends on the type of building blocks used. According to Strategy I, synthesis starts with the preparation of an alkyne betulin derivative, which is then clicked with a 1-azido sugar. The second option is to modify the betulin with an azide moiety and click it with propargyl *O*-glucoside.

Unfortunately, a preliminary cytotoxicity assay of the obtained betulin glycoconjugates showed that the addition of a sugar unit to the native betulin structure via a few selected linkers containing a 1,2,3-triazole ring is not a significant way for biological activity. Monosubstituted betulin glycoconjugates show comparable or slightly lower activity than that of betulin alone while betulin derivatives containing two sugar units protected with acetyl groups increase cell viability, which may result from the release of glucose from the tested glycoconjugates under physiological conditions. In turn, the observed lack of biological activity of deprotected betulin glycoconjugates may be due to the insufficient affinity of these compounds for glucose transporters.

3,28-*O,O′*-Di(2-(4-(hydroxymethyl-*1H*-1,2,3-triazol-1-yl)acetyl)betulin **21**, modified only by introducing a fragment containing the 1,2,3-triazole system without the sugar unit, turned out to be surprisingly active. Unfortunately, it also turned out to be very toxic. This indicates that modifying the betulin backbone by introducing a 1,2,3-triazole ring significantly improves its cytotoxicity. However, further work should be done on increasing the selectivity of the obtained connections.

The obtained results show that despite the fact that the obtained betulin glycoconjugates do not show interesting antitumor activity, the idea of adding a sugar unit to the betulin backbone may, after some modifications, turn out to be correct and allow for the targeted transport of biologically active compounds into the tumor cells. The position used in the sugar unit for conjugation with the betulin derivative appears to be crucial for maintaining its affinity for GLUT transporters. Based on literature reports [59,60], it can be assumed that the 6-OH group (the group least involved in binding to the GLUT transporter) is the most neutral position that can be modified in sugar, so as not to reduce its affinity to the transport protein on the surface of cancer cells. Due to the above, it seems advisable to undertake further research on the chemical modifications of betulin by attaching sugar through the functionalization of this position. Glycoconjugation of betulin with the use of sugar derivatives modified in the C-6 position will be carried out both with the use of 1,3-dipolar azide-alkyne cycloaddition, as well as by creating bonds that can be broken under the action of intracellular enzymes (ester, amide, or carbamate).

Supplementary Materials: The following are available online. Supporting information includes experimental procedures and spectroscopic properties of synthesized compounds (**4**, **5**, **9–14**), ^{1}H NMR, ^{13}C NMR spectra of all synthesized compounds (**1–24**) and gHSQC and FT-IR for selected compounds.

Author Contributions: Conceptualization and methodology, M.G. and G.P.-G.; synthesis and characterization of chemical compounds, M.Z., S.B. and M.K.; cytotoxicity tests, A.L.; mass spectra, K.E.; supervision, M.G. and G.P.-G.; analysis and interpretation of the results, M.G., G.P.-G., M.Z. and S.B.; writing-original draft preparation, M.G. and G.P.-G.; writing-review and editing, M.G. and G.P.-G. All authors have read and agreed to the published version of the manuscript.

Funding: This research was supported by Grant BK No. 04/020/BK/20/0122 and BKM No. 04/020/BKM20/0138 (BKM-611/RCH2/2020) as part of a targeted subsidy for conducting scientific research or development works and related tasks for the development of young scientists and participants of doctoral studie granted by Ministry of Science and Higher Education, Poland. The cytotoxicity studies was supported by BK-274/RAu1/2020.

Conflicts of Interest: The authors declare no conflict of interest.

References

1. Krasutsky, P.A. Birch bark research and development. *Nat. Prod. Rep.* **2006**, *23*, 919–942. [CrossRef] [PubMed]
2. Safe, S.; Kasiappan, R. Natural Products as Mechanism-based Anticancer Agents: Sp Transcription Factors as Targets. *Phytother. Res.* **2016**, *30*, 1723–1732. [CrossRef] [PubMed]
3. Hordyjewska, A.; Ostapiuk, A.; Horecka, A. Betulin and betulinic acid in cancer research. *J. Pre-Clin. Clin. Res.* **2018**, *12*, 72–75. [CrossRef]
4. Dutta, D.; Chakraborty, B.; Sarkar, A.; Chowdhury, C.; Das, P. A potent betulinic acid analogue ascertains an antagonistic mechanism between autophagy and proteasomal degradation pathway in HT-29 cells. *BMC Cancer* **2016**, *16*, 23. [CrossRef] [PubMed]
5. Boryczka, S.; Bębenek, E.; Wietrzyk, J.; Kempińska, K.; Jastrzębska, M.; Kusz, J.; Nowak, M. Synthesis, Structure and Cytotoxic Activity of New Acetylenic Derivatives of Betulin. *Molecules* **2013**, *18*, 4526–4543. [CrossRef]
6. Baratto, L.C.; Porsani, M.V.; Pimentel, I.C.; Netto, A.B.P.; Paschke, R.; Oliveira, B.H. Preparation of betulinic acid derivatives by chemical and biotransformation methods and determination of cytotoxicity against selected cancer cell lines. *Eur. J. Med. Chem.* **2013**, *68*, 121–131. [CrossRef]
7. Kommera, H.; Kaluđerović, G.N.; Kalbitz, J.; Dräger, B.; Paschke, R. Small structural changes of pentacyclic lupane type triterpenoid derivatives lead to significant differences in their anticancer properties. *Eur. J. Med. Chem.* **2010**, *45*, 3346–3353. [CrossRef]
8. Kommera, H.; Kaluđerović, G.N.; Dittrich, S.; Kalbitz, J.; Dräger, B.; Mueller, T.; Paschke, R. Carbamate derivatives of betulinic acid and betulin with selective cytotoxic activity. *Bioorg. Med. Chem. Lett.* **2010**, *20*, 3409–3412. [CrossRef]
9. Dangroo, N.A.; Singh, J.; Rath, S.K.; Gupta, N.; Qayum, A.; Singh, S.; Sangwan, P.L. A convergent synthesis of novel alkyne–azide cycloaddition congeners of betulinic acid as potent cytotoxic agent. *Steroids* **2017**, *123*, 1–12. [CrossRef]
10. Gauthier, C.; Legault, J.; Lebrun, M.; Dufour, P.; Pichette, A. Glycosidation of lupane-type triterpenoids as potent in vitro cytotoxic agents. *Bioorg. Med. Chem.* **2006**, *14*, 6713–6725. [CrossRef]

11. Drąg-Zalesińska, M.; Kulbacka, J.; Saczko, J.; Wysocka, T.; Zabel, M.; Surowiak, P.; Drag, M. Esters of betulin and betulinic acid with amino acids have improved water solubility and are selectively cytotoxic toward cancer cells. *Bioorg. Med. Chem. Lett.* **2009**, *19*, 4814–4817. [CrossRef] [PubMed]

12. Yang, S.-J.; Liu, M.-C.; Xiang, H.-M.; Zhao, Q.; Xue, W.; Yang, S. Synthesis and in vitro antitumor evaluation of betulin acid ester derivatives as novel apoptosis inducers. *Eur. J. Med. Chem.* **2015**, *102*, 249–255. [CrossRef] [PubMed]

13. Zhang, D.-M.; Xu, H.-G.; Wang, L.; Li, Y.-J.; Sun, P.; Wu, X.-M.; Wang, G.; Chen, W.-M.; Ye, W.-C. Betulinic Acid and its Derivatives as Potential Antitumor Agents. *Med. Res. Rev.* **2015**, *35*, 1127–1155. [CrossRef] [PubMed]

14. Zhao, H.; Liu, Z.; Liu, W.; Han, X.; Zhao, M. Betulin attenuates lung and liver injuries in sepsis. *Int. Immunopharmacol.* **2016**, *30*, 50–56. [CrossRef] [PubMed]

15. Moghaddam, M.G.; Ahmad, F.B.H.; Samzadeh-Kermani, A. Biological Activity of Betulinic Acid: A Review. *Pharmacol. Pharm.* **2012**, *3*, 119–123. [CrossRef]

16. Zhao, H.; Zheng, Q.; Hu, X.; Shen, H.; Li, F. Betulin attenuates kidney injury in septic rats through inhibiting TLR4/NF-κB signaling pathway. *Life Sci.* **2016**, *144*, 185–193. [CrossRef]

17. Fulda, S.; Kroemer, G. Targeting mitochondrial apoptosis by betulinic acid in human cancers. *Drug Discov. Today* **2009**, *14*, 885–890. [CrossRef]

18. Suman, P.; Patel, A.; Solano, L.; Jampana, G.; Gardner, Z.S.; Holt, C.M.; Jonnalagadda, S.C. Synthesis and cytotoxicity of Baylis-Hillman template derived betulinic acid-triazole conjugates. *Tetrahedron* **2017**, *73*, 4214–4226. [CrossRef]

19. Jonnalagadda, S.; Suman, P.; Morgan, D.; Seay, J. Recent Developments on the Synthesis and Applications of Betulin and Betulinic Acid Derivatives as Therapeutic Agents. *Bioact. Nat. Prod. (Part O)* **2017**, *53*, 45–84. [CrossRef]

20. Baltina, L.A.; Kazakova, O.B.; Nigmatullina, L.; Boreko, E.I.; Pavlova, N.; Nikolaeva, S.; Savinova, O.; Tolstikov, G. Lupane triterpenes and derivatives with antiviral activity. *Bioorg. Med. Chem. Lett.* **2003**, *13*, 3549–3552. [CrossRef]

21. Bori, I.D.; Hung, H.-Y.; Qian, K.; Chen, C.-H.; Morris-Natschke, S.L.; Lee, K.-H. Anti-AIDS agents 88. Anti-HIV conjugates of betulin and betulinic acid with AZT prepared via click chemistry. *Tetrahedron Lett.* **2012**, *53*, 1987–1989. [CrossRef]

22. Amiri, S.; Dastghaib, S.; Ahmadi, M.; Mehrbod, P.; Khadem, F.; Behrouj, H.; Aghanoori, M.-R.; Machaj, F.; Ghamsari, M.; Rosik, J.; et al. Betulin and its derivatives as novel compounds with different pharmacological effects. *Biotechnol. Adv.* **2020**, *38*, 107409. [CrossRef]

23. Zdzisińska, B. Właściwości lecznicze betuliny i kwasu betulinowego, składników ekstraktu z kory brzozy. *Farm. Przegl. Nauk.* **2010**, *3*, 33–39.

24. Grymel, M.; Zawojak, M.; Adamek, J. Triphenylphosphonium Analogues of Betulin and Betulinic Acid with Biological Activity: A Comprehensive Review. *J. Nat. Prod.* **2019**, *82*, 1719–1730. [CrossRef] [PubMed]

25. Pokorný, J.; Horka, V.; Sidova, V.; Urban, M. Synthesis and characterization of new conjugates of betulin diacetate and bis(triphenysilyl)betulin with substituted triazoles. *Mon. Für Chem.—Chem. Mon.* **2018**, *149*, 839–845. [CrossRef]

26. Bhunia, D.; Pallavi, P.M.C.; Bonam, S.R.; Reddy, S.A.; Verma, Y.K.; Halmuthur, S.K.M. Design, Synthesis, and Evaluation of Novel 1,2,3-Triazole-Tethered Glycolipids as Vaccine Adjuvants. *Arch. Pharm.* **2015**, *348*, 689–703. [CrossRef] [PubMed]

27. Ye, Y.; Zhang, T.; Yuan, H.; Li, D.; Lou, H.; Fan, P. Mitochondria-Targeted Lupane Triterpenoid Derivatives and Their Selective Apoptosis-Inducing Anticancer Mechanisms. *J. Med. Chem.* **2017**, *60*, 6353–6363. [CrossRef] [PubMed]

28. Dalvie, D.K.; Kalgutkar, A.S.; Khojasteh-Bakht, S.C.; Obach, R.S.; O'Donnell, J.P. Biotransformation Reactions of Five-Membered Aromatic Heterocyclic Rings. *Chem. Res. Toxicol.* **2002**, *15*, 269–299. [CrossRef]

29. Valverde, I.E.; Bauman, A.; Kluba, C.A.; Vomstein, S.; Walter, M.A.; Mindt, T.L. 1,2,3-Triazoles as Amide Bond Mimics: Triazole Scan Yields Protease-Resistant Peptidomimetics for Tumor Targeting. *Angew. Chem. Int. Ed.* **2013**, *52*, 8957–8960. [CrossRef]

30. Dheer, D.; Singh, V.; Shankar, R. Medicinal attributes of 1,2,3-triazoles: Current developments. *Bioorganic Chem.* **2017**, *71*, 30–54. [CrossRef]

31. Thirumurugan, P.; Matosiuk, D.; Jozwiak, K. Click Chemistry for Drug Development and Diverse Chemical–Biology Applications. *Chem. Rev.* **2013**, *113*, 4905–4979. [CrossRef] [PubMed]

32. Pokorný, L.B.A.M.U.J.; Borkova, L.; Urban, M. Click Reactions in Chemistry of Triterpenes—Advances Towards Development of Potential Therapeutics. *Curr. Med. Chem.* **2018**, *25*, 636–658. [CrossRef] [PubMed]

33. Antimonova, A.N.; Petrenko, N.I.; Shakirov, M.M.; Rybalova, T.V.; Frolova, T.S.; Shul'Ts, E.E.; Kukina, T.P.; Sinitsyna, O.I.; Tolstikov, G.A. Synthesis and study of mutagenic properties of lupane triterpenoids containing 1,2,3-triazole fragments in the C-30 position. *Chem. Nat. Compd.* **2013**, *49*, 657–664. [CrossRef]

34. Bębenek, E.; Jastrzębska, M.; Kadela-Tomanek, M.; Chrobak, E.; Orzechowska, B.; Zwolińska, K.; Latocha, M.; Mertas, A.; Czuba, Z.P.; Boryczka, S. Novel Triazole Hybrids of Betulin: Synthesis and Biological Activity Profile. *Molecules* **2017**, *22*, 1876. [CrossRef] [PubMed]

35. Shi, W.; Tang, N.; Yan, W. Synthesis and cytotoxicity of triterpenoids derived from betulin and betulinic acid via click chemistry. *J. Asian Nat. Prod. Res.* **2015**, *17*, 159–169. [CrossRef] [PubMed]

36. Krawczyk, M.; Pastuch-Gawołek, G.; Mrozek-Wilczkiewicz, A.; Kuczak, M.; Skonieczna, M.; Musiol, R. Synthesis of 8-hydroxyquinoline glycoconjugates and preliminary assay of their β1,4-GalT inhibitory and anti-cancer properties. *Bioorg. Chem.* **2019**, *84*, 326–338. [CrossRef] [PubMed]

37. Krawczyk, M.; Pastuch-Gawołek, G.; Pluta, A.; Erfurt, K.; Domiński, A.; Kurcok, P. 8-Hydroxyquinoline Glycoconjugates: Modifications in the Linker Structure and Their Effect on the Cytotoxicity of the Obtained Compounds. *Molecules* **2019**, *24*, 4181. [CrossRef]

38. Krawczyk, M.; Pastuch-Gawołek, G.; Hadasik, A.; Erfurt, K. 8-Hydroxyquinoline Glycoconjugates Containing Sulfur at the Sugar Anomeric Position—Synthesis and Preliminary Evaluation of Their Cytotoxicity. *Molecules* **2020**, *25*, 4174. [CrossRef]

39. Domiński, A.; Krawczyk, M.; Konieczny, T.; Kasprów, M.; Foryś, A.; Pastuch-Gawołek, G.; Kurcok, P. Biodegradable pH-responsive micelles loaded with 8-hydroxyquinoline glycoconjugates for Warburg effect based tumor targeting. *Eur. J. Pharm. Biopharm.* **2020**, *154*, 317–329. [CrossRef]

40. Křen, V.; Martinkova, L. Glycosides in Medicine: "The Role of Glycosidic Residue in Biological Activity". *Curr. Med. Chem.* **2001**, *8*, 1303–1328. [CrossRef]

41. A Smith, T. Facilitative glucose transporter expression in human cancer tissue. *Br. J. Biomed. Sci.* **1999**, *56*, 285–292. [PubMed]

42. Heiden, M.G.V.; Cantley, L.C.; Thompson, C.B. Understanding the Warburg Effect: The Metabolic Requirements of Cell Proliferation. *Science* **2009**, *324*, 1029–1033. [CrossRef] [PubMed]

43. Calvaresi, E.C.; Hergenrother, P.J. Glucose conjugation for the specific targeting and treatment of cancer. *Chem. Sci.* **2013**, *4*, 2319–2333. [CrossRef] [PubMed]

44. Zhao, F.-Q. Functional Properties and Genomics of Glucose Transporters. *Curr. Genom.* **2007**, *8*, 113–128. [CrossRef]

45. Barnett, J.E.G.; Holman, G.D.; Munday, K.A. Structural requirements for binding to the sugar-transport system of the human erythrocyte. *Biochem. J.* **1973**, *131*, 211–221. [CrossRef]

46. Ma, J.; Liu, H.; Xi, Z.; Hou, J.; Li, Y.; Niu, J.; Liu, T.; Bi, S.; Wang, X.; Wang, C.; et al. Protected and De-protected Platinum(IV) Glycoconjugates With GLUT1 and OCT2-Mediated Selective Cancer Targeting: Demonstrated Enhanced Transporter-Mediated Cytotoxic Properties in vitro and in vivo. *Front. Chem.* **2018**, *6*, 386. [CrossRef]

47. Khan, I.; Guru, S.K.; Rath, S.K.; Chinthakindi, P.K.; Singh, B.; Koul, S.; Bhushan, S.; Sangwan, P.L. A novel triazole derivative of betulinic acid induces extrinsic and intrinsic apoptosis in human leukemia HL-60 cells. *Eur. J. Med. Chem.* **2016**, *108*, 104–116. [CrossRef]

48. Komissarova, N.G.; Dubovitskii, S.N.; Shitikova, O.V.; Vyrypaev, E.M.; Spirikhin, L.V.; Eropkina, E.M.; Lobova, T.G.; Eropkin, M.; Yunusov, M.S. Synthesis of Conjugates of Lupane-Type Pentacyclic Triterpenoids with 2-Aminoethane- and N-Methyl-2-Aminoethanesulfonic Acids. Assessment of in vitro Toxicity. *Chem. Nat. Compd.* **2017**, *53*, 907–914. [CrossRef]

49. Yang, Q.-R.; Qiao, W.-H.; Zhang, S.-M.; Qu, J.-P.; Liu, D.-L. Synthesis and Characterization of a New Cationic Galactolipid with Carbamate for Gene Delivery. *Tenside Surfactants Deterg.* **2010**, *47*, 294–299. [CrossRef]

50. Rajaram, H.; Palanivelu, M.K.; Arumugam, T.V.; Rao, V.M.; Shaw, P.N.; McGeary, R.P.; Ross, B.P. 'Click' assembly of glycoclusters and discovery of a trehalose analogue that retards Aβ40 aggregation and inhibits Aβ40-induced neurotoxicity. *Bioorg. Med. Chem. Lett.* **2014**, *24*, 4523–4528. [CrossRef]

51. Lancuški, A.; Bossard, F.; Fort, S. Carbohydrate-Decorated PCL Fibers for Specific Protein Adhesion. *Biomacromolecules* **2013**, *14*, 1877–1884. [CrossRef] [PubMed]

52. Zemplén, G.; Pacsu, E. Über die Verseifung acetylierter Zucker und verwandter Substanzen. *Berichte Dtsch. Chem. Ges. (A B Ser.)* **1929**, *62*, 1613–1614. [CrossRef]

53. Brown, R.S.; Wahl, R.L. Overexpression of Glut-1 glucose transporter in human breast cancer. An im-munohistochemical study. *Cancer* **1993**, *72*, 2979–2985. [CrossRef]

54. Haber, R.S.; Rathan, A.; Weiser, K.R.; Pritsker, A.; Itzkowitz, S.H.; Bodian, C.; Slater, G.; Weiss, A.; Burstein, D.E. GLUT1 glucose transporter expression in colorectal carcinoma: A marker for poor progno-sis. *Cancer* **1998**, *83*, 34–40. [CrossRef]

55. Kumamoto, K.; Goto, Y.; Sekikawa, K.; Takenoshita, S.; Ishida, N.; Kawakita, M.; Kannagi, R. Increased expression of UDP-galactose transporter messenger RNA in human colon cancer tissues and its implica-tion in synthesis of Thomsen-Friedenreich antigen and sialyl Lewis A/X determinants. *Cancer Res.* **2001**, *61*, 4620–4627.

56. Lutter, A.-H.; Scholka, J.; Richter, H.; Anderer, U. Applying XTT, WST-1, and WST-8 to human chondrocytes: A comparison of membrane-impermeable tetrazolium salts in 2D and 3D cultures. *Clin. Hemorheol. Microcirc.* **2017**, *67*, 327–342. [CrossRef]

57. Van De Waterbeemd, H.; Smith, D.A.; Jones, B.C. Lipophilicity in PK design: Methyl, ethyl, futile. *J. Comput. Mol. Des.* **2001**, *15*, 273–286. [CrossRef]

58. Lombardo, F.; Obach, R.; Shalaeva, M.Y.; Gao, F. Prediction of Volume of Distribution Values in Hu-mans for Neutral and Basic Drugs Using Physicochemical Measurements and Plasma Protein Binding Data. *J. Med. Chem.* **2002**, *45*, 2867–2876. [CrossRef]

59. Deng, D.; Xu, C.; Sun, P.; Wu, J.; Yan, C.; Hu, M.; Yan, N. Crystal structure of the human glucose transporter GLUT1. *Nat. Cell Biol.* **2014**, *510*, 121–125. [CrossRef]

60. Fernández, C.; Nieto, O.; Rivas, E.; Montenegro, G.; Fontenla, J.A.; Fernández-Mayoralas, A. Synthesis and biological studies of glycosyl dopamine derivatives as potential antiparkinsonian agents. *Carbohydr. Res.* **2000**, *327*, 353–365. [CrossRef]

Sample Availability: Samples of the compounds **15–24** are available from the authors.

Publisher's Note: MDPI stays neutral with regard to jurisdictional claims in published maps and institutional affiliations.

Review

Bioactive Compounds and Metabolites from Grapes and Red Wine in Breast Cancer Chemoprevention and Therapy

Danielly C. Ferraz da Costa [1], Luciana Pereira Rangel [2], Julia Quarti [3], Ronimara A. Santos [1], Jerson L. Silva [4],* and Eliane Fialho [3],*

[1] Departamento de Nutrição Básica e Experimental, Instituto de Nutrição, Universidade do Estado do Rio de Janeiro, Rio de Janeiro 20550-013, Brazil; daniellyferraz@ymail.com (D.C.F.d.C.); ronimaras@gmail.com (R.A.S.)

[2] Faculdade de Farmácia, Universidade Federal do Rio de Janeiro, Rio de Janeiro 21941-902, Brazil; lprangel@pharma.ufrj.br

[3] Departamento de Nutrição Básica e Experimental, Instituto de Nutrição Josué de Castro, Universidade Federal do Rio de Janeiro, Rio de Janeiro 21941-902, Brazil; julia_quarti@hotmail.com

[4] Programa de Biologia Estrutural, Instituto de Bioquímica Médica Leopoldo de Meis, Instituto Nacional de Ciência e Tecnologia de Biologia Estrutural e Bioimagem, Universidade Federal do Rio de Janeiro, Rio de Janeiro 21941-902, Brazil

* Correspondences: jerson@bioqmed.ufrj.br (J.L.S.); fialho@nutricao.ufrj.br (E.F.); Tel.: +55-21-3938-6756 (J.L.S.); +55-21-3938-6799 (E.F.)

Academic Editors: José Antonio Lupiáñez, Amalia Pérez-Jiménez, Eva E. Rufino-Palomares and Luca Rolle
Received: 15 May 2020; Accepted: 28 July 2020; Published: 1 August 2020

Abstract: Phytochemicals and their metabolites are not considered essential nutrients in humans, although an increasing number of well-conducted studies are linking their higher intake with a lower incidence of non-communicable diseases, including cancer. This review summarizes the current findings concerning the molecular mechanisms of bioactive compounds from grapes and red wine and their metabolites on breast cancer—the most commonly occurring cancer in women—chemoprevention and treatment. Flavonoid compounds like flavonols, monomeric catechins, proanthocyanidins, anthocyanins, anthocyanidins and non-flavonoid phenolic compounds, such as resveratrol, as well as their metabolites, are discussed with respect to structure and metabolism/bioavailability. In addition, a broad discussion regarding in vitro, in vivo and clinical trials about the chemoprevention and therapy using these molecules is presented.

Keywords: bioactive compounds; metabolites; wine; grapes; breast cancer; chemoprevention; chemotherapy

---

1. Introduction

Breast cancer was ranked as the fifth leading cause of death (627,000 deaths, 6.6%) worldwide in 2018 [1]. During the last decade, new strategies based on the use of dietary chemopreventive agents for breast cancer management have been developed. Several in vitro and in vivo studies have reported the beneficial effects promoted by bioactive compounds from grapes and its derivative products [2]. The positive impact on health has been attributed to its phenolic compounds, such as flavonoids, stilbenes, anthocyanins and other molecules. This review aims to summarize the current findings regarding the role of bioactive compounds from grapes and red wine and their metabolites on breast cancer chemoprevention and treatment by exploring its molecular targets and mechanisms of action [3].

In cancer research, a wide variety of established breast cancer cell lines are used as experimental models. Most of them resemble the different subtypes of breast cancer seen in the clinic. These cell lines offer an infinite supply of a relatively homogeneous cell population that is capable of self-replication in standard cell culture medium and are available through commercial cell banks. The most commonly used breast cancer cell line in the world, MCF-7, was established in 1973 at the Michigan Cancer Foundation. With different molecular characteristics, MDA-MB-231, MDA-MB-453, MDA-MB-468, TD47D, among others, are frequently used in studies, as will be described below [4].

2. Anticancer Effects Produced by Grapes and Seed Extracts

Grapes and their derivative products are a rich source of bioactive molecules, including flavonoid compounds (flavonols, monomeric catechins, proanthocyanidins, anthocyanins, anthocyanidins) and non-flavonoid phenolic compounds (resveratrol), as well as their metabolites. Several molecular pathways involved in breast cancer cell signaling and differentiation, cell cycle arrest, apoptosis, and metastasis can be modulated by these compounds, as described below (Figure 1).

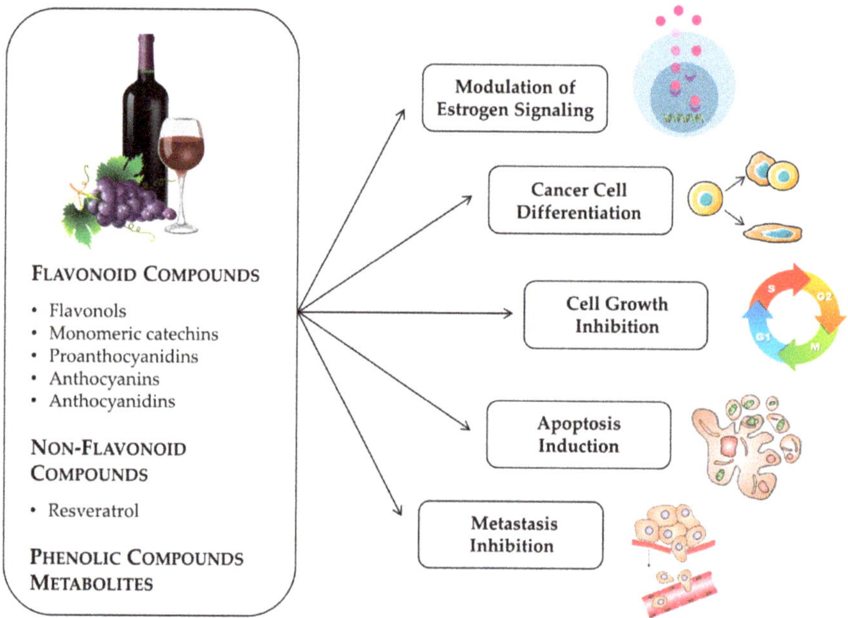

Figure 1. Anticancer activities promoted by phenolic compounds from grapes and red wine and their metabolites.

A polyphenolic fraction isolated from grape seeds (GSP) containing 50% of procyanidins, 5% of catechin and 6% of epicatechin that has been described to cause irreversible inhibition growth of MDA-MB-468, a metastatic breast cancer cell line, by a mechanism involving activation of MAPK/ERK1/2 and MAPK/p38, the two MAPK pathways associated with cell growth and differentiation. GSP also promoted the induction of CDKI Cip1/p21 and a decrease in CDK4, resulting in G1 arrest [5]. Polyphenols obtained by hydroalcoholic extraction from grape seeds promoted a selective inhibition of cell viability and induction of apoptotic cell death on MCF-7 cells. The authors hypothesize that this effect is mediated by gap-junction-mediated cell-cell communications improvement via re-localization of Cx43 proteins and up-regulation of CX43 gene, since gap junctions have been associated with the apoptotic process [6].

Extracellular matrix remodeling, which is influenced by urokinase-type plasminogen activator (uPA) and matrix metalloproteinases (MMPs), is a critical event in the metastasizing process. Dinicola et al. [7] have studied the effect of grape seed extract (GSE) containing 6.2 mg/g of catechins and 5,6 mg/g of procyanidins on metastatic human breast carcinoma cell line focusing on migration and invasion processes. Low GSE concentrations (25 µg/mL) were reported to strongly inhibit MDA-MB-231 cell migration and invasion by decreasing uPA, MMP-2 and MMP-9 activity, as well as down-regulating β-catenin, fascin and NF-κB expression. On the other hand, high GSE concentration (50 and 100 µg/mL) triggered proliferation arrest and apoptosis.

Another target in the treatment of several cancers is the inhibition of angiogenesis, which is supported by vascular endothelial growth factor (VEGF). A study by Lu et al. [8] showed that GSE (85% procyanidins) reduced VEGF expression in both U251 human glioma cells and MDA-MB-231 human breast cancer cells, supporting the hypothesis that GSE may be a natural anti-angiogenesis source of compounds.

Previous studies described many phytochemicals found in grapes and wine as aromatase inhibitors [9]. Polyphenols were demonstrated to modulate estrogen signaling and to compete for steroid-binding sites. Kijima et al. [10] reported that GSP (74–78% proanthocyanidins and <6% catechin, epicatechin, and their gallic acid esters) suppressed aromatase expression and activity on MCF-7aro (aromatase transfected MCF-7 cells) and SK-BR-3 cells. Aromatase catalyzes critical reactions of estrogen synthesis, converting androgen to estrogen, which is known to stimulate breast cancer cell growth by binding to the estrogen receptor (ER). Another approach includes the combination of therapeutic compounds, like doxorubicin (Dox), and phytochemicals for cancer management. GSE (95% procyanidins) increases the efficacy of Dox in human breast cancer MCF-7, MDA-MB468, and MDA-MB231 cells suggesting a strong possibility of a synergistic effect of GSE and Dox combination, independent of the estrogen receptor status of cells [11].

Grape seed proanthocyanidin extract (GSPE) showed also a promising therapeutic role against adverse effects of the chemotherapeutic agents carboplatin and thalidomide. Administration of these agents in rats led to an enhancement in the TNF-α and IL-6 cytokine levels, which could be partially reversed by administration of GSPE. In addition, GSPE reduced free radicals like thiobarbituric acid-reactive substances and nitric oxide and increased glutathione and antioxidant enzymes in liver and heart [12].

3. Flavonoid Compounds

3.1. Flavonols

Flavonols are a subgroup of the flavonoids group, structurally resembling flavones, with the presence of an additional hydroxyl in position 3 of the A ring of the flavones general backbone (3-hidroxyflavone) (Figure 2). The most abundant flavonols present in wine are quercetin, a majoritary compound in this beverage, kaempferol, myricetin, isorhamnetin (quercetin 3′-methylether), and rutin, a glucoside derivative of quercetin [13].

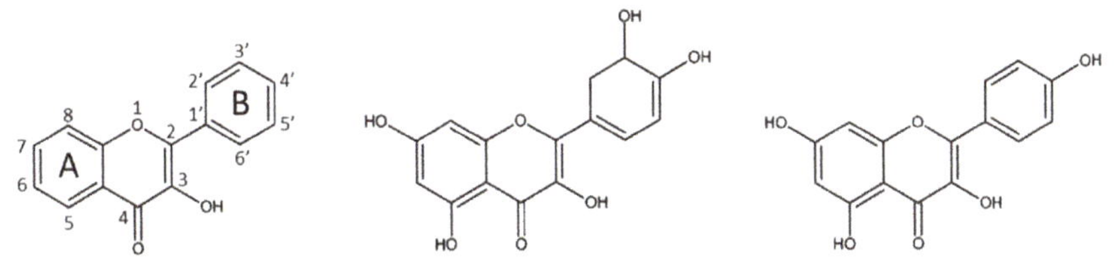

Figure 2. (a) Basic structure of flavonols. (b) Kaempferol. (c) Quercetin.

In red wine, flavonol (aglycones and glycosides) concentration ranges from around 3 to 50 mg/L [14,15], while quercetin varies from around 1 to 10 mg/L in different wines from different regions of the world [16,17]. Flavonol concentration appears to be related to the degree of sun exposure of the grapes while cultivated and the degradation is both related to UV exposure and temperature [13,18]. Moreover, flavonol content is dramatically altered during the processing of grapes, wine production, and storage [13,19,20].

Flavonol bioavailability may vary according to the substitutions found in the molecule, e.g., sugars (mostly glycosides). A relevant example is rutin, a glycosidic disaccharide conjugate of quercetin [13]. The glycosylated forms are the most frequently found in wine and these compounds appear to be more resistant to degradation than their correspondent aglycones [21].

We discuss below the particular effects of quercetin and kaempferol, the most abundant flavonols in wine, on breast cancer. Their effects encompass both chemoprevention and therapy. Although their aglycone forms share a high structural resemblance, the targets, mechanisms of action and bioavailability of these compounds may vary according to the metabolites produced and substitutions found [22,23].

The anticancer effects of kaempferol and quercetin have been described in several different cancer types, such as bladder, breast, prostate, ovarian, liver, and colon. A special feature is given to breast cancer due to a superior number of published works. The few epidemiological studies available using flavonoids such as quercetin or kaempferol on breast cancer involve the observation of their dietary contribution. However, their findings vary from null to positive effects in cancer prevention [24–26]. It is important to mention that, in spite of the larger number of basic research papers published so far, there is a small number of pre-clinical and clinical studies using kaempferol or quercetin as anticancer agents [27,28]. This is due to their low bioavailability, which is around 2% for kaempferol [29,30] and 20% for quercetin, while only 1% was found in the free form in serum [31,32]. This has inspired several works describing drug delivery systems including quercetin and other drugs, such as doxorubicin [33–35]. Likewise, the blending of kaempferol and other flavonoids, either in a complex mixture [36] or simply in combination with quercetin, has been described to increase their anticancer properties [37].

In a general way, the anticancer activities of these two flavonols in breast cancer can be organized in three groups: apoptosis induction, growth inhibition (cell cycle arrest), and inhibition of the metastatic behavior: invasion, migration and epithelial-mesenchymal transition (EMT). Several different papers describe the effects of kaempferol and quercetin in the induction of apoptosis in breast cancer cells, such as MCF-7, MDA-MB-453, SK-BR-3, and MDA-MB-231 [38–40]. Kaempferol and quercetin decrease the growth of MCF-7 and MDA-MB-231 cells with micromolar concentrations. However, the MCF-7 cell line appeared to be more sensitive to quercetin than MDA-MB-231 [41–44]. The same was observed for kaempferol in an MCF-7 3-D model, with ERK signaling being responsible for the apoptotic death [45] and cell cycle arrest at the sub-G1 phase [40].

Kaempferol was found to inhibit the RhoA and RacA signaling pathways, leading to cell migration and invasion inhibition in MDA-MB-231 and MDA-MB-453, triple negative breast cancer (TNBC) cells [41]. This compound was also found to promote PARP cleavage for apoptosis induction through the downregulation of Bcl-2 and BAX induction [38]. On the other hand, kaempferol demonstrated a strong antioxidant activity that was able to attenuate ROS-induced hemolysis and to promote an antiproliferative effect on different tumor cell lines, including MCF-7 cells [46].

Quercetin has been widely reported for its activity against breast cancer cell lines such as MCF-7 and MDA-MB-231 [47–49]. The mechanisms of action involved apoptosis induction through different pathways, such as caspase activation through the mitochondrial pathway [50,51], inhibition of the Akt signaling pathways [52,53], and cell cycle arrest in the G2/M phase [39,44,50,54]. These effects have been shown in vitro, but also in vivo [22]. Quercetin induces necroptosis, with an increase in BAX expression and Bcl-2 inhibition [55]. In combination with chloramphenicol, isorhamnetin, the main metabolite of quercetin in mammals, was shown to induce mitochondrial fission through CaMKII/Drp1, leading to apoptosis [56]. Also, quercetin-3-*O*-d-galactopyranoside induced apoptosis via ROS through the inhibition of NFκβ signaling pathway and activation of the BAX-caspase 3 axis [57]. Dietary quercetin has demonstrated an inverted "U"-shaped dose-dependent curve on the C3(1)/SV40 Tag breast cancer mouse model [58].

The antimetastatic potential of these compounds involves their ability to inhibit the expression of metalloproteinases such as MMP-3 and MMP-9 [53,59,60]. EMT and angiogenesis in breast cancer cells are also inhibited by quercetin [61,62].

Fatty acid synthase inhibition by flavonoids has been reported and associated with modulation of cell growth and promotion of apoptosis [63]. Quercetin led MDA-MB-231 and MCF-7 cells to EGFR reduction and this signaling promoted fatty acid alterations, including fatty acid isomerization and free radicals production [64].

The structure similarity of estradiol and flavones confers on them a potential to interfere in tumor growth and development through the interaction with ERs [65–68] and aromatase [69,70]. Preliminary reports show an anti-estrogenic activity for quercetin [71,72], but kaempferol stood out in a panel of phytoestrogens as the one with the most affinity with ERα, which enhances estrogen-dependent cell proliferation [68], displaying estrogenic affinity at 5 μM in a luciferase model in MCF-7 cells. Also, in this cell line, kaempferol revealed a dual effect, according to the concentration used: at the 10 μM range, it was described as an ER competitor with estrogen, since the increase in estrogen concentration was able to impair its functions. On the other hand, at a higher concentration of kaempferol, 100 μM, increments in estrogen concentration were unable to block the kaempferol effect, suggesting different pathways for activation by this compound [73].

In this sense, kaempferol is able to overcome the effects produced by triclosan and bisphenol A, exogenous xenoestrogenic compounds considered endocrine-disrupting chemicals (EDCs) with anti-apoptosis effects [74]. Kaempferol was shown to reverse these effects, increasing BAX levels and reducing Bcl-2 levels, leading VM7Luc4E2 cells to apoptosis [75]. Kaempferol was also able to suppress the EMT and metastatic-related behaviors of MCF-7 cells induced by triclosan [76] and to reverse triclosan-induced phosphorylation of IRS-1, AKT, MEK1/2 and ERK. In a 17β-estradiol (E2) or triclosan tumor growth-induced in vivo xenograft mouse model, co-treatment with kaempferol inhibited tumor growth [74].

Quercetin has been described as a multidrug resistance (MDR) inhibitor in several different works. It is defined as an inhibitor of p-glycoprotein by direct binding to this efflux pump, but also through the downregulation of p-gp expression [77–79]. Furthermore, quercetin was shown to potentiate the doxorubicin effect and to reduce its toxicity and side effects, both in vitro and in vivo [33,80]. Similar effects were seen with other drugs such as docetaxel [81], tamoxifen [82], paclitaxel and vincristine [79] using different drug delivery systems. Kaempferol is able to reverse drug resistance promoted by ABCG2 [83,84] and to inhibit quercetin efflux by this transporter [85].

Multidrug transporters are also acknowledged in the metabolization/elimination of quercetin and kaempferol. ABCG2 and ABCC2 participate in kaempferol-3-glucuronide (the major metabolite of kaempferol) elimination in vivo [86]. Finally, the cooperation between ABC transporters and UDP-glucuronosyltransferases appears to regulate kaempferol glucuronidation, thus regulating its accumulation in cells (in comparison to the glucuronide forms) with effects on the pharmacological properties of this compound [87].

3.2. Monomeric Catechins and Proanthocyanidins

Catechins, epicatechins, and proanthocyanidins are naturally occurring flavan-3-ols, typically found in tea, cocoa, grape, and wine [88]. Proanthocyanidins are the major phenolic compounds in grape seed and skin [89] and catechins are present in large amounts in green and black teas [90] and in red wine [91]. Proantocyanidins, also known as condensed tannins, are phenolic compounds that take the form of dimers, trimers, and highly polymerized oligomers of flavan-3-ol units [92,93]. Therefore, proanthocyanidins are metabolized to catechins and catechin derivatives [94].

Previous studies indicated that (+)-catechin and (−)-epicatechin (Figure 3) are rapidly absorbed from the upper portion of the small intestine in both human and animal organisms [95]. Catechin bioavailability is inversely proportional to its molecular masses. For example, although the (−)-epigallocatechin-3-gallate (EGCG) content in tea is much higher than other catechins, the peak plasma levels for EGCG (458 Da), (−)-epigallocatechin (306 Da), and (−)-epicatechin (290 Da) are 0.26, 0.48, and 0.19 µM, respectively. Plasma (+)-catechin concentrations increased in response to the ingestion of a single serving of reconstituted red wine. A maximum level of (+)-catechin at 76.7 nmol/L was detected in humans at 1.4 h after intake of both dealcoholized and reconstituted wine [91]. The bioavailability of procyanidins closely resembles that of flavan-3-ol monomers. Different studies, following the ingestion of GSE and GSPE, have shown that during digestion, the oligomers are fragmented into monomeric units of (+)-catechin and (−)-epicatechin and free forms of dimers and trimers have been detected in rat plasma [96,97]. Procyanidin B1 was also detected in human serum 2 h after intake of GSE [98]. Biotransformation of catechins directly undergoes phase II of metabolism, where they can be methylated by catechol-*O*-methyltransferase (COMT), glucuronidated by UDP-glucuronosyltransferase (UGT) or sulfated by sulfotransferase (SULT) [99]. Catechins can also be degraded in the intestinal tract by microorganisms to ring fission metabolites M4, M6, and M6′ [100].

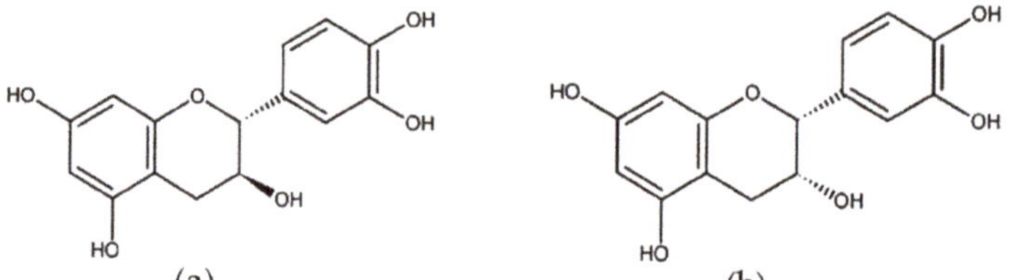

Figure 3. Structures of (**a**) (+)-catechin and (**b**) (−)-epicatechin.

The biological activities of these polyphenols that exert an effect on breast cancer were obtained from a variety of studies, including in vitro and in vivo data. Isolated catechin decreased cell viability and proliferation of MCF-7 human breast cancer cells at 30 and 60 µg/mL [101]. Alshatwi et al. [102] demonstrated that catechin hydrate (150 µg/mL and 300 µg/mL) effectively induced apoptosis in MCF-7 cells through increased expression levels of caspases -3, -8, -9 and p53.

Interestingly, inhibition of cell proliferation by purified (+)-catechin and (−)-epicatechin was more effective in hormone-sensitive breast cancer cells (MCF-7 and T47D), also demonstrating a possible implication of steroid hormone receptors in the action of polyphenols, and in fact, a competition of epicatechin for ER was reported [103].

The anti-carcinogenic activity of wine polyphenols is related to the protection of DNA damages by chemically reactive molecules, such as ROS (reactive oxygen species). The antioxidant effect of purified polyphenols was investigated in three types of breast cancer cells. The treatment with catechin and epicatechin decreased about 80% of ROS production on T47D cells, while no effect was noticed on MDA-MB-231 and MCF-7 cells. The authors attributed the results to different constitutive ROS production between cell lines, hormone receptor spectrum (considering interaction of ROS and these molecules) and limitations of the method [103].

The chemopreventive activity of GSP was demonstrated in an established carcinogen-induced animal model of breast cancer. Adult rats that received 5% GSE (86% proanthocyanidins) showed 44% reduction in the number of DMBA (7,12-dimethylbenz(a)anthracene)-induced mammary tumors [94]. The same effect was observed on a xenograft model using BALB/c nu/nu, athymic, ovariectomized mice carrying MCF-7aro tumors. Mice gavaged with GSE (74–78% proanthocyanidins and <6% catechin, epicatechin and their gallic acid esters) had a 70% reduction in tumor growth, indicating that GSE could suppress aromatase-positive tumors in vivo [10]. On the other hand, diets supplemented with 0.1%, 0.5% and 1.0% of grape seed proanthocyanidins (3.8% of catechin and epicatechin, 96.2% of oligomers and polymers) were not effective in reducing DMBA-induced rat mammary carcinogenesis. The authors attributed the results to the poor absorption of the components and, thus, insufficient amounts being available in the mammary gland to modulate tumorigenesis [104].

In humans, a pilot study with daily doses of grape GSPE failed to decrease plasma estrogens in postmenopausal women [105], despite the same extract exhibit inhibition in cell growth on MCF-7 cells in culture [106]. A double-blind placebo-controlled randomized phase II trial investigated the efficacy of IH636 GSPE in patients with tissue induration, considered a late adverse effect of curative radiotherapy for early breast cancer. In this study, 44 volunteers were given 100 mg of GSPE three times a day orally for six months. The authors considered that the study failed to demonstrate the efficacy of orally-administered GSPE, since there was no significant difference between the groups in terms of external assessments (tissue hardness, breast appearance) or patient self-assessments (breast hardness, pain or tenderness), 12 months post-randomization [107], although the same extract exhibited inhibition in cell growth of MCF-7 cells in culture at 25 mg/L [106].

Additionally, other health effects attributed to GSE and GSPE demonstrated potential to improve antioxidant cell defenses and modulate proinflammatory cytokines, which possibly complement the antitumoral functions of these matrices [108].

3.3. Anthocyanins and Anthocyanidins

Anthocyanins are the most abundant flavonoid pigments in young red wines, being responsible for their intense red color [109]. To date, the number of reported types of anthocyanins exceeds 600 [110], but the most common anthocyanidins, aglycone forms of anthocyanins, are cyanidin, pelargonidin, delphinidin, peonidin, petunidin, and malvidin [111] (Figure 4). Anthocyanidins could be immediately metabolized after ingestion of anthocyanins since the β-glucosidase found in intestinal bacteria can easily hydrolyze respective anthocyanins (glycosides) to anthocyanidins (aglycones) [112]. Anthocyanins are known for their apparent poor bioavailability (less to 1–2%). However, presystemic metabolism of these compounds may underestimate their bioavailability if only parent compounds and/or phenolic acid breakdown products are targeted in bioassays. Taking into account the original parent compounds, generated metabolites (from phase I and phase II metabolism and from microbiota-generated) and conjugated products, total bioavailability is much higher than previously credited, after all, anthocyanins are very influential to health [113].

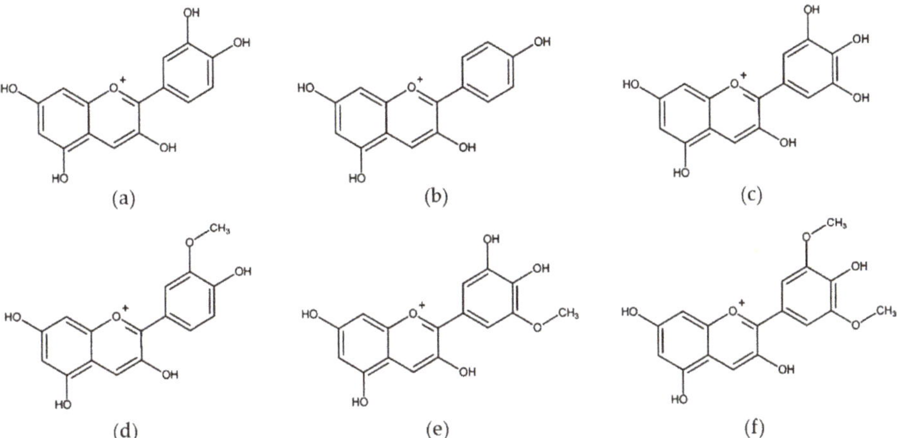

Figure 4. The most frequent anthocyanidins. (**a**) cyanidin, (**b**) pelargonidin, (**c**) delphinidin, (**d**) peonidin, (**e**) petunidin, (**f**) malvidin.

Anthocyanins show a range of antitumor activity in vitro and in vivo, from chemoprevention to chemotherapy. Their potential antitumor effects include antioxidant activities, anti-inflammatory effects, anti-mutagenesis, induction of differentiation and cell cycle arrest, stimulation of apoptosis, autophagy modulation, anti-metastasis, reversion of drug resistance and increasing the sensitivity of cancer cells to chemotherapy [114].

The effects of anthocyanins on breast cancer are some of the most studied, both concerning prevention and treatment of this disease. One of the proposed mechanisms of carcinogenesis is the formation of carcinogen-DNA adducts in target tissues, which is essential to the initiation of chemically induced breast cancer [115]. Singletary et al. [116] evaluated an anthocyanin-rich extract from Concord grapes and the major anthocyanins detected in this extract, delphinidin aglycone and its glucoside, for their capacity to inhibit DNA adduct formation due to the environmental carcinogen benzo[a]pyrene (BP) in a noncancerous, immortalized human breast epithelial cell line (MCF-10F). These authors observed that both grape extract (10 and 20 µg/mL) and isolated compounds (0.6 µM) inhibited BP–DNA adduct formation, through enhancing phase II metabolizing enzymes (GST and NQO1) activities, and suppressed reactive metabolites such as ROS.

Syed et al. [117] also showed that delphinidin, the most common anthocyanidin monomer, could prevent tumor development and malignant progression through inhibition of breast oncogenesis. These authors also assessed experiments with the MCF-10A cell line, a non-tumorigenic mammary epithelial cell line for studying normal breast cell function and transformation, and demonstrated that delphinidin (5 to 40 µM) inhibited HGF-induced early biochemical effects, blocking proliferation and migration of this cell line.

Hepatocyte growth factor (HGF) is produced mainly by mesenchymal cells and acts primarily through its only receptor, c-Met [118]. A variety of cellular responses are activated by c-Met/HGF signaling and mediate critical physiological processes for tumor growth and metastasis in human cancers, including angiogenesis [119], cellular invasion [120–122], and morphogenic differentiation [123]. In addition to observing effects on non-tumor breast cells, Syed et al. [117] showed that delphinidin treatment caused growth inhibition of breast cancer cells that express HGF, suggesting that this compound could prevent HGF-mediated activation of signaling pathways implicated in breast cancer.

The chemopreventive effects of delphinidin-3-glucoside were also evaluated on breast carcinogenesis [124]. Yang et al. [124] described that this phytochemical effectively suppressed carcinogenic transformation of MCF-10A cells induced by carcinogen treatment (NNK and BP). After that, these authors investigated the molecular mechanism related to lncRNA HOX transcript antisense RNA (HOTAIR) modulation. Long non-coding RNAs (lncRNA) are usually related to a group of RNAs with more than 200 nucleotides and are not involved in protein generation, despite being involved in different regulatory processes, such as modulation of gene expression [125]. HOTAIR, which is over-expressed in different types of cancers, is a lncRNA that plays a role in carcinogenesis and cancer progression by promoting cancer cell viability, growth, and metastasis [126]. In its turn, HOTAIR is regulated by the interferon regulatory factor-1 (IRF1) protein, which decreases HOTAIR expression. On the other hand, Akt activation decreases IRF1 expression and, consequently, elevates HOTAIR levels [127,128]. Yang et al. [124] observed that delphinidin-3-glucoside treatment (40 µM) inhibited HOTAIR expression in breast carcinogenesis and breast cancer cells. Besides that, these researchers also found the same results in xenografted breast tumors in athymic mice. Mechanistically, in this study delphinidin-3-glucoside down-regulates HOTAIR by inhibiting Akt activation and promoting IRF1.

The first report of tumor cell proliferation inhibitory activity of anthocyanidins from grape skin was published by Zhang et al. [129]. These authors tested the cell proliferation inhibitory activity of five anthocyanidins (cyanidin, delphinidin, pelargonidin, petunidin, and malvidin) and four anthocyanins (cyanidin-3-glucoside, cyanidin-3-galactoside, delphinidin-3-galactoside, and pelargonidin-3-galactoside) against diverse human cancer cell lines. Although anthocyanins did not inhibit proliferation of any cell line tested, even at the highest concentration (200 µg/mL), anthocyanidins inhibited cancer cell proliferation, with malvidin and pelargonidin being the most promising compounds, since they affected many different cancer cells at the same time.

Afaq et al. [130] evaluated the effect of delphinidin (5–40 µM) on epidermal growth factor receptor (EGFR)-positive breast cancer AU-565 cells and non-tumorigenic MCF-10A cells. EGFR is overexpressed in about 20% of invasive breast carcinoma [131] and has been implicated in tumor progression since it promotes a loss of balance between proliferation and apoptosis. The authors showed that this compound is an inhibitor of EGFR and its downstream signaling, the PI3K/AKT and MAPK pathways, both of which play a significant role in the mitogenic and cell survival responses mediated by EGFR. This same study also demonstrated that delphinidin treatment caused more dramatic inhibition of growth of AU-565 cells than MCF-10A cells and had minimal effect on normal mammary epithelial 184A1 cells, which express very low levels of EGFR, suggesting an important contribution of EGFR in delphinidin action. Moreover, delphinidin treatment of AU-565 cells resulted in induction of caspase 3-dependent apoptosis.

Delphinidin has also been shown to induce apoptosis and autophagy in MDA-MB-453 (concentrations of 20, 40 and 80 µM) and BT474 (concentrations of 60, 100 and 140 µM) cell lines [132]. In this study, the autophagy inhibitors, 3-methyladenine (3-MA) or bafilomycin A1 (BA1), enhanced the delphinidin-induced apoptosis in both breast cancer cell lines, suggesting that autophagy might exert a protective effect in this experimental model. In addition, these authors showed that delphinidin induced autophagy via the mTOR and AMPK signaling pathways.

The effect of cyanidin-3-glucoside, the main anthocyanin studied in breast cancer cells, was evaluated on breast cancer-induced angiogenesis [133]. This anthocyanin attenuated breast cancer-induced angiogenesis via inhibiting the expression and secretion of VEGF, the most important angiogenic cytokine, in a dose-dependent manner (concentrations up to 20 µM). The mechanism proposed by these authors involves the downregulation of STAT3, at both mRNA and protein level, via inducing miR-124, resulting in VEGF inhibition. Thus, the inhibitory effect of cyanidin-3-glucoside on the endogenous STAT3 may occur in a non-canonical way, via miRNAs, which could downregulate target gene expression with mRNA degradation.

Recently, Liang et al. [134] reported that cyanidin-3-glucoside (20 µM) decreased the migratory and invasive nature of triple-negative breast cancer cell lines through reversion of the EMT, which is highly associated with cancer metastasis. Cyanidin-3-glucoside increased epithelial markers (E-cadherin and zonula occludens-1), decreased mesenchymal markers (vimentin and N-cadherin) and EMT-associated transcription factors (Snail1, Snail2). Mechanistically, this phytochemical also attenuated the pivotal factor for EMT, NF-κB, and induced the inhibitor Sirt1 in triple-negative breast cancer cell lines.

There is currently available evidence that endogenous estrogens play a critical role in the development of breast cancer [135]. Despite the fact that triple-negative breast cancer is characterized by a lack of ERα expression [136], data suggest that estrogen still plays a critical role in the etiology of this type of cancer since a 36-kDa variant of ERα, known as ER alpha 36 (ERα36), is highly expressed in triple-negative breast cancer [137] and is involved in rapid estrogen signaling [138]. Studies also emphasize the causal link between ERα36 and EGFR, since this receptor is one of the most critical downstream targets of activated ERα36 signaling [138].

Wang et al. [139] found that cyanidin-3-glucoside (150 µM) preferentially promotes cell death, by the extrinsic apoptosis pathway of triple-negative breast cancer cells (MDA-MB-231) that co-express ERα36 and EGFR. Cyanidin-3-glucoside directly binds to the ERα36 receptor, which in turn inhibits its downstream signaling, the EGFR/AKT pathway leading to EGFR degradation through the proteasome system. A xenograft mouse model also confirmed these properties of cyanidin-3-glucoside.

Fernandes et al. [140] evaluated the effect of cyanidin-3-glucoside, delphinidin-3-glucoside, and vinylpyranoanthocyanin-catechins (portisins) on MCF-7 cells. Overall, the studied compounds inhibited, in a dose-dependent manner (12.5–100 µM), the growth of MCF-7 cells, however, delphinidin-3-glucoside and its respective portisin presented the highest cytotoxic effect. This same study also highlighted a structural requirement for a more potent cytotoxicity effect on MCF-7 cells, characterized by an ortho- trihydroxylated substituent attached to the phenolic ring. Nevertheless, this study was unable to elucidate whether anthocyanins antiproliferative effect could be dependent or independent of ERs or other molecular pathways involved.

Although the monoclonal antibody trastuzumab improves survival of patients with HER2-positive breast cancers [141], the majority of patients who initially respond to this therapy demonstrate disease progression within 12–24 months [142]. Thereby, identifying alternative strategies to overcome trastuzumab resistance targeting HER2 may improve treatment response in breast cancer. Li et al. [143] investigated the antitumor properties of anthocyanins, peonidin-3-glucoside, and cyanidin-3-glucoside, on parental HER2-positive cells and their trastuzumab-resistant cell lines. Treatment with cyanidin-3-glucoside and peonidin-3-glucoside significantly inhibited cell growth in the parental and trastuzumab-resistant cells in a dose-dependent manner. The authors also observed that the mechanisms of action of cyanidin-3-glucoside (5 μg/mL) and peonidin-3-glucoside (5 μg/mL) involve the inhibition of HER2 phosphorylation, the induction of apoptosis in both sensitive and trastuzumab-resistant cell lines, and in a higher anthocyanins concentration (1 mg/mL), an inhibition of trastuzumab-resistant cells migration and invasion was also observed. Besides that, treatment with both anthocyanins (6 mg/Kg/twice weekly, intraperitoneal) reduced trastuzumab-resistant cell-mediated tumor growth in vivo.

As described above, cyanidin-3-glucoside can act alone against breast cancer. However, it has also been shown to be effective in combination with trastuzumab in three representative HER2-positive breast cancer cell lines [144]. These researchers demonstrated that HER2 inactivation seems to represent a central role in the synergistic effect between cyanidin-3-glucoside and trastuzumab in all HER2-positive breast cancer cells tested. Moreover, cyanidin-3-glucoside (5 μg/mL) alone and in combination with trastuzumab (5 μg/mL) induced cell apoptosis in HER2-positive cell lines. These authors also evaluated, in an in vivo xenograft model in mice, the effect of 6 mg/mL cyanidin-3-glucoside in association with 6 mg/mL trastuzumab intraperitoneally twice a week for 25 days. These results demonstrated that anthocyanins were able to significantly enhance trastuzumab-induced tumor growth inhibition.

4. Non-Flavonoid Phenolic Compounds

Resveratrol

Natural stilbenes are an important group of non-flavonoid polyphenols characterized by the presence of a 1,2-diphenylethylene nucleus in their structure [145]. Among them, resveratrol (3,4′,5-trihydroxy-*trans*-stilbene) is a phenolic compound derived from grapes, berries, peanuts, and other plant sources. The molecule consists of two aromatic rings that are connected through a methylenic bridge and exists as *cis*- and *trans*-resveratrol isomers (Figure 5), and their glucosides, *cis*- and *trans*-piceid [146]. Resveratrol was originally identified as a phytoalexin by Langcake and Pryce [147] and is produced by a wide range of plant species under stressful environmental conditions, such as pathogen infection and ultraviolet radiation. Grapes and their derivative products, particularly red wine, are the most important natural sources of resveratrol. The resveratrol composition of wines depends on the grape varieties used, as well as the growing conditions and the wine-making methods, which may vary. In fresh grape skin, the concentration of this compound is in the range of 50–100 μg per gram, and red wine contains about 1.9 ± 1.7 mg/L of *trans*-resveratrol [148–152].

In humans, resveratrol is extensively metabolized and rapidly eliminated. When consumed orally, the molecule is absorbed via passive diffusion or by membrane transporters in the intestine, and then conjugated into glucuronides and sulfates. Although oral absorption is around 75%, only a small fraction of resveratrol ingested from dietary sources reaches the bloodstream and body tissues. It was previously described that metabolism in the liver and intestine results in oral bioavailability of about 1–2% of *trans*-resveratrol [153–155]. Rapid conjugation and low bioavailability are some of the major limitations and challenges of the in vivo use of this compound. Different methodological approaches, such as encapsulation in liposomes, emulsions, micelles, insertion into polymeric nanoparticles, solid dispersions, and nanocrystals, have been developed to improve the low aqueous solubility and the poor bioavailability of resveratrol [156]. Furthermore, the use of naturally occurring or synthetic resveratrol derivatives, with a better pharmacokinetic profile, low toxicity, less side effects, and improved biological activities, are promising strategies for clinical applications of stilbene compounds [157].

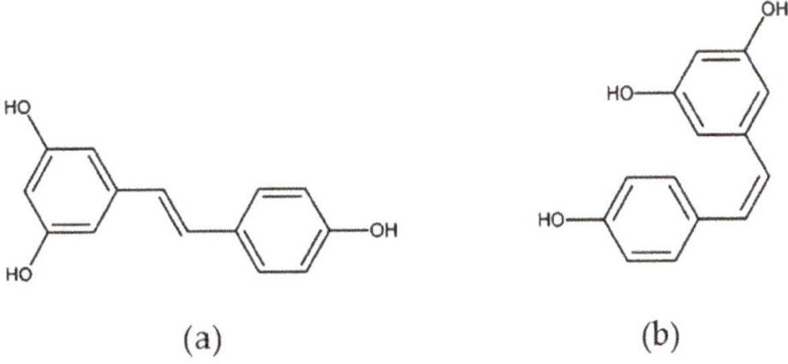

(a)   (b)

Figure 5. Structures of stilbenes. (**a**) Trans- resveratrol and (**b**) cis-resveratrol.

Resveratrol exhibits multiple bioactivities, including anti-oxidative, anti-inflammatory, cardioprotective, neuroprotective, anti-aging and anticancer properties. Accumulated experimental and clinical evidence clearly shows the chemopreventive and chemotherapeutic potential of resveratrol, as reviewed in our recent publication [158]. Scientific interest in this molecule has grown considerably during the last 23 years, since Jang and colleagues first demonstrated the ability of resveratrol to inhibit in vivo carcinogenesis in a mouse skin cancer model [159,160]. Resveratrol is reported to act as a multi-target suppressor of all three carcinogenesis stages (initiation, promotion, and progression), by regulating signal transduction pathways that control cell division and growth, apoptosis, inflammation, angiogenesis, and metastasis. Furthermore, resveratrol increases the efficacy of traditional chemotherapy and radiotherapy by reducing drug resistance and sensitizing tumor cells to a chemotherapeutic agent [160–163].

A plethora of studies, including in vitro and in vivo investigations, have suggested that resveratrol triggers chemopreventive and therapeutic responses against several tumor types, such as skin, breast, prostate, lung, colon, and liver cancer [163,164]. As indicated by a recent search on PubMed (accessed in April 2020), most of these studies (570 of 3524 hits) have been reported in breast cancer models. In 2005, it was shown for the first time that resveratrol from grape consumption is inversely related to breast cancer risk, as reported in a case-control study conducted between 1993 and 2003 in the Swiss Canton of Vaud on 369 cases and 602 controls [165]. Among its wide range of biological properties, resveratrol has attracted considerable attention in breast carcinogenesis because of its role as a phytoestrogen. This compound can compete with natural estrogens for binding to ERs, thus modulating its biological responses [146,155,166,167].

Hormone-dependent tumors may be prevented by regular exposition to selective estrogen receptor modulators (SERMs). These compounds exhibit different levels of estrogen agonism or antagonism, depending on the cell type and gene expression targeted by ERs [168]. Gehm and colleagues were the first to investigate whether resveratrol would have estrogenic activity due to its structural similarity to the synthetic estrogen diethylstilbestrol (DES; 4,4′-dihydroxy-*trans*-α, β-diethylstilbene). Based on its ability to compete with E2 for binding to and modulating the activity of ERα, resveratrol was characterized as a phytoestrogen [169]. It binds to ER at a low micromolar range (3–10 μM) and with lower affinity than estradiol. Despite this, resveratrol may act as a superagonist in activating hormone receptor-mediated gene transcription in MCF-7 cells [169,170]. In contrast, the antiestrogenic activity of resveratrol in breast cancer was also reported, being related to pathways that inhibit estrogen-induced cellular outcomes, such as proliferation, tumoral transformation, and progression [146]. Lu and Serrero reported ER antagonism of resveratrol (5 μM) in the presence of E2 and partial agonism in its absence [171]. It was also demonstrated that resveratrol exerted a mixed agonist/antagonist action in cells transiently transfected with ER, and mediated higher transcriptional activity when bound to ERβ than to ERα. Moreover, resveratrol showed antagonist activity with ERα, but not with ERβ [172]. Based on these reports, resveratrol may be categorized as a natural SERM, since it behaves as both agonist and antagonist of ERs. These opposite responses vary according to cell type, resveratrol concentration, hormone competition and ERs expression [155,173]. Resveratrol also modulates the expression of the progesterone receptor (PR). It was previously reported that resveratrol produces greater transcriptional activation of PR than estradiol. In MCF-7 cells, resveratrol was as effective as a maximal dose of estradiol in activating PR gene expression [169].

In tumors, expression of aromatase is upregulated compared to that of surrounding noncancerous tissue. The suppression of in situ estrogen formation by using aromatase inhibitors is a promising route for chemoprevention against breast cancer. In SK-BR-3 cells, resveratrol significantly reduced aromatase mRNA and protein expression in a dose-dependent manner [174]. Resveratrol also inhibits the expression and enzyme activity of aromatase, thus reducing localized estrogen production in breast cancer cells [175]. When tested in a co-culture system of T47D breast cancer cells with human breast adipose fibroblasts (BAFs), resveratrol (20 μM) promoted an aromatase inhibitory effect as potent as 20 nM of letrozole, which is a clinically used anti-aromatase drug in breast cancer treatment [176].

As reviewed by different authors, several experimental approaches have been used to describe the molecular mechanisms of resveratrol in breast carcinogenesis [155,158,162,177]. In addition to the phytoestrogenic action, resveratrol modulates xenobiotic metabolism by altering ABCG2 and CYP1A1 activities [178]; decreases the production of prostaglandins by inhibiting COX-2 expression and activity at multiple levels [177]; suppresses the growth of different breast cancer cell lines and induces a number of biological pathways, thus leading to cell growth arrest and apoptosis [155,165,177,179,180]; modulates the p53 tumor suppressor protein by inducing post-translational modifications [158,180]; prevents mutant p53 aggregation in breast cancer cells and in breast tumor xenografts [181]; regulates extracellular growth factors and receptor tyrosine kinases [162]; induces epigenetic mechanisms by modulation of histone acetylation/methylation [182]; inhibits angiogenesis, EMT, and metastasis [155]; acts as an MDR reversion molecule [183] and sensitizes breast cancer cells toward chemotherapy [161]. In animal studies, resveratrol inhibits chemically-induced breast carcinogenesis; it reduces tumor growth, decreases angiogenesis and increases the apoptotic index in xenograft breast cancer models; delays the tumor development, reduces the mean number and size of tumors and diminished the number of lung metastases in spontaneous breast tumor models [155]. In recent years, accumulating evidence also suggests that resveratrol may be effective in breast cancer management when given in combination with other naturally occurring and chemotherapeutic agents, thus suggesting that resveratrol can enhance the efficacy of other compounds [184].

Although the antitumor activity of resveratrol in in vitro and animal breast cancer models is well established, the clinical evidence regarding its therapeutic effect against breast cancer is still limited. Considering that preclinical and clinical studies suggested that resveratrol may modulate several hormone-related factors involved in breast cancer risk, a pilot phase I clinical study was conducted in a group of forty postmenopausal women with high body mass index, to determine the clinical effect of resveratrol on systemic sex steroid hormones. The resveratrol intervention (1 g daily, for 12 weeks) did not result in significant changes in serum concentrations of estradiol, estrone, or testosterone, but had favorable effects on estrogen metabolism and steroid hormone-binding globulin (SHBG) [185]. Further clinical trials are required to ascertain and validate the efficacy of resveratrol on breast cancer.

5. Grape and Wine Metabolites and Breast Cancer (In Vitro and In Vivo Studies)

The health-promoting effect of wine can be focused on consumption, bioavailability, metabolism and microbiota influence on bioactive compounds. There is now strong evidence that the molecules responsible for those effects are probably not the ingested ones but rather their metabolites that occur after the action of microbiota and absorption process. The identification and quantification of these metabolites has not been an easy task, but improvement of analytical methods and sensitivity has allowed some advances in metabolomics area [186].

The WinMet database contains 2030 putative compounds present in oenological matrices covering 10 different families, such as phenols, organic acids, biogenic amines, sugars, polyols, fatty acids, higher alcohols, aldehydes, lignans, and ketones [187]. These molecules can be divided into primary metabolites (e.g., sugars, amino acids, and short chain organic acids) and secondary metabolites (flavonoids and phenol compounds) and are well documented in the literature about wines [188].

Wine is a complex mixture of many different molecules and several factors interfere in its composition, such as grape type, fermentation process, aging, among others. For example, catechin and epicatechin decrease during aging in all wines, while gallic acid increases in almost all red wines [189]. Thus, the purpose of this section will be to discuss in vitro and in vivo studies related to the metabolites of the flavonoid and non-flavonoid compounds present in red wine described previously in this review, and the relationship between these molecules and breast cancer.

The majority of phenolic compounds from grapes and wine are metabolized in the gastrointestinal tract, where they are broken down by gut microbiota and typically involve deglycosylation, followed by breakdown of ring structures to produce phenolic acids and aldehydes. These metabolites can be detected in bloodstream, urine, and fecal samples by using sophisticated instrumentation methods for quantitation and identification at low concentrations [190].

An intervention study with red wine offered to eight healthy adults for 20 days revealed significant changes in eight metabolites: 3,5-dihydroxybenzoic acid, 3-O-methylgallic acid, p-coumaric acid, phenylpropionic acid, protocatechuic acid, vanillic acid, syringic acid and 4-hydroxy-5-(phenyl)valeric acid without any influence of ethanol on the microbial action [191]. The same research group characterized the metabolome of human feces after moderate consumption of red wine by healthy subjects during four weeks and showed 37 metabolites related to wine intake, from which 20 could be tentatively or completely identified, including the following: wine compounds, microbial-derived metabolites of wine polyphenols and endogenous metabolites and/or others derived from different nutrient pathways. After wine consumption, fecal metabolome is usually enriched in flavan-3-ols metabolites [192].

To determine which compounds in grapes and wine are the most bioactive, their effects in disease models must be known, including absorption and metabolism. Rats that consume a red wine extract have elevated levels of the microbial phenolic acid metabolites 3-hydroxyphenylpropionic, 3-hydroxybenzoic, 3-hydroxyhippuric, hippuric, p-coumaric, vanillic, 4-hydroxybenzoic, and 3-hydroxyphenylacetic acids in urine. These urine metabolites account for roughly 10% of the administered red wine polyphenols [193]. Most grape and wine flavonoids and others are rapidly metabolized in the human body, making it difficult to determine whether these compounds are effective against disease.

Based on these metabolites, the combination of hippuric acid (HA) nanocomposite (intercalation of hippuric acid into a zinc-layered hydroxide) with doxorubicin and oxaliplatin induced cytotoxicity in MDA-MB-231 and MCF-7 cell lines [194]. 4-Hydroxybenzoic acid (4-HBA) and a histone deacetylase 6 (HDAC6) inhibitor could successfully reverse adryamicin (ADM) resistance in human breast cancer cells. 4-HBA significantly promoted the anticancer effect of ADM on apoptosis induction, as evidenced by the increased expressions of caspase-3 and PARP cleavage, which were associated with the promotion of p53 and homeodomain interacting protein kinase-2 (HIPK2) expressions in ADM-resistant breast cancer cells. Therefore, 4-HBA could be applied as an effective HDAC6 inhibitor to reverse human breast cancer resistance. Herein, the 4-HBA and ADM combination might represent a useful therapeutic strategy to prevent human breast cancer [195].

Apoptotic effects of protocatechuic acid (PCA), another metabolite of wine, were examined on MCF-7 cells. Results showed that PCA concentration-dependently decreased cell viability, increased lactate dehydrogenase leakage, enhanced DNA fragmentation, reduced mitochondrial membrane potential and lowered Na^+-K^+-ATPase activity. PCA also concentration-dependently elevated caspases-3 and -8 activities and significantly inhibited cell adhesion. These findings suggest that PCA is a potent anticancer agent to cause apoptosis or retard invasion and metastasis in breast cancer and other cells [196].

The metabolites gallic acid, 4-*O*-methylgallic acid and 3-*O*-methylgallic acid are detected in the plasma of humans who consume 300 mL of red wine [197]. In fact, the metabolites gallic acid and 4-*O*-methylgallic acid are well correlated with wine consumption and may be used as urinary biomarkers for wine intake in health-related studies [198]. Phenolic acid metabolites are mainly formed from gut microbiota metabolism and could be responsible for much of the disease reduction associated with consuming wine and grape phenolics.

Gallic acid (GA) possesses potential for antitumoral activity on different types of malignancies. GA treatment significantly decreased the cell viability of MDA-MB-231 and HS578T cells in a dose-dependent manner. In addition, GA exerted relatively lower cytotoxicity on non-cancer breast fibroblast MCF-10F. The changes in cell cycle distribution in response to GA treatment led to an increase of G0/G1 and sub-G1 phase ratio in MDA-MB-231 cells. GA also downregulated cyclin D1/CDK4 and cyclin E/CDK2, upregulated p21Cip1 and p27Kip1 and induced activation of caspase-9 and caspase-3. In addition, it modulated p38 mitogen-activated protein kinase that was involved in the GA-mediated cell-cycle arrest and apoptosis. GA inhibited the cell viability of TNBC cells, which may be related to the G1 phase arrest and cellular apoptosis via p38 mitogen-activated protein kinase/p21/p27 axis. Thus, GA could be beneficial for TNBC treatment [199]. GA also promoted inhibition of proliferation and induction of apoptosis in MCF-7 cells. The results revealed that GA induced apoptosis by triggering the extrinsic or Fas/FasL pathway as well as the intrinsic or mitochondrial pathway. Furthermore, the apoptotic signaling induced by GA was amplified by a cross-link between the two pathways. Taken together, these findings may be useful for understanding the mechanism of action of GA on breast cancer cells and provide new insights into the possible application of such a compound and its derivatives in breast cancer therapy [200].

5.1. Resveratrol Metabolites

Resveratrol is a minor component of red wines and, following its ingestion, it is converted to glucuronide and sulfate metabolites, which are present in the circulatory system in nanomolar concentrations [201]. Nevertheless, the by far most commonly studied form of resveratrol is the aglycone, often at concentrations largely exceeding those attainable in vivo. By contrast, very little is known about the biological activity of the resveratrol metabolites formed upon intestinal absorption, which represent the major circulating forms of resveratrol; in particular, the glucuronic acid and the sulfate conjugates of trans-resveratrol, which are produced at the enterocyte and hepatocyte level [202]. Besides dihydroresveratrol, Bode et al. [203] found, in vivo and in vitro, bacterial trans-resveratrol metabolites: 3′,4-dihydroxy-trans-stilbene and 3′,4′-dihydroxybibenzyl (lunularin). In estrogen-sensitive cancer cells, like MCF-7, 3′,4-dihydroxy-trans-stilbene showed agonist properties [204].

Resveratrol-3-*O*-sulfate (R3S), but no other resveratrol derivative, exerted a pronounced antiestrogenic activity on both receptors (α and β), with a marked preference for ER. R3S, the main resveratrol metabolite accumulating in human plasma after ingestion of dietary amounts of resveratrol, is an effective ER-preferential anti-estrogen in both yeast and mammalian cells [205]. A significant increase in MCF-7 cancer cells growth rates was shown in the presence of picomolar concentrations of dihydroresveratrol (DH-RSV) because this polyphenol has a profound proliferative effect on hormone-sensitive tumor cell lines such as MCF-7.

The proliferative effect of DH-RSV was not observed in cell lines that do not express hormone receptors (MDA-MB-231, BT-474 and K-562) [206]. Human MCF-7 (wild-type p53), MDA-MB-231 (mutant p53) and nontumorigenic MCF-10A cells are treated with resveratrol and physiological-derived metabolites (RSV-3-*O*-glucuronide, RSV-3-*O*-sulfate, RSV-4′-*O*-sulfate, DH-RSV and DH-RSV-3-*O*-glucuronide). Cellular senescence is measured by SA-β-gal activity and senescence-associated markers (p53, p21Cip1/Waf1, p16INK4a and phosphorylation status of retinoblastoma (pRb/tRb). While no effect is observed in MDA-MB-231 and normal cells, resveratrol metabolites induce cellular senescence in MCF-7 cells by reducing their clonogenic capacity and arresting cell cycle at the G2M/S phase, but do not induce apoptosis. Senescence is induced through the p53/p21Cip1/Waf1 and p16INK4a/Rb pathways, depending on the resveratrol metabolite, and requires ABC transporters, but not ERs. Recent evidence demonstrates that resveratrol metabolites, but not free resveratrol, reach malignant tumors (MT) in breast cancer (BC) patients. Since these metabolites, as detected in MT, do not exert short-term antiproliferative or estrogenic/antiestrogenic activities, long-term tumor senescent chemoprevention has been hypothesized. These data suggest that resveratrol metabolites, as found in MT from BC patients, are not deconjugated to release free resveratrol, but enter the cells and may exert long-term tumor-senescent chemoprevention [207].

5.2. Catechins Metabolites

Catechin appears to be metabolized only if absorbed from the small intestinal lumen. Both 3′-*O*-methylcatechin-glucuronide and catechin-glucuronide are produced in intestinal cells and methylation and sulfation of catechin metabolites occur in the liver [208]. Catechin glucuronide and 3′-*O*-methylcatechin glucuronide are mainly found in plasma of rats after ingestion of catechins [208,209]. Large amounts of the 3′-*O*-methyl metabolite are also found to be glucuronidated and sulfated on the same compound, presumably produced in the liver, and are only detected in the bile [208]. In humans, between 3.0 and 10.3% of ingested catechins from red wine are accounted for in urine, mostly as catechin and its 39-*O*-methyl-glucuronide and sulfate metabolites [210].

Aside from metabolism that occurs in intestinal cells and liver, catechins can also be metabolized by gut microbiota to produce phenolic acid metabolites. In rats, these metabolites can be found in urine, being 3-hydroxyphenylpropionic acid, 3-hydroxybenzoic acid and 3-hydroxyhippuric acid present in the highest concentrations [193]. When catechin is incubated with human gut microbiota, it is metabolized to 4-hydroxybenzoic acid, 2,4,6-trihydroxybenzaldehyde, phloroglucinol and 4-methoxysalicylic acid [211], again emphasizing the effects of individual microbiota profiles on gut metabolism. We have not found much research showing the association of catechin metabolites with breast cancer, only with the use of phloroglucinol, as can be seen above.

Metastasis is a challenging clinical problem and the primary cause of death in breast cancer patients. Treatment with phloroglucinol (PG) effectively inhibited mesenchymal phenotypes of basal type breast cancer cells through downregulation of SLUG without causing a cytotoxic effect. Importantly, PG decreased SLUG through inhibition of PI3K/AKT and RAS/RAF-1/ERK signaling. Treatment with PG sensitized breast cancer cells to anticancer drugs such as cisplatin, etoposide, and taxol, as well as to ionizing radiation. Taken together, these data indicate PG to be a good candidate to target breast cancer stem-like cells (BCSCs) and to prevent disease relapse [212,213].

5.3. Anthocyanins Metabolites

In humans, nanomolar plasma concentrations of anthocyanins are found after they are consumed. Anthocyanins such as cyanidin-3-glucoside and pelargonidin-3-glucoside could be absorbed in their intact form into the gastrointestinal wall, undergo extensive first-pass metabolism, and enter the systemic circulation as metabolites. Phenolic acid metabolites were found in the bloodstream in much higher concentrations than their parent compounds. These metabolites could be responsible for the health benefits associated with anthocyanins [214].

After rats ate cyanidin-3-glucoside, the aglycone was only found in the small intestine, cyanidin-3-glucoside was found in the plasma, and methylated cyanidin-3-glucoside was found in the liver and kidney [215]. Cyanidin-3-glucoside attenuates the angiogenesis of breast cancer via inhibiting STAT3/VEGF pathway [133].

In humans and Caco-2 cells, cyanidin-3-*O*-glucoside's major metabolites are protocatechuic acid (PCA) and phloroglucinaldehyde which are also subjected to entero-hepatic recycling, although caffeic acid and peonidin-3-glucoside seem to be strictly produced in the large bowel and renal tissues [216].

Previous studies evaluated the bioavailability of anthocyanins using red wine and dealcoholized red wine [217,218]. One of the first studies is the work of Bub and co-workers who only detected the main native anthocyanin in plasma and urine with no effect of ethanol on the amount quantified [217]. Ethanol enhances cyanidin-3-*O*-glucoside's metabolism potentiating its conversion into methylated and glucuronidated derivatives, showing an increase in the two main anthocyanin conjugates, methyl-cyanidin-glucuronide and 3′-methyl-cyanidin-3-*O*-glucoside, of 59% and 57%, respectively. But in this case, the food matrix used was blackberry puree with or without ethanol, and not wine or grapes [219].

The accumulation of multiple phenolic metabolites might ultimately be responsible for reported anthocyanin bioactivity, with the gut microbiota apparently playing an important role in the biotransformation process. Nevertheless, phase II conjugates of cyanidin-3-*O*-glucoside and cyanidin (cyanidin-glucuronide, methyl cyanidin and methyl-cyanidin-glucuronide) were also detected in plasma and urine [186]. The most important metabolites corresponded to products of anthocyanin degradation (i.e., benzoic, phenylacetic and phenylpropenoic acids, phenolic aldehydes and hippuric acid) and their phase II conjugates, which were found at 60- and 45-fold higher concentrations than their parent compounds in urine and plasma, respectively [220].

Delphinidin-3-glucoside, cyanidin-3-glucoside and petunidin-3-glucoside methylated metabolites were obtained by enzymatic hemi-synthesis and decreased or did not alter the antiproliferative effect of the original anthocyanin in MCF-7 cells [221]. The methylation reaction alters the number of hydroxyl and methoxyl groups in ring B, so these metabolites are likely to have different antioxidant activities in comparison with the original anthocyanins. Generally, the health effects of anthocyanins are associated with an increase in the endogenous antioxidant defenses. In a paper by Fernandes et al. [221] the synthetized methylated metabolites still displayed some antiproliferative activities for the three cell lines although not as intense as parental anthocyanin. The biological studies conducted with the metabolites in comparison with the native compounds allow understanding of the real contribution of methylation towards the antioxidant and antiproliferative effects of anthocyanins. However, this subject is new and needs more publications for a good discussion, especially from methylated anthocyanin-derived metabolites.

5.4. Quercetin Metabolites

Quercetin or its metabolites may have cytotoxic activities [222]. Studies on the metabolism of quercetin suggest degradation by intestinal microbiota and relatively low absorption [223], limiting its use as a biomarker. Metabolism of quercetin includes 3,4-dihydroxyphenylacetic acid (DHPAA) as homoprotocatechuic acid, m-hydroxyphenylacetic acid (mHPAA), and 3-methoxy-4-hydroxyphenylacetic acid as homovanillic acid (HVA) [224]. These three metabolites are excreted in the urine of rats, rabbits, and humans [224,225].

Recently, Yamazaki et al. [226,227] investigated the effects of quercetin and its main circulating metabolite quercetin-3-*O*-glucuronide on MCF-10A and MDA-MB-231 cells and suggested that these flavonoids may suppress invasion of these cells by controlling β_2-adrenergic signaling, and may be a dietary chemopreventive factor for stress-related breast cancer.

5.5. Metabolites and Breast Cancer Patients

In relation to breast cancer patients, previous reports have shown that glycolysis, lipogenesis and the production of volatile organic metabolites were increased in the serum of these patients compared to healthy women [228]. The serum levels of choline, tyrosine, valine, lactate, isoleucine are up-regulated, and glutamate levels are downregulated in patients with early-stage breast cancer [229]. These studies reveal that metabolic alterations are important indications for breast cancer. There is evidence that metabolic changes are correlated with metastasis and metabolism of tumors [230–232]. Metabolism changes are often associated with resistance to chemotherapy and therapeutic sensitivity in clinical chemotherapy. Breast cancer cells not only show significant differences in metabolism compared with healthy breast cells, but also show differences in drug resistance [233,234]. Cancer and metabolism are deeply interconnected, studies indicate that cancer evolution is associated with abnormal glucose metabolism that is related to high proliferation, metastasis and clinical characteristics and is allied to the action of a particular drug. In this context, chemoresistance enables cancer cells to survive drug action and proliferate uncontrollably, which may lead to strong metastatic potential and cancer progression [230–234].

A recent clinical trial has reported resveratrol accumulation, mainly as sulfates and glucuronides, in normal and malignant human breast tissues. Although phase-II conjugation might hamper a direct anticancer activity, long-term tumor-senescent chemoprevention cannot be discarded [235]. Metabolites of wine bioactive compounds have been positively related to in vitro and in vivo breast anticancer properties and this evidence was associated with the ingestion of several flavonoids present in large amounts in red wine. However, the concentration required to trigger a biological event is dependent not only on the amount ingested, but also on critical variables that include bioaccessibility, bioavailability, stability under in vivo conditions, and so on. Many studies are still required to clarify the role of many of these metabolites with regard to the health-promoting properties of wine.

Table 1 summarizes the data collected from the literature about the metabolite dosage used or found in the different in vitro and in vivo models mentioned in this review.

Table 1. Summary of metabolites found or used in different in vitro and in vivo models.

Dietary Factor/Isolated Compound	Model (Human/Animal/Cancer Cell Lines)	Sample Type	Discriminating Metabolites	Average of Metabolites (Found or Used)	Primary Reference
Red wine	Human (healthy subjects)	Plasma	catechin gallic acid 4-*O*-methylgallic acid 3-*O*-methylgallic acid caffeic acid	0.13–1.5 µmol/L	[197]
Red wine resveratrol in capsules	Human (healthy subjects)	Plasma	*trans*-resveratrol *trans*-resveratrol-4'-*O*-glucuronide *trans*-resveratrol-3'-*O*-glucuronide resveratrol resveratrol-*O*-glucuronide resveratrol-*O*-sulfate	0–0.03 µM 0–3.9 µM 0–2.7 µM 0.04–0.23 µM 0.56–2.90 µM 0.75–4.78 µM	[201]
Red wine and red grape juice	Human (male healthy subjects)	Plasma Urine	cyanidin 3-glucoside, delphinidin 3-glucoside, malvidin 3-glucoside, peonidin 3-glucoside, petunidin 3-glucoside	0.42–48.8 ng/mL (maximum) 0.66–86.7 µg/h	[218]
Red wine, dealcoholized red wine and grape juice	Human (male healthy subjects)	Plasma Urine	malvidin-3-glucoside	1.38 nM (maximum) 13.3–27.0 µg	[217]
Habitual diets	Human (healthy subjects)	Urine	3,4-dihydroxyphenylacetic acid *m*-hydroxyphenylacetic acid homovanillic acid	0.7 µg/mL 4.8 µg/mL 2.8 µg/mL	[225]
Red wine and dealcoholized red wine	Human (healthy subjects)	Urine	catechin (unmethylated conjugates) catechin (methylated conjugates)	5.32 µmol 1.27 µmol	[210]
Wine	Human (healthy subjects)	Urine	gallic acid 4-*O*-methylgallic acid	1.6–6.1 µmol/d	[198]
Red wine and dealcoholized red wine	Human (healthy adults)	Human feces	3,5-dihydroxybenzoic acid, protocatechuic acid, 3-*O*-methylgallic acid, vanillic acid, syringic acid, p-coumaric acid, phenylpropionic acid, 4-hydroxy-5-(phenyl)valeric acid, 2-hydroxyglutaric acid, 2-methylbutyric acid, 2,3-pentanedione, diethylmalonate, 2-phenethyl butyrate, 2-phenylethyl hexanoate, 5-(3',4'-dihydroxyphenyl)gamma-valerolactone, 3-(3'-hydroxyphenyl)propionic acid, 4-hydroxy-5-(3'-hydroxyphenyl)valeric acid, benzoic acid, 4-hydroxy-5-(phenyl)valeric acid	0.2–50 µg/g	[191]
Catechin	Human (healthy subjects)	Feces	4-hydroxybenzoic acid 2,4,6-trihydroxybenzoic acid phloroglucinol 4-methoxysalicylic acid	Not determined	[211]
Trans-resveratrol	Human (healthy subjects)	Feces	dihydroresveratrol 3,4'-dihydroxy-*trans*-stilbene 3,4'-dihydroxybibenzyl (lunularin)	0–86.9 µmol/L 0–11.1 µmol/L 0–79.8 µmol/L	[203]

Table 1. *Cont.*

Dietary Factor/Isolated Compound	Model (Human/Animal/Cancer Cell Lines)	Sample Type	Discriminating Metabolites	Average of Metabolites (Found or Used)	Primary Reference
Fried onions, quercetin rutinoside, quercetin aglycone	Human (healthy ileostomy subjects)	Ileostomy effluent urine	quercetin	37–72 mg 73–275 µg	[223]
Isotopically labeled cyanidin-3-glucoside (6,8,10,3′,5′-13C$_5$-C3G)	Human (male healthy subjects)	Serum Urine	24 labeled metabolites were identified (cyanidin-glucuronide, methyl cyanidin-glucuronide, methyl C3G-glucuronide, protocatechuic acid (PCA), phloroglucinaldehyde, phloroglucinaldehyde, PCA-3-glucuronide, PCA-4-glucuronide, PCA-3-sulfate, PCA-4-sulfate, vanillic acid, isovanillic acid, vanillic acid-glucuronide, isovanillic acid-glucuronide, vanillic acid-sulfate, isovanillic acid-sulfate, methyl 3,4-dihydroxybenzoate, 2-hydroxy-4-methoxybenzoic acid, methyl vanillate, 3,4-dihydroxyphenylacetic acid, 4-hydroxyphenylacetic acid, caffeic acid, ferulic acid, hippuric acid)	6.11 µmol/L (maximum) 15.69 µmol/L (maximum)	[220]
Red wine powder	Animal (male Wistar rats)	Urine Plasma	aromatic acids catechins hippuric acid hippuric acid catechin glucuronide	4.7–2790 µg/d 0–8 mg/d 0.6–3 mg/d 60–110 µmol/L 0.2–2.8 µmol/L	[193]
(+)-Catechin	Animal (male Wistar rats)	Plasma	catechin glucuronide + sulfate 3′-O-methyl catechin-glucuronide 3′-O-methyl catechin-glucuronide + sulfate	0.1–0.8 µmol/L 0.3–19.3 µmol/L 16.8–38.3 µmol/L	[208]
(+)-Catechin (−)-Epicatechin (+)-Catechin + (−)-Epicatechin	Animal (male Sprague-Dawley rats)	Plasma Urine	catechin epicatechin 3′-O-methyl-catechin 3′-O-methyl-epicatechin catechin epicatechin 3′-O-methyl-catechin 3′-O-methyl-epicatechin	0.15–44.2 µmol.h.L^{-1} 0–41.9 µmol.h.L^{-1} 0–23.0 µmol.h.L^{-1} 0.82–78.3 µmol.h.L^{-1} 0.01–8.85 µmol.h.L^{-1} 0.03–16.6 µmol.h.L^{-1} 0–3.60 µmol.h.L^{-1} 0–9.45 µmol.h.L^{-1}	[209]
Cyanidin 3-O-β-D-glucoside	Animal (male Wistar rats)	Plasma Kidney Liver	cyanidin 3-O-β-D-glucoside protocatechuic acid cyanidin 3-O-β-D-glucoside methylated cyanidin 3-O-β-D-glucoside cyanidin 3-O-β-D-glucoside methylated cyanidin 3-O-β-D-glucoside	0–0.31 µmol/L 0–2.56 µmol/L 0–3.20 µmol/L 0–1.32 µmol/L 0 µmol/L 0–0.64 µmol/L	[215]
Rutin Quercetin	Animal (rabbits)	Urine	3,4-dihydroxyphenylacetic acid m-hydroxyphenylacetic acid p-hydroxyphenylacetic acid homovanillic acid	Not determined	[224]
Phloroglucinol	Animal (athymic Balb/c female nude mice)	Mice	phloroglucinol	25 mg of phloroglucinol/kg of body	[213]
Hippuric acid associated with doxorubicin or oxaliplatin	Cancer cell lines (MDA-MB-231, MCF-7, Caco-2)	Cells	Hippuric acid associated with doxorubicin or oxaliplatin	0.13–20 µg/mL (IC$_{50}$)	[194]
4-hydroxybenzoic acid	Cancer cell lines (MCF-7, adriamycin-resistant cells MCF-7/ADM, MDA-MB-231, MDA-MB-468, 4T1) Animal (BALB/c mice)	Cells Tumor	4-hydroxybenzoic acid	0–20 µM 2 mg/Kg	[195]

Table 1. *Cont.*

Dietary Factor/Isolated Compound	Model (Human/Animal/Cancer Cell Lines)	Sample Type	Discriminating Metabolites	Average of Metabolites (Found or Used)	Primary Reference
Protocatechuic acid	Cancer cell lines (MCF-7, A549, HepG2, HeLa, LNCap)	Cells	protocatechuic acid	1–8 µmol/L	[196]
Gallic acid	Cancer cell lines (MDA-MB-231, HS578T, MCF-7)	Cells	gallic acid	5–400 µM	[199]
Resveratrol Hydrosystilbenes Dihydroresveratrol	Cancer cell lines (MCF-7, MDA-MB-231, BT-474, K-562)	Cells	resveratrol hydrosystilbenes dihydroresveratrol	1 nM–10 µM	[206]
Resveratrol-3-*O*-sulfate Resveratrol Resveratrol-3′-*O*-glucuronide	Cancer cell line (MCF-7)	Cells	resveratrol-3-*O*-sulfate resveratrol resveratrol-3′-*O*-glucuronide	500 nM–100 µM	[205]
Resveratrol 3′-*O*-sulfate Resveratrol 4′-*O*-sulfate Dihydroresveratrol Dihydroresveratrol-3′-*O*-glucuronide	Cancer cell lines (MCF-7, MDA-MB-231)	Cells	resveratrol 3′-*O*-sulfate resveratrol 4′-*O*-sulfate dihydroresveratrol dihydroresveratrol-3′-*O*-glucuronide	0.4–10 µmol/L	[207]
Phloroglucinol	Cancer cell lines (BT549, MDA-MB-231, MCF-7, SK-BR3, BT549)	Cells	phloroglucinol	0–100 µM	[212]
Delphinidin-3-glucuronide Cyanidin-3-glucuronide Petunidin-3-glucuronide	Cancer cell lines (MKN-28, Caco-2, MCF-7)	Cells	delphinidin-3-glucuronide cyanidin-3-glucuronide petunidin-3-glucuronide	6.3–100 µM	[221]
Quercetin-3-*O*-glucuronide	Non-tumorigenic cell line (MCF-10A) and cancer cell line (MDA-MB-231)	Cells	quercetin-3-*O*-glucuronide	0.01–µM	[226]

6. Conclusions

As can be seen in this compilation, grapes and wines are rich and complex sources of bioactive molecules with multiple targets and effects. The natural polyphenols from these dietary products belong to different classes of compounds, both flavonoid and non-flavonoid, and have been studied in different models of breast cancer, both in vivo and in vitro. The major anticancer activities promoted by these compounds are summarized in Figure 1 and include modulation of estrogen cell signaling, cancer cell differentiation, cell growth inhibition, apoptosis induction and suppression of the metastatic behavior.

Based on dietary source, bioactive compounds or their metabolites used in different in vitro and in vivo studies for breast cancer, we conclude that there is a great variation of doses utilized or found. When the studies utilize wine or grape as a bioactive compound source, it is possible to observe a great variation on metabolite quality and quantity. On the other hand, when the isolated metabolite or its precursor were used, mainly in cancer cell lines, variations from 1 nM until 100 μM were used, and some authors justify the use of these concentrations to approximate the physiological concentrations. It is also important to point out that the effects produced by the glycosidic forms and the aglycones might lead to different routes of absorption and/or metabolization, leading to important variations in bioavailability and global effects produced.

The bioavailability of these compounds is another important issue that must be circumvented to improve local biological effects. In this way, grape and wine have long been used as sources of lead compounds in the search for breast cancer chemotherapy candidates and should be further explored in clinical studies, along with the biotechnological improvements necessary for their application.

Author Contributions: Writing—original draft preparation—D.C.F.d.C., L.P.R., J.Q., R.A.S., E.F.; writing and editing—D.C.F.d.C., L.P.R., J.L.S., E.F.; All authors have read and agreed to the published version of the manuscript.

Funding: This study was financed in part by the Coordenação de Aperfeiçoamento de Pessoal de Nível Superior—Brasil (CAPES)—Finance Code 001, Conselho Nacional de Desenvolvimento Científico e Tecnológico (CNPq) and Fundação Carlos Chagas Filho de Amparo à Pesquisa do Estado do Rio de Janeiro (FAPERJ).

Conflicts of Interest: The authors declare no conflict of interest.

References

1. The International Agency for Research on Cancer (IARC) Report, World Health Organization (WHO). *Latest Global Cancer Data: Cancer Burden Rises to 18.1 Million New Cases and 9.6 Million Cancer Deaths in 2018*; IARC: Lyon, France, 2018.
2. Teixeira, A.; Baenas, N.; Dominguez-Perles, R.; Barros, A.; Rosa, E.; Moreno, D.A.; Garcia-Viguera, C. Natural bioactive compounds from winery by-products as health promoters: A review. *Int. J. Mol. Sci.* **2014**, *15*, 15638–15678. [CrossRef]
3. Amor, S.; Châlons, P.; Aires, V.; Delmas, D. Polyphenol Extracts from Red Wine and Grapevine: Potential Effects on Cancers. *Diseases* **2018**, *6*, 106. [CrossRef] [PubMed]
4. Holliday, D.L.; Speirs, V. Choosing the right cell line for breast cancer research. *Breast Cancer Res.* **2011**, *13*, 1–7. [CrossRef] [PubMed]
5. Agarwal, C.; Sharma, Y.; Zhao, J.; Agarwal, R. A polyphenolic fraction from grape seeds causes irreversible growth inhibition of breast carcinoma MDA-MB468 cells by inhibiting mitogen- activated protein kinases activation and inducing G1 arrest and differentiation. *Clin. Cancer Res.* **2000**, *6*, 2921–2930. [PubMed]
6. Leone, A.; Longo, C.; Gerardi, C.; Trosko, J.E. Pro-apoptotic effect of grape seed extract on MCF-7 involves transient increase of gap junction intercellular communication and Cx43 up-regulation: A mechanism of chemoprevention. *Int. J. Mol. Sci.* **2019**, *20*, 3244. [CrossRef]
7. Dinicola, S.; Pasqualato, A.; Cucina, A.; Coluccia, P. Grape seed extract suppresses MDA-MB231 breast cancer cell migration and invasion. *Eur. J. Nutr.* **2014**, *53*, 421–431. [CrossRef]
8. Lu, J.; Zhang, K.; Chen, S.; Wen, W. Grape seed extract inhibits VEGF expression via reducing HIF-1α protein expression. *Carcinogenesis* **2009**, *30*, 636–644. [CrossRef]

9. Eng, E.T.; Williams, D.; Mandava, U.; Kirma, N.; Tekmal, R.R.; Chen, S. Anti-Aromatase Chemicals in Red Wine. *Ann. N. Y. Acad. Sci.* **2006**, *963*, 239–246. [CrossRef]

10. Kijima, I.; Phung, S.; Hur, G.; Kwok, S.L.; Chen, S. Grape seed extract is an aromatase inhibitor and a suppressor of aromatase expression. *Cancer Res.* **2006**, *66*, 5960–5967. [CrossRef]

11. Sharma, G.; Tyagi, A.K.; Singh, R.P.; Chan, D.C.F.; Agarwal, R. Synergistic anti-cancer effects of grape seed extract and conventional cytotoxic agent doxorubicin against human breast carcinoma cells. *Breast Cancer Res. Treat.* **2004**, *85*, 1–12. [CrossRef]

12. Yousef, M.I.; Mahdy, M.A.; Abdou, H.M. The potential protective role of grape seed proanthocyanidin extract against the mixture of carboplatin and thalidomide induced hepatotoxicity and cardiotoxicity in male rats. *Prev. Med. Commun. Health* **2020**, *2*, 1–7. [CrossRef]

13. Makris, D.P.; Kallithraka, S.; Kefalas, P. Flavonols in grapes, grape products and wines: Burden, profile and influential parameters. *J. Food Compos. Anal.* **2006**, *19*, 396–404. [CrossRef]

14. Simonetti, P.; Pietta, P.; Testolin, G. Polyphenol content and total antioxidant potential of selected Italian wines. *J. Agric. Food Chem.* **1997**, *45*, 1152–1155. [CrossRef]

15. Gardner, P.T.; McPhail, D.B.; Crozier, A.; Duthie, G.G. Electron spin resonance (ESR) spectroscopic assessment of the contribution of quercetin and other flavonols to the antioxidant capacity of red wines. *J. Sci. Food Agric.* **1999**, *79*, 1011–1014. [CrossRef]

16. McDonald, M.S.; Hughes, M.; Burns, J.; Lean, M.E.J.; Matthews, D.; Crozier, A. Survey of the Free and Conjugated Myricetin and Quercetin Content of Red Wines of Different Geographical Origins. *J. Agric. Food Chem.* **1998**, *46*, 368–375. [CrossRef]

17. Dabeek, W.M.; Marra, M.V. Dietary quercetin and kaempferol: Bioavailability and potential cardiovascular-related bioactivity in humans. *Nutrients* **2019**, *11*, 2288. [CrossRef]

18. Kennedy, J.A. Grape and wine phenolics: Observations and recent findings. *Ciencia e Investigación Agraria* **2008**, *35*, 107–120. [CrossRef]

19. Fernández de Simón, B.; Hernández, T.; Cadahía, E.; Dueñas, M.; Estrella, I. Phenolic compounds in a Spanish red wine aged in barrels made of Spanish, French and American oak wood. *Eur. Food Res. Technol.* **2003**, *216*, 150–156. [CrossRef]

20. Castellari, M.; Piermattei, B.; Arfelli, G.; Amati, A. Influence of aging conditions on the quality of red Sangiovese wine. *J. Agric. Food Chem.* **2001**, *49*, 3672–3676. [CrossRef]

21. Makris, D.P.; Rossiter, J.T. Heat-induced, metal-catalyzed oxidative degradation of quercetin and rutin (quercetin 3-*O*-rhamnosylglucoside) in aqueous model systems. *J. Agric. Food Chem.* **2000**, *48*, 3830–3838. [CrossRef]

22. Hashemzaei, M.; Far, A.D.; Yari, A.; Heravi, R.E.; Tabrizian, K.; Taghdisi, S.M.; Sadegh, S.E.; Tsarouhas, K.; Kouretas, D.; Tzanakakis, G.; et al. Anticancer and apoptosis-inducing effects of quercetin in vitro and in vivo. *Oncol. Rep.* **2017**, *38*, 819–828. [CrossRef] [PubMed]

23. Imran, M.; Salehi, B.; Sharifi-Rad, J.; Gondal, T.A.; Saeed, F.; Imran, A.; Shahbaz, M.; Fokou, P.V.T.; Arshad, M.U.; Khan, H.; et al. Kaempferol: A key emphasis to its anticancer potential. *Molecules* **2019**, *24*, 2277. [CrossRef] [PubMed]

24. Franke, A.A.; Custer, L.J.; Wilkens, L.R.; Le Marchand, L.; Nomura, A.M.Y.; Goodman, M.T.; Kolonel, L.N. Liquid chromatographic-photodiode array mass spectrometric analysis of dietary phytoestrogens from human urine and blood. *J. Chromatogr. B Anal. Technol. Biomed. Life Sci.* **2002**, *777*, 45–59. [CrossRef]

25. Wang, L.; Lee, I.M.; Zhang, S.M.; Blumberg, J.B.; Buring, J.E.; Sesso, H.D. Dietary intake of selected flavonols, flavones, and flavonoid-rich foods and risk of cancer in middle-aged and older women. *Am. J. Clin. Nutr.* **2009**, *89*, 905–912. [CrossRef] [PubMed]

26. Mokbel, K.; Mokbel, K. Chemoprevention of Breast Cancer with Vitamins and Micronutrients: A Concise Review. *In Vivo* **2019**, *33*, 983–997. [CrossRef]

27. Wang, X.; Yang, Y.; An, Y.; Fang, G. The mechanism of anticancer action and potential clinical use of kaempferol in the treatment of breast cancer. *Biomed. Pharmacother.* **2019**, *117*, 109086. [CrossRef]

28. Tang, S.M.; Deng, X.T.; Zhou, J.; Li, Q.P.; Ge, X.X.; Miao, L. Pharmacological basis and new insights of quercetin action in respect to its anti-cancer effects. *Biomed. Pharmacother.* **2020**, *121*, 109604. [CrossRef]

29. Barve, A.; Chen, C.; Hebbar, V.; Desiderioa, J.; Constance Lay-Lay Saw, A.-N.K. Metabolism, oral bioavailability and pharmacokinetics of chemopreventive kaempferol in rats. *Biopharm. Drug Dispos.* **2009**, *30*, 356–365. [CrossRef]

30. Zabela, V.; Sampath, C.; Oufir, M.; Moradi-Afrapoli, F.; Butterweck, V.; Hamburger, M. Pharmacokinetics of dietary kaempferol and its metabolite 4-hydroxyphenylacetic acid in rats. *Fitoterapia* **2016**, *115*, 189–197. [CrossRef]

31. Lamson, D.W.; Brignall, M.S. Antioxidants and cancer III: Quercetin. *Altern. Med. Rev.* **2000**, *5*, 196–208.

32. Cai, X.; Fang, Z.; Dou, J.; Yu, A.; Zhai, G. Bioavailability of Quercetin: Problems and Promises. *Curr. Med. Chem.* **2013**, *20*, 2572–2582. [CrossRef] [PubMed]

33. Liu, S.; Li, R.; Qian, J.; Sun, J.; Li, G.; Shen, J.; Xie, Y. The combination therapy of doxorubicin and quercetin on multidrug resistant breast cancer and their sequential delivery by reduction-sensitive hyaluronic acid-based conjugate/D-α-tocopheryl polyethylene glycol 1000 succinate mixed micelles. *Mol. Pharm.* **2020**, *17*, 1415–1427. [CrossRef] [PubMed]

34. Liu, Z.; Balasubramanian, V.; Bhat, C.; Vahermo, M.; Mäkilä, E.; Kemell, M.; Fontana, F.; Janoniene, A.; Petrikaite, V.; Salonen, J.; et al. Quercetin-Based Modified Porous Silicon Nanoparticles for Enhanced Inhibition of Doxorubicin-Resistant Cancer Cells. *Adv. Healthc. Mater.* **2017**, *6*, 1–11. [CrossRef] [PubMed]

35. Lv, L.; Liu, C.; Chen, C.; Yu, X.; Chen, G.; Shi, Y.; Qin, F.; Ou, J.; Qiu, K.; Li, G. Quercetin and doxorubicin co-encapsulated biotin receptortargeting nanoparticles for minimizing drug resistance in breast cancer. *Oncotarget* **2016**, *7*, 32184–32199. [CrossRef]

36. Kubatka, P.; Kapinová, A.; Kello, M.; Kruzliak, P.; Kajo, K.; Výbohová, D.; Mahmood, S.; Murin, R.; Viera, T.; Mojžiš, J.; et al. Fruit peel polyphenols demonstrate substantial anti-tumour effects in the model of breast cancer. *Eur. J. Nutr.* **2015**, *55*, 955–965. [CrossRef]

37. Leigh Ackland, M.; Van De Waarsenburg, S.; Jones, R. Synergistic antiproliferation action of the flavonols quercetin and kaempferol in cultured human cancer cell lines. *In Vivo* **2005**, *19*, 69–76.

38. Yi, X.; Zuo, J.; Tan, C.; Xian, S.; Luo, C.; Chen, S.; Yu, L.; Luo, Y. Kaempferol, a flavonoid compound from gynura medica induced apoptosis and growth inhibition in MCF-7 breast cancer cell. *Afr. J. Tradit. Complement. Altern. Med.* **2016**, *13*, 210–215. [CrossRef]

39. Choi, E.J.; Ahn, W.S. Kaempferol induced the apoptosis via cell cycle arrest in human breast cancer MDA-MB-453 cells. *Nutr. Res. Pract.* **2008**, *2*, 322–325. [CrossRef]

40. Kang, G.Y.; Lee, E.R.; Kim, J.H.; Jung, J.W.; Lim, J.; Kim, S.K.; Cho, S.G.; Kim, K.P. Downregulation of PLK-1 expression in kaempferol-induced apoptosis of MCF-7 cells. *Eur. J. Pharmacol.* **2009**, *611*, 17–21. [CrossRef]

41. Li, S.; Yan, T.; Deng, R.; Jiang, X.; Xiong, H.; Wang, Y.; Yu, Q.; Wang, X.; Chen, C.; Zhu, Y. Low dose of kaempferol suppresses the migration and invasion of triple-negative breast cancer cells by downregulating the activities of RhoA and Rac1. *OncoTargets Ther.* **2017**, *10*, 4809–4819. [CrossRef]

42. Zhu, L.; Xue, L. Kaempferol suppresses proliferation and induces cell cycle arrest, apoptosis, and DNA damage in breast cancer cells. *Oncol. Res.* **2019**, *27*, 629–634. [CrossRef] [PubMed]

43. Ranganathan, S.; Halagowder, D.; Sivasithambaram, N.D. Quercetin suppresses twist to induce apoptosis in MCF-7 breast cancer cells. *PLoS ONE* **2015**, *10*, 1–21. [CrossRef] [PubMed]

44. Kabała-Dzik, A.; Rzepecka-Stojko, A.; Kubina, R.; Iriti, M.; Wojtyczka, R.D.; Buszman, E.; Stojko, J. Flavonoids, bioactive components of propolis, exhibit cytotoxic activity and induce cell cycle arrest and apoptosis in human breast cancer cells MDA-MB-231 and MCF-7—A comparative study. *Cell. Mol. Biol.* **2018**, *64*, 1–10. [CrossRef] [PubMed]

45. Kim, B.W.; Lee, E.R.; Min, H.; Jeong, H.S.; Ahn, J.Y.; Kim, J.H.; Choi, H.Y.; Choi, H.; Eun, Y.K.; Se, P.P.; et al. Sustained ERK activation is involved in the kaempferol-induced apoptosis of breast cancer cells and is more evident under 3-D culture condition. *Cancer Biol. Ther.* **2008**, *7*, 1080–1089. [CrossRef]

46. Liao, W.; Chen, L.; Ma, X.; Jiao, R.; Li, X.; Wang, Y. Protective effects of kaempferol against reactive oxygen species-induced hemolysis and its antiproliferative activity on human cancer cells. *Eur. J. Med. Chem.* **2016**, *114*, 24–32. [CrossRef]

47. Zhang, H.; Zhang, M.; Yu, L.; Zhao, Y.; He, N.; Yang, X. Antitumor activities of quercetin and quercetin-5′,8-disulfonate in human colon and breast cancer cell lines. *Food Chem. Toxicol.* **2012**, *50*, 1589–1599. [CrossRef]

48. Wu, Q.; Needs, P.W.; Lu, Y.; Kroon, P.A.; Ren, D.; Yang, X. Different antitumor effects of quercetin, quercetin-3′-sulfate and quercetin-3-glucuronide in human breast cancer MCF-7 cells. *Food Funct.* **2018**, *9*, 1736–1746. [CrossRef]

49. Niazvand, F.; Orazizadeh, M.; Khorsandi, L.; Abbaspour, M.; Mansouri, E.; Khodadadi, A. Effects of Quercetin-Loaded Nanoparticles on MCF-7 Human Breast Cancer Cells. *Medicina* **2019**, *55*, 114. [CrossRef]

50. Chou, C.C.; Yang, J.S.; Lu, H.F.; Ip, S.W.; Lo, C.; Wu, C.C.; Lin, J.P.; Tang, N.Y.; Chung, J.G.; Chou, M.J.; et al. Quercetin-mediated cell cycle arrest and apoptosis involving activation of a caspase cascade through the mitochondrial pathway in human breast cancer MCF-7 cells. *Arch. Pharm. Res.* **2010**, *33*, 1181–1191. [CrossRef]

51. Chien, S.Y.; Wu, Y.C.; Chung, J.G.; Yang, J.S.; Lu, H.F.; Tsou, M.F.; Wood, W.; Kuo, S.J.; Chen, D.R. Quercetin-induced apoptosis acts through mitochondrial- and caspase-3-dependent pathways in human breast cancer MDA-MB-231 cells. *Hum. Exp. Toxicol.* **2009**, *28*, 493–503. [CrossRef]

52. Balakrishnan, S.; Mukherjee, S.; Das, S.; Bhat, F.A.; Raja Singh, P.; Patra, C.R.; Arunakaran, J. Gold nanoparticles–conjugated quercetin induces apoptosis via inhibition of EGFR/PI3K/Akt–mediated pathway in breast cancer cell lines (MCF-7 and MDA-MB-231). *Cell Biochem. Funct.* **2017**, *35*, 217–231. [CrossRef] [PubMed]

53. Lin, C.W.; Hou, W.C.; Shen, S.C.; Juan, S.H.; Ko, C.H.; Wang, L.M.; Chen, Y.C. Quercetin inhibition of tumor invasion via suppressing PKCδ/ERK/AP-1-dependent matrix metalloproteinase-9 activation in breast carcinoma cells. *Carcinogenesis* **2008**, *29*, 1807–1815. [CrossRef] [PubMed]

54. Li, J.; Zhu, F.; Lubet, R.A.; De Luca, A.; Grubbs, C.; Ericson, M.E.; D'Alessio, A.; Normanno, N.; Dong, Z.; Bode, A.M. Quercetin-3-methyl ether inhibits lapatinib-sensitive and -resistant breast cancer cell growth by inducing G2/M arrest and apoptosis. *Mol. Carcinog.* **2013**, *52*, 134–143. [CrossRef] [PubMed]

55. Khorsandi, L.; Orazizadeh, M.; Niazvand, F.; Abbaspour, M.R.; Mansouri, E.; Khodadadi, A. Quercetin induces apoptosis and necroptosis in MCF-7 breast cancer cells. *Bratislava Med. J.* **2017**, *118*, 123–128. [CrossRef]

56. Hu, J.; Zhang, Y.; Jiang, X.; Zhang, H.; Gao, Z.; Li, Y.; Fu, R.; Li, L.; Li, J.; Cui, H.; et al. ROS-mediated activation and mitochondrial translocation of CaMKII contributes to Drp1-dependent mitochondrial fission and apoptosis in triple-negative breast cancer cells by isorhamnetin and chloroquine. *J. Exp. Clin. Cancer Res.* **2019**, *38*, 1–16. [CrossRef]

57. Qiu, J.; Zhang, T.; Zhu, X.; Yang, C.; Wang, Y.; Zhou, N.; Ju, B.; Zhou, T.; Deng, G.; Qiu, C. Hyperoside induces breast cancer cells apoptosis via ROS-mediated NF-κB signaling pathway. *Int. J. Mol. Sci.* **2020**, *21*, 131. [CrossRef]

58. Steiner, J.L.; Davis, J.M.; McClellan, J.L.; Enos, R.T.; Carson, J.A.; Fayad, R.; Nagarkatti, M.; Nagarkatti, P.S.; Altomare, D.; Creek, K.E.; et al. Dose-dependent benefits of quercetin on tumorigenesis in the C3(1)/SV40Tag transgenic mouse model of breast cancer. *Cancer Biol. Ther.* **2014**, *15*, 1456–1467. [CrossRef]

59. Phromnoi, K.; Yodkeeree, S.; Anuchapreeda, S.; Limtrakul, P. Inhibition of MMP-3 activity and invasion of the MDA-MB-231 human invasive breast carcinoma cell line by bioflavonoids. *Acta Pharmacol. Sin.* **2009**, *30*, 1169–1176. [CrossRef]

60. Li, C.; Yang, D.; Zhao, Y.; Qiu, Y.; Cao, X.; Yu, Y.; Guo, H.; Gu, X.; Yin, X. Inhibitory Effects of Isorhamnetin on the Invasion of Human Breast Carcinoma Cells by Downregulating the Expression and Activity of Matrix Metalloproteinase-2/9. *Nutr. Cancer* **2015**, *67*, 1191–1200. [CrossRef]

61. Balakrishnan, S.; Bhat, F.A.; Raja Singh, P.; Mukherjee, S.; Elumalai, P.; Das, S.; Patra, C.R.; Arunakaran, J. Gold nanoparticle–conjugated quercetin inhibits epithelial–mesenchymal transition, angiogenesis and invasiveness via EGFR/VEGFR-2-mediated pathway in breast cancer. *Cell Prolif.* **2016**, *49*, 678–697. [CrossRef]

62. Oh, S.J.; Kim, O.; Lee, J.S.; Kim, J.A.; Kim, M.R.; Choi, H.S.; Shim, J.H.; Kang, K.W.; Kim, Y.C. Inhibition of angiogenesis by quercetin in tamoxifen-resistant breast cancer cells. *Food Chem. Toxicol.* **2010**, *48*, 3227–3234. [CrossRef] [PubMed]

63. Brusselmans, K.; Vrolix, R.; Verhoeven, G.; Swinnen, J.V. Induction of cancer cell apoptosis by flavonoids is associated with their ability to inhibit fatty acid synthase activity. *J. Biol. Chem.* **2005**, *280*, 5636–5645. [CrossRef] [PubMed]

64. Hanikoglu, A.; Kucuksayan, E.; Hanikoglu, F.; Ozben, T.; Menounou, G.; Sansone, A.; Chatgilialoglu, C.; Di Bella, G.; Ferreri, C. Effects of somatostatin, curcumin, and quercetin on the fatty acid profile of breast cancer cell membranes. *Can. J. Physiol. Pharmacol.* **2020**, *98*, 131–138. [CrossRef] [PubMed]

65. Kuiper, G.G.J.M.; Lemmen, J.G.; Carlsson, B.; Corton, J.C.; Safe, S.H.; Van Der Saag, P.T.; Van Der Burg, B.; Gustafsson, J.Å. Interaction of estrogenic chemicals and phytoestrogens with estrogen receptor β. *Endocrinology* **1998**, *139*, 4252–4263. [CrossRef] [PubMed]

66. Hung, H. Inhibition of Estrogen Receptor Alpha Expression and Function in MCF-7 Cells by Kaempferol. *J. Cell. Physiol.* **2004**, *198*, 197–208. [CrossRef] [PubMed]

67. Yiğitaslan, S.; Erol, K.; Özatik, F.Y.; Özatik, O.; Şahin, S.; Çengelli, Ç. Estrogen-like activity of quercetin in female rats. *Erciyes Tip Derg.* **2016**, *38*, 53–58. [CrossRef]

68. Puranik, N.V.; Srivastava, P.; Bhatt, G.; John Mary, D.J.S.; Limaye, A.M.; Sivaraman, J. Determination and analysis of agonist and antagonist potential of naturally occurring flavonoids for estrogen receptor (ERα) by various parameters and molecular modelling approach. *Sci. Rep.* **2019**, *9*, 1–11. [CrossRef] [PubMed]

69. Park, Y.J.; Choo, W.H.; Kim, H.R.; Chung, K.H.; Oh, S.M. Inhibitory aromatase effects of flavonoids from ginkgo biloba extracts on estrogen biosynthesis. *Asian Pac. J. Cancer Prev.* **2015**, *16*, 6317–6325. [CrossRef]

70. Kao, Y.C.; Zhou, C.; Sherman, M.; Laughton, C.A.; Chen, S. Molecular basis of the inhibition of human aromatase (estrogen synthetase) by flavone and isoflavone phytoestrogens: A site-directed mutagenesis study. *Environ. Health Perspect.* **1998**, *106*, 85–92. [CrossRef]

71. van Meeuwen, J.A.; ter Burg, W.; Piersma, A.H.; van den Berg, M.; Sanderson, J.T. Mixture effects of estrogenic compounds on proliferation and pS2 expression of MCF-7 human breast cancer cells. *Food Chem. Toxicol.* **2007**, *45*, 2319–2330. [CrossRef]

72. Resende, F.A.; de Oliveira, A.P.S.; de Camargo, M.S.; Vilegas, W.; Varanda, E.A. Evaluation of Estrogenic Potential of Flavonoids Using a Recombinant Yeast Strain and MCF7/BUS Cell Proliferation Assay. *PLoS ONE* **2013**, *8*, 1–7. [CrossRef] [PubMed]

73. Seung, M.O.; Yeon, P.K.; Kyu, H.C. Biphasic effects of kaempferol on the estrogenicity in human breast cancer cells. *Arch. Pharm. Res.* **2006**, *29*, 354–362.

74. Kim, S.H.; Hwang, K.A.; Choi, K.C. Treatment with kaempferol suppresses breast cancer cell growth caused by estrogen and triclosan in cellular and xenograft breast cancer models. *J. Nutr. Biochem.* **2016**, *28*, 70–82. [CrossRef] [PubMed]

75. Lee, G.A.; Choi, K.C.; Hwang, K.A. Treatment with phytoestrogens reversed triclosan and bisphenol A-induced anti-apoptosis in breast cancer cells. *Biomol. Ther.* **2018**, *26*, 503–511. [CrossRef] [PubMed]

76. Lee, G.A.; Choi, K.C.; Hwang, K.A. Kaempferol, a phytoestrogen, suppressed triclosan-induced epithelial-mesenchymal transition and metastatic-related behaviors of MCF-7 breast cancer cells. *Environ. Toxicol. Pharmacol.* **2017**, *49*, 48–57. [CrossRef] [PubMed]

77. Mei, Y.Q.; Kai, L.; Qun, S.; Qingyue, L.; Jinghui, H.; Fangxuan, H.; Ren-Wang, J. Reversal of multidrug resistance in cancer by multi-functional flavonoids. *Front. Oncol.* **2019**, *9*, 487.

78. Soo, Y.C.; Min, K.S.; Na, H.K.; Jung, O.J.; Eun, J.G.; Hwa, J.L. Inhibition of P-glycoprotein by natural products in human breast cancer cells. *Arch. Pharm. Res.* **2005**, *28*, 823–828.

79. Li, S.; Yuan, S.; Zhao, Q.; Wang, B.; Wang, X.; Li, K. Quercetin enhances chemotherapeutic effect of doxorubicin against human breast cancer cells while reducing toxic side effects of it. *Biomed. Pharmacother.* **2018**, *100*, 441–447. [CrossRef]

80. Li, S.; Zhao, Q.; Wang, B.; Yuan, S.; Wang, X.; Li, K. Quercetin reversed MDR in breast cancer cells through down-regulating P-gp expression and eliminating cancer stem cells mediated by YB-1 nuclear translocation. *Phyther. Res.* **2018**, *32*, 1530–1536. [CrossRef]

81. Desale, J.P.; Swami, R.; Kushwah, V.; Katiyar, S.S.; Jain, S. Chemosensitizer and docetaxel-loaded albumin nanoparticle: Overcoming drug resistance and improving therapeutic efficacy. *Nanomedicine* **2018**, *13*, 2759–2776. [CrossRef]

82. Kumar, M.; Sharma, G.; Misra, C.; Kumar, R.; Singh, B.; Katare, O.P.; Raza, K. N-desmethyl tamoxifen and quercetin-loaded multiwalled CNTs: A synergistic approach to overcome MDR in cancer cells. *Mater. Sci. Eng. C* **2018**, *89*, 274–282. [CrossRef] [PubMed]

83. Imai, Y.; Tsukahara, S.; Asada, S.; Sugimoto, Y. Phytoestrogens/flavonoids reverse breast cancer resistance protein/ABCG2-mediated multidrug resistance. *Cancer Res.* **2004**, *64*, 4346–4352. [CrossRef] [PubMed]

84. Fan, X.; Bai, J.; Zhao, S.; Hu, M.; Sun, Y.; Wang, B.; Ji, M.; Jin, J.; Wang, X.; Hu, J.; et al. Evaluation of inhibitory effects of flavonoids on breast cancer resistance protein (BCRP): From library screening to biological evaluation to structure-activity relationship. *Toxicol. Vitr.* **2019**, *61*, 104642. [CrossRef] [PubMed]

85. An, G.; Gallegos, J.; Morris, M.E. The bioflavonoid kaempferol is an Abcg2 substrate and inhibits Abcg2-mediated quercetin efflux. *Drug Metab. Dispos.* **2011**, *39*, 426–432. [CrossRef]

86. Zheng, L.; Zhu, L.; Zhao, M.; Shi, J.; Li, Y.; Yu, J.; Jiang, H.; Wu, J.; Tong, Y.; Liu, Y.; et al. In Vivo Exposure of Kaempferol Is Driven by Phase II Metabolic Enzymes and Efflux Transporters. *AAPS J.* **2016**, *18*, 1289–1299. [CrossRef]

87. Li, Y.; Lu, L.; Wang, L.; Qu, W.; Liu, W.; Xie, Y.; Zheng, H.; Wang, Y.; Qi, X.; Hu, M.; et al. Interplay of Efflux Transporters with Glucuronidation and Its Impact on Subcellular Aglycone and Glucuronide Disposition: A Case Study with Kaempferol. *Mol. Pharm.* **2018**, *15*, 5602–5614. [CrossRef]

88. Yao, L.H.; Jiang, Y.M.; Shi, J.; Tomás-Barberán, F.A.; Datta, N.; Singanusong, R.; Chen, S.S. Flavonoids in food and their health benefits. *Plant Foods Hum. Nutr.* **2004**, *59*, 113–122. [CrossRef]

89. Pastrana-Bonilla, E.; Akoh, C.C.; Sellappan, S.; Krewer, G. Phenolic Content and Antioxidant Capacity of Muscadine Grapes. *J. Agric. Food Chem.* **2003**, *51*, 5497–5503. [CrossRef]

90. Graham, H.N. Green tea composition, consumption, and polyphenol chemistry. *Prev. Med.* **1992**, *21*, 334–350. [CrossRef]

91. Bell, J.R.C.; Donovan, J.L.; Wong, R.; Waterhouse, A.L.; German, J.B.; Walzem, R.L. (+)-Catechin in human plasma after ingestion of a single serving of reconstituted red wine. *Am. J. Clin. Nutr.* **2000**, *71*, 103–108. [CrossRef]

92. Silva, J.M.R.; Rigaud, J.; Cheynier, V.; Cheminat, A.; Moutounet, M. Procyanidin dimers and trimers from grape seeds. *Phytochemistry* **1991**, *30*, 1259–1264. [CrossRef]

93. Prieur, C.; Rigaud, J.; Cheynier, V.; Moutounet, M. Oligomeric and polymeric procyanidins from grape seeds. *Phytochemistry* **1994**, *36*, 781–784. [CrossRef]

94. Kim, H.; Hall, P.; Smith, M.; Kirk, M.; Prasain, J.K.; Barnes, S.; Grubbs, C. Chemoprevention by Grape Seed Extract and Genistein in Carcinogen- induced Mammary Cancer in Rats Is Diet Dependent 1, 2. *Int. Res. Conf. Food Nutr. Cancer* **2004**, 3445–3452. [CrossRef] [PubMed]

95. Ou, K.; Gu, L. Absorption and metabolism of proanthocyanidins. *J. Funct. Foods* **2014**, *7*, 43–53. [CrossRef]

96. Tsang, C.; Auger, C.; Mullen, W.; Bornet, A.; Rouanet, J.-M.; Crozier, A.; Teissedre, P.-L. The absorption, metabolism and excretion of flavan-3-ols and procyanidins following the ingestion of a grape seed extract by rats. *Br. J. Nutr.* **2005**, *94*, 170–181. [CrossRef]

97. Serra, A.; MacI, A.; Romero, M.P.; Valls, J.; Bladé, C.; Arola, L.; Motilva, M.J. Bioavailability of procyanidin dimers and trimers and matrix food effects in in vitro and in vivo models. *Br. J. Nutr.* **2010**, *103*, 944–952. [CrossRef]

98. Sano, A.; Yamakoshi, J.; Tokutake, S.; Tobe, K.; Kubota, Y.; Kikuchi, M. Procyanidin B1 is detected in human serum after intake of proanthocyanidin-rich grape seed extract. *Biosci. Biotechnol. Biochem.* **2003**, *67*, 1140–1143. [CrossRef]

99. Yang, C.S.; Sang, S.; Lambert, J.D.; Lee, M.J. Bioavailability issues in studying the health effects of plant polyphenolic compounds. *Mol. Nutr. Food Res.* **2008**, *52*, 139–151. [CrossRef]

100. Meng, X.; Sang, S.; Zhu, N.; Lu, H.; Sheng, S.; Lee, M.J.; Ho, C.T.; Yang, C.S. Identification and characterization of methylated and ring-fission metabolites of tea catechins formed in humans, mice, and rats. *Chem. Res. Toxicol.* **2002**, *15*, 1042–1050. [CrossRef]

101. Faria, A.; Calhau, C.; De Freitas, V.; Mateus, N. Procyanidins as antioxidants and tumor cell growth modulators. *J. Agric. Food Chem.* **2006**, *54*, 2392–2397. [CrossRef]

102. Alshatwi, A.A. Catechin hydrate suppresses MCF-7 proliferation through TP53/Caspase-mediated apoptosis. *J. Exp. Clin. Cancer Res.* **2010**, *29*, 1–9. [CrossRef] [PubMed]

103. Damianaki, A.; Bakogeorgou, E.; Kampa, M.; Notas, G.; Hatzoglou, A.; Panagiotou, S.; Gemetzi, C.; Kouroumalis, E.; Martin, P.M.; Castanas, E. Potent inhibitory action of red wine polyphenols on human breast cancer cells. *J. Cell. Biochem.* **2000**, *78*, 429–441. [CrossRef]

104. Singletary, K.W.; Meline, B. Effect of Grape Seed Proanthocyanidins on Colon Aberrant Crypts and Breast Tumors in a Rat Dual-Organ Tumor Model. *Nutr. Cancer* **2001**, *39*, 252–258. [CrossRef] [PubMed]

105. Wahner-roedler, D.L.; Bauer, B.A.; Loehrer, L.L.; Cha, S.S.; Vera, J.; Hoskin, T.L.; Olson, J.E. The Effect of Grape Seed Extract on Estrogen Levels of Postmenopausal Women—A Pilot Study. *J. Diet. Suppl.* **2016**, *11*, 184–197. [CrossRef]

106. Ye, X.; Krohn, R.L.; Liu, W.; Joshi, S.S.; Kuszynski, C.A.; McGinn, T.R.; Bagchi, M.; Preuss, H.G.; Stohs, S.J.; Bagchi, D. The cytotoxic effects of a novel IH636 grape seed proanthocyanidin extract on cultured human cancer cells. *Stress Adapt. Prophyl. Treat.* **1999**, *196*, 99–108.

107. Brooker, S.; Martin, S.; Pearson, A.; Bagchi, D.; Earl, J.; Gothard, L.; Hall, E.; Porter, L.; Yarnold, J. Double-blind, placebo-controlled, randomised phase II trial of IH636 grape seed proanthocyanidin extract (GSPE) in patients with radiation-induced breast induration. *Radiother. Oncol.* **2006**, *79*, 45–51. [CrossRef]

108. Ma, Z.F.; Zhang, H. Phytochemical constituents, health benefits, and industrial applications of grape seeds: A mini-review. *Antioxidants* **2017**, *6*, 71. [CrossRef]
109. Tang, K.; Liu, T.; Han, Y.; Xu, Y.; Li, J.M. The Importance of Monomeric Anthocyanins in the Definition of Wine Colour Properties. *S. Afr. J. Enol. Vitic.* **2017**, *38*, 1–10. [CrossRef]
110. Zhang, Y.; Butelli, E.; Martin, C. Engineering anthocyanin biosynthesis in plants. *Curr. Opin. Plant Biol.* **2014**, *19*, 81–90. [CrossRef]
111. Khoo, H.E.; Azlan, A.; Tang, S.T.; Lim, S.M. Anthocyanidins and anthocyanins: Colored pigments as food, pharmaceutical ingredients, and the potential health benefits. *Food Nutr. Res.* **2017**, *61*, 1361779. [CrossRef]
112. Tsuda, T.; Watanabe, M.; Ohshima, K.; Norinobu, S.; Choi, S.W.; Kawakishi, S.; Osawa, T. Antioxidative Activity of the Anthocyanin Pigments Cyanidin 3-*O*-β-d-Glucoside and Cyanidin. *J. Agric. Food Chem.* **1994**, *42*, 2407–2410. [CrossRef]
113. Lila, M.A.; Burton-Freeman, B.; Grace, M.; Kalt, W. Unraveling Anthocyanin Bioavailability for Human Health. *Annu. Rev. Food Sci. Technol.* **2016**, *7*, 375–393. [CrossRef] [PubMed]
114. Lin, B.W.; Gong, C.C.; Song, H.F.; Cui, Y.Y. Effects of anthocyanins on the prevention and treatment of cancer. *Br. J. Pharmacol.* **2017**, *174*, 1226–1243. [CrossRef] [PubMed]
115. Hemmlnki, K. DNA adducts, mutations and cancer. *Carcinogenesis* **1993**, *14*, 2007–2012. [CrossRef]
116. Singletary, K.W.; Jung, K.J.; Giusti, M. Anthocyanin-rich grape extract blocks breast cell DNA damage. *J. Med. Food* **2007**, *10*, 244–251. [CrossRef]
117. Syed, D.N.; Afaq, F.; Sarfaraz, S.; Khan, N.; Kedlaya, R.; SAetaluri, V.; Mukhtar, H. Delphinidin inhibits cell proliferation and invasion via modulation of Met receptor phosphorylation. *Toxicol. Appl. Pharmacol.* **2008**, *231*, 52–60. [CrossRef]
118. Jeffers, M.; Schmidt, L.; Nakaigawa, N.; Webb, C.P.; Weirich, G.; Kishida, T.; Zbar, B.; Vande Woude, G.F. Activating mutations for the Met tyrosine kinase receptor in human cancer. *Proc. Natl. Acad. Sci. USA* **1997**, *94*, 11445–11450. [CrossRef]
119. Ding, S.; Merkulova-Rainon, T.; Han, Z.C.; Tobelem, G. HGF receptor up-regulation contributes to the angiogenic phenotype of human endothelial cells and promotes angiogenesis in vitro. *Blood* **2003**, *101*, 4816–4822. [CrossRef]
120. Bae-Jump, V.; Segreti, E.M.; Vandermolen, D.; Kauma, S. Hepatocyte growth factor (HGF) induces invasion of endometrial carcinoma cell lines in vitro. *Gynecol. Oncol.* **1999**, *73*, 265–272. [CrossRef]
121. Jeffers, M.; Rong, S.; Woude, G.F. Enhanced tumorigenicity and invasion-metastasis by hepatocyte growth factor/scatter factor-met signalling in human cells concomitant with induction of the urokinase proteolysis network. *Mol. Cell. Biol.* **1996**, *16*, 1115–1125. [CrossRef]
122. Murray, D.W.; Didier, S.; Chan, A.; Paulino, V.; Van Aelst, L.; Ruggieri, R.; Tran, N.L.; Byrne, A.T.; Symons, M. Guanine nucleotide exchange factor Dock7 mediates HGF-induced glioblastoma cell invasion via Rac activation. *Br. J. Cancer* **2014**, *110*, 1307–1315. [CrossRef]
123. Yang, Y.; Spitzer, E.; Meyer, D.; Sachs, M.; Niemann, C.; Hartmann, G.; Weidner, K.M.; Birchmeier, C.; Birchmeier, W. Sequential requirement of hepatocyte growth factor and neuregulin in the morphogenesis and differentiation of the mammary gland. *J. Cell Biol.* **1995**, *131*, 215–226. [CrossRef] [PubMed]
124. Yang, X.; Luo, E.; Liu, X.; Han, B.; Yu, X.; Peng, X. Delphinidin-3-glucoside suppresses breast carcinogenesis by inactivating the Akt/HOTAIR signaling pathway. *BMC Cancer* **2016**, *16*, 423. [CrossRef] [PubMed]
125. Khorkova, O.; Hsiao; Wahlestedt, C. Basic Biology and Therapeutic Implications of lncRNA. *Physiol. Behav.* **2015**, *87*, 15–24. [CrossRef] [PubMed]
126. Hajjari, M.; Salavaty, A. HOTAIR: An oncogenic long non-coding RNA in different cancers. *Cancer Biol. Med.* **2015**, *12*, 1–9. [PubMed]
127. Yang, G.; Zhang, S.; Gao, F.; Liu, Z.; Lu, M.; Peng, S.; Zhang, T.; Zhang, F. Osteopontin enhances the expression of HOTAIR in cancer cells via IRF1. *Biochim. Biophys. Acta Gene Regul. Mech.* **2014**, *1839*, 837–848. [CrossRef]
128. Chen, J.; Lin, C.; Yong, W.; Ye, Y.; Huang, Z. Calycosin and genistein induce apoptosis by inactivation of HOTAIR/p-akt signaling pathway in human breast cancer MCF-7 cells. *Cell. Physiol. Biochem.* **2015**, *35*, 722–728. [CrossRef]
129. Zhang, Y.; Vareed, S.K.; Nair, M.G. Human tumor cell growth inhibition by nontoxic anthocyanidins, the pigments in fruits and vegetables. *Life Sci.* **2005**, *76*, 1465–1472. [CrossRef]

130. Afaq, F.; Zaman, N.; Khan, N.; Syed, D.N.; Sarfaraz, S.; Zaid, M.A.; Mukhtar, H. Inhibition of epidermal growth factor receptor signaling pathway by delphinidin, an anthocyanidin in pigmented fruits and vegetables. *Int. J. Cancer* **2008**, *123*, 1508–1515. [CrossRef]

131. Abd El-Rehim, D.M.; Pinder, S.E.; Paish, C.E.; Bell, J.A.; Rampaul, R.S.; Blamey, R.W.; Robertson, J.F.R.; Nicholson, R.I.; Ellis, I.O. Expression and co-expression of the members of the epidermal growth factor receptor (EGFR) family in invasive breast carcinoma. *Br. J. Cancer* **2004**, *91*, 1532–1542. [CrossRef]

132. Chen, J.; Zhu, Y.; Zhang, W.; Peng, X.; Zhou, J.; Li, F.; Han, B.; Liu, X.; Ou, Y.; Yu, X. Delphinidin induced protective autophagy via mTOR pathway suppression and AMPK pathway activation in HER-2 positive breast cancer cells. *BMC Cancer* **2018**, *18*, 342. [CrossRef] [PubMed]

133. Ma, X.; Ning, S. Cyanidin-3-glucoside attenuates the angiogenesis of breast cancer via inhibiting STAT3/VEGF pathway. *Phyther. Res.* **2019**, *33*, 81–89. [CrossRef] [PubMed]

134. Liang, L.; Liu, X.; He, J.; Shao, Y.; Liu, J.; Wang, Z.; Xia, L.; Han, T.; Wu, P. Cyanidin-3-glucoside induces mesenchymal to epithelial transition via activating Sirt1 expression in triple negative breast cancer cells. *Biochimie* **2019**, *162*, 107–115. [CrossRef] [PubMed]

135. Samavat, H.; Kurzer, M.S. Estrogen Metabolism and Breast Cancer. *Cancer Lett.* **2015**, *356*, 231–243. [CrossRef]

136. Reis-Filho, J.S.; Tutt, A.N.J. Triple negative tumours: A critical review. *Histopathology* **2008**, *52*, 108–118. [CrossRef]

137. Zhang, X.T.; Kang, L.G.; Ding, L.; Vranic, S.; Gatalica, Z.; Wang, Z.Y. A positive feedback loop of ER-α36/EGFR promotes malignant growth of ER-negative breast cancer cells. *Oncogene* **2011**, *30*, 770. [CrossRef]

138. Wang, Z.Y.; Zhang, X.T.; Shen, P.; Loggie, B.W.; Chang, Y.C.; Deuel, T.F. A variant of estrogen receptor-α, hER-α36: Transduction of estrogen- and antiestrogen-dependent membrane-initiated mitogenic signaling. *Proc. Natl. Acad. Sci. USA* **2006**, *103*, 9063–9068. [CrossRef]

139. Wang, L.; Li, H.; Yang, S.; Ma, W.; Liu, M.; Guo, S.; Zhan, J.; Zhang, H.; Tsang, S.Y.; Zhang, Z.; et al. Cyanidin-3-o-glucoside directly binds to ERa36 and inhibits EGFR-positive triple-negative breast cancer. *Oncotarget* **2016**, *7*, 68864. [CrossRef]

140. Fernandes, I.; Faria, A.; Azevedo, J.; Soares, S.; Calhau, C.; De Freitas, V.; Mateus, N. Influence of anthocyanins, derivative pigments and other catechol and pyrogallol-type phenolics on breast cancer cell proliferation. *J. Agric. Food Chem.* **2010**, *58*, 3785–3792. [CrossRef]

141. Swain, S.M.; Baselga, J.; Kim, S.B.; Ro, J.; Semiglazov, V.; Campone, M.; Ciruelos, E.; Ferrero, J.; Schneeweiss, A.; Heeson, S.; et al. Pertuzumab, trastuzumab, and docetaxel in HER2-positive metastatic breast cancer. *N. Engl. J. Med.* **2015**, *372*, 724–734. [CrossRef]

142. Witzel, I.; Müller, V.; Abenhardt, W.; Kaufmann, M.; Schoenegg, W.; Schneeweis, A.; Jänicke, F. Long-term tumor remission under trastuzumab treatment for HER2 positive metastatic breast cancer—Results from the HER-OS patient registry. *BMC Cancer* **2014**, *14*, 806. [CrossRef] [PubMed]

143. Li, X.; Xu, J.; Tang, X.; Liu, Y.; Yu, X.; Wang, Z.; Liu, W. Anthocyanins inhibit trastuzumab-resistant breast cancer in vitro and in vivo. *Mol. Med. Rep.* **2016**, *13*, 4007–4013. [CrossRef] [PubMed]

144. Liu, W.; Xu, J.; Liu, Y.; Yu, X.; Tang, X.; Wang, Z.; Li, X. Anthocyanins potentiate the activity of trastuzumab in human epidermal growth factor receptor 2-positive breast cancer cells in vitro and in vivo. *Mol. Med. Rep.* **2014**, *10*, 1921–1926. [CrossRef] [PubMed]

145. Shen, T.; Wang, X.N.; Lou, H.X. Natural stilbenes: An overview. *Nat. Prod. Rep.* **2009**, *26*, 916–935. [CrossRef]

146. Signorelli, P.; Ghidoni, R. Resveratrol as an anticancer nutrient: Molecular basis, open questions and promises. *J. Nutr. Biochem.* **2005**, *16*, 449–466. [CrossRef]

147. Langcake, P.; Pryce, R.J. The production of resveratrol by Vitis vinifera and other members of the Vitaceae as a response to infection or injury. *Physiol. Plant Pathol.* **1976**, *9*, 77–86. [CrossRef]

148. Jeandet, P.; Bessis, R.; Gautheron, B. The production of resveratrol (3, 5, 4′-trihydroxystilbene) by grape berries in different developmental stages. *Am. J. Enol. Vitic.* **1991**, *42*, 41–46.

149. Goldberg, D.M.; Yan, J.; Ng, E.; Diamandis, E.P.; Karumanchiri, A.; Soleas, G.; Waterhouse, A.L. A global survey of trans-resveratrol concentrations in commercial wines. *Am. J. Enol. Vitic.* **1995**, *46*, 159–165.

150. Romero-Pérez, A.I.; Lamuela-Raventós, R.M.; Waterhouse, A.L.; De La Torre-Boronat, M.C. Levels of cis- and trans-Resveratrol and Their Glucosides in White and Rosé Vitis vinifera Wines from Spain. *J. Agric. Food Chem.* **1996**, *44*, 2124–2128. [CrossRef]

151. Pace-Asciak, C.; Rounova, O.; Hahn, S.E.; Diamandis, E.P.; Goldberg, D.M. Wines and grape juices as modulators of platelet aggregation in healthy human subjects. *Clin. Chim. Acta* **1996**, *246*, 163–182. [CrossRef]

152. Weiskirchen, S.; Weiskirchen, R. Resveratrol: How Much Wine Do You Have to Drink to Stay Healthy? *Adv. Nutr. An Int. Rev. J.* **2016**, *7*, 706–718. [CrossRef] [PubMed]

153. Goldberg, D.M.; Yan, J.; Soleas, G.J. Absorption of three wine-related polyphenols in three different matrices by healthy subjects. *Clin. Biochem.* **2003**, *36*, 79–87. [CrossRef]

154. Walle, T. Bioavailability of resveratrol. *Ann. N. Y. Acad. Sci.* **2011**, *1215*, 9–15. [CrossRef] [PubMed]

155. Sinha, D.; Sarkar, N.; Biswas, J.; Bishayee, A. Resveratrol for breast cancer prevention and therapy: Preclinical evidence and molecular mechanisms. *Semin. Cancer Biol.* **2016**, *40*, 209–232. [CrossRef] [PubMed]

156. Chimento, A.; De Amicis, F.; Sirianni, R.; Sinicropi, M.S.; Puoci, F.; Casaburi, I.; Saturnino, C.; Pezzi, V. Progress to improve oral bioavailability and beneficial effects of resveratrol. *Int. J. Mol. Sci.* **2019**, *20*, 1381. [CrossRef] [PubMed]

157. Nawaz, W.; Zhou, Z.; Deng, S.; Ma, X.; Ma, X.; Li, C.; Shu, X. Therapeutic versatility of resveratrol derivatives. *Nutrients* **2017**, *9*, 1188. [CrossRef] [PubMed]

158. da Costa, D.C.F.; Fialho, E.; Silva, J.L. Cancer chemoprevention by resveratrol: The P53 tumor suppressor protein as a promising molecular target. *Molecules* **2017**, *22*, 1014. [CrossRef]

159. Jang, M.; Cai, L.; Udeani, G.O.; Slowing, K.V.; Thomas, C.F.; Beecher, C.W.W.; Fong, H.H.S.; Farnsworth, N.R.; Kinghorn, A.D.; Mehta, R.G.; et al. Cancer chemopreventive activity of resveratrol, a natural product derived from grapes. *Science* **1997**, *275*, 218–220. [CrossRef]

160. Pezzuto, J.M. Resveratrol: Twenty years of growth, development and controversy. *Biomol. Ther.* **2019**, *27*, 1. [CrossRef]

161. Casanova, F.; Quarti, J.; Da Costa, D.C.F.; Ramos, C.A.; Da Silva, J.L.; Fialho, E. Resveratrol chemosensitizes breast cancer cells to melphalan by cell cycle arrest. *J. Cell. Biochem.* **2012**, *113*, 2586–2596. [CrossRef]

162. Varoni, E.M.; Lo Faro, A.F.; Sharifi-Rad, J.; Iriti, M. Anticancer Molecular Mechanisms of Resveratrol. *Front. Nutr.* **2016**, *3*, 8. [CrossRef] [PubMed]

163. Ko, J.H.; Sethi, G.; Um, J.Y.; Shanmugam, M.K.; Arfuso, F.; Kumar, A.P.; Bishayee, A.; Ahn, K.S. The role of resveratrol in cancer therapy. *Int. J. Mol. Sci.* **2017**, *18*, 2589. [CrossRef] [PubMed]

164. Ndiaye, M.; Kumar, R.; Ahmad, N. Resveratrol in cancer management: Where are we and where we go from here? *Ann. N. Y. Acad. Sci.* **2011**, *1215*, 144–149. [CrossRef] [PubMed]

165. Levi, F.; Pasche, C.; Lucchini, F.; Ghidoni, R.; Ferraroni, M.; La Vecchia, C. Resveratrol and breast cancer risk. *Eur. J. Cancer Prev.* **2005**, *14*, 139–142. [CrossRef] [PubMed]

166. Chakraborty, S.; Levenson, A.S.; Biswas, P.K. Structural insights into Resveratrol's antagonist and partial agonist actions on estrogen receptor alpha. *BMC Struct. Biol.* **2013**, *13*, 27. [CrossRef]

167. Poschner, S.; Maier-Salamon, A.; Thalhammer, T.; Jäger, W. Resveratrol and other dietary polyphenols are inhibitors of estrogen metabolism in human breast cancer cells. *J. Steroid Biochem. Mol. Biol.* **2019**, *190*, 11–18. [CrossRef]

168. Jordan, V.C. Antiestrogens and Selective Estrogen Receptor Modulators as Multifunctional Medicines. 2. Clinical Considerations and New Agents. *J. Med. Chem.* **2003**, *46*, 1081–1111. [CrossRef]

169. Gehm, B.D.; McAndrews, J.M.; Chien, P.Y.; Jameson, J.L. Resveratrol, a polyphenolic compound found in grapes and wine, is an agonist for the estrogen receptor. *Proc. Natl. Acad. Sci. USA* **1997**, *94*, 14138–14143. [CrossRef]

170. Levenson, A.S.; Gehm, B.D.; Pearce, S.T.; Horiguchi, J.; Simons, L.A.; Ward, J.E.; Jameson, J.L.; Jordan, V.C. Resveratrol acts as an estrogen receptor (ER) agonist in breast cancer cells stably transfected with ERα. *Int. J. Cancer* **2003**, *104*, 587–596. [CrossRef]

171. Lu, R.; Serrero, G. Resveratrol, a natural product derived from grape, exhibits antiestrogenic activity and inhibits the growth of human breast cancer cells. *J. Cell. Physiol.* **1999**, *179*, 297–304. [CrossRef]

172. Bowers, J.L.; Tyulmenkov, V.V.; Jernigan, S.C.; Klinge, C.M. Resveratrol acts as a mixed agonist/antagonist for estrogen receptors α and β. *Endocrinology* **2000**, *141*, 3657–3667. [CrossRef] [PubMed]

173. Gehm, B.D.; Levenson, A.S.; Liu, H.; Lee, E.J.; Amundsen, B.M.; Cushman, M.; Jordan, V.C.; Jameson, J.L. Estrogenic effects of resveratrol in breast cancer cells expressing mutant and wild-type estrogen receptors: Role of AF-1 and AF-2. *J. Steroid Biochem. Mol. Biol.* **2004**, *88*, 223–234. [CrossRef] [PubMed]

174. Simpson, E.R.; Michael, M.D.; Agarwal, V.R.; Hinshelwood, M.M.; Bulun, S.E.; Zhao, Y. Cytochromes P450 11: Expression of the CYP19 (aromatase) gene: An unusual case of alternative promoter usage. *FASEB J.* **1997**, *11*, 29–36. [CrossRef]

175. Wang, Y.; Lee, K.W.; Chan, F.L.; Chen, S.; Leung, L.K. The red wine polyphenol resveratrol displays bilevel inhibition on aromatase in breast cancer cells. *Toxicol. Sci.* **2006**, *92*, 71–77. [CrossRef]

176. Chottanapund, S.; Van Duursen, M.B.M.; Navasumrit, P.; Hunsonti, P.; Timtavorn, S.; Ruchirawat, M.; Van den Berg, M. Anti-aromatase effect of resveratrol and melatonin on hormonal positive breast cancer cells co-cultured with breast adipose fibroblasts. *Toxicol. In Vitro* **2014**, *28*, 1215–1221. [CrossRef]

177. Le Corre, L.; Chalabi, N.; Delort, L.; Bignon, Y.J.; Bernard-Gallon, D.J. Resveratrol and breast cancer chemoprevention: Molecular mechanisms. *Mol. Nutr. Food Res.* **2005**, *49*, 462–471. [CrossRef] [PubMed]

178. Sinha, R.; Caporaso, N. Diet, Genetic Susceptibility and Human Cancer Etiology. *J. Nutr.* **1999**, *129*, 556S–559S. [CrossRef]

179. Pozo-Guisado, E.; Alvarez-Barrientos, A.; Mulero-Navarro, S.; Santiago-Josefat, B.; Fernandez-Salguero, P.M. The antiproliferative activity of resveratrol results in apoptosis in MCF-7 but not in MDA-MB-231 human breast cancer cells: Cell-specific alteration of the cell cycle. *Biochem. Pharmacol.* **2002**, *64*, 1375–1386. [CrossRef]

180. Da Costa, D.C.F.; Casanova, F.A.; Quarti, J.; Malheiros, M.S.; Sanches, D.; dos Santos, P.S.; Fialho, E.; Silva, J.L. Transient Transfection of a Wild-Type p53 Gene Triggers Resveratrol-Induced Apoptosis in Cancer Cells. *PLoS ONE* **2012**, *7*, 1–12.

181. Da Costa, D.C.F.; Campos, N.P.C.; Santos, R.A.; Guedes-da-Silva, F.H.; Martins-Dinis, M.M.D.C.; Zanphorlin, L.; Ramos, C.; Rangel, L.P.; Silva, J.L. Resveratrol prevents p53 aggregation in vitro and in breast cancer cells. *Oncotarget* **2018**, *9*, 29112–29122. [CrossRef]

182. Chang, L.C.; Yu, Y.L. Dietary components as epigenetic-regulating agents against cancer. *Biomedicine* **2016**, *6*, 1–8. [CrossRef]

183. Alamolhodaei, N.S.; Tsatsakis, A.M.; Ramezani, M.; Hayes, A.W.; Karimi, G. Resveratrol as MDR reversion molecule in breast cancer: An overview. *Food Chem. Toxicol.* **2017**, *103*, 223–232. [CrossRef]

184. Singh, C.K.; George, J.; Ahmad, N. Resveratrol-based combinatorial strategies for cancer management. *Ann. N. Y. Acad. Sci.* **2014**, *1290*, 113–121. [CrossRef] [PubMed]

185. Chow, H.H.S.; Garland, L.L.; Heckman-Stoddard, B.M.; Hsu, C.H.; Butler, V.D.; Cordova, C.A.; Chew, W.M.; Cornelison, T.L. A pilot clinical study of resveratrol in postmenopausal women with high body mass index: Effects on systemic sex steroid hormones. *J. Transl. Med.* **2014**, *12*, 1–7. [CrossRef]

186. Fernandes, I.; Pérez-Gregorio, R.; Soares, S.; Mateus, N.; De Freitas, V.; Santos-Buelga, C.; Feliciano, A.S. Wine flavonoids in health and disease prevention. *Molecules* **2017**, *22*, 292. [CrossRef] [PubMed]

187. Arbulu, M.; Sampedro, M.C.; Gómez-Caballero, A.; Goicolea, M.A.; Barrio, R.J. Untargeted metabolomic analysis using liquid chromatography quadrupole time-of-flight mass spectrometry for non-volatile profiling of wines. *Anal. Chim. Acta* **2015**, *858*, 32–41. [CrossRef] [PubMed]

188. Alves Filho, E.G.; Silva, L.M.A.; Ribeiro, P.R.V.; de Brito, E.S.; Zocolo, G.J.; Souza-Leão, P.C.; Marques, A.T.B.; Quintela, A.L.; Larsen, F.H.; Canuto, K.M. 1 H NMR and LC-MS-based metabolomic approach for evaluation of the seasonality and viticultural practices in wines from São Francisco River Valley, a Brazilian semi-arid region. *Food Chem.* **2019**, *289*, 558–567. [CrossRef] [PubMed]

189. Cassino, C.; Tsolakis, C.; Bonello, F.; Gianotti, V.; Osella, D. Wine evolution during bottle aging, studied by 1H NMR spectroscopy and multivariate statistical analysis. *Food Res. Int.* **2019**, *116*, 566–577. [CrossRef]

190. Forester, S.C.; Waterhouse, A.L. Metabolites Are Key to Understanding Health Effects of Wine Polyphenolics. *J. Nutr.* **2009**, *139*, 1824S–1831S. [CrossRef]

191. Jiménez-Girón, A.; Queipo-Ortuño, M.; Boto-Ordóñez, M.; Muñoz-Gonzáles, I.; Sánches-Patán, F.; Monagas, M.; Martín-Álvarez, P.; Murri, M.; Tinahones, J.F.; Abdrés-Lacueva, C.; et al. Comparative Study of Microbial-Derived Phenolic Metabolites in Human Feces after Intake of Gin, Red Wine, and Dealcoholized Red Wine. *J. Agric. Food Chem.* **2013**, *61*, 3909–3915. [CrossRef]

192. Jiménez-Girón, A.; Ibáñez, C.; Cifuentes, A.; Simó, C.; Muñoz-González, I.; Martín-Álvarez, P.J.; Bartolomé, B.; Moreno-Arribas, M.V. Faecal metabolomic fingerprint after moderate consumption of red wine by healthy subjects. *J. Proteome Res.* **2015**, *14*, 897–905. [CrossRef] [PubMed]

193. Gonthier, M.-P.; Cheynier, V.; Donovan, J.L.; Manach, C.; Morand, C.; Mila, I.; Lapierre, C.; Rémésy, C.; Scalbert, A. Microbial Aromatic Acid Metabolites Formed in the Gut Account for a Major Fraction of the Polyphenols Excreted in Urine of Rats Fed Red Wine Polyphenols. *J. Nutr.* **2003**, *133*, 461–467. [CrossRef] [PubMed]

194. Al Ali, S.H.H.; Al-Qubaisi, M.; Hussein, M.Z.; Ismail, M.; Bullo, S. Hippuric acid nanocomposite enhances doxorubicin and oxaliplatin-induced cytotoxicity in MDA-MB231, MCF-7 and Caco2 cell lines. *Drug Des. Devel. Ther.* **2013**, *7*, 25. [CrossRef]

195. Wang, X.N.; Wang, K.Y.; Zhang, X.S.; Yang, C.; Li, X.Y. 4-Hydroxybenzoic acid (4-HBA) enhances the sensitivity of human breast cancer cells to adriamycin as a specific HDAC6 inhibitor by promoting HIPK2/p53 pathway. *Biochem. Biophys. Res. Commun.* **2018**, *504*, 812–819. [CrossRef] [PubMed]

196. Yin, M.C.; Lin, C.C.; Wu, H.C.; Tsao, S.M.; Hsu, C.K. Apoptotic effects of protocatechuic acid in human breast, lung, liver, cervix, and prostate cancer cells: Potential mechanisms of action. *J. Agric. Food Chem.* **2009**, *57*, 6468–6473. [CrossRef]

197. Cartron, E.; Fouret, G.; Carbonneau, M.A.; Lauret, C.; Michel, F.; Monnier, L.; Descomps, B.; Léger, C.L. Red-wine beneficial long-term effect on lipids but not on antioxidant characteristics in plasma in a study comparing three types of wine—Description of two O-methylated derivatives of gallic acid in humans. *Free Radic. Res.* **2003**, *37*, 1021–1035. [CrossRef]

198. Mennen, L.I.; Sapinho, D.; Ito, H.; Bertrais, S.; Galan, P.; Hercberg, S.; Scalbert, A. Urinary flavonoids and phenolic acids as biomarkers of intake for polyphenol-rich foods. *Br. J. Nutr.* **2006**, *96*, 191–198. [CrossRef]

199. Lee, H.L.; Lin, C.S.; Kao, S.H.; Chou, M.C. Gallic acid induces G1 phase arrest and apoptosis of triple-negative breast cancer cell MDA-MB-231 via p38 mitogen-activated protein kinase/p21/p27 axis. *Anticancer Drugs* **2017**, *28*, 1150–1156. [CrossRef]

200. Wang, K.; Zhu, X.; Zhang, K.; Zhu, L.; Zhou, F. Investigation of Gallic Acid Induced Anticancer Effect in Human Breast Carcinoma MCF-7 Cells. *J. Biochem. Mol. Toxicol.* **2014**, *28*, 387–393. [CrossRef]

201. Vitaglione, P.; Sforza, S.; Del Rio, D. Occurrence, Bioavailability, and Metabolism of Resveratrol. In *Flavonoids and Related Compounds: Bioavailability and Function*, 1st ed.; Spencer, J.P.E., Crozier, A., Eds.; CRC Press: Boca Raton, FL, USA, 2012; pp. 167–182.

202. Henry-Vitrac, C.; Desmoulière, A.; Girard, D.; Mérillon, J.M.; Krisa, S. Transport, deglycosylation, and metabolism of trans-piceid by small intestinal epithelial cells. *Eur. J. Nutr.* **2006**, *45*, 376–382. [CrossRef]

203. Bode, L.M.; Bunzel, D.; Huch, M.; Cho, G.S.; Ruhland, D.; Bunzel, M.; Bub, A.; Franz, C.M.A.P.; Kulling, S.E. In vivo and in vitro metabolism of trans-resveratrol by human gut microbiota. *Am. J. Clin. Nutr.* **2013**, *97*, 295–309. [CrossRef] [PubMed]

204. Lappano, R.; Rosano, C.; Madeo, A.; Albanito, L.; Plastina, P.; Gabriele, B.; Forti, L.; Stivala, L.A.; Iacopetta, D.; Dolce, V.; et al. Structure-activity relationships of resveratrol and derivatives in breast cancer cells. *Mol. Nutr. Food Res.* **2009**, *53*, 845–858. [CrossRef] [PubMed]

205. Ruotolo, R.; Calani, L.; Fietta, E.; Brighenti, F.; Crozier, A.; Meda, C.; Maggi, A.; Ottonello, S.; Del Rio, D. Anti-estrogenic activity of a human resveratrol metabolite. *Nutr. Metab. Cardiovasc. Dis.* **2013**, *23*, 1086–1092. [CrossRef] [PubMed]

206. Gakh, A.A.; Anisimova, N.Y.; Kiselevsky, M.V.; Sadovnikov, S.V.; Stankov, I.N.; Yudin, M.V.; Rufanov, K.A.; Krasavin, M.Y.; Sosnov, A.V. Dihydro-resveratrol—A potent dietary polyphenol. *Bioorganic Med. Chem. Lett.* **2010**, *20*, 6149–6151. [CrossRef]

207. Giménez-Bastida, J.A.; Ávila-Gálvez, M.Á.; Espín, J.C.; González-Sarrías, A. Conjugated Physiological Resveratrol Metabolites Induce Senescence in Breast Cancer Cells: Role of p53/p21 and p16/Rb Pathways, and ABC Transporters. *Mol. Nutr. Food Res.* **2019**, *63*, 1900629. [CrossRef]

208. Donovan, J.L.; Crespy, V.; Manach, C.; Morand, C.; Besson, C.; Scalbert, A.; Rémésy, C. Catechin Is Metabolized by Both the Small Intestine and Liver of Rats. *J. Nutr.* **2001**, *131*, 1753–1757. [CrossRef]

209. Baba, S.; Osakabe, N.; Natsume, M.; Muto, Y.; Takizawa, T.; Terao, J. In Vivo Comparison of the Bioavailability of (+)-Catechin, (−)-Epicatechin and Their Mixture in Orally Administered Rats. *J. Nutr.* **2001**, *131*, 2885–2891. [CrossRef]

210. Donovan, J.L.; Kasim-Karakas, S.; German, J.B.; Waterhouse, A.L. Urinary excretion of catechin metabolites by human subjects after red wine consumption. *Br. J. Nutr.* **2002**, *87*, 31–37. [CrossRef]

211. Kim, D.H.; Jung, E.A.; Sohng, I.S.; Han, J.A.; Kim, T.H.; Han, M.J. Intestinal bacterial metabolism of flavonoids and its relation to some biological activities. *Arch. Pharm. Res.* **1998**, *21*, 17–23. [CrossRef]

212. Kim, R.K.; Suh, Y.; Yoo, K.C.; Cui, Y.H.; Hwang, E.; Kim, H.J.; Kang, J.S.; Kim, M.J.; Lee, Y.Y.; Lee, S.J. Phloroglucinol suppresses metastatic ability of breast cancer cells by inhibition of epithelial-mesenchymal cell transition. *Cancer Sci.* **2015**, *106*, 94–101. [CrossRef]

213. Kim, R.K.; Uddin, N.; Hyun, J.W.; Kim, C.; Suh, Y.; Lee, S.J. Novel anticancer activity of phloroglucinol against breast cancer stem-like cells. *Toxicol. Appl. Pharmacol.* **2015**, *286*, 143–150. [CrossRef] [PubMed]

214. Fang, J. Bioavailability of anthocyanins. *Drug Metab. Rev.* **2014**, *46*, 508–520. [CrossRef] [PubMed]

215. Tsuda, T.; Horio, F.; Osawa, T. Absorption and metabolism of cyanidin 3-*O*-β-D-glucoside in rats. *FEBS Lett.* **1999**, *449*, 179–182. [CrossRef]

216. Olivas-Aguirre, F.J.; Rodrigo-García, J.; Martínez-Ruiz, N.D.R.; Cárdenas-Robles, A.I.; Mendoza-Díaz, S.O.; Álvarez-Parrilla, E.; González-Aguilar, G.A.; De La Rosa, L.A.; Ramos-Jiménez, A.; Wall-Medrano, A. Cyanidin-3-*O*-glucoside: Physical-chemistry, foodomics and health effects. *Molecules* **2016**, *21*, 1264. [CrossRef] [PubMed]

217. Bub, A.; Watzl, B.; Heeb, D.; Rechkemmer, G.; Briviba, K. Malvidin-3-glucoside bioavailability in humans after ingestion of red wine, dealcoholized red wine and red grape juice. *Eur. J. Nutr.* **2001**, *40*, 113–120. [CrossRef]

218. Bitsch, R.; Netzel, M.; Frank, T.; Strass, G.; Bitsch, I. Bioavailability and biokinetics of anthocyanins from red grape juice and red wine. *J. Biomed. Biotechnol.* **2004**, *2004*, 293–298. [CrossRef]

219. Marques, C.; Fernandes, I.; Norberto, S.; Sá, C.; Teixeira, D.; de Freitas, V.; Mateus, N.; Calhau, C.; Faria, A. Pharmacokinetics of blackberry anthocyanins consumed with or without ethanol: A randomized and crossover trial. *Mol. Nutr. Food Res.* **2016**, *60*, 2319–2330. [CrossRef]

220. Czank, C.; Cassidy, A.; Zhang, Q.; Morrison, D.J.; Preston, T.; Kroon, P.A.; Botting, N.P.; Kay, C.D. Human metabolism and elimination of the anthocyanin, cyanidin-3-glucoside: A ^{13}C-tracer study. *Am. Clin. Nutr.* **2013**, *97*, 995–1003. [CrossRef]

221. Fernandes, I.; Marques, F.; De Freitas, V.; Mateus, N. Antioxidant and antiproliferative properties of methylated metabolites of anthocyanins. *Food Chem.* **2013**, *141*, 2923–2933. [CrossRef]

222. Geahlen, R.L.; Koonchanok, N.M.; McLaughlin, J.L.; Pratt, D.E. Inhibition of protein-tyrosine kinase activity by flavanoids and related compounds. *J. Nat. Prod.* **1989**, *52*, 982–986. [CrossRef]

223. Hollman, P.C.; de Vries, J.H.; van Leeuwen, S.D.; Mengelers, M.J.; Katan, Martijn, K.B. Absorption of dietary quercetin glycosides and quercetin healthy ileostomy volunteers. *Am. J. Clin. Nutr.* **1995**, *62*, 1276–1282. [CrossRef] [PubMed]

224. Booth, A.N.; Murray, C.W.; Jones, F.T.; Deeds, F. The Metabolic Fate of Rutin and Quercetin in the Animal Body. *J. Biol. Chem.* **1956**, *233*, 251–257.

225. Gross, M.; Pfeiffer, M.; Martini, M.; Campbell, D.; Slavin, J.; Potter, J. The quantitation of metabolites of quercetin flavonols in human urine. *Cancer Epidemiol. Biomarkers Prev.* **1996**, *5*, 711–720. [PubMed]

226. Yamazaki, S.; Sakakibara, H.; Takemura, H.; Yasuda, M.; Shimoi, K. Quercetin-3-*O*-glucronide inhibits noradrenaline binding to α2-adrenergic receptor, thus suppressing DNA damage induced by treatment with 4-hydroxyestradiol and noradrenaline in MCF-10A cells. *J. Steroid Biochem. Mol. Biol.* **2014**, *143*, 122–129. [CrossRef] [PubMed]

227. Yamazaki, S.; Miyoshi, N.; Kawabata, K.; Yasuda, M.; Shimoi, K. Quercetin-3-*O*-glucuronide inhibits noradrenaline-promoted invasion of MDA-MB-231 human breast cancer cells by blocking β2-adrenergic signaling. *Arch. Biochem. Biophys.* **2014**, *557*, 18–27. [CrossRef] [PubMed]

228. Hadi, N.I.; Jamal, Q.; Iqbal, A.; Shaikh, F.; Somroo, S.; Musharraf, S.G. Serum Metabolomic Profiles for Breast Cancer Diagnosis, Grading and Staging by Gas Chromatography-Mass Spectrometry. *Sci. Rep.* **2017**, *7*, 1–11. [CrossRef]

229. Jové, M.; Collado, R.; Quiles, J.L.; Ramírez-Tortosa, M.C.; Sol, J.; Ruiz-Sanjuan, M.; Fernandez, M.; Cabrera, C.d.l.T.; Ramírez-Tortosa, C.; Granados-Principal, S.; et al. A plasma metabolomic signature discloses human breast cancer. *Oncotarget* **2017**, *8*, 19522. [CrossRef]

230. Liu, H.; Liu, Y.; Zhang, J.T. A new mechanism of drug resistance in breast cancer cells: Fatty acid synthase overexpression-mediated palmitate overproduction. *Mol. Cancer Ther.* **2008**, *7*, 263–270. [CrossRef]

231. Wang, J.B.; Erickson, J.W.; Fuji, R.; Ramachandran, S.; Gao, P.; Dinavahi, R.; Wilson, K.F.; Ambrosio, A.L.B.; Dias, S.M.G.; Dang, C.V.; et al. Targeting mitochondrial glutaminase activity inhibits oncogenic transformation. *Cancer Cell* **2010**, *18*, 207–219. [CrossRef]

232. Wang, G.; Li, Y.; Yang, Z.; Xu, W.; Yang, Y.; Tan, X. ROS mediated EGFR/MEK/ERK/HIF-1α Loop Regulates Glucose metabolism in pancreatic cancer. *Biochem. Biophys. Res. Commun.* **2018**, *500*, 873–878. [CrossRef]

233. Cardoso, M.R.; Santos, J.C.; Ribeiro, M.L.; Talarico, M.C.R.; Viana, L.R.; Derchain, S.F.M. A metabolomic approach to predict breast cancer behavior and chemotherapy response. *Int. J. Mol. Sci.* **2018**, *19*, 617. [CrossRef] [PubMed]

234. Gomes, L.; Viana, L.; Silva, J.L.; Mermelstein, C.; Atella, G.; Fialho, E. Resveratrol Modifies Lipid Composition of Two Cancer Cell Lines. *Biomed Res. Int.* **2020**, *2020*, 1–10. [CrossRef] [PubMed]

235. Ávila-Gálvez, M.Á.; García-Villalba, R.; Martínez-Díaz, F.; Ocaña-Castillo, B.; Monedero-Saiz, T.; Torrecillas-Sánchez, A.; Abellán, B.; González-Sarrías, A.; Espín, J.C. Metabolic Profiling of Dietary Polyphenols and Methylxanthines in Normal and Malignant Mammary Tissues from Breast Cancer Patients. *Mol. Nutr. Food Res.* **2019**, *63*, 1801239. [CrossRef] [PubMed]

 *molecules*

Review

Role of Photoactive Phytocompounds in Photodynamic Therapy of Cancer

Kasipandi Muniyandi [1,2], Blassan George [1], Thangaraj Parimelazhagan [2] and Heidi Abrahamse [1,*]

[1] Laser Research Centre, Faculty of Health Sciences, University of Johannesburg, 17011, Doornfontein 2028, South Africa; kasim@uj.ac.za (K.M.); blassang@uj.ac.za (B.G.)
[2] Bioprospecting Laboratory, Department of Botany, School of Life Sciences, Bharathiar University, Coimbatore, Tamil Nadu 641046, India; drparimel@buc.edu.in
* Correspondence: habrahamse@uj.ac.za

Academic Editor: José Antonio Lupiáñez
Received: 30 July 2020; Accepted: 4 September 2020; Published: 8 September 2020

Abstract: Cancer is one of the greatest life-threatening diseases conventionally treated using chemo- and radio-therapy. Photodynamic therapy (PDT) is a promising approach to eradicate different types of cancers. PDT requires the administration of photosensitisers (PSs) and photoactivation using a specific wavelength of light in the presence of molecular oxygen. This photoactivation exerts an anticancer effect via apoptosis, necrosis, and autophagy of cancer cells. Recently, various natural compounds that exhibit photosensitising potentials have been identified. Photoactive substances derived from medicinal plants have been found to be safe in comparison with synthetic compounds. Many articles have focused on PDT mechanisms and types of PSs, but limited attention has been paid to the phototoxic activities of phytocompounds. The reduced toxicity and side effects of natural compounds inspire the researchers to identify and use plant extracts or phytocompounds as a potent natural PS candidate for PDT. This review focusses on the importance of common photoactive groups (furanocoumarins, polyacetylenes, thiophenes, curcumins, alkaloids, and anthraquinones), their phototoxic effects, anticancer activity and use as a potent PS for an effective PDT outcome in the treatment of various cancers.

Keywords: photodynamic therapy; cancer; photosensitiser; natural compounds

1. Introduction

Cancer is one of the deadliest diseases reported in developed as well as developing countries [1]. It is mainly characterised by the uncontrolled cell growth and development of normal cells due to genetic alterations or exposure to the carcinogenic substances. The mutation of normal cells leads to abnormal cellular proliferation and develops into tumour [2]; this can be either benign, premalignant (non-cancerous) or malignant (cancerous) [3,4]. Presently surgery, radiotherapy, and chemotherapy either as monotherapy or as combined treatments are used in the treatment of cancer. However, these treatments frequently stimulate redundant side effects [2]. Many of the current chemotherapeutic drugs are of low molecular weight with high pharmacokinetic profiles [4]. Hence, in order to achieve the bioavailability and cytotoxicity induction, the drugs are administrated in high concentrations. In photodynamic therapy (PDT), photoactive drugs are generally administered systemically, but because of the precise application of light from the laser source, the cytotoxicity is attained in the tumour location. Due to the lesser drug specificity and toxicity to healthy cells, the chemotherapeutic drugs used in cancer treatments need to be improved.

Photodynamic therapy (PDT) is a promising minimally invasive therapy for the treatment of cancer. This involves the administration of photosensitiser (PS) and subsequent excitation of PS by light irradiation at a specific wavelength. The excited PS then reacts with cellular oxygen and produces reactive oxygen species (ROS). This reaction results in oxidising the cellular macromolecules surrounding tumour cells [5]. This remedial method has been developed over the last few years [6,7], and has not only been utilised in cancer treatment, but also in dermatological [8] and ophthalmic [9] conditions, including psoriasis and age-related diseases [10–12]. The use of photodynamic therapy to treat cancers has gained attention around the world [13,14]. The mechanism of PDT is based on various photocatalytic reactions that induce the destruction of cancer cells, and it has been clinically used for the treatment of cancer for over a decade [5]. In the first clinical PDT study reported by Granelli et al. [15], hematoporphyrin was used as a potent photosensitiser (PS) against glioma cancer cells. PDT destroys cancer cells through three fundamentally different pathways, namely, by damaging cancer cells over time, damaging vascular tissues that supply oxygen to cells, and finally by activating host immune response systems [13,16]. Combining PDT with chemotherapy, radiotherapy, and herbal therapy could be an emerging future methodology in cancer treatment. The combination therapy has more of a tendency to reduce the side effects when compared to monotherapy regimes and can significantly lower cancer cell proliferation by improving the drug uptake [17].

Since ancient times, herbal medicine from natural products has been utilised for treating various human ailments [18]. Most current medicines are derived from various medicinal plants, and it is evident that herbal extracts and their compounds should be examined as possible active lead components in cancer drug discoveries [19,20]. Nature is a valuable reserve for medicinal plants, and many of the pharmaceutically active compounds isolated from medicinal plants have not been tested for photoactive properties. There have been few studies attempting to identify new chemical compounds with photoactivity from plant extracts that can be used as potent natural PSs [21–25]. Hypericin (isolated from *Hypericum perforatum*) is a recognised plant-based PS used in PDT. The in vitro and in vivo studies reported that hypericin PS activated at 594 nm could destroy cancer cell proliferation effectively. The researchers already demonstrated that the effect of herbal extracts combined with illumination could significantly reduce cancer development by prohibiting metabolic viability and proliferation cancer cells [26–29].

Due to the low or no adverse side effects, herbal products have been used for the treatment of many more ailments than synthetic drugs. Studies have shown that plant-based compounds could be used in the treatment of various cancers [30]. Many phototoxic substances were subsequently reported in various plant species that are equally efficient as of conventional PSs [31]. These studies recommend that natural compounds with photosensitising abilities can be isolated from plants and used as alternatives for conventional PSs used in PDT. In this review, the underlying principles of PDT, PSs and plant-based photoactive compounds were addressed. This review mainly focused on the anticancer activity of furanocoumarins, polyacetylenes, thiophenes, curcumins, alkaloids and anthraquinones in relation to the light-absorbing properties.

2. Basic Principles of Photodynamic Therapy

Photodynamic therapy involves coordination with three individual factors, namely, the photosensitiser, oxygen, and light [7]. These components are not toxic to cells individually, but when irradiated, these can initiate a photochemical reaction that generates highly reactive singlet oxygen ($^1O_2$) and cause significant toxicity, leading to cell death. PDT is normally described in two stages, the first is administration of the PS and the second stage is the irradiation. Generally, the effect of PDT is affected by the PS type, dosage, light fluence, as well as exposure time. PDT can be used either before or after chemotherapy, radiotherapy, or surgery without compromise. The clinically approved PS should not accumulate in the body and does not develop resistant cancer cells. Pain during administration and continuous photosensitisation are the major drawbacks of PDT treatment. There are three types of lights ranges from 600 to 800 nm that are commonly used in PDT, namely, blue, red, and infrared lights. Among them, blue light penetrates the tissue the least when compared to red and infrared lights. The wavelengths below 800 nm are mostly used in PDT than higher wavelengths (above 800 nm) due to their lack of photodynamic reactions. The choice of light source is commonly based on PS nature, absorption spectra of PS, location, and size and characteristics of the infected tissue [7,32].

More than 300 chemical compounds have already been identified as potential candidates to be used as PSs. Amongst these, a few were authorised for clinical application in PDT, and others were medically evaluated, whereas some are still under examination [33,34]. We have tabulated some PSs which are used in various cancer treatments in Table 1. Photosensitisers are naturally or chemically produced compound conjugated with a visible light-absorbing chromophore group with a strong chemical absorbance. Choosing the correct PS is the most important phase in PDT for a successful outcome [33,34]. The purity and the presence of a tetrapyrrole structure with good storage stability are the preferable properties of most PSs used in PDT. The potent and effective PS should have the ability to initiate a photodynamic reaction after irradiation with 600–800 nm lights and should not cause any toxicity under dark conditions. It should be easily distinguishable from the body with no or minimum phototoxic side effects [35]. The better diffusion of PS through the cells after long administration might contribute to the effectiveness of PDT [36]. The production of a significant amount of ROS after irradiation that induces apoptosis with less inflammation is most likely to be a suitable PS for PDT application [37,38].

When a PS is subjected to a particular wavelength light, the electron of the outermost orbital will be shifted from the ground state (S0) to the first excited state (S1). Subsequently, the electromagnetic propulsion switches the molecule to an excited triplet state (T1) with a longer life span (Figure 1). In each of these excited states, PSs are quite unstable and lose their energy in the form of fluorescence, phosphorescence, and internal heat conversion. PSs in the T1 state may react photochemically in any of the two pathways. In the type 1 pathway, the excited PS reacts through an electron transfer process with the surrounding oxygen, which ultimately leads to generation of reactive oxygen species (ROS). Such free radicals communicate readily with the biomolecules (lipids, peptides, proteins, and nucleic acids) and destroy them [39,40]. In contrast, in the type 2 pathway, the energy is directly transferred from the T1 state of the PS to the S0-state oxygen. This results in the ground-state PS transformation and excited-state singlet reactive oxygen. The disruption caused by PDT is local because both singlet oxygen as well as free radicals have a short half-life between 10–300 nanoseconds and a small diffusion distance of 10–55 nm [41].

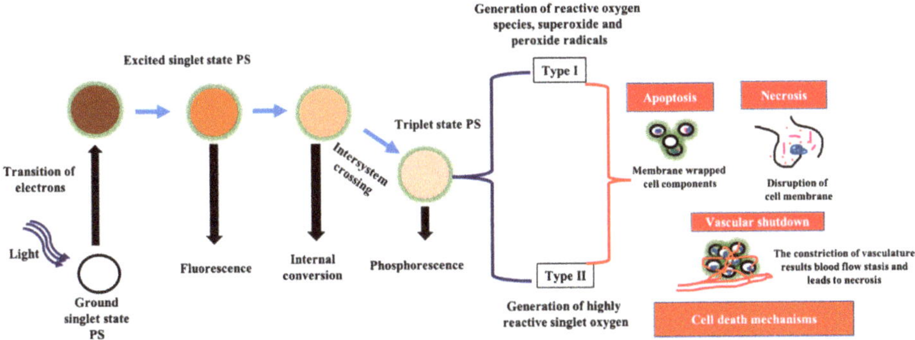

Figure 1. The general mechanism of photodynamic therapy.

Table 1. List of photosensitisers used in photodynamic therapy of various cancers.

Photosensitiser	Commercial Name	λ max (nm)	Structure	Type of Cancer	Reference
First-Generation Photosensitiser					
Hematoporphyrin derivatives	Photofrin Photoheme	630	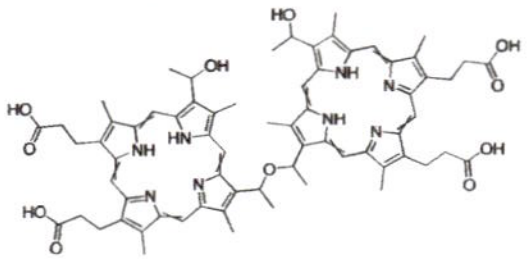	Lung, bladder, skin, cervical, breast cancer.	[17,42]
Second-Generation Photosensitisers					
5-Aminolevulinic acid	Levulan Alasens	635	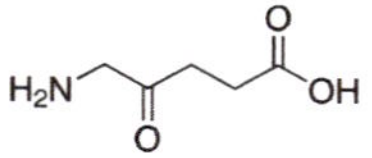	Bladder, skin, lung, ovary and gastrointestinal cancer.	[43–45]
Meta-tetra(hydroxyphenyl) chlorin	Foscan	652	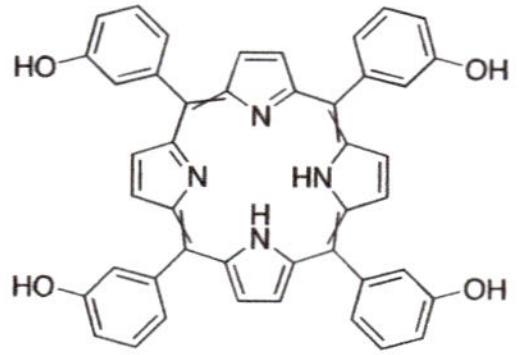	Approved drug for the treatment of bronchial and oesophageal cancers.	[46–48]
Chlorin e6	MACEDACEPhotoditazine	664	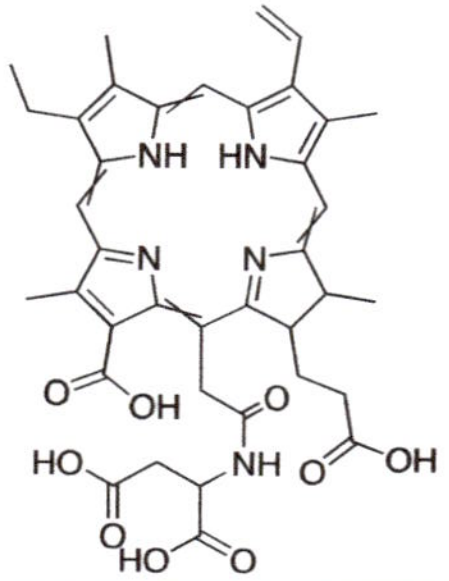	Gynaecological diseases, prostate cancer, fibrosarcoma, Liver, brain, lung, and oral cancers.	[49–51]

Table 1. Cont.

Photosensitiser	Commercial Name	λ max (nm)	Structure	Type of Cancer	Reference
Benzoporphyrin	Visudyne	690		Prostate and skin cancer.	[52,53]
Texaphyrins	Lutrin, Antrin, Optrin, Xcytrin	720–760		Hepatocellular cancer, leukaemia, nasopharyngeal carcinoma, colon, prostate, bronchial and oesophageal cancers.	[54–58]
Phthalocyanines	Photosense	640–690		Breast, cervical, skin, lung, liver, colon and gastrointestinal cancers.	[17,59–61]
Purpurins	Purlytin	660		Breast cancer, prostate cancer and Kaposi's sarcoma.	[62–64]

3. PDT's Cancer Cell Death Mechanism

PDT's cancer cell death mechanism starts after the activation of administrated PS by a specific wavelength of light. The PS's hydrophilic, hydrophobic, and ionic charge-related interaction nature plays an important role in the targeting of particular cancer cell receptor (globulins and Low-Density Lipoprotein (LDL) receptors) [34]. After the activation of PSs, the cancer cell death mechanism might occur in three main pathways (Figure 1), namely, apoptosis, necrosis, and autophagy [33,65,66]. However, the level of cell death induced by PDT may be affected by various aspects, including subcellular localisation, bioavailability, the physicochemical nature of the PS, the cellular oxygen concentration, as well as the applied light intensity and wavelength [67]. In general, the light-absorbed PS interacts with cellular oxygen and highly produce ROS (hydroperoxides, superoxide, or hydroxyl radicals) as well as singlet oxygen ($^1O_2$). These produced ROS can induce cancerous cell death via the above-mentioned mechanisms. Both type 1 and 2 reactions may occur separately or in combination, but type 1 (generation of ROS followed by the apoptotic cell death mechanism) is commonly exhibited by most approved PSs [67].

4. PS from Natural Resources

The effectiveness of PDT is mainly based on the PS; it should possess all the properties of the PS as previously explained. The PS can be divided into first- and second-generation types. Hematoporphyrin and its derivative Photofrin®® were classified as first-generation PSs. After extensive studies, new and improved second-generation PSs, such as Levulan®®, Alasens®®, and Foscan have been introduced for PDT application (Table 1). Although these are widely used for various cancer treatments, their clinical usage is limited by various drawbacks such as lack of chemical purity, a longer half-life, accumulation in tissues and poor ability in relation to depth of tissue penetration [31–39].

Subsequently, there are some research reports on PSs with potent pharmaceutical properties to overcome the shortcomings of first- (Porphyrin based sensitisers) and second-generation (non-porphyrin derivatives) PSs [35–37]. These drawbacks of current PSs specifically imply the need for new PSs as anticancer agents from natural resources. The discovery of new PS compounds with anticipated pharmacological properties and clinical application is an inspiring task. Recently, a greater number of plant-based compounds have been reported for their anticancer activity, and these compounds are pharmaceutically very important for the development of potent drugs. The use of light to activate the bioactivities of natural products is generally called photopharmacology (a combination of photophysics and photochemistry). The absorption of lights ($\lambda < 350$ nm) by a molecule mainly depends on the chromophore compound attached (Figure 2) [36,37]. This review presents an overview of natural photoactive compounds as potent third-generation photosensitisers in the improvement of PSs in relation to their prospective application in cancer treatments.

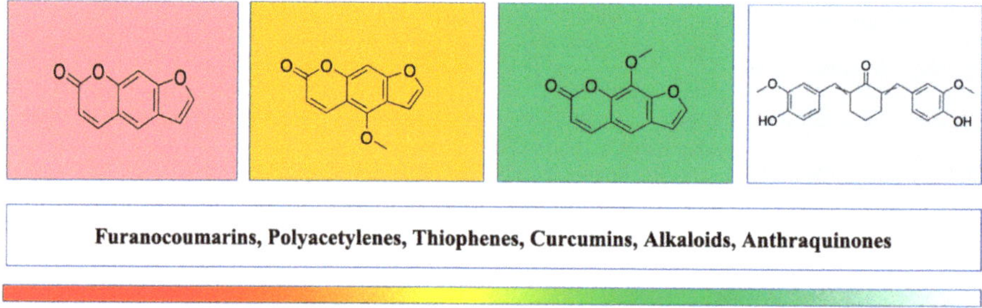

Figure 2. Phototherapeutic window of natural compounds.

5. Natural Photoactive Compounds from Plants

The search for the natural compounds as efficient PSs has been progressively moving forward because of the side effects caused by current synthetic drugs. The advanced isolation, identification and characterisation techniques improved the extraction of desirable compounds from plants. Recently, using these advanced techniques, the isolation of natural photoactive compounds has become easy. Although there have been few studies attempting to identify new chemical compounds with photoactivity from plant extracts, this review discusses the photoactivity as well as the anticancer activity of some plant-based compounds such as furanocoumarins, polyacetylenes, thiophenes, curcumins, alkaloids and anthraquinones (Table 2).

Table 2. List of plant-based natural photoactive compounds with known photoactivity.

Name	Absorption Maxima	Chemical Property and Groups	Natural Sources	Possible Mode of Action	Reference
Furanocoumarins	333 nm	Aromatic compounds possessing a furan ring.	*Angelicae dahuricae, Tetradium daniellii, Glehnia littoralis, Heracleum persicum, Syzygium Sps, Ruta graveolens, Ficus sps.*	DNA intercalation under dark type 2 PDT reaction. Crosslinking and adduct formation with DNA and RNA. Cell membrane damage.	[68–70]
Polyacetylenes and Thiophenes	488 nm	Furanoacetylenes thiarubrines, thiophenes, polyacetylene (aliphatic compounds with more than three conjugated triple bonds), thiophenes (aromatic acetylenes; e.g., phenylheptatriyne).	*Asteraceae spp, Heliopsisa, Rudbeckia spp, Arnica, Centaurea scabiosa, Tagetes erecta, Porophyllum obscurum, Echinops, Bidens, Ambrosia chamissonis, T. minuta, E. latifolius, E. sgrijissi, Rhaponticum uniflorum.*	Membrane damage or erythrocyte leakage; type 1 and type 2 PDT reaction, as well as type 1 and 2 PDT mixed reaction.	[71–75]
Curcumins	420–480 nm	Dicinnamoylmethane, curcumin, curcuminoids, demethoxycurcumin, bisdemethoxycurcumin.	*Curcuma longa.*	Cell membrane is the primary target of curcuminoids. Induction of caspase-mediated cell death.	[76–78]
Alkaloids	360 nm	Chinolin alkaloids, pterins, benzylisoquinolines, beta-carbolines, harmine.	*Guatteria blepharophylla, Berberis vulgaris, Sanguinaria Canadensis, Mahonia aquifolium Peganum harmala, Indigofera tinctoria.*	Photo-oxidises histidine and tryptophan, resulting in DNA crosslinking. Photooxidation, type 1 PDT mechanism and targets mitochondria.	[79–87]
Anthraquinones	437 nm	Hydroxyanthraquinones, rhein, physcion, emodin, rubiadin, damnacanthol, soranjidiol, alizarin, purpurin, rubiadin, aloe-emodin, 1,5-dihydroxy przewalsquinone B, ziganein, uredinorubellins, caeruleoramularin, hypericin, cercosporin, elsinochromes A-C pleichrome, hypocrellin.	*Polygonum cuspidatum, Heterophyllaea pustulata, H. lycioides Aloe vera, Rheum palmatum, Rumex crispus Polyathia suberosa, Dactylopius coccus, Xanthoria parietina, Drechslera avenae, Ramularia collo-cygni. H. perforatum, Fagopyrum esculentum.*	Type 1 and 2 PDT action.	[88–91]

5.1. Furanocoumarins

The secondary metabolites, furanocoumarins (FC; Figure 3), are mostly present in higher plants. The photoactive furanocoumarins were mainly composed of a linear core, and the biological distribution, photochemistry and phototoxicity mechanisms of FC after PUVA (psoralen and long-wave ultraviolet radiation) irradiation were reported in previous study [92]. In terms of phototherapy, psoralen is activated in the wavelength range of 300–400 nm ultraviolet radiation to treat psoriasis, dermatitis, eczema, and other skin problems [92]. Over a few years, many researchers reported anticancer activity of FCs against various types of cancer such as breast, skin, and leukaemia. FCs modulate several pathways inducing cancer cell death by inhibiting signal transducer and activator of transcription 3 (STAT3), nuclear factor-κB (NF-κB), phosphatidylinositol-3-kinase and AKT protein expression (Figure 4). These pathways play a key role in tumour development through regular activation of several inflammatory genes. Studies show that FC displayed potent activity against breast cancer development by inhibiting STAT3 protein expression [93]. Panno et al. [92,93], demonstrated inhibition of breast cancer cell growth in a dose-dependent manner through activation of p53 and Bax, leading to the cleavage of caspase 9. In contrast, in leukaemia cells, FC inactivated the JAK (Janus-activated kinase), protein c-Src, and STAT3, and downregulated Bcl-xl and Bcl-2 proteins which are responsible for apoptosis [93–96]. The enhanced activity against the malignant melanoma cell line (A375) after UV irradiation of plant extracts containing FC also supported the possible photoactive nature of FCs [93–96]. The linear forms of furanocoumarins like psoralen and its derivatives 5-methoxypsoralen (5-MOP) and 8-methoxypsoralen (8-MOP) are reported to increase the cytotoxicity after irradiation by ultraviolet light in the 320–400 nm wavelength range against cutaneous T-cell lymphoma [97,98], and photoactivated psoralens induce apoptosis by forming adducts with DNA. This leads to the activation of p21waf/Cip and p53 and subsequently leads to cell death by the release of mitochondrial cytochrome c. The photoactivation of psoralen can also cause cell death by blocking oncogenic receptor tyrosine kinase signalling and the PI3K pathway by interfering with efficient recruitment of effector Akt kinase to the activated plasma membrane [92–104]. PUVA treatment was found effectively against B16F10 murine melanoma cells by cell cycle arrest in G2/M phases [93–96].

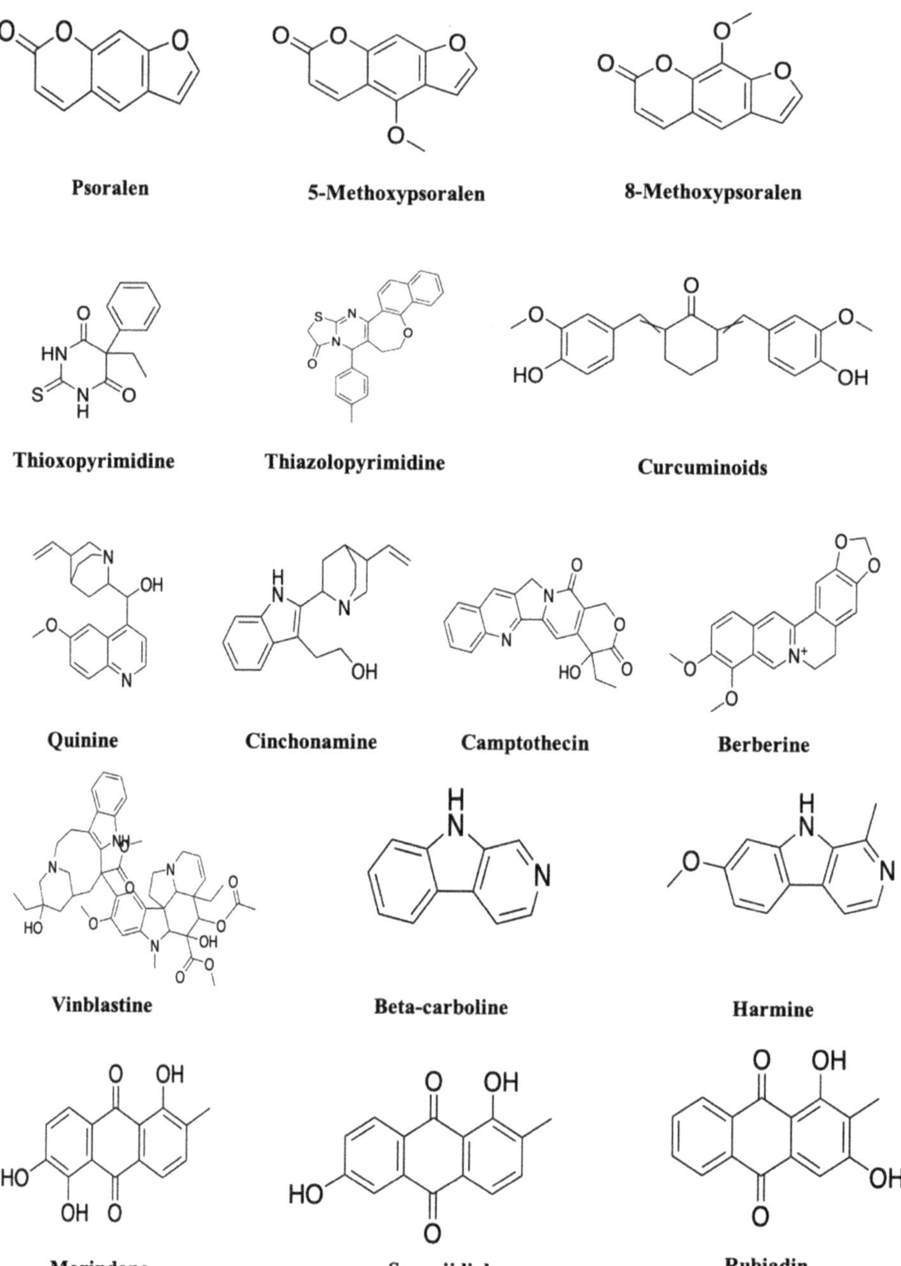

Figure 3. Natural photoactive compounds presented in this review.

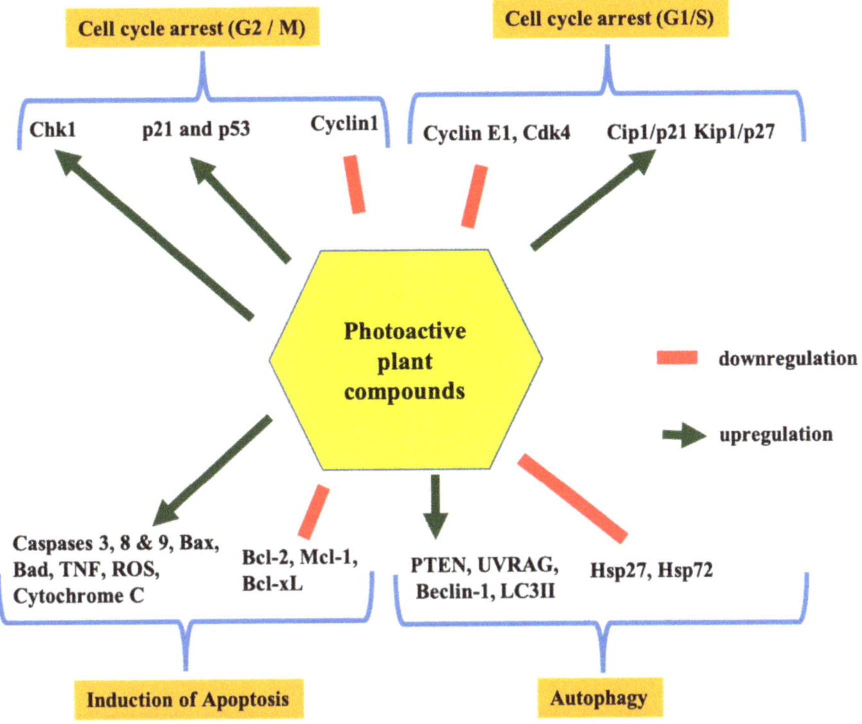

Figure 4. Common molecular targets of major photoactive compounds.

5.2. Polyacetylene and Thiophenes

These group compounds are characterised by a triple bond carbon–carbon molecule [75] and thiophenes compounds. Generally, the aliphatic compounds conjugated with three or more acetylenic bonds are considered phototoxic in nature. Among these, polyacetylenes compounds can produce $^1O_2$ under irradiation and thiophenes can provide high photo yield, leading to type 2 PDT reaction yields [105]. The polyacetylene and thiophenes compounds were reported to be activated or excited at a wavelength range of 314–350 nm absorbance maximum for the relevant photobiological effects [75]. The derivatives of these compounds were reported with a variety of potent biological activities, including analgesic, anti-inflammatory, antitumour, and antimicrobial activities. Some of the derivatives substituted by pyrimidines show antimicrobial, anti-inflammatory, and antitumour activities. A few numbers of thiophenes were reported for their cytotoxic effects against human cancer cell lines. *Echinops grijisii* root-derived thiophenes exhibited cytotoxicity against HL-60, K562 and MCF-7 cells [106]. Notably, derivatives such as thioxopyrimidine and thiazolopyrimidine were reported to possess anticancer activities against MCF-7 (breast adenocarcinoma), NCI-H460 (non-small cell lung cancer), and SF-268 (CNS cancer) cells. These acetylenic compounds and derivatives when combined with PDT might improve the efficacy of various cancer treatments [107–112]. The UV irradiation of some thiophenes also showed increased cytotoxic activities [113]; this might be due to their instable nature under UV radiation. Hence, the UV irradiation of polyacetylene and thiophene compounds can form free radicals that would induce cell death. Due to the ability to produce ROS after irradiation, these compounds can be used as an alternative PS from natural sources.

5.3. Curcumins

Curcumin (CU) is a plant-based therapeutic compound isolated from rhizome of *Curcuma longa* of the Zingiberaceae family. Curcuminoid is one of the most extensively studied plant-derived bioactive compounds [114]. Since the 1980s, the photobiological potential of CU was of great interest [115,116], and studies described CU as a desirable, highly promising photosensitiser [115,117–121]. The foremost property of CU is that it is biologically safe even at higher doses, and it can be easily produced on a large scale [117,122]. The photobleaching analysis reported the degradation profile of curcumin derivatives and its ability to produce singlet oxygen species [123]. CU was characterised by an absorption spectrum of 300–500 nm with a high extinction coefficient. This suggests that CU can induce a strong phototoxic reaction even at lower concentrations [124–126]. Curcumin is considered as a potential anticancer agent and inhibits cancer cell proliferation in breast, lung, colon, kidney, ovary, and liver cancers [114]. The in vitro and in vivo anticancer activity of curcumin has been proved by inhibition of various transcription factors such as NF-κB, AP-1, VEGF, iNOS, COX-2, 5-LOX, MMP-2, MMP-9 and IL-8, which are mainly responsible for angiogenesis and tumour growth [127,128]. The administration of curcumin significantly reduced the expression of the CDK4/cylin D1 complex by inhibiting p53 expression and causing the apoptotic process by inducing ROS generation. Furthermore, the enhanced antitumour activity was noted after UVB irradiation of curcumin by caspase activation on HaCaT (human keratinocyte cell) cells. It was also efficient against MCF-7 breast cancer cells at 30 J/cm^2 [77,78]. These data suggested that curcumin may act as a potent anticancer agent by preventing cancer progression, migration and invasion [129–131]. Dovigo et al. [132] found the light absorption ability of CU in the range of 300 and 500 nm with a maximum absorption at 430 nm, which might support its usage as a PS. The ROS-inducing and anticancer ability of CU makes it a potent candidate as a natural PS [133]. Nevertheless, Chan and Wu [134] observed that the photoactive nature of CU on human epidermal A431 carcinoma cells and the higher amount of CU also affect the irradiation penetration [132,134,135]. The irradiation of CU under a 290–320 nm UVB light source with the fluence of 100 mJ/cm^2 induced apoptosis in HaCaT keratinocyte cells [78]. Based on the above reports, CU can be used as a natural PS, and it can achieve high efficacy at a low concentration when combined with PDT. The existing PDT and photoactive reports on CU suggest that CU can be used as a potential and promising natural PS in PDT. In conclusion, CU can be a potent photosensitiser in the treatment of cancer and skin infections. Therefore, investigating the photodynamic potential of CU derivatives in terms of higher absorption and extinction coefficient will contribute to the increased efficacy of photodynamic toxicity.

5.4. Alkaloids

Alkaloids, a diverse secondary metabolites group from higher plants, contain a heterocyclic structure with a nitrogen atom in the ring [136]. Nitrogen-containing alkaloids are normally photoactive in nature, e.g., quinine and cinchonamine. The alkaloids were reported for many significant properties, such as analgesic and anticancer activity [136–139]. The alkaloids camptothecin and vinblastine are few alkaloids were successfully utilised as chemotherapeutic drugs [138,140]. The anticancer activity of alkaloids was proved by different studies by means of disturbing tumour progression by induction of cell cycle arrest at the G1 or G2/M phases, regulating cyclin-dependent kinase (CDK) and promoting apoptosis as well as autophagy in tumour cells. Furthermore, these compounds induce apoptosis by regulating Bax, Bcl-2, Bcl-xL, NF-κB and various caspase proteins [140–143]. In addition, the combination of alkaloids with chemotherapeutic drugs and irradiation also enhanced the biological activities [144,145]. Furthermore, alkaloids induce the formation of intracellular ROS in cancer cells, which leads to the destruction of cancer cell metabolism [140–143]. The photochemically best-known alkaloid is berberine; Luiza Andreazza et al. [146], Bhattacharyya et al. [147] and Inbaraj et al. [81] reported the antitumour activity of berberine upon UV and blue light irradiation. The irradiation of berberine at 410 nm proved to be effective in controlling brain cancer cell growth [146]. Beta-carboline and harmine are also a noticeable alkaloid with a photoactive nature and are reported to produce a significant amount of ROS after irradiation [148], which is considered an important feature of potent PSs. The photoactivity of harmine was proved by the UVA (long-wave ultraviolet radiation) irradiation against tumour cell lines [148]. Berberine was extensively investigated as a potential photosensitising agent for PDT [149–151]. The fluorescent active nature of berberine is indicated for its efficiency in PDT [149]; thus, berberine and its associated alkaloids can be used as a new candidate for photodynamic therapy [150]. Different studies have proved the photosensitising as well as ROS generation ability of alkaloids in the presence of a light source [151]. Therefore, berberine can be studied as a natural photosensitiser in PDT applications with minimal side effects.

5.5. Anthraquinones (AQ)

Anthraquinone are the largest group among natural quinones from higher plants, which, including naphthoquinones and benzoquinones, includes over 700 compounds, including emodin, physcion, catenarin and rhein [152,153]. The hydroxylation pattern, however, dictates the possibility of AQs' photopharmacological properties. Notably, AQs' aminoanthraquinone derivatives were studied extensively for their photoactive properties among the plant compounds due to their UV/vis absorption and photosensitising nature [154,155]. The AQs were reported as kinase and tyrosinase inhibitors as well as cytotoxicity agents. The *M. elliptica* AQs such as morindone, soranjidiol and rubiadin were also reported for their antitumour activity against lymphocytic leukaemia (P-388) cells [156]. The anthraquinones isolated from *H. pustulata* leaves and stem exhibited photosensitising properties by generation of singlet oxygen and/or superoxide anion radicals [157]. Comini et al. [158] reported that irradiation of AQs (soranjidiol and rubiadin) under visible radiation of 380–480 nm can promote the anti-proliferative effect on MCF-7 breast cancer cells. In addition, Montoya et al. [157] and Vittar et al. [159] also reported photosensitisation effects of AQs in Balb/c mice and their leukocyte-inhibiting ability in a dose-dependent manner by inducing apoptosis, necrosis, or autophagy. These study results show the photoactive nature of AQs to inhibit the proliferation of cancerous cells. Based on the previous studies and the above data, molecular targets responsible for the anticancer activity of AQs and major phytocompounds are summarised in Figure 4.

6. Theorical Studies for Assessing the Photoactivity of Natural Compounds

The development of various antitumor compounds with different molecular targets initiated an exciting field of investigation with recently developed theoretical studies. The theoretical studies including density functional theory (DFT) and time-dependent density functional theory (TD-DFT) were used to assess a series of photophysical properties, including absorption spectra, excitation energies (singlet and triplet) and spin–orbit matrix elements. All the reported compounds are potential UVA chemotherapeutic agents which require the lowest triplet-state energy for producing highly cytotoxic ROS [160,161].

7. Advantages and Scope of Natural PSs

The anticancer property of many plant extracts and bioactive compounds have been analysed, but not so much in terms of as sources of photosensitisers. Selecting proper PSs is the first step in PDT, and, to date, only a few PSs are clinically approved, such as Photofrin, Foscan and Levulan. The present study explored the common photoactive nature of various phytocompounds. Many of the natural photoactive compounds were reported for their non-toxicity against normal cells and toxicity towards cancer cells. The important property of a PS is the nontoxic nature during the absence of light. The increasing activity of extracts or phytocompounds after irradiation by light makes them good photosensitising candidates for PDT. Another important feature that makes photoactive plant compounds suitable photosensitisers is their absorption maxima at 400–700 nm, which is biologically compatible. The selective nature of these compounds is important in clinical PDT to overcome side effects. Future studies are warranted to isolate and evaluate these specific photoactive compounds from plants to be used as a potent PSs for PDT for cancer and related disorders [162,163].

8. Conclusions and Future Perspectives

As discussed in this review, plant-based photoactive compounds can be used as a natural PSs in PDT application. There are wide range of unknown natural compounds with different photoactive and phototoxic properties. This review summarises and encourages researchers to identify and elucidate natural photoactive plant-based compounds and to use them as alternatives for the synthesis PSs for a better PDT outcome. Furthermore, discovering natural phototoxic agents as PSs will be helpful to reduce toxicity and side effects and improve selectivity. In conclusion, use the plant-based PSs in PDT typically causes less and minimal adverse effects than other treatments that are commonly used in cancer therapies.

Author Contributions: Conceptualisation and writing, K.M.; review and editing, K.M., B.G., T.P. and H.A.; supervision, B.G., T.P. and H.A. The final version of the submitted manuscript was read and agreed by all the authors. All authors have read and agreed to the published version of the manuscript.

Funding: This work is supported by Science and Engineering Research Board (SERB), Department of Science and Technology (DST), Government of India in the form of SERB—Overseas Visiting Doctoral Fellowship (ODF/2018/000072). This work is also based on the research supported by the South African Research Chairs Initiative of the Department of Science and Technology and National Research Foundation of South Africa (Grant No. 98337).

Acknowledgments: The authors sincerely thank the Science and Engineering Research Board (SERB), Department of Science and Technology (DST), Government of India and the Laser research centre, University of Johannesburg, South Africa for their support.

Conflicts of Interest: The authors declare no conflict of interest.

References

1. El-Hussein, A.; Harith, M.; Abrahamse, H. Assessment of DNA Damage after Photodynamic Therapy Using a Metallophthalocyanine Photosensitizer. *Int. J. Photoenergy* **2012**, *2012*, 1–10. [CrossRef]
2. Klug, W.S.; Cummings, M.R.; Spencer, C.A. *Concepts of Genetics*, 8th ed.; Pearson Education International: Upper Saddle River, NJ, USA, 2006.
3. Matés, J.M.; Segura, J.A.; Alonso, F.J.; Márquez, J.D. Intracellular redox status and oxidative stress: Implications for cell proliferation, apoptosis, and carcinogenesis. *Arch. Toxicol.* **2008**, *82*, 273–299. [CrossRef] [PubMed]
4. Santiago-Montero, R.; Sossa-Azuela, H.; Gutiérrez-Hernández, D.; Zamudio, V.; Hernández-Bautista, I.; Valadez-Godínez, S. Novel Mathematical Model of Breast Cancer Diagnostics Using an Associative Pattern Classification. *Diagnostics* **2020**, *10*, 136. [CrossRef] [PubMed]
5. Zhou, Z.; Song, J.; Nie, L.; Chen, X.S. Reactive oxygen species generating systems meeting challenges of photodynamic cancer therapy. *Chem. Soc. Rev.* **2016**, *45*, 6597–6626. [CrossRef] [PubMed]
6. Baskaran, R.; Lee, J.; Yang, S.-G. Clinical development of photodynamic agents and therapeutic applications. *Biomater. Res.* **2018**, *22*, 25. [CrossRef] [PubMed]
7. De Almeida, D.R.Q.; Terra, L.F.; Labriola, L.; Dos Santos, A.F.; Baptista, M.S. Photodynamic therapy in cancer treatment—An update review. *J. Cancer Metastasis Treat.* **2019**, *2019*, 10–20517. [CrossRef]
8. Nguyen, K.; Khachemoune, A. An update on topical photodynamic therapy for clinical dermatologists. *J. Dermatol. Treat.* **2019**, *30*, 732–744. [CrossRef]
9. Blasi, M.A.; Pagliara, M.M.; Lanza, A.; Sammarco, M.G.; Caputo, C.G.; Grimaldi, G.; Scupola, A. Photodynamic Therapy in Ocular Oncology. *Biomedicines* **2018**, *6*, 17. [CrossRef]
10. Choi, Y.M.; Adelzadeh, L.; Wu, J.J. Photodynamic therapy for psoriasis. *J. Dermatol. Treat.* **2014**, *26*, 202–207. [CrossRef]
11. Silva, A.M.; Siopa, J.R.; Martins-Gomes, C.; Teixeira, M.D.C.; Santos, D.J.; Pires, M.D.A.; Andreani, T. New strategies for the treatment of autoimmune diseases using nanotechnologies. *Emerg. Nanotechnol. Immunol.* **2018**, 135–163. [CrossRef]
12. Hatz, K.; Schneider, U.; Henrich, P.B.; Braun, B.; Sacu, S. Ranibizumab plus Verteporfin Photodynamic Therapy in Neovascular Age-Related Macular Degeneration: 12 Months of Retreatment and Vision Outcomes from a Randomized Study. *Ophthalmologia* **2014**, *233*, 66–73. [CrossRef] [PubMed]

13. Oniszczuk, A.; Wojtunik-Kulesza, K.A.; Oniszczuk, T.; Kasprzak, K. The potential of photodynamic therapy (PDT)—Experimental investigations and clinical use. *Biomed. Pharmacother.* **2016**, *83*, 912–929. [CrossRef] [PubMed]

14. Zhang, J.; Jiang, C.; Longo, J.P.F.; Azevedo, R.B.; Zhang, H.; Muehlmann, L.A. An updated overview on the development of new photosensitizers for anticancer photodynamic therapy. *Acta Pharm. Sin. B* **2017**, *8*, 137–146. [CrossRef] [PubMed]

15. Granelli, S.G.; Diamond, I.; McDonagh, A.F.; Wilson, C.B.; Nielsen, S.L. Photochemotherapy of glioma cells by visible light and hematoporphyrin. *Cancer Res.* **1975**, *35*, 2567–2570.

16. Abrahamse, H.; Hamblin, M.R. *Photomedicine and Stem Cells: The Janus Face of Photodynamic Therapy (PDT) to Kill Cancer Stem Cells, and Photobiomodulation (PBM) to Stimulate Normal Stem Cells*; Morgan & Claypool Publishers: Bristol, UK, 2017.

17. Moreira, L.M.; Dos Santos, F.V.; Lyon, J.P.; Maftoum-Costa, M.; Soares, C.P.; Da Silva, N.S. Photodynamic Therapy: Porphyrins and Phthalocyanines as Photosensitizers. *Aust. J. Chem.* **2008**, *61*, 741–754. [CrossRef]

18. Mohammadi, A.; Mansoori, B.; Baradaran, B. Regulation of miRNAs by herbal medicine: An emerging field in cancer therapies. *Biomed. Pharmacother.* **2017**, *86*, 262–270. [CrossRef]

19. Mohammadi, A.; Mansoori, B.; Aghapour, M.; Baradaran, B. Urtica dioica dichloromethane extract induce apoptosis from intrinsic pathway on human prostate cancer cells (PC3). *Cell. Mol. Boil.* **2016**, *62*, 78–83.

20. Mohammadi, A.; Mansoori, B.; Goldar, S.; Shanehbandi, D.; Khaze, V.; Mohammadnejad, L.; Baghbani, E.; Baradaran, B. Effects of Urtica dioica dichloromethane extract on cell apoptosis and related gene expression in human breast cancer cell line (MDA-MB-468). *Cell. Mol. Boil.* **2016**, *62*, 62–67.

21. Alali, F.Q.; Tawaha, K. Dereplication of bioactive constituents of the genus hypericum using LC-(+,−)-ESI-MS and LC-PDA techniques: Hypericum triquterifolium as a case study. *Saudi Pharm. J.* **2009**, *17*, 269–274. [CrossRef]

22. Bailly, C. Ready for a comeback of natural products in oncology. *Biochem. Pharmacol.* **2009**, *77*, 1447–1457. [CrossRef]

23. Mishra, B.B.; Tiwari, V.K. Natural products: An evolving role in future drug discovery. *Eur. J. Med. Chem.* **2011**, *46*, 4769–4807. [CrossRef] [PubMed]

24. Rodrigues, M.C. Photodynamic Therapy Based on Arrabidaea chica (Crajiru) Extract Nanoemulsion: In vitro Activity against Monolayers and Spheroids of Human Mammary Adenocarcinoma MCF-7 Cells. *J. Nanomed. Nanotechnol.* **2015**, *6*, 1–6. [CrossRef]

25. Tan, P.J.; Appleton, D.R.; Mustafa, M.R.; Lee, H.B. Rapid Identification of Cyclic Tetrapyrrolic Photosensitisers for Photodynamic Therapy Using On-line Hyphenated LC-PDA-MS Coupled with Photo-cytotoxicity Assay. *Phytochem. Anal.* **2011**, *23*, 52–59. [CrossRef]

26. Skalkos, D.; Gioti, E.; Stalikas, C.; Meyer, H.; Papazoglou, T.; Filippidis, G.; Papazoglou, T.G. Photophysical properties of Hypericum perforatum L. extracts—Novel photosensitizers for PDT. *J. Photochem. Photobiol. B Boil.* **2006**, *82*, 146–151. [CrossRef] [PubMed]

27. Zeisser-Labouèbe, M.; Lange, N.; Gurny, R.; Delie, F. Hypericin-loaded nanoparticles for the photodynamic treatment of ovarian cancer. *Int. J. Pharm.* **2006**, *326*, 174–181. [CrossRef] [PubMed]

28. Mirmalek, S.A.; Azizi, M.A.; Jangholi, E.; Yadollah-Damavandi, S.; Javidi, M.A.; Parsa, Y.; Parsa, T.; Salimi-Tabatabaee, S.A.; Kolagar, H.G.; Alizadeh-Navaei, R. Cytotoxic and apoptogenic effect of hypericin, the bioactive component of Hypericum perforatum on the MCF-7 human breast cancer cell line. *Cancer Cell Int.* **2016**, *16*, 3. [CrossRef]

29. Yonar, D.; Süloğlu, A.K.; Selmanoğlu, G.; Sünnetçioğlu, M.M. An Electron paramagnetic resonance (EPR) spin labeling study in HT-29 Colon adenocarcinoma cells after Hypericin-mediated photodynamic therapy. *BMC Mol. Cell Boil.* **2019**, *20*, 16. [CrossRef]

30. Aggarwal, B.B.; Ichikawa, H.; Garodia, P.; Weerasinghe, P.; Sethi, G.; Bhatt, I.D.; Pandey, M.K.; Shishodia, S.; Nair, M.G. From traditional Ayurvedic medicine to modern medicine: Identification of therapeutic targets for suppression of inflammation and cancer. *Expert Opin. Ther. Targets* **2006**, *10*, 87–118. [CrossRef]

31. Chaturvedi, D.; Singh, K.; Singh, V.K. Therapeutic and pharmacological aspects of photodynamic product chlorophyllin. *Eur. J. Biol. Res.* **2019**, *9*, 64–76.

32. Juzeniene, A.; Nielsen, K.P.; Moan, J. Biophysical Aspects of Photodynamic Therapy. *J. Environ. Pathol. Toxicol. Oncol.* **2006**, *25*, 7–28. [CrossRef]

33. George, B.P.; Abrahamse, H. A Review on Novel Breast Cancer Therapies: Photodynamic Therapy and Plant Derived Agent Induced Cell Death Mechanisms. *Anti-Cancer Agents Med. Chem.* **2016**, *15*, 1. [CrossRef]

34. Aniogo, E.C.; George, B.P.; Abrahamse, H. The role of photodynamic therapy on multidrug resistant breast cancer. *Cancer Cell Int.* **2019**, *19*, 91. [CrossRef] [PubMed]

35. Allison, R.R.; Sibata, C.H. Oncologic photodynamic therapy photosensitizers: A clinical review. *Photodiagn. Photodyn. Ther.* **2010**, *7*, 61–75. [CrossRef] [PubMed]

36. Chen, B.; Roskams, T.; De Witte, P.A.M. Antivascular tumor eradication by hypericin-mediated photodynamic therapy. *Photochem. Photobiol.* **2002**, *76*, 509. [CrossRef]

37. Ascencio, M.; Collinet, P.; Farine, M.; Mordon, S. Protoporphyrin IX fluorescence photobleaching is a useful tool to predict the response of rat ovarian cancer following hexaminolevulinate photodynamic therapy. *Lasers Surg. Med.* **2008**, *40*, 332–341. [CrossRef]

38. Garg, A.D.; Nowis, D.; Golab, J.; Vandenabeele, P.; Krysko, D.V.; Agostinis, P. Immunogenic cell death, DAMPs and anticancer therapeutics: An emerging amalgamation. *Biochim. Biophys. Acta (BBA) Rev. Cancer* **2010**, *1805*, 53–71. [CrossRef]

39. Castano, A.P.; Demidova, T.N.; Hamblin, M.R. Mechanisms in photodynamic therapy: Part three-Photosensitizer pharmacokinetics, biodistribution, tumor localization and modes of tumor destruction. *Photodiagn. Photodyn. Ther.* **2005**, *2*, 91–106. [CrossRef]

40. Ogilby, P.R. Singlet oxygen: There is indeed something new under the sun. *Chem. Soc. Rev.* **2010**, *39*, 3181. [CrossRef]

41. Oseroff, A.R.; Blumenson, L.R.; Wilson, B.D.; Mang, T.S.; Bellnier, D.A.; Parsons, J.C.; Frawley, N.; Cooper, M.; Zeitouni, N.; Dougherty, T.J. A dose ranging study of photodynamic therapy with porfimer sodium (Photofrin®) for treatment of basal cell carcinoma. *Lasers Surg. Med.* **2006**, *38*, 417–426. [CrossRef]

42. Juzeniene, A.; Juzenas, P.; Ma, L.-W.; Iani, V.; Moan, J. Effectiveness of different light sources for 5-aminolevulinic acid photodynamic therapy. *Lasers Med. Sci.* **2004**, *19*, 139–149. [CrossRef]

43. Lang, P. Methyl aminolaevulinate–photodynamic therapy: A review of clinical trials in the treatment of actinic keratoses and nonmelanoma skin cancer. *Yearb. Dermatol. Dermatol. Surg.* **2008**, *2008*, 322–323. [CrossRef]

44. Jeffes, E.W.; McCullough, J.L.; Weinstein, G.D.; Kaplan, R.; Glazer, S.D.; Taylor, J. Photodynamic therapy of actinic keratoses with topical aminolevulinic acid hydrochloride and fluorescent blue light. *J. Am. Acad. Dermatol.* **2001**, *45*, 96–104. [CrossRef] [PubMed]

45. Lu, K.; He, C.; Lin, W. A Chlorin-Based Nanoscale Metal–Organic Framework for Photodynamic Therapy of Colon Cancers. *J. Am. Chem. Soc.* **2015**, *137*, 7600–7603. [CrossRef] [PubMed]

46. Moore, C.M.; Nathan, T.; Lees, W.; Mosse, C.; Freeman, A.; Emberton, M.; Bown, S. Photodynamic therapy using meso tetra hydroxy phenyl chlorin (mTHPC) in early prostate cancer. *Lasers Surg. Med.* **2006**, *38*, 356–363. [CrossRef]

47. Grosjean, P.; Savary, J.-F.; Wagnières, G.; Mizeret, J.; Woodtli, A.; Theumann, J.-F.; Fontolliet, C.; Bergh, H.V.D.; Monnier, P. Tetra(m-hydroxyphenyl)chlorin clinical photodynamic therapy of early bronchial and oesophageal cancers. *Lasers Med. Sci.* **1996**, *11*, 227–235. [CrossRef]

48. Kessel, D. Pharmacokinetics of N-aspartyl chlorin e6 in cancer patients. *J. Photochem. Photobiol. B Boil.* **1997**, *39*, 81–83. [CrossRef]

49. Taber, S.W.; Fingar, V.H.; Coots, C.T.; Wieman, T.J. Photodynamic therapy using mono-L-aspartyl chlorin e6 (Npe6) for the treatment of cutaneous disease: A Phase I clinical study. *Clin. Cancer Res.* **1998**, *4*, 2741–2746.

50. Lagudaev, D.M. Sorokatyĭ Photodynamic therapy of prostatic adenoma. *Urologiia* **2007**, *4*, 34–37.

51. Momma, T.; Hamblin, M.R.; Wu, H.C.; Hasan, T. Photodynamic therapy of orthotopic prostate cancer with benzoporphyrin derivative: Local control and distant metastasis. *Cancer Res.* **1998**, *58*, 5425–5431.

52. Levy, J.G.; Waterfield, E.; Richter, A.M.; Smits, C.; Lui, H.; Hruza, L.; Anderson, R.R.; Salvatori, V. Photodynamic therapy of malignancies with benzoporphyrin derivative monoacid ring A. *Europto Biomedical Optics '93* **1994**, *2078*, 91–101. [CrossRef]

53. Young, S.W.; Woodburn, K.W.; Wright, M.; Mody, T.D.; Fan, Q.; Sessler, J.L.; Dow, W.C.; Miller, R.A. Lutetium Texaphyrin (PCI-0123): A Near-Infrared, Water-Soluble Photosensitizer. *Photochem. Photobiol.* **1996**, *63*, 892–897. [CrossRef]

54. Sessler, J.L.; Miller, R.A. Texaphyrins: New drugs with diverse clinical applications in radiation and photodynamic therapy. *Biochem. Pharmacol.* **2000**, *59*, 733–739. [CrossRef]

55. Du, K.; Mick, R.; Busch, T.; Zhu, T.C.; Finlay, J.; Yu, G.; Yodh, A.; Malkowicz, S.; Smith, D.; Whittington, R.; et al. Preliminary results of interstitial motexafin lutetium-mediated PDT for prostate cancer. *Lasers Surg. Med.* **2006**, *38*, 427–434. [CrossRef] [PubMed]

56. Rockson, S.G.; Lorenz, D.P.; Cheong, W.-F.; Woodburn, K.W. Photoangioplasty: An emerging clinical cardiovascular role for photodynamic therapy. *Circulation* **2000**, *102*, 591–596. [CrossRef] [PubMed]

57. Kogias, E.; Vougioukas, V.; Hubbe, U.; Halatsch, M.-E. Minimally Invasive Approach for the Treatment of Lateral Lumbar Disc Herniations. Technique and Results. *Minim. Invasive Neurosurg.* **2007**, *50*, 160–162. [CrossRef]

58. Wood, S.R.; Holroyd, J.A.; Brown, S.B. The Subcellular Localization of Zn(ll) Phthalocyanines and Their Redistribution on Exposure to Light. *Photochem. Photobiol.* **1997**, *65*, 397–402. [CrossRef] [PubMed]

59. Stuchinskaya, T.; Moreno, M.; Cook, M.J.; Edwards, D.R.; Russell, D.A. Targeted photodynamic therapy of breast cancer cells using antibody–phthalocyanine–gold nanoparticle conjugates. *Photochem. Photobiol. Sci.* **2011**, *10*, 822. [CrossRef]

60. Sekhejane, P.R.; Houreld, N.N.; Abrahamse, H. Multiorganelle Localization of Metallated Phthalocyanine Photosensitizer in Colorectal Cancer Cells (DLD-1 and CaCo-2) Enhances Efficacy of Photodynamic Therapy. *Int. J. Photoenergy* **2014**, *2014*, 1–10. [CrossRef]

61. Hunt, D.W.C. Rostaporfin (Miravant Medical Technologies). *IDrugs* **2002**, *5*, 180–186.

62. Kaplan, M.; Somers, R.H.; Greenburg, R.; Ackler, J. Photodynamic therapy in the management of metastatic cutaneous adenocarcinomas: Case reports from phase 1/2 studies using tin ethyl etiopurpurin (SnET2). *J. Surg. Oncol.* **1998**, *67*, 121–125. [CrossRef]

63. Selman, S.H.; Keck, R.W.; Hampton, J.A. Transperineal Photodynamic Ablation of the Canine Prostate. *J. Urol.* **1996**, *156*, 258–260. [CrossRef]

64. Agostinis, P.; Berg, K.; Cengel, K.A.; Foster, T.H.; Girotti, A.W.; Gollnick, S.O.; Hahn, S.M.; Hamblin, M.R.; Juzeniene, A.; Kessel, D.; et al. Photodynamic therapy of cancer: An update. *CA Cancer J. Clin.* **2011**, *61*, 250–281. [CrossRef] [PubMed]

65. Reiners, J.J.; Agostinis, P.; Berg, K.; Oleinick, N.L.; Kessel, D. Assessing autophagy in the context of photodynamic therapy. *Autophagy* **2010**, *6*, 7–18. [CrossRef] [PubMed]

66. Kruger, C.; Abrahamse, H. Utilisation of Targeted Nanoparticle Photosensitiser Drug Delivery Systems for the Enhancement of Photodynamic Therapy. *Molecules* **2018**, *23*, 2628. [CrossRef]

67. Redmond, R.W.; Gamlin, J.N. A compilation of singlet oxygen yields from biologically relevant molecules. *Photochem. Photobiol.* **1999**, *70*, 391–475. [CrossRef]

68. Kitamura, N.; Kohtani, S.; Nakagaki, R. Molecular aspects of furocoumarin reactions: Photophysics, photochemistry, photobiology, and structural analysis. *J. Photochem. Photobiol. C Photochem. Rev.* **2005**, *6*, 168–185. [CrossRef]

69. Fracarolli, L.; Rodrigues, G.B.; Pereira, A.C.; Júnior, N.S.M.; Silva-Junior, G.J.; Bachmann, L.; Wainwright, M.; Bastos, J.K.; Braga, G.U. Inactivation of plant-pathogenic fungus Colletotrichum acutatum with natural plant-produced photosensitizers under solar radiation. *J. Photochem. Photobiol. B Boil.* **2016**, *162*, 402–411. [CrossRef]

70. Chobot, V.; Vytlačilová, J.; Kubicová, L.; Opletal, L.; Jahodář, L.; Laakso, I.; Vuorela, P. Phototoxic activity of a thiophene polyacetylene from Leuzea carthamoides. *Fitoterapia* **2006**, *77*, 194–198. [CrossRef]

71. Lima, B.; Agüero, M.B.; Zygadlo, J.; Tapia, A.; Solís, C.; De Arias, A.R.; Yaluff, G.; Zacchino, S.; Feresin, G.E.; Schmeda-Hirschmann, G. Antimicrobial activity of extracts, essential oil and metabolites obtained from tagetes mendocina. *J. Chil. Chem. Soc.* **2009**, *54*, 68–72. [CrossRef]

72. Jin, Q.; Lee, J.W.; Jang, H.; Choi, J.E.; Kim, H.S.; Lee, N.; Hong, J.T.; Lee, M.K.; Hwang, B.Y. Dimeric sesquiterpene and thiophenes from the roots of Echinops latifolius. *Bioorg. Med. Chem. Lett.* **2016**, *26*, 5995–5998. [CrossRef]

73. Postigo, A.; Funes, M.; Petenatti, E.; Bottai, H.; Pacciaroni, A.; Sortino, M. Antifungal photosensitive activity of Porophyllum obscurum (Spreng.) DC.: Correlation of the chemical composition of the hexane extract with the bioactivity. *Photodiagn. Photodyn. Ther.* **2017**, *20*, 263–272. [CrossRef] [PubMed]

74. Ibrahim, S.R.M.; Abdallah, H.M.; El Halawany, A.M.; Mohamed, G.A. Naturally occurring thiophenes: Isolation, purification, structural elucidation, and evaluation of bioactivities. *Phytochem. Rev.* **2015**, *15*, 197–220. [CrossRef]

75. Park, K.; Lee, J.-H. Photosensitizer effect of curcumin on UVB-irradiated HaCaT cells through activation of caspase pathways. *Oncol. Rep.* **2007**, *17*, 537–540. [CrossRef] [PubMed]

76. Lin, H.-Y.; Lin, J.-N.; Ma, J.-W.; Yang, N.-S.; Ho, C.-T.; Kuo, S.-C.; Way, T.-D. Demethoxycurcumin induces autophagic and apoptotic responses on breast cancer cells in photodynamic therapy. *J. Funct. Foods* **2015**, *12*, 439–449. [CrossRef]

77. Randazzo, W.; Aznar, R.; Sánchez, G. Curcumin-Mediated Photodynamic Inactivation of Norovirus Surrogates. *Food Environ. Virol.* **2016**, *8*, 244–250. [CrossRef]

78. Lee, H.-J.; Kang, S.-M.; Jeong, S.-H.; Chung, K.-H.; Kim, B.-I. Antibacterial photodynamic therapy with curcumin and Curcuma xanthorrhiza extract against Streptococcus mutans. *Photodiagnosis Photodyn. Ther.* **2017**, *20*, 116–119. [CrossRef]

79. Bhavya, M.; Hebbar, H.U. Efficacy of blue LED in microbial inactivation: Effect of photosensitization and process parameters. *Int. J. Food Microbiol.* **2019**, *290*, 296–304. [CrossRef]

80. Morten, A.G.; Martinez, L.J.; Holt, N.; Sik, R.H.; Reszka, K.; Chignell, C.F.; Tonnesen, H.H.; Roberts, J.E. Photophysical Studies on Antimalariai Drugs. *Photochem. Photobiol.* **2008**, *69*, 282–287. [CrossRef]

81. Inbaraj, J.J.; Kukielczak, B.M.; Bilski, P.; Sandvik, S.L.; Chignell, C.F. Photochemistry and Photocytotoxicity of Alkaloids from Goldenseal (Hydrastis canadensis L.) 1. Berberine. *Chem. Res. Toxicol.* **2001**, *14*, 1529–1534. [CrossRef]

82. Flors, C.; Prat, C.; Suau, R.; Najera, F.; Nonell, S. Photochemistry of Phytoalexins Containing Phenalenone-like Chromophores: Photophysics and Singlet Oxygen Photosensitizing Properties of the Plant Oxoaporphine Alkaloid Oxoglaucine. *Photochem. Photobiol.* **2005**, *81*, 120. [CrossRef]

83. Lorente, C.; Thomas, A.H. Photophysics and photochemistry of pterins in aqueous solution. *Acc. Chem. Res.* **2006**, *39*, 395–402. [CrossRef] [PubMed]

84. Phillipson, J.D.; Roberts, M.F.; Zenk, M.H. *The Chemistry and Biology of Isoquinoline Alkaloids*; Springer Science & Business Media: Berlin, Germany, 2012.

85. Vignoni, M.; Erra-Balsells, R.; Epe, B.; Cabrerizo, F.M. Intra- and extra-cellular DNA damage by harmine and 9-methyl-harmine. *J. Photochem. Photobiol. B Boil.* **2014**, *132*, 66–71. [CrossRef] [PubMed]

86. Reid, L.O.; Roman, E.A.; Thomas, A.H.; Dántola, M.L. Photooxidation of Tryptophan and Tyrosine Residues in Human Serum Albumin Sensitized by Pterin: A Model for Globular Protein Photodamage in Skin. *Biochemistry* **2016**, *55*, 4777–4786. [CrossRef] [PubMed]

87. Yañuk, J.G.; Denofrio, M.P.; Rasse-Suriani, F.A.O.; Villarruel, F.D.; Fassetta, F.; Einschlag, F.S.G.; Erra-Balsells, R.; Epe, B.; Cabrerizo, F.M. DNA damage photo-induced by chloroharmine isomers: Hydrolysis versus oxidation of nucleobases. *Org. Biomol. Chem.* **2018**, *16*, 2170–2184. [CrossRef]

88. Daub, M.E.; Herrero, S.; Chung, K.-R. Photoactivated perylenequinone toxins in fungal pathogenesis of plants. *FEMS Microbiol. Lett.* **2005**, *252*, 197–206. [CrossRef]

89. Montoya, S.C.N.; Comini, L.R.; Sarmiento, M.; Becerra, C.; Albesa, I.; Argüello, G.A.; Cabrera, J.L. Natural anthraquinones probed as Type I and Type II photosensitizers: Singlet oxygen and superoxide anion production. *J. Photochem. Photobiol. B Boil.* **2005**, *78*, 77–83. [CrossRef]

90. Comini, L.R.; Montoya, S.C.N.; Sarmiento, M.; Cabrera, J.L.; Argüello, G.A. Characterizing some photophysical, photochemical and photobiological properties of photosensitizing anthraquinones. *J. Photochem. Photobiol. A Chem.* **2007**, *188*, 185–191. [CrossRef]

91. Mastrangelopoulou, M.; Grigalavicius, M.; Berg, K.; Ménard, M.; Theodossiou, T.A. Cytotoxic and Photocytotoxic Effects of Cercosporin on Human Tumor Cell Lines. *Photochem. Photobiol.* **2018**, *95*, 387–396. [CrossRef]

92. Panno, M.L.; Giordano, F.; Palma, M.G.; Bartella, V.; Rago, V.; Maggiolini, M.; Sisci, D.; Lanzino, M.; De Amicis, F.; Ando, S. Evidence that bergapten, independently of its photoactivation, enhances p53 gene expression and induces apoptosis in human breast cancer cells. *Curr. Cancer Drug Targets* **2009**, *9*, 469–481. [CrossRef]

93. Panno, M.L.; Giordano, F.; Rizza, P.; Pellegrino, M.; Zito, D.; Giordano, C.; Mauro, L.; Catalano, S.; Aquila, S.; Sisci, D.; et al. Bergapten induces ER depletion in breast cancer cells through SMAD4-mediated ubiquitination. *Breast Cancer Res. Treat.* **2012**, *136*, 443–455. [CrossRef]

94. Kim, S.-M.; Lee, J.H.; Sethi, G.; Kim, C.; Baek, S.H.; Nam, D.; Chung, W.-S.; Shim, B.S.; Ahn, K.S.; Kim, S.-H. Bergamottin, a natural furanocoumarin obtained from grapefruit juice induces chemosensitization and apoptosis through the inhibition of STAT3 signaling pathway in tumor cells. *Cancer Lett.* **2014**, *354*, 153–163. [CrossRef] [PubMed]

95. Kim, S.-M.; Lee, E.-J.; Lee, J.H.; Yang, W.M.; Nam, D.; Lee, J.H.; Lee, S.-G.; Um, J.-Y.; Shim, B.S.; Ahn, K.S. Simvastatin in combination with bergamottin potentiates TNF-induced apoptosis through modulation of NF-κB signalling pathway in human chronic myelogenous leukaemia. *Pharm. Boil.* **2016**, *54*, 1–11. [CrossRef] [PubMed]

96. Ge, Z.-C.; Qu, X.; Yu, H.-F.; Zhang, H.-M.; Wang, Z.-H.; Zhang, Z.-T. Antitumor and apoptotic effects of bergaptol are mediated via mitochondrial death pathway and cell cycle arrest in human breast carcinoma cells. *Bangladesh J. Pharmacol.* **2016**, *11*, 489. [CrossRef]

97. Nagatani, T.; Matsuzaki, T.; Kim, S.; Baba, N.; Ichiyama, S.; Miyamoto, H.; Nakajima, H. Treatment of cutaneous T-cell lymphoma (CTCL) by extracorporeal photochemotherapy. *J. Dermatol. Sci.* **1990**, *1*, 226. [CrossRef]

98. Bethea, D.; Fullmer, B.; Syed, S.; Seltzer, G.; Tiano, J.; Rischko, C.; Gillespie, L.; Brown, D.; Gasparro, F.P. Psoralen photobiology and photochemotherapy: 50 years of science and medicine. *J. Dermatol. Sci.* **1999**, *19*, 78–88. [CrossRef]

99. McKenna, K.E. PUVA, Psoralens and Skin Cancer. *Skin Cancer UV Radiat.* **1997**, 416–424.

100. El-Domyati, M.; Moftah, N.H.; Nasif, G.A.; Abdel-Wahab, H.M.; Barakat, M.T.; Abdel-Aziz, R.T. Evaluation of apoptosis regulatory proteins in response to PUVA therapy for psoriasis. *Photodermatol. Photoimmunol. Photomed.* **2013**, *29*, 18–26. [CrossRef]

101. Holtick, U.; Wang, X.N.; Marshall, S.R.; Von Bergwelt-Baildon, M.; Scheid, C.; Dickinson, A.M. In Vitro PUVA Treatment Preferentially Induces Apoptosis in Alloactivated T Cells. *Transplantation* **2012**, *94*, e31–e34. [CrossRef]

102. Schmitt, I.M.; Chimenti, S.; Gasparro, F.P. Psoralen-protein photochemistry—A forgotten field. *J. Photochem. Photobiol. B Boil.* **1995**, *27*, 101–107. [CrossRef]

103. Van Aelst, B.; Devloo, R.; Zachee, P.; T'Kindt, R.; Sandra, K.; Vandekerckhove, P.; Compernolle, V.; Feys, H.B. Psoralen and Ultraviolet A Light Treatment Directly Affects Phosphatidylinositol 3-Kinase Signal Transduction by Altering Plasma Membrane Packing. *J. Boil. Chem.* **2016**, *291*, 24364–24376. [CrossRef]

104. Xia, W.; Gooden, D.; Liu, L.; Zhao, S.; Soderblom, E.J.; Toone, E.J.; Beyer, W.F.; Walder, H.; Spector, N. Photo-Activated Psoralen Binds the ErbB2 Catalytic Kinase Domain, Blocking ErbB2 Signaling and Triggering Tumor Cell Apoptosis. *PLoS ONE* **2014**, *9*, e88983. [CrossRef] [PubMed]

105. Ghosh, G.; Colón, K.L.; Fuller, A.; Sainuddin, T.; Bradner, E.; McCain, J.; Monro, S.M.A.; Yin, H.; Hetu, M.W.; Cameron, C.G.; et al. Cyclometalated Ruthenium(II) Complexes Derived from α-Oligothiophenes as Highly Selective Cytotoxic or Photocytotoxic Agents. *Inorg. Chem.* **2018**, *57*, 7694–7712. [CrossRef] [PubMed]

106. Zhang, P.; Jin, W.-R.; Shi, Q.; He, H.; Ma, Z.; Qu, H.-B. Two novel thiophenes from Echinops grijissi Hance. *J. Asian Nat. Prod. Res.* **2008**, *10*, 977–981. [CrossRef] [PubMed]

107. Galushko, S.; Shishkina, I.; Alekseeva, I. Relationship between retention parameters in reversed-phase high-performance liquid chromatography and antitumour activity of some pyrimidine bases and nucleosides. *J. Chromatogr. A* **1991**, *547*, 161–166. [CrossRef]

108. Wang, Y.D.; Johnson, S.; Powell, D.; McGinnis, J.P.; Miranda, M.; Rabindran, S.K. Inhibition of tumor cell proliferation by thieno[2,3-d]pyrimidin-4(1H)-one-based analogs. *Bioorg. Med. Chem. Lett.* **2005**, *15*, 3763–3766. [CrossRef]

109. Alagarsamy, V.; Meena, S.; Ramseshu, K.; Solomon, V.; Thirumurugan, K.; Dhanabal, K.; Murugan, M. Synthesis, analgesic, anti-inflammatory, ulcerogenic index and antibacterial activities of novel 2-methylthio-3-substituted-5,6,7,8-tetrahydrobenzo (b) thieno[2,3-d]pyrimidin-4(3H)-ones. *Eur. J. Med. Chem.* **2006**, *41*, 1293–1300. [CrossRef]

110. Starcevic, K.; Kralj, M.; Piantanida, I.; Šuman, L.; Pavelic, K.; Karminski-Zamola, G. Synthesis, photochemical synthesis, DNA binding and antitumor evaluation of novel cyano- and amidino-substituted derivatives of naphtho-furans, naphtho-thiophenes, thieno-benzofurans, benzo-dithiophenes and their acyclic precursors. *Eur. J. Med. Chem.* **2006**, *41*, 925–939. [CrossRef]

111. Galindo, M.A.; Romero, M.A.; Navarro, J.A. Cyclic assemblies formed by metal ions, pyrimidines and isogeometrical heterocycles: DNA binding properties and antitumour activity. *Inorg. Chim. Acta* **2009**, *362*, 1027–1030. [CrossRef]

112. Prachayasittikul, S.; Worachartcheewan, A.; Nantasenamat, C.; Chinworrungsee, M.; Sornsongkhram, N.; Ruchirawat, S.; Prachayasittikul, V. Synthesis and structure–activity relationship of 2-thiopyrimidine-4-one analogs as antimicrobial and anticancer agents. *Eur. J. Med. Chem.* **2011**, *46*, 738–742. [CrossRef]

113. Jin, W.; Shi, Q.; Hong, C.; Cheng, Y.; Ma, Z.; Qu, H. Cytotoxic properties of thiophenes from Echinops grijissi Hance. *Phytomedicine* **2008**, *15*, 768–774. [CrossRef]

114. Karunagaran, D.; Rashmi, R.; Kumar, T.R.S. Induction of Apoptosis by Curcumin and Its Implications for Cancer Therapy. *Curr. Cancer Drug Targets* **2005**, *5*, 117–129. [CrossRef] [PubMed]

115. Tønnesen, H.H.; De Vries, H.; Karlsen, J.; Van Henegouwen, G.B. Studies on Curcumin and Curcuminoids IX: Investigation of the Photobiological Activity of Curcumin Using Bacterial Indicator Systems. *J. Pharm. Sci.* **1987**, *76*, 371–373. [CrossRef]

116. Dahl, T.A.; McGowan, W.M.; Shand, M.A.; Srinivasan, V.S. Photokilling of bacteria by the natural dye curcumin. *Arch. Microbiol.* **1989**, *151*, 183–185. [CrossRef] [PubMed]

117. Haukvik, T.; Bruzell, E.; Kristensen, S.; Tønnesen, H.H. Photokilling of bacteria by curcumin in selected polyethylene glycol 400 (PEG 400) preparations. Studies on curcumin and curcuminoids, XLI. *Die Pharm.* **2010**, *65*, 600–606.

118. Nardo, L.; Andreoni, A.; Másson, M.; Haukvik, T.; Tønnesen, H.H. Studies on Curcumin and Curcuminoids. XXXIX. Photophysical Properties of Bisdemethoxycurcumin. *J. Fluoresc.* **2010**, *21*, 627–635. [CrossRef] [PubMed]

119. Bruzell, E.M.; Morisbak, E.; Tønnesen, H.H. Studies on curcumin and curcuminoids. XXIX. Photoinduced cytotoxicity of curcumin in selected aqueous preparations. *Photochem. Photobiol. Sci.* **2005**, *4*, 523. [CrossRef]

120. Nardo, L.; Andreoni, A.; Bondani, M.; Másson, M.; Haukvik, T.; Tønnesen, H.H. Studies on Curcumin and Curcuminoids. XLVI. Photophysical Properties of Dimethoxycurcumin and Bis-dehydroxycurcumin. *J. Fluoresc.* **2011**, *22*, 597–608. [CrossRef]

121. Nardo, L.; Andreoni, A.; Bondani, M.; Másson, M.; Tønnesen, H.H. Studies on curcumin and curcuminoids. XXXIV. Photophysical properties of a symmetrical, non-substituted curcumin analogue. *J. Photochem. Photobiol. B Boil.* **2009**, *97*, 77–86. [CrossRef]

122. Araújo, N.C.; Fontana, C.R.; Gerbi, M.E.M.; Bagnato, V.S. Overall-Mouth Disinfection by Photodynamic Therapy Using Curcumin. *Photomed. Laser Surg.* **2012**, *30*, 96–101. [CrossRef]

123. Rego-Filho, F.D.A.; De Araujo, M.T.; De Oliveira, K.T.; Bagnato, V.S. Validation of Photodynamic Action via Photobleaching of a New Curcumin-Based Composite with Enhanced Water Solubility. *J. Fluoresc.* **2014**, *24*, 1407–1413. [CrossRef]

124. Sreedhar, A.; Sarkar, I.; Rajan, P.; Pai, J.; Malagi, S.; Kamath, V.; Barmappa, R. Comparative evaluation of the efficacy of curcumin gel with and without photo activation as an adjunct to scaling and root planing in the treatment of chronic periodontitis: A split mouth clinical and microbiological study. *J. Nat. Sci. Boil. Med.* **2015**, *6*, 102–S109. [CrossRef] [PubMed]

125. MacRobert, A.J.; Komerik, N. Photodynamic Therapy as an Alternative Antimicrobial Modality for Oral Infections. *J. Environ. Pathol. Toxicol. Oncol.* **2006**, *25*, 487–504. [CrossRef] [PubMed]

126. Priyadarsini, K.I. Photophysics, photochemistry and photobiology of curcumin: Studies from organic solutions, bio-mimetics and living cells. *J. Photochem. Photobiol. C Photochem. Rev.* **2009**, *10*, 81–95. [CrossRef]

127. Maheshwari, R.K.; Singh, A.K.; Gaddipati, J.; Srimal, R.C. Multiple biological activities of curcumin: A short review. *Life Sci.* **2006**, *78*, 2081–2087. [CrossRef] [PubMed]

128. Yance, D.R.; Sagar, S.M. Targeting Angiogenesis with Integrative Cancer Therapies. *Integr. Cancer Ther.* **2006**, *5*, 9–29. [CrossRef] [PubMed]

129. Salvioli, S.; Sikora, E.; Cooper, E.L.; Franceschi, C. Curcumin in Cell Death Processes: A Challenge for CAM of Age-Related Pathologies. *Evidence-Based Complement Altern. Med.* **2007**, *4*, 181–190. [CrossRef]

130. Vallianou, N.; Evangelopoulos, A.; Schizas, N.; Kazazis, C. Potential anticancer properties and mechanisms of action of curcumin. *Anticancer Res.* **2015**, *35*, 645–651.

131. Yang, J.-Y.; Zhong, X.; Yum, H.-W.; Lee, H.-J.; Kundu, J.K.; Na, H.-K.; Surh, Y.-J. Curcumin Inhibits STAT3 Signaling in the Colon of Dextran Sulfate Sodium-treated Mice. *J. Cancer Prev.* **2013**, *18*, 186–191. [CrossRef]

132. Dovigo, L.N.; Pavarina, A.C.; Carmello, J.C.; Machado, A.L.; Brunetti, I.L.; Bagnato, V.S. Susceptibility of clinical isolates of Candida to photodynamic effects of curcumin. *Lasers Surg. Med.* **2011**, *43*, 927–934. [CrossRef]

133. Bernd, A. Visible light and/or UVA offer a strong amplification of the anti-tumor effect of curcumin. *Phytochem. Rev.* **2013**, *13*, 183–189. [CrossRef]

134. Chan, W.-H.; Wu, H.-J. Anti-apoptotic effects of curcumin on photosensitized human epidermal carcinoma A431 cells. *J. Cell. Biochem.* **2004**, *92*, 200–212. [CrossRef] [PubMed]

135. Araújo, N.C.; Fontana, C.R.; Bagnato, V.S.; Gerbi, M.E.M. Photodynamic Effects of Curcumin Against Cariogenic Pathogens. *Photomed. Laser Surg.* **2012**, *30*, 393–399. [CrossRef]

136. Benyhe, S. Morphine: New aspects in the study of an ancient compound. *Life Sci.* **1994**, *55*, 969–979. [CrossRef]

137. Li, W.; Shao, Y.; Hu, L.; Zhang, X.; Chen, Y.; Tong, L.; Li, C.; Shen, X.; Ding, J. BM6, a new semi-synthetic vinca alkaloid, exhibits its potent in vivo anti-tumor activities via its high binding affinity for tubulin and improved pharmacokinetic profiles. *Cancer Boil. Ther.* **2007**, *6*, 787–794. [CrossRef] [PubMed]

138. Huang, M.; Gao, H.; Chen, Y.; Zhu, H.; Cai, Y.; Zhang, X.; Miao, Z.; Jiang, H.; Zhang, J.; Shen, H.; et al. Chimmitecan, a Novel 9-Substituted Camptothecin, with Improved Anticancer Pharmacologic Profiles In vitro and In vivo. *Clin. Cancer Res.* **2007**, *13*, 1298–1307. [CrossRef] [PubMed]

139. Sun, Y.; Xun, K.; Wang, Y.; Chen, X. A systematic review of the anticancer properties of berberine, a natural product from Chinese herbs. *Anti-Cancer Drugs* **2009**, *20*, 757–769. [CrossRef]

140. Eom, K.S.; Kim, H.-J.; So, H.-S.; Park, R.; Kim, T.Y. Berberine-induced apoptosis in human glioblastoma T98G cells is mediated by endoplasmic reticulum stress accompanying reactive oxygen species and mitochondrial dysfunction. *Boil. Pharm. Bull.* **2010**, *33*, 1644–1649. [CrossRef]

141. Diogo, C.V.; Machado, N.G.; Barbosa, I.A.; Serafim, T.L.; Burgeiro, A.; Oliveira, P.J. Berberine as a Promising Safe Anti-Cancer Agent- Is there a Role for Mitochondria? *Curr. Drug Targets* **2011**, *12*, 850–859. [CrossRef]

142. Tan, W.; Lu, J.-J.; Huang, M.; Li, Y.; Chen, M.; Wu, G.; Gong, J.; Zhong, Z.; Xu, Z.; Dang, Y.; et al. Anti-cancer natural products isolated from chinese medicinal herbs. *Chin. Med.* **2011**, *6*, 27. [CrossRef]

143. Burgeiro, A.A.C.; Gajate, C.; Dakir, E.H.; Villa-Pulgarin, J.A.; Oliveira, P.J.; Mollinedo, F. Involvement of mitochondrial and B-RAF/ERK signaling pathways in berberine-induced apoptosis in human melanoma cells. *Anti-Cancer Drugs* **2011**, *22*, 507–518. [CrossRef]

144. Youn, M.-J.; So, H.-S.; Cho, H.-J.; Kim, H.-J.; Kim, Y.; Lee, J.-H.; Sohn, J.S.; Kim, Y.K.; Chung, S.-Y.; Park, R. Berberine, a Natural Product, Combined with Cisplatin Enhanced Apoptosis through a Mitochondria/Caspase-Mediated Pathway in HeLa Cells. *Boil. Pharm. Bull.* **2008**, *31*, 789–795. [CrossRef] [PubMed]

145. Hur, J.-M.; Hyun, M.-S.; Lim, S.; Lee, W.-Y.; Kim, N. The combination of berberine and irradiation enhances anti-cancer effects via activation of p38 MAPK pathway and ROS generation in human hepatoma cells. *J. Cell. Biochem.* **2009**, *107*, 955–964. [CrossRef] [PubMed]

146. Andreazza, N.L.; Vevert-Bizet, C.; Bourg-Heckly, G.; Sureau, F.; Salvador, M.J.; Bonneau, S. Berberine as a Photosensitizing Agent for Antitumoral Photodynamic Therapy: Insights into its Association to Low Density Lipoproteins. *Int. J. Pharm.* **2016**, *510*, 240–249. [CrossRef] [PubMed]

147. Bhattacharyya, R.; Gupta, P.; Bandyopadhyay, S.K.; Patro, B.S.; Chattopadhyay, S. Coralyne, a protoberberine alkaloid, causes robust photosenstization of cancer cells through ATR-p38 MAPK-BAX and JAK2-STAT1-BAX pathways. *Chem. Interact.* **2018**, *285*, 27–39. [CrossRef] [PubMed]

148. Martín, J.P.; Labrador, V.; Freire, P.F.; Molero, M.L.; Hazen, M. Ultrastructural changes induced in HeLa cells after phototoxic treatment with harmine. *J. Appl. Toxicol.* **2004**, *24*, 197–201. [CrossRef] [PubMed]

149. Arnason, J.T.; Towers, G.H.N.; Abramowski, Z.; Campos, F.; Champagne, D.; McLachlan, D.; Philogène, B.J.R. Berberine: A naturally occurring phototoxic alkaloid. *J. Chem. Ecol.* **1984**, *10*, 115–123. [CrossRef]

150. Cheng, L.-L.; Wang, M.; Zhu, H.; Li, K.; Zhu, R.-R.; Sun, X.-Y.; Yao, S.-D.; Wu, Q.-S.; Wang, S.-L. Characterization of the transient species generated by the photoionization of Berberine: A laser flash photolysis study. *Spectrochim. Acta Part A Mol. Biomol. Spectrosc.* **2009**, *73*, 955–959. [CrossRef]

151. Jantova, S.; Letašiová, S.; Brezová, V.; Cipak, L.; Lábaj, J. Photochemical and phototoxic activity of berberine on murine fibroblast NIH-3T3 and Ehrlich ascites carcinoma cells. *J. Photochem. Photobiol. B Boil.* **2006**, *85*, 163–176. [CrossRef]

152. Dave, H.; Ledwani, L. A review on anthraquinones isolated from Cassia species and their applications. *Indian J. Nat. Prod. Resour.* **2012**, *3*, 291–319.

153. Seigler, D.S. *Plant Secondary Metabolism*; Springer Science & Business Media: Berlin, Germany, 1998.
154. Gutiérrez, I.; Bertolotti, S.G.; Biasutti, M.; Soltermann, A.T.; García, N.A. Quinones and hydroxyquinones as generators and quenchers of singlet molecular oxygen. *Can. J. Chem.* **1997**, *75*, 423–428. [CrossRef]
155. Pawłowska, J.; Tarasiuk, J.; Wolf, C.R.; Paine, M.J.I.; Borowski, E. Differential Ability of Cytostatics From Anthraquinone Group to Generate Free Radicals in Three Enzymatic Systems: NADH Dehydrogenase, NADPH Cytochrome P450 Reductase, and Xanthine Oxidase. *Oncol. Res. Featur. Preclin. Clin. Cancer Ther.* **2003**, *13*, 245–252. [CrossRef] [PubMed]
156. Singh, A. *Herbal Drugs as Therapeutic Agents*; CRC Press: Boca Raton, FL, USA, 2014.
157. Montoya, S.C.N.; Comini, L.; Vittar, B.R.; Fernández, I.M.; Rivarola, V.A.; Cabrera, J.L. Phototoxic effects of Heterophyllaea pustulata (Rubiaceae). *Toxicon* **2008**, *51*, 1409–1415. [CrossRef] [PubMed]
158. Comini, L.; Fernandez, I.; Vittar, N.R.; Montoya, S.N.; Cabrera, J.L.; Rivarola, V.A. Photodynamic activity of anthraquinones isolated from Heterophyllaea pustulata Hook f. (Rubiaceae) on MCF-7c3 breast cancer cells. *Phytomedicine* **2011**, *18*, 1093–1095. [CrossRef] [PubMed]
159. Vittar, N.B.R.; Awruch, J.; Azizuddin, K.; Rivarola, V.A. Caspase-independent apoptosis, in human MCF-7c3 breast cancer cells, following photodynamic therapy, with a novel water-soluble phthalocyanine. *Int. J. Biochem. Cell Boil.* **2010**, *42*, 1123–1131. [CrossRef]
160. Pirillo, J.; De Simone, B.C.; Russo, N. Photophysical properties prediction of selenium- and tellurium-substituted thymidine as potential UVA chemotherapeutic agents. *Theor. Chem. Accounts* **2015**, *135*, 8. [CrossRef]
161. Mazzone, G.; Alberto, M.E.; De Simone, B.C.; Marino, T.; Russo, N. Can Expanded Bacteriochlorins Act as Photosensitizers in Photodynamic Therapy? Good News from Density Functional Theory Computations. *Molecules* **2016**, *21*, 288. [CrossRef] [PubMed]
162. Mansoori, B.; Mohammadi, A.; Doustvandi, M.A.; Mohammadnejad, F.; Kamari, F.; Gjerstorff, M.F.; Baradaran, B.; Hamblin, M.R. Photodynamic therapy for cancer: Role of natural products. *Photodiagnosis Photodyn. Ther.* **2019**, *26*, 395–404. [CrossRef] [PubMed]
163. Pandey, R.K.; Goswami, L.N.; Chen, Y.; Gryshuk, A.; Missert, J.R.; Oseroff, A.; Dougherty, T.J. Nature: A rich source for developing multifunctional agents. tumor-imaging and photodynamic therapy. *Lasers Surg. Med.* **2006**, *38*, 445–467. [CrossRef]

 *molecules*

Review

Mitocanic Di- and Triterpenoid Rhodamine B Conjugates

Sophie Hoenke [1], **Immo Serbian** [1], **Hans-Peter Deigner** [2] and **René Csuk** [1,*]

[1] Organic Chemistry, Martin-Luther University Halle-Wittenberg, Kurt-Mothes Street 2, D-06120 Halle, Germany; sophie.hoenke@chemie.uni-halle.de (S.H.); immo.serbian@chemie.uni-halle.de (I.S.)

[2] Medical and Life Science Faculty, Institute of Precision Medicine, Furtwangen University, Jakob–Kienzle–Street 17, D-78054 Villigen–Schwenningen, Germany; dei@hs-furtwangen.de

* Correspondence: rene.csuk@chemie.uni-halle.de; Tel.: +49-345-5525660

Academic Editor: José Antonio Lupiáñez

Received: 29 October 2020; Accepted: 16 November 2020; Published: 20 November 2020

Abstract: The combination of the "correct" triterpenoid, the "correct" spacer and rhodamine B (**RhoB**) seems to be decisive for the ability of the conjugate to accumulate in mitochondria. So far, several triterpenoid rhodamine B conjugates have been prepared and screened for their cytotoxic activity. To obtain cytotoxic compounds with EC_{50} values in a low nano-molar range combined with good tumor/non-tumor selectivity, the Rho B unit has to be attached via an amine spacer to the terpenoid skeleton. To avoid spirolactamization, secondary amines have to be used. First results indicate that a homopiperazinyl spacer is superior to a piperazinyl spacer. Hybrids derived from maslinic acid or tormentic acid are superior to those from oleanolic, ursolic, glycyrrhetinic or euscaphic acid. Thus, a tormentic acid-derived **RhoB** conjugate **32**, holding a homopiperazinyl spacer can be regarded, at present, as the most promising candidate for further biological studies.

Keywords: triterpenoic acid; maslinic acid; tormentic acid; betulinic acid; oleanolic acid; rhodamine B; cytotoxicity

---

1. Introduction

Cancer remains one of the leading causes of death worldwide, and the incidence is increasing. Cancer is the second-leading cause of death globally, accounting for 9.6 million deaths in 2018 [1]. It is expected that in 2030, 21 million people worldwide will suffer from cancer [2]. Tremendous progress, however, has been made in the treatment of individual cancers [3–6]. This is due, on one hand, to improved early detection and prophylaxis and, on the other, to the development of highly efficient drugs in a wide range of different substance classes. Thus, the probability of premature death from cancer per year decreased from 8.3% in the year 2000 to 6.9% in 2015. It is expected to be as low as 5.3% in 2030, saving approximately 1.1 million lives per year [7].

Both proteins and small molecules have proven their worth in therapy and are in use in the clinic. However, low-molecular-weight drugs often bear the stigma of reduced selectivity in which the cytotoxic drug not only targets cancerous cells, but it also damages healthy tissue. This increases serious side effects and stress symptoms in patients, including nausea, heart or brain damage, impairments to the central nervous system and damage to cells of the inner ear; losses of fertility, hearing and hair have also been noted [8].

These serious side effects and impairments limit the use and acceptance of a drug, as they reduce patient compliance due to a significantly reduced quality of life. This not only endangers the chances of successful therapy, but it often also leads to discontinuation of the therapy [9,10].

Since the problems of reconversion of cancer cells into normal cells ("reprogramming", for example, of terminally differentiated cancer cells into cancer cells of benign phenotypes) [11] have only remotely been solved today, cancer cells have to be removed either by surgery or destroyed by radiation or chemotherapy. Cell death by chemotherapy can be induced in many different ways [12–17], but the mitochondria play a major role in the life or death of a cell. Thus, agents that target mitochondria and induce a controlled cell death, so called "mitocans", have received increased attention in recent years [17–38]. This seems even more significant inasmuch as cancer cells are closely linked to dysregulated apoptosis of the cells; as a consequence, drug resistance of the cancer cells can develop [39].

Mitocans (as well as other cytotoxic agents) are often able to induce apoptosis; however, the death of a cell, irrespective of whether this cell is malignant, is not random at all. Triggering of controlled cell death is always preferable to an unselective rupture of membranes following the application of extreme but locally applied heat, freeze–thaw cycles or steep osmotic gradients. Controlled cell death can be triggered on a cellular level from nuclear, reticular, cytoskeletal, lysosomal, membrane or, most important, mitochondrial origins [16].

Usually, the cells in a living organism closely cooperate, and cells are constantly in an equilibrium between life and death. Triggering programmed cell death routines removes damaged, infected and out-of-control cells from the organism. The problem arises from the latter cells, since most cancer cells do not respond to extrinsic apoptotic triggers. Thus, mitochondria present a target of emerging interest for cancer therapy as they can trigger apoptosis through an intrinsic pathway. Apoptosis usually starts with loss of mitochondrial membrane potential, followed by the release of cytochrome c and activation of caspase 3 [40]. Furthermore, permeabilization of the outer mitochondrial membrane and the release of cytochrome c are required in many cell death stimuli [41]. This release of cytochrome c can be regarded as a "point of no return" finally leading to the death of the cell. This highlights the importance of mitochondrial compartmentation and explains the devastating effect following permeabilization of the mitochondrial membrane [42–44]. In addition to membrane permeabilization, the opening of the mitochondrial permeability transition pore is also considered an important event resulting in mitochondrial depolarization and the release of apoptotic factors [29].

In recent years, triterpenes have repeatedly and increasingly entered the focus of scientific interest. Extensive studies on their apoptotic and cytotoxic properties have been performed. A major concern in dealing with cancer is the MDR (multiple drug resistance) phenotype [39]. These cancer cells overexpress ATP-dependent transporters that eject toxic compounds from the cell before they cause harm to the cell. Some triterpenes are known inhibitors of the efflux pump MDR1, but they are also known to downregulate the transcription factor NF-κB. For cancer, it is widely accepted [45] that NF-κB promotes tumor migration and tumor proliferation.

2. Results

Mitochondrial membranes of malignant cells hold an increased membrane potential compared to non–malignant cells [46]. This effect fosters the accumulation of cationic molecules [17,47,48], hence inducing high selectivity for mitocans holding a (more or less) lipophilic cation such as a rhodamine scaffold. The same effect applies for triphenylphosphonium cations [49–57] and to a small extent for quaternary ammonium ions [58–60], zwitterionic *N*-oxides [60,61] and triterpenes substituted with BODIPYs [62–66] or a safirinium moiety [67]. However, the presence of a cationic center is not alone decisive for achieving high cytotoxic effects [60].

Rhodamine B (**RhoB**) seems to be a privileged scaffold. This fluorescent dye, also known as rhodamine 610, C.I. Pigment violet 1, basic violet 10, and C.I. 45170 [68], was invented in 1888 ("Tetraethyl-rhodamine") by M. Cérésole [69,70], and since then it has been widely used in biology, biotechnology and as a biosensor [71,72]. **RhoB** exists in an equilibrium [73–77] between an "open" positively charged form A (Figure 1) that is fluorescent and a "closed", non–fluorescent form B. Under acidic conditions, pink-colored A dominates, while colorless B dominates under basic conditions. Further, in less polar organic solvents, the zwitterionic form C undergoes a rapid reversible conversion to B [78–81].

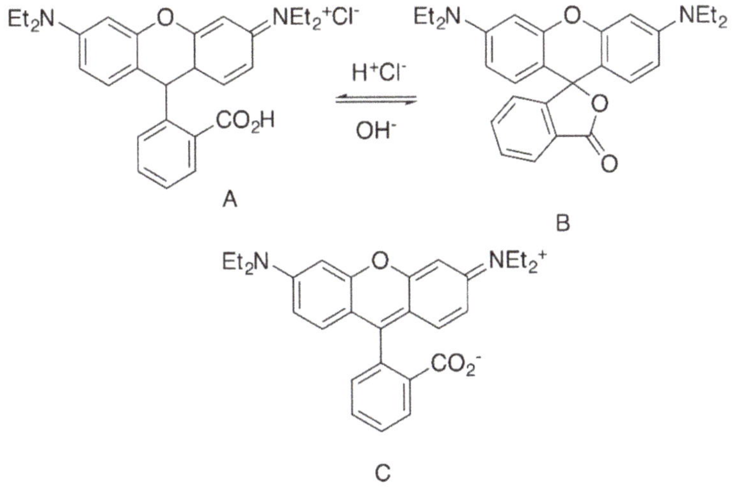

Figure 1. Structure of Rhodamine B (**RhoB**) in its "open" form **A**, "closed" lactone form **B** and the zwitterion **C**.

RhoB is suspected to be carcinogenic [82–85]. The LD_{50} value for orally administered RhoB in rats is >500 mg/kg, and an older report classified **RhoB** (as well as Rho6G) as possibly carcinogenic in rats [85]. **RhoB**, however, seems not to be mutagenic in Chinese hamster ovary cells [86], but it presents a genotoxic hazard for mammalian organisms [87]. As far as the **RhoB**–triterpene conjugates are concerned, two types of compounds have been accessed so far: triterpenes with a **RhoB** moiety directly attached to the skeleton of the triterpene, and compounds wherein these two units are separated by a suitable spacer.

To date, hybrid molecules have been prepared from oleanolic acid (**OA**, Figure 2), ursolic acid (**UA**), glycyrrhetinic acid (**GA**), betulinic acid (**BA**), maslinic acid (**MA**), augustic acid (**AU**), 11-keto-β-boswellic acid (**KBA**), asiatic acid (**AA**), tormentic acid (**TA**) and euscaphic acid (**EA**).

By means of suitable double-staining experiments, it could be shown that these hybrids are actually effective as mitocans [88], and preliminary molecular modeling studies suggest these compounds might target the mitochondrial NADH dehydrogenase and mitochondrial succinate dehydrogenase [89]. Both enzymes are part of the mitochondrial electron transport chain; this also suggests an increased production of reactive oxygen species (ROS). An increased production of ROS would lead to an oxidative damage of the cell and trigger apoptosis through an intrinsic pathway. Therefore, the integrity of the **RhoB** basic structure seems to be of crucial importance. It has been shown that derivatives from the triphenylmethane dye malachite green still exhibit increased cytotoxicity as compared to their parent compounds [90]. The cytotoxicity, however, of these hybrids was much lower than those observed for the **RhoB** derivatives (vide infra).

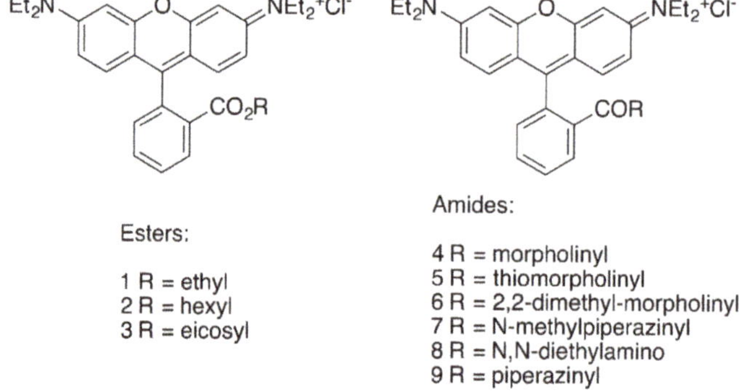

oleanolic acid (OA)

R1 = CH₃, R2 = H ursolic acid (UA)
R1 = CH₂OH, R2 = OH, asiatic acid (AA)

X = CH₂, betulinic acid (BA)
X = O, platanic acid (PA)

maslinic acid (MA)

augustic acid (AU)

glycyrrhetinic acid (GA)

11-keto-β-boswellic acid (KBA)

euscaphic acid (EA)

tormentic acid (TA)

Figure 2. Structure of some important pentacyclic triterpenoic acids.

The triterpenoid skeleton is equally important. Here, too, it was shown that "simple" RhoB conjugates **1–9** (Figure 3) also had lower cytotoxicity than the corresponding triterpenoid analogs, but their tumor cell/non-tumor cell selectivity was also diminished (Table 1) [91].

Esters:

1 R = ethyl
2 R = hexyl
3 R = eicosyl

Amides:

4 R = morpholinyl
5 R = thiomorpholinyl
6 R = 2,2-dimethyl-morpholinyl
7 R = N-methylpiperazinyl
8 R = N,N-diethylamino
9 R = piperazinyl

Figure 3. Structures of "simple" **RhoB** conjugates **1–9**.

Table 1. Cytotoxicity of selected "simple" **RhoB** conjugates.

Compound	A375	HT29	MCF-7	A2780	FaDu	NIH 3T3
RhoB	>30	>30	>30	>30	>30	>30
1	0.38	0.41	0.23	0.21	0.30	0.96
2	0.19	0.19	0.14	0.17	0.15	0.32
3	>30	>30	>30	>30	>30	>30
4	7.09	5.46	1.54	1.66	4.53	>30
5	1.79	1.54	0.44	0.52	1.12	5.09
6	3.05	1.74	0.49	0.70	1.52	7.92
7	16.05	17.34	3.74	3.62	11.78	>30
8	1.03	0.54	0.32	0.27	0.64	3.27
9	>30	>30	17.80	26.40	>30	>30

EC_{50} in µM from SRB assays; cut-off 30 µM.

Of special interest seems the morpholinyl derivative **4** inasmuch as this compound held the highest selectivity of this series with respect to MCF-7 carcinoma cells (S = ($EC_{50, \text{NIH 3T3}}$ / $EC_{50, \text{MCF-7}}$) > 19.5) and A2780 ovarian cancer cells (S = ($EC_{50, \text{NIH 3T3}}$ / $EC_{50, \text{A27807}}$) > 18.1) [90].

The highest cytotoxicity was observed for the hexyl ester **2** (EC_{50} = 0.15–0.19 µM) for the different tumor cell lines. Interestingly, an eicosyl ester **3** with a lipophilicity similar to that of triterpenoids did not show even moderate cytotoxicity [90], while hydroxycinnamic acid rhodamine B conjugates displayed good cytotoxicity in the low µM range [92].

The importance of the presence of a triterpenoid backbone is also evident from studies concerning RhoB steroid conjugates (Figure 4) [93]. In these studies, the reaction of the steroids cholesterol, testosterone, prednisone and abiraterone with an activated **RhoB** chloride furnished ester conjugates holding low EC_{50} values (SRB assays with several human tumor cell lines, Table 2). Thus, a testosterone conjugate **10** held EC_{50} = 60 nm for MCF-7 cells, but acted by necrosis (20%, A2780 cells). A prednisone conjugate **11** was less cytotoxic (0.2 µM for MCF-7 cells) but acted in A2780 cells mainly by apoptosis (48%) and late apoptosis (14%). In addition, this compound showed a higher selectivity for the A2780 tumor cells (S = 73) than for NIH 3T3 fibroblasts. For comparison, an abiratone conjugate **12** was less cytotoxic and also less selective [93].

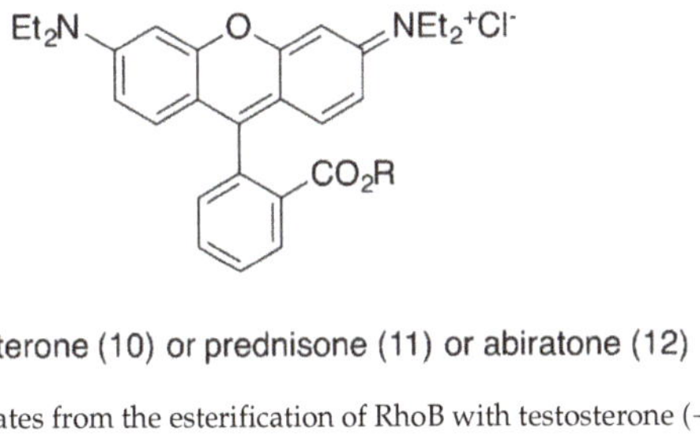

Et$_2$N — O — NEt$_2$$^+Cl^-$

CO$_2$R

R = testosterone (10) or prednisone (11) or abiratone (12)

Figure 4. RhoB steroid conjugates from the esterification of RhoB with testosterone (→ **10**), prednisone (→ **11**) or abiratone (→ **12**), respectively.

Table 2. Cytotoxicity of selected steroidal **RhoB** conjugates.

Compound	A375	HT29	MCF-7	A2780	FaDu	NIH 3T3
10	0.16	0.12	0.06	0.08	0.26	0.25
11	0.11	0.64	0.21	0.31	0.40	1.81
12	0.22	0.21	0.23	0.13	0.24	0.37

EC_{50} in µM from SRB assays.

A closer look at the cell cycle by FACS (with A2780 cells) showed a decrease of the G1 and G2/M peak with an increase of cells in the S phase. For cells treated with **11**, the S phase peak and the subG1/apoptosis peak increased significantly. However, for all compounds the selectivity between tumor cells and non-malignant fibroblasts NIH 3T3 was small and never exceeded 7.3 (**11**, for MCF-7 cells) [93].

A similar behavior was observed for dehydroabietylamine (DHAA) derivatives **13–16** (Figure 5, Table 3). These products were easily obtained from dehydroabietylamine by the microwave-assisted multicomponent Ugi reaction using paraformaldehyde, an isocyanide and **RhoB** with yields between 47 and 50% [94].

Table 3. Cytotoxicity of selected **DHAA**-derived **RhoB** conjugates.

Compound	A375	HT29	MCF-7	A2780	FaDu	NIH 3T3
13	3.2	0.18	0.10	0.37	0.23	0.28
14	0.23	0.32	0.16	0.57	0.35	0.41
15	0.20	0.28	0.12	0.66	0.32	0.44
16a/16b	>30	>30	>30	>30	>30	>30

EC_{50} in µM from SRB assays.

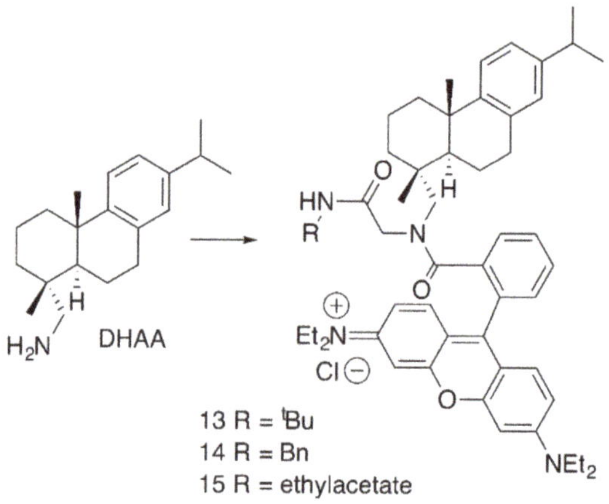

Figure 5. Dehydroabiethylamine (**DHAA**)-derived **RhoB** conjugates **13–15** obtained by Ugi-multi component reactions.

Although the cytotoxicity of these compounds was good, their pharmacological potential was restricted by low selectivity values. Interestingly enough, products **16a/16b** (Figure 6), having been obtained from a simple Schotten–Baumann reaction with **DHAA** and **RhoB,** were not cytotoxic at all [94]. As mentioned above, **RhoB** conjugates derived from primary amines are able to form intramolecular non-fluorescent spirolactams (here **16a**). From a photo-induced ring opening reaction, **16b** was obtained from **16a** very quickly within 10 s of irradiation either with visible light or with UV light (λ = 254 or 366 nm). This equilibrium is also strongly influenced by changes in temperature, and at room temperature **16a** dominates the equilibrium [94].

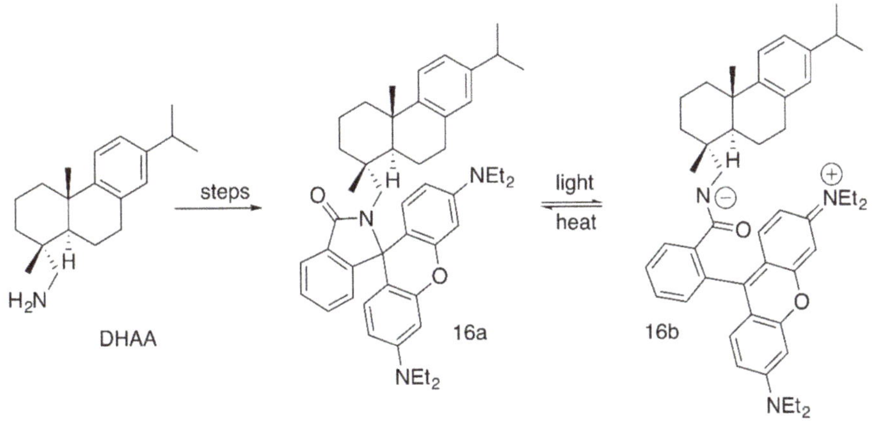

Figure 6. Synthesis of **DHAA**-derived **RhoB** conjugates **16a/16b** and their equilibrium.

As far as the triterpene **RhoB** conjugates are concerned, the **RhoB** moiety can be attached to the triterpenoid scaffold either directly (e.g., in form of a triterpene **RhoB** ester) or with the aid of a suitable spacer. Pentacyclic triterpenoic acids (Figure 2) holding an **RhoB** moiety without an extra spacer have been prepared by esterification of **UA**, **OA**, **GA** and **BA** with **RhoB**, respectively, (Figure 7) [95].

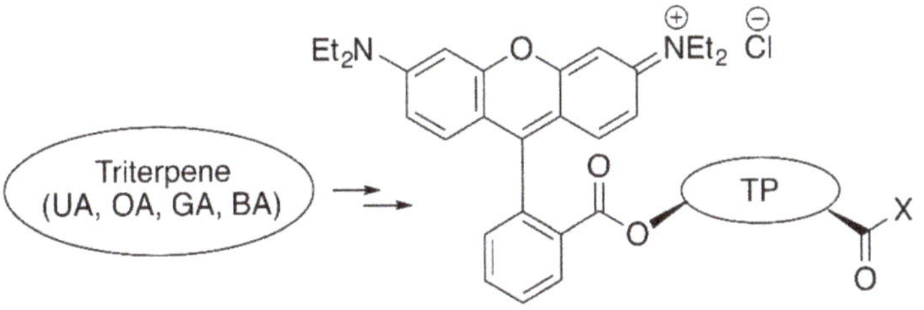

Figure 7. Un-spaced **UA-, OA-, GA-** and **BA-**derived esters **17–24** and amides **25–28**.

All of these compounds had EC$_{50}$ values between 0.02 and 15.8 μM (Table 4); thereby, the cytotoxicity of benzyl esters **21–24** was lower than the cytotoxicity of the methyl esters **17–20**, while the benzyl amides **25–28** were the most cytotoxic compounds of this series. The presence of a benzyl ester group as in **21–24** seems to be disadvantageous, while the opposite is true for the benzyl amides **25–28**. Compound **27** was the most cytotoxic compound (EC$_{50}$ = 0.02–0.08 μM), but it was not selective for human tumor cells. Extra staining experiments showed this compound to be accumulated in the mitochondria of A2780 cells and to act mainly by apoptosis [95].

Table 4. Cytotoxicity of un-spaced esters **UA-**, **OA-**, **GA-** and **BA**-derived esters **17–24** and amides **25–28**.

Compound	TP	R	FaDu	A2780	HT29	MCF-7	SW1736	NIH 3T3
17	UA	OMe	1.96	1.75	1.85	1.83	1.72	1.84
18	OA	OMe	1.99	1.14	2.75	2.31	1.76	2.63
19	GA	OMe	0.19	0.08	0.15	0.18	0.15	0.20
20	BA	OMe	1.29	0.42	0.61	0.81	0.74	1.77
21	UA	OBn	15.79	10.10	11.41	13.75	12.66	15.42
22	OA	OBn	9.12	3.35	8.90	9.40	9.05	11.25
23	GA	OBn	1.54	0.90	1.42	1.47	1.13	1.28
24	BA	OBn	7.59	3.36	5.33	5.05	6.43	8.04
25	UA	NBn	0.44	0.34	0.45	0.30	0.24	0.37
26	OA	NBn	0.50	0.32	0.46	0.36	0.27	0.40
27	GA	NBn	0.06	0.02	0.06	0.04	0.04	0.08
28	BA	NBn	0.54	0.31	0.53	0.47	0.45	0.54

EC_{50} in µM from SRB assays.

Noteworthy in this context is the higher cytotoxicity of the glycyrrhetinic acid derivatives as compared to analogs derived from **OA**, **UA** or **BA**. Extensions in the design of these compounds led to the synthesis of triterpene conjugates with further modifications in the backbone ($\rightarrow$ tormentic acid (**TA**) and euscaphic acid (**EA**)) as well as to changes in the ring size of the heterocyclic spacer between the backbone of the triterpene and the **RhoB** moiety (Figure 8).

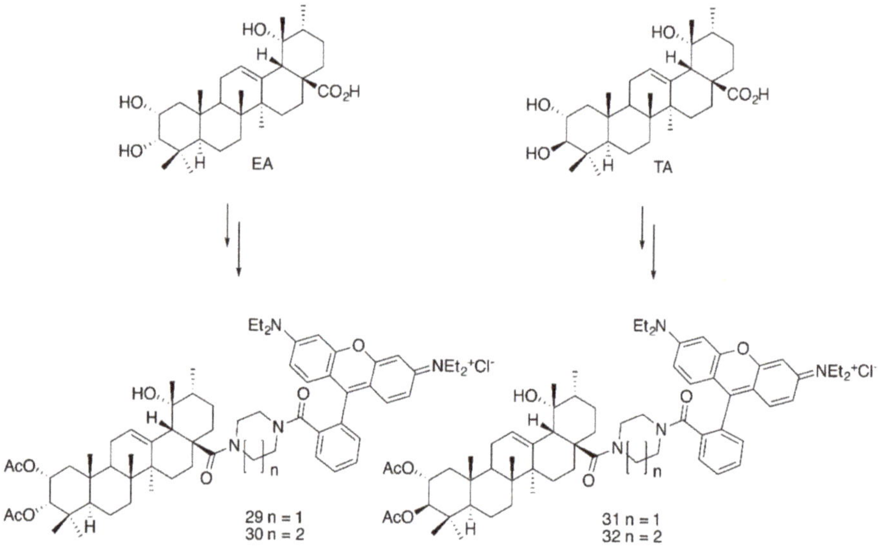

Figure 8. Synthesis of euscaphic (**EA**)- or tormentic acid (**TA**)-derived **RhoB** conjugates **29–32**.

The significantly higher cytotoxicity (Table 5) of **TA**-derived **32** seems particularly noteworthy when comparing the different spacers: Thereby, the presence of a homopiperazinyl spacer [96] (as in **32**) proved to be clearly superior to the piperazinyl moiety (as in **31**). A similar trend was also noted for **EA**-derived compounds **29** and **30**. On the other hand, **TA**-derived compounds were more cytotoxic than the corresponding **EA** derivatives. Interestingly, the absolute configuration at C–2 and C–3 in **TA** corresponds exactly to the configuration found in maslinic acid (**MA**). Several **MA** derivatives (for example [97,98], a diacetylated benzylamide **EM2**, Figure 9) were of higher cytotoxicity and better selectivity than their corresponding **OA** or **UA** derivatives.

Table 5. Cytotoxicity of euscaphic (**EA**)- or tormentic acid (**TA**)-derived **RhoB** conjugates **29–32**, asiatic acid (**AA**)-derived **33** and maslinic acid (**MA**)-derived amide **EM2**.

Compound	A375	HT29	MCF-7	A2780	FaDu	NIH 3T3
29	0.19	0,19	0.094	0.066	0.074	0.21
30	0.012	0.012	0.022	0.004	0.004	0.164
31	0.14	0.16	0.0084	0.037	0.041	0.25
32	0.06	0.005	0.008	0.001	0.001	0.19
33	n.d.	0.017	0.012	0.008	n.d.	0.178
EM2	n.d.	4.70	7.70	0.50	n.d.	33.8

EC_{50} in µM from SRB assays.

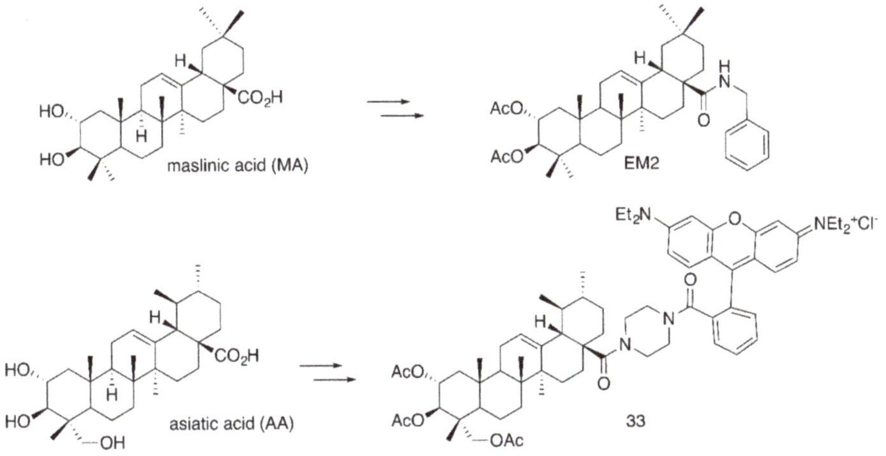

Figure 9. Synthesis of maslinic acid (**MA**)-derived **EM2** holding the same absolute configuration of hydroxyl groups in ring A as asiatic acid (**AA**)-derived conjugate **33**.

The same configuration is found in asiatic acid (**AA**). Again, its acetylated piperazinyl-rhodamine B conjugate **33** was most cytotoxic to many human tumor cell lines, being accumulated in the mitochondria, and it also acted as a mitocan [99]. However, for this compound an unusual non-linear rate of growth was detected for some human tumor cell lines (e.g., colorectal carcinoma HT29 and melanoma 518A2). In a bimodal manner at two different concentrations the tumor cells were killed, a phenomenon that might be due to an accelerated recovery of the mitochondrial membrane potential or due to a modulation of the mitochondrial permeability pores. However, at present a concentration-triggered activation of a metabolizing enzyme cannot completely be ruled out [99].

A graphical comparison of all derivatives (using the target line A2780 as an example) is given in Figure 10 including a comparison of tumor cell/non-tumor cell selectivity (A2780 vs. NIH 3T3) of all compounds.

From Figure 10 the high potential of compound **32** (selectivity for A2780 or FaDu cells, ca. 190) becomes clearly visible, making this compound an interesting drug for advanced testing and biological screening.

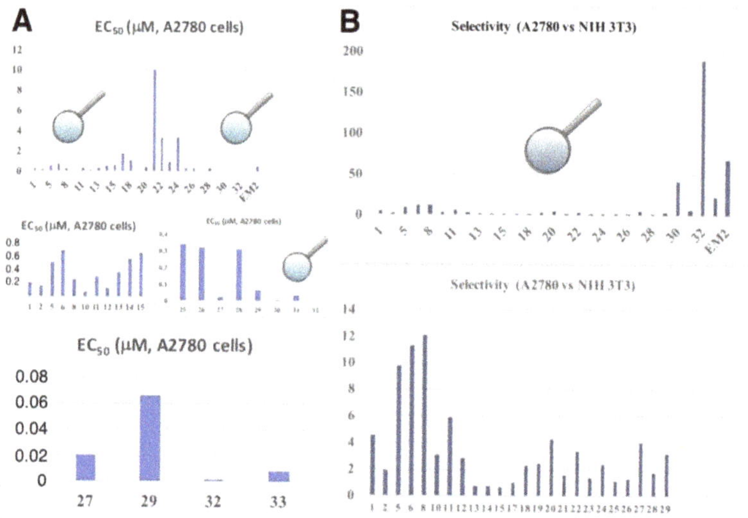

Figure 10. Graphical representation of the cytotoxicity of all compounds (EC$_{50}$ in µM) for the cell line A2780 (**A**) combined with a comparison of tumor cell/non-tumor cell selectivity (A2780 vs. NIH 3T3, selectivity = (EC$_{50,\,NIH3T3}$/EC$_{50,\,A2780}$)) of all compounds (**B**).

3. Conclusions

OA-derived **RhoB** conjugates appear to be superior to analog **UA**-derived compounds in the majority of cases with respect to their cytotoxicity. Although **AKBA**-derived derivatives have good cytotoxicity properties, they were found to be less cytotoxic compared to other triterpene carboxylic acid derivatives, but they often showed better tumor cell/non-tumor cell selectivity. So far, the best cytotoxicity properties have been found for **MA-**, **EA-** and **TA-**derived derivatives. These allowed the transition to compounds of nano-molar activity, while many other triterpene carboxylic acid derivatives were cytotoxic only on a micro-molar concentration range. **MA-** derived derivatives seem to be approximately equivalent to **EA-**derived compounds. They are currently only surpassed in many tumor cell lines only by the analogous derivatives from **TA**. From results available so far, it can be concluded that compounds holding a homopiperazinyl spacer are superior to those with a piperazinyl spacer. This underlines the importance of the spacer for obtaining good cytotoxicity properties. Replacement of the secondary amide derived spacer by a primary amine like ethylenediamine has invariably led to **RhoB** conjugates of insignificant cytotoxicity (EC$_{50}$ > 30 µM) due to the formation of a spirolactam holding no positive charge in the RhoB part.

However, the presence of a distal cation is not sufficient to obtain compounds with excellent cytotoxicity, as has been shown for several quaternary ammonium compounds or compounds where the **RhoB** part has been replaced by, for example, malachite green, a BODIPY residue or a safirinium group. In addition, the latter compounds do not act as mitocans, since their primary target is the endoplasmic reticulum.

A statement on the extent to which the replacement of the **RhoB** group with another rhodamine has a positive effect on biological activity cannot be made at present. The cytotoxic properties of these compounds, other spacers and other triterpene carboxylic acids are currently the subject of further investigation. The combination of the "correct" parent structure, the "correct" spacer and the "correct" **RhoB** seems to be decisive for the ability of the conjugate to accumulate in mitochondria. So far, a tormentic acid acid-derived **RhoB** conjugate **32** holding a homopiperazinyl spacer can be regarded as the most promising candidate for further biological studies. At present, no extended investigations have been carried out on the precise mode of action of these molecules.

Author Contributions: Conceptualization, R.C. and H.-P.D.; validation, S.H., I.S. and R.C.; investigation, S.H.; writing—original draft preparation, R.C. writing—review and editing, R.C., S.H., I.S., H.-P.D. All authors have read and agreed to the published version of the manuscript.

Funding: We acknowledge the financial support within the funding program Open Access Publishing by the German Research Foundation (DFG).

Acknowledgments: We like to thank all members of our groups for their continuing support and skillful work in the laboratories.

Conflicts of Interest: The authors declare no conflict of interest.

References

1. Available online: https://www.who.int/health-topics/cancer-tab=tab_1 (accessed on 20 October 2020).
2. Bray, F.; Jemal, A.; Grey, N.; Ferlay, J.; Forman, D. Global cancer transitions according to the Human Development Index (2008–2030): A population-based study. *Lancet Oncol.* **2012**, *13*, 790–801. [CrossRef]
3. Siegel, R.L.; Mph, K.D.M.; Sauer, A.G.; Fedewa, S.A.; Butterly, L.F.; Anderson, J.C.; Cercek, A.; Smith, R.A.; Jemal, A. Colorectal cancer statistics, 2020. *CA: A Cancer J. Clin.* **2020**, *70*, 145–164. [CrossRef] [PubMed]
4. Markham, M.J.; Wachter, K.; Agarwal, N.; Bertagnolli, M.M.; Chang, S.M.; Dale, W.; Diefenbach, C.S.M.; Rodriguez-Galindo, C.; George, D.J.; Gilligan, T.D.; et al. Clinical Cancer Advances 2020: Annual Report on Progress Against Cancer From the American Society of Clinical Oncology. *J. Clin. Oncol.* **2020**, *38*, 1081. [CrossRef] [PubMed]
5. Oostveen, M.; Pritchard-Jones, K. Pharmacotherapeutic Management of Wilms Tumor: An Update. *Pediatr. Drugs* **2019**, *21*, 1–13. [CrossRef]
6. Di Paolo, A.; Arrigoni, E.; Luci, G.; Cucchiara, F.; Danesi, R.; Galimberti, S. Precision Medicine in Lymphoma by Innovative Instrumental Platforms. *Front. Oncol.* **2019**, *9*, 1417. [CrossRef]
7. Available online: https://www.paho.org/hq/index.php?option=com_docman&view=download&category_slug=4-cancer-country-profiles-2020&alias=51561-global-cancer-profile-2020&Itemid=270&lang=fr (accessed on 20 October 2020).
8. Green, D.R. Cancer and Apoptosis: Who Is Built to Last? *Cancer Cell* **2017**, *31*, 2–4. [CrossRef]
9. Pretta, A.; Trevisi, E.; Bregni, G.; Deleporte, A.; Hendlisz, A.; Sclafani, F. Treatment compliance in early-stage anal cancer. *Ann. Oncol.* **2020**, *31*, 1282–1284. [CrossRef]
10. Muzumder, S.; Srikantia, N.; Vashishta, G.D.; Udayashankar, A.H.; Raj, J.M.; Sebastian, M.G.J.; Kainthaje, P.B. Compliance, toxicity and efficacy in weekly versus 3-weekly cisplatin concurrent chemoradiation in locally advanced head and neck cancer. *J. Radiother. Pr.* **2018**, *18*, 21–25. [CrossRef]
11. Gong, L.; Yan, Q.; Zhang, Y.; Fang, X.; Liu, B.; Guan, X.-Y. Cancer cell reprogramming: A promising therapy converting malignancy to benignity. *Cancer Commun.* **2019**, *39*, 1–13. [CrossRef]
12. Fuchs, Y.; Steller, H. Live to die another way: Modes of programmed cell death and the signals emanating from dying cells. *Nat. Rev. Mol. Cell Biol.* **2015**, *16*, 329–344. [CrossRef]
13. Wang, C.X.; Youle, R.J. The Role of Mitochondria in Apoptosis. *Annu. Rev. Genet.* **2009**, *43*, 95–118. [CrossRef] [PubMed]
14. Johnstone, R.W.; Ruefli, A.A.; Lowe, S.W. Apoptosis: A link between cancer genetics and chemotherapy. *Cell* **2002**, *108*, 153–164. [CrossRef]
15. Fiers, W.; Beyaert, R.; Declercq, W.; Vandenabeele, P. More than one way to die: Apoptosis, necrosis and reactive oxygen damage. *Oncogene* **1999**, *18*, 7719–7730. [CrossRef] [PubMed]
16. Galluzzi, L.; Pedro, J.M.B.-S.; Kroemer, G. Organelle-specific initiation of cell death. *Nat. Cell Biol.* **2014**, *16*, 728–736. [CrossRef] [PubMed]
17. Bing, Y.; Dong, L.; Neuzil, J. Mitochondria: An intriguing target for killing tumour-initiating cells. *Mitochondrion* **2016**, *26*, 86–93. [CrossRef]
18. Pei, J.; Panina, S.B.; Kirienko, N.V. An Automated Differential Nuclear Staining Assay for Accurate Determination of Mitocan Cytotoxicity. *J. Vis. Exp.* **2020**, *159*, e61295. [CrossRef]
19. Panina, S.B.; Pei, J.; Baran, N.; Konopleva, M.; Kirienko, N.V. Utilizing Synergistic Potential of Mitochondria-Targeting Drugs for Leukemia Therapy. *Front. Oncol.* **2020**, *10*, 435. [CrossRef]
20. Ralph, S.J.; Neuzil, J. Mitochondria as Targets for Cancer Therapy. In *Mitochondria and Cancer*; Costello, L., Singh, K., Eds.; Springer: New York, NY, USA, 2009; pp. 211–249.

21. Ralph, S.J.; Rodríguez-Enríquez, S.; Neuzil, J.; Moreno-Sánchez, R. Bioenergetic pathways in tumor mitochondria as targets for cancer therapy and the importance of the ROS-induced apoptotic trigger. *Mol. Asp. Med.* **2010**, *31*, 29–59. [CrossRef]

22. Ralph, S.J.; Rodríguez-Enríquez, S.; Neuzil, J.; Saavedra, E.; Moreno-Sánchez, R. The causes of cancer revisited: "Mitochondrial malignancy" and ROS-induced oncogenic transformation – Why mitochondria are targets for cancer therapy. *Mol. Asp. Med.* **2010**, *31*, 145–170. [CrossRef] [PubMed]

23. Chiu, H.Y.; Tay, E.X.Y.; Ong, D.S.T.; Taneja, R. Mitochondrial Dysfunction at the Center of Cancer Therapy. *Antioxidants Redox Signal.* **2020**, *32*, 309–330. [CrossRef]

24. Singh, Y.; Viswanadham, K.K.D.R.; Pawar, V.K.; Meher, J.; Jajoriya, A.K.; Omer, A.; Jaiswal, S.; Dewangan, J.; Bora, H.K.; Singh, P.; et al. Induction of Mitochondrial Cell Death and Reversal of Anticancer Drug Resistance via Nanocarriers Composed of a Triphenylphosphonium Derivative of Tocopheryl Polyethylene Glycol Succinate. *Mol. Pharm.* **2019**, *16*, 3744–3759. [CrossRef] [PubMed]

25. Nguyen, C.; Pandey, S. Exploiting Mitochondrial Vulnerabilities to Trigger Apoptosis Selectively in Cancer Cells. *Cancers* **2019**, *11*, 916. [CrossRef] [PubMed]

26. Olivas-Aguirre, M.; Pottosin, I.; Dobrovinskaya, O. Mitochondria as emerging targets for therapies against T cell acute lymphoblastic leukemia. *J. Leukoc. Biol.* **2019**, *105*, 935–946. [CrossRef]

27. Sassi, N.; Biasutto, L.; Mattarei, A.; Carraro, M.; Giorgio, V.; Citta, A.; Bernardi, P.; Garbisa, S.; Szabò, I.; Paradisi, C.; et al. Cytotoxicity of a mitochondriotropic quercetin derivative: Mechanisms. *Biochim. et Biophys. Acta (BBA) - Gen. Subj.* **2012**, *1817*, 1095–1106. [CrossRef] [PubMed]

28. Sassi, N.; Mattarei, A.; Azzolini, M.; Bernardi, P.; Szabò, I.; Paradisi, C.; Zoratti, M.; Biasutto, L. Mitochondria-targeted resveratrol derivatives act as cytotoxic pro-oxidants. *Curr. Pharm. Des.* **2014**, *20*, 172–179. [CrossRef]

29. Chen, G.; Wang, F.; Trachootham, D.; Huang, P. Preferential killing of cancer cells with mitochondrial dysfunction by natural compounds. *Mitochondrion* **2010**, *10*, 614–625. [CrossRef]

30. Guzman-Villanueva, D.; Weissig, V. Mitochondria-Targeted Agents: Mitochondriotropics, Mitochondriotoxics, and Mitocans. *Muscarinic Receptors* **2016**, *240*, 423–438. [CrossRef]

31. Rohlena, J.; Dong, L.-F.; Neuzil, J. Targeting the mitochondrial electron transport chain complexes for the induction of apoptosis and cancer treatment. *Curr. Pharm. Biotechnol.* **2013**, *14*, 377–389. [CrossRef]

32. Rohlena, J.; Dong, L.; Ralph, S.J.; Neuzil, J. Anticancer Drugs Targeting the Mitochondrial Electron Transport Chain. *Antioxidants Redox Signal.* **2011**, *15*, 2951–2974. [CrossRef]

33. Kornblihtt, L.I.; Carreras, M.C.A.; Blanco, G. Targeting Mitophagy in Combined Therapies of Haematological Malignancies. *Autophagy Curr. Trends Cell. Physiol. Pathol.* **2016**, *19*, 411–431. [CrossRef]

34. Neuzil, J.; Dong, L.-F.; Rohlena, J.; Truksa, J.; Ralph, S.J. Classification of mitocans, anti-cancer drugs acting on mitochondria. *Mitochondrion* **2013**, *13*, 199–208. [CrossRef] [PubMed]

35. Neuzil, J.; Dyason, J.C.; Freeman, R.; Dong, L.; Prochazka, L.; Wang, X.-F.; Scheffler, I.; Ralph, S.J. Mitocans as anti-cancer agents targeting mitochondria: Lessons from studies with vitamin E analogues, inhibitors of complex II. *J. Bioenerg. Biomembr.* **2007**, *39*, 65–72. [CrossRef] [PubMed]

36. Neuzil, J.; Wang, X.-F.; Dong, L.-F.; Low, P.; Ralph, S.J. Molecular mechanism of 'mitocan'-induced apoptosis in cancer cells epitomizes the multiple roles of reactive oxygen species and Bcl-2 family proteins. *FEBS Lett.* **2006**, *580*, 5125–5129. [CrossRef] [PubMed]

37. Morrison, B.J.; Andera, L.; Reynolds, B.A.; Ralph, S.J.; Neuzil, J. Future use of mitocans against tumour-initiating cells? *Mol. Nutr. Food Res.* **2009**, *53*, 147–153. [CrossRef] [PubMed]

38. Hahn, T.; Polanczyk, M.J.; Borodovsky, A.; Ramanathapuram, L.V.; Akporiaye, E.T.; Ralph, S.J. Use of anti-cancer drugs, mitocans, to enhance the immune responses against tumors. *Curr. Pharm. Biotechnol.* **2013**, *14*, 357–376. [CrossRef] [PubMed]

39. Amaral, L.; Spengler, G.; Molnar, J. Identification of Important Compounds Isolated from Natural Sources that Have Activity Against Multidrug-resistant Cancer Cell Lines: Effects on Proliferation, Apoptotic Mechanism and the Efflux Pump Responsible for Multi-resistance Phenotype. *Anticancer. Res.* **2016**, *36*, 5665–5672. [CrossRef]

40. Abdelmageed, N.; Morad, S.A.; Elghoneimy, A.A.; Syrovets, T.; Simmet, T.; El-Zorba, H.; El-Banna, H.A.; Cabot, M.; Abdel-Aziz, M.I. Oleanolic acid methyl ester, a novel cytotoxic mitocan, induces cell cycle arrest and ROS-Mediated cell death in castration-resistant prostate cancer PC-3 cells. *Biomed. Pharmacother.* **2017**, *96*, 417–425. [CrossRef]

41. Bhola, P.D.; Letai, A. Mitochondria—Judges and Executioners of Cell Death Sentences. *Mol. Cell* **2016**, *61*, 695–704. [CrossRef]

42. Jacotot, E.E.; Ferri, K.K.; El Hamel, C.C.; Brenner, C.C.; Druillennec, S.S.; Hoebeke, J.; Rustin, P.P.; Métivier, D.D.; Lenoir, C.C.; Geuskens, M.; et al. Control of Mitochondrial Membrane Permeabilization by Adenine Nucleotide Translocator Interacting with HIV-1 Viral Protein R and Bcl-2. *J. Exp. Med.* **2001**, *193*, 509–520. [CrossRef]

43. Zamzami, N.; Kroemer, G. The mitochondrion in apoptosis: How Pandora's box opens. *Nat. Rev. Mol. Cell Biol.* **2001**, *2*, 67–71. [CrossRef]

44. Zamzami, N.; Maisse, C.; Métivier, D.; Kroemer, G. Chapter 8 Measurement of membrane permeability and permeability transition of mitochondria. *Echinoderms Part B* **2001**, *65*, 147–158. [CrossRef]

45. Yang, H.; Dou, Q.P. Targeting apoptosis pathway with natural terpenoids: Implications for treatment of breast and prostate cancer. *Curr. Drug Targets* **2010**, *11*, 733–744. [CrossRef] [PubMed]

46. Fulda, S.; Kroemer, G. Mitochondria as Therapeutic Targets for the Treatment of Malignant Disease. *Antioxidants Redox Signal.* **2011**, *15*, 2937–2949. [CrossRef] [PubMed]

47. Serafim, T.L.; Carvalho, F.S.; Pereira, G.; Greene, A.; Perkins, E.; Holy, J.; Krasutsky, D.A.; Kolomitsyna, O.N.; Krasutsky, P.A.; Oliveira, P.J. Lupane triterpenoids as mitocans against breast cancer cells. *Eur. J. Clin. Invest.* **2013**, *43*, 48–49.

48. Serafim, T.L.; Carvalho, F.S.; Bernardo, T.C.; Pereira, G.C.; Perkins, E.; Holy, J.; Krasutsky, D.A.; Kolomitsyna, O.N.; Krasutsky, P.A.; Oliveira, P.J. New derivatives of lupine triterpenoids disturb breast cancer mitochondria and induce cell death. *Bioorg. Med. Chem.* **2014**, *22*, 6270–6287. [CrossRef] [PubMed]

49. Grymel, M.; Zawojak, M.; Adamek, J. Triphenylphosphonium Analogues of Betulin and Betulinic Acid with Biological Activity: A Comprehensive Review. *J. Nat. Prod.* **2019**, *82*, 1719–1730. [CrossRef]

50. Kolevzon, N.; Kuflik, U.; Shmuel, M.; Benhamron, S.; Ringel, I.; Yavin, E. Multiple Triphenylphosphonium Cations as a Platform for the Delivery of a Pro-Apoptotic Peptide. *Pharm. Res.* **2011**, *28*, 2780–2789. [CrossRef]

51. Spivak, A.Y.; Nedopekina, D.A.; Khalitova, R.R.; Gubaidullin, R.; Odinokov, V.N.; Bel'Skii, Y.P.; Bel'Skaya, N.V.; Khazanov, V.A. Triphenylphosphonium cations of betulinic acid derivatives: Synthesis and antitumor activity. *Med. Chem. Res.* **2017**, *26*, 518–531. [CrossRef]

52. Akhmetova, V.R.; Nedopekina, D.A.; Shakurova, E.R.; Khalitova, R.R.; Gubaidullin, R.R.; Odinokov, V.N.; Dzhemilev, U.M.; Bel'Skii, Y.P.; Bel'Skaya, N.V.; Stankevich, S.A.; et al. Synthesis of lupane triterpenoids with triphenylphosphonium substituents and studies of their antitumor activity. *Russ. Chem. Bull.* **2013**, *62*, 188–198. [CrossRef]

53. Huang, H.; Wu, H.; Huang, Y.; Zhang, S.; Lam, Y.-W.; Ao, N. Antitumor activity and antitumor mechanism of triphenylphosphonium chitosan against liver carcinoma. *J. Mater. Res.* **2018**, *33*, 2586–2597. [CrossRef]

54. Strobykina, I.Y.; Andreeva, O.V.; Belenok, M.G.; Semenova, M.N.; Semenov, V.V.; Chuprov-Netochin, R.N.; Sapunova, A.S.; Voloshina, A.D.; Dobrynin, A.B.; Semenov, V.E.; et al. Triphenylphosphonium conjugates of 1,2,3-triazolyl nucleoside analogues. Synthesis and cytotoxicity evaluation. *Med. Chem. Res.* **2020**, *29*, 1–15. [CrossRef]

55. Jara, J.A.; Castro-Castillo, V.; Saavedra-Olavarría, J.; Peredo, L.; Pavanni, M.; Jaña, F.; Letelier, M.E.; Parra, E.; Becker, M.I.; Morello, A.; et al. Antiproliferative and Uncoupling Effects of Delocalized, Lipophilic, Cationic Gallic Acid Derivatives on Cancer Cell Lines. Validation in Vivo in Singenic Mice. *J. Med. Chem.* **2014**, *57*, 2440–2454. [CrossRef] [PubMed]

56. Kafkova, A.; Trnka, J. Mitochondria-targeted compounds in the treatment of cancer. *Neoplasma* **2020**, *67*, 450–460. [CrossRef] [PubMed]

57. Ye, Y.; Zhang, T.; Yuan, H.; Li, D.; Lou, H.; Fan, P. Mitochondria-Targeted Lupane Triterpenoid Derivatives and Their Selective Apoptosis-Inducing Anticancer Mechanisms. *J. Med. Chem.* **2017**, *60*, 6353–6363. [CrossRef]

58. Sarek, J.; Biedermann, D.; Eignerova, B.; Hajduch, M. Synthesis and Evaluation of Biological Activity of the Quaternary Ammonium Salts of Lupane-, Oleanane-, and Ursane-Type Acids. *Synthesis* **2010**, *22*, 3839–3848. [CrossRef]

59. Kataev, V.E.; Strobykina, I.Y.; Zakharova, L.Y. Quaternary ammonium derivatives of natural terpenoids. Synthesis and properties. *Russ. Chem. Bull.* **2014**, *63*, 1884–1900. [CrossRef]

60. Brandes, B.; Koch, L.; Hoenke, S.; Deigner, H.-P.; Csuk, R. The presence of a cationic center is not alone decisive for the cytotoxicity of triterpene carboxylic acid amides. *Steroids* **2020**, *163*, 108713. [CrossRef]

61. Perreault, M.; Maltais, R.; Dutour, R.; Poirier, D. Explorative study on the anticancer activity, selectivity and metabolic stability of related analogs of aminosteroid RM-133. *Steroids* **2016**, *115*, 105–113. [CrossRef]

62. Brandes, B.; Hoenke, S.; Fischer, L.; Csuk, R. Design, synthesis and cytotoxicity of BODIPY FL labelled triterpenoids. *Eur. J. Med. Chem.* **2020**, *185*, 111858. [CrossRef]

63. Barut, B.; Çoban, Ö.; Özgür, Y.C.; Baş, H.; Sari, S.; Biyiklioglu, Z.; Demirbaş, Ü.; Özel, A. Synthesis, DNA interaction, in vitro/in silico topoisomerase II inhibition and photodynamic therapy activities of two cationic BODIPY derivatives. *Dye. Pigment.* **2020**, *174*, 108072. [CrossRef]

64. Barut, B.; Özgür, Y.C.; Sari, S.; Çoban, Ö.; Keleş, T.; Biyiklioglu, Z.; Abudayyak, M.; Demirbaş, Ü.; Özel, A. Novel water soluble BODIPY compounds: Synthesis, photochemical, DNA interaction, topoisomerases inhibition and photodynamic activity properties. *Eur. J. Med. Chem.* **2019**, *183*, 11685. [CrossRef] [PubMed]

65. Taki, S.; Ardestani, M.S. Novel nanosized AS1411-chitosan-BODIPY conjugate for molecular fluorescent imaging. *Int. J. Nanomed.* **2019**, *14*, 3543–3555. [CrossRef] [PubMed]

66. Li, M.; Li, X.; Cao, Z.; Wu, Y.; Chen, J.-A.; Gao, J.; Wang, Z.; Guo, W.; Gu, X. Mitochondria-targeting BODIPY-loaded micelles as novel class of photosensitizer for photodynamic therapy. *Eur. J. Med. Chem.* **2018**, *157*, 599–609. [CrossRef] [PubMed]

67. Kraft, O.; Kozubek, M.; Hoenke, S.; Serbian, I.; Major, D.; Csuk, R. Cytotoxic triterpenoid–safirinium conjugates target the endoplasmic reticulum. *Eur. J. Med. Chem.* **2020**, 112920. [CrossRef] [PubMed]

68. Cooksey, C. Quirks of dye nomenclature. 5. Rhodamines. *Biotech. Histochem.* **2015**, *91*, 1–6. [CrossRef]

69. Ceresole, M. Rhodamin Dye. U.S. Patent 516589 A 18940313, 13 March 1894.

70. Ceresole, M. Red Dyestuff. U.S. Patent 456081 18910714, 14 July 1891.

71. Kim, H.N.; Lee, M.H.; Kim, H.J.; Kim, J.S.; Yoon, J. A new trend in rhodamine-based chemosensors: Application of spirolactam ring-opening to sensing ions. *Chem. Soc. Rev.* **2008**, *37*, 1465–1472. [CrossRef]

72. Wang, D.; Wang, Z.; Li, Y.; Song, Y.; Song, Y.; Zhang, M.; Yu, H. A single rhodamine spirolactam probe for localization and pH monitoring of mitochondrion/lysosome in living cells. *New J. Chem.* **2018**, *42*, 11102–11108. [CrossRef]

73. Adamovich, L.P.; Mel'nik, V.V.; McHedlov-Petrosyan, N.O. Rhodamine B equilibriums in water-salt solutions. *Zh. Fiz. Khim.* **1979**, *53*, 356–359.

74. Hinckley, D.A.; Seybold, P.G. Thermodynamics of the rhodamine B lactone zwitterion equilibrium: An undergraduate laboratory experiment. *J. Chem. Educ.* **1987**, *64*, 362. [CrossRef]

75. Ramart-Lucas, P. Structure of rhodamine from its absorption spectrum. *Compt. Rend.* **1938**, *207*, 1416–1418.

76. Ramette, R.W.; Blackburn, T.R. Specific effects of cations on rhodamine B equilibriums. *J. Phys. Chem.* **1958**, *62*, 1601–1603. [CrossRef]

77. Ramette, R.W.; Sandell, E.B. Rhodamine B equilibriums. *J. Am. Chem. Soc.* **1956**, *78*, 4872–4878. [CrossRef]

78. Hinckley, D.A.; Seybold, P.G.; Borris, D.P. Solvatochromism and thermochromism of rhodamine solutions. *Spectrochim. Acta Part A Mol. Spectrosc.* **1986**, *42*, 747–754. [CrossRef]

79. Beija, M.; Afonso, C.A.M.; Martinho, J.M.G. Synthesis and applications of Rhodamine derivatives as fluorescent probes. *Chem. Soc. Rev.* **2009**, *38*, 2410–2433. [CrossRef]

80. Arbeloa, F.L.; Estevez, M.J.T. Photophysics of rhodamines: Molecular structure and solvent effects. *J. Phys. Chem.* **1991**, *95*, 2203–2208. [CrossRef]

81. Arbeloa, T.L.; Estévez, M.T.; López-Arbeloa, F.; Aguirresacona, I.U.; Arbeloa, I.L. Luminescence properties of rhodamines in water/ethanol mixtures. *J. Lumin* **1991**, *48*, 400–404. [CrossRef]

82. Milvy, P.; Kay, K. Mutagenicity of 19 major graphic arts and printing dyes. *J. Toxicol. Environ. Heal. Part A* **1978**, *4*, 31–36. [CrossRef]

83. Smart, P.L. A review of the toxicity of twelve fluorescent dyes used for water tracing. *NSS Bull.* **1984**, *46*, 21–33.

84. Umeda, M. Experimental production of sarcoma in rats by injection of rhodamine B. I. *Gann* **1952**, *43*, 120–122.

85. Umeda, M. Experimental study of xanthene dyes as carcinogenic agents. *Gann* **1956**, *47*, 51–78.

86. Nestmann, E.R.; Douglas, G.R.I.; Matula, T.E.; Grant, C.; Kowbel, D.J. Mutagenic activity of rhodamine dyes and their impurities as detected by mutation induction in Salmonella and DNA damage in Chinese hamster ovary cells. *Cancer Res.* **1979**, *39*, 4412–4417. [PubMed]

87. Elliott, G.S.; Mason, R.W.; Edwards, I.R. Studies on the pharmacokinetics and mutagenic potential of rhodamine b. *J. Toxicol. Clin. Toxicol.* **1990**, *28*, 45–59. [CrossRef] [PubMed]

88. Sommerwerk, S.; Heller, L.; Kerzig, C.; Kramell, A.E.; Csuk, R. Rhodamine B conjugates of triterpenoic acids are cytotoxic mitocans even at nanomolar concentrations. *Eur. J. Med. Chem.* **2017**, *127*, 1–9. [CrossRef] [PubMed]

89. Macasoi, I.; Mioc, M.; Vaduva, D.B.; Ghiulai, R.; Mioc, A.; Soica, C.; Muntean, D.; Dumitrascu, V. In silico Evaluation of the Antiproliferative Mithocondrial Targeted Mechanism of Action of Some Pentacyclic Triterpene Derivatives. *Rev. Chim.* **2019**, *69*, 3361–3363. [CrossRef]

90. Friedrich, S.; Serbian, I.; Hoenke, S.; Wolfram, R.K.; Csuk, R. Synthesis and cytotoxic evaluation of malachite green derived oleanolic and ursolic acid piperazineamides. *Med. Chem. Res.* **2020**, *29*, 926–933. [CrossRef]

91. Serbian, I.; Hoenke, S.; Kraft, O.; Csuk, R. Ester and amide derivatives of rhodamine B exert cytotoxic effects on different human tumor cell lines. *Med. Chem. Res.* **2020**, *29*, 1655–1661. [CrossRef]

92. Kozubek, M.; Serbian, I.; Hoenke, S.; Kraft, O.; Csuk, R. Synthesis and cytotoxic evaluation of hydroxycinnamic acid rhodamine B conjugates. *Results Chem.* **2020**, *2*, 100057. [CrossRef]

93. Serbian, I.; Hoenke, S.; Csuk, R. Synthesis of some steroidal mitocans of nanomolar cytotoxicity acting by apoptosis. *Eur. J. Med. Chem.* **2020**, *199*, 112425. [CrossRef]

94. Wiemann, J.; Fischer, L.; Kessler, J.; Ströhl, D.; Csuk, R. Ugi Multicomponent-reaction: Syntheses of cytotoxic dehydroabietylamine derivatives. *Bioorganic Chem.* **2018**, *81*, 567–576. [CrossRef]

95. Wolfram, R.K.; Heller, L.; Csuk, R. Targeting mitochondria: Esters of rhodamine B with triterpenoids are mitocanic triggers of apoptosis. *Eur. J. Med. Chem.* **2018**, *152*, 21–30. [CrossRef]

96. Wolfram, R.K.; Fischer, L.; Kluge, R.; Ströhl, D.; Al-Harrasi, A.; Csuk, R. Homopiperazine-rhodamine B adducts of triterpenoic acids are strong mitocans. *Eur. J. Med. Chem.* **2018**, *155*, 869–879. [CrossRef]

97. Siewert, B.; Pianowski, E.; Obernauer, A.; Csuk, R. Towards cytotoxic and selective derivatives of maslinic acid. *Bioorg. Med. Chem.* **2014**, *22*, 594–615. [CrossRef]

98. Sommerwerk, S.; Heller, L.; Kuhfs, J.; Csuk, R. Urea derivates of ursolic, oleanolic and maslinic acid induce apoptosis and are selective cytotoxic for several human tumor cell lines. *Eur. J. Med. Chem.* **2016**, *119*, 1–16. [CrossRef]

99. Kahnt, M.; Wiemann, J.; Fischer, L.; Sommerwerk, S.; Csuk, R. Transformation of asiatic acid into a mitocanic, bimodal-acting rhodamine B conjugate of nanomolar cytotoxicity. *Eur. J. Med. Chem.* **2018**, *159*, 143–148. [CrossRef] [PubMed]

Sample Availability: Samples of the compounds are not available from the authors.

Review

Biophysical Characterization and Anticancer Activities of Photosensitive Phytoanthraquinones Represented by Hypericin and Its Model Compounds

Valéria Verebová [1], Jiří Beneš [2] and Jana Staničová [1,2,*]

[1] Department of Chemistry, Biochemistry and Biophysics, University of Veterinary Medicine & Pharmacy, Komenského 73, 041 81 Košice, Slovakia; valeria.verebova@uvlf.sk
[2] Institute of Biophysics and Informatics, First Faculty of Medicine, Charles University, Kateřinská 1, 121 08 Prague, Czech Republic; jiri.benes@lf1.cuni.cz
* Correspondence: jana.stanicova@uvlf.sk; Tel.: +421-915-984-613

Academic Editors: José Antonio Lupiáñez, Amalia Pérez-Jiménez and Eva E. Rufino-Palomares
Received: 27 October 2020; Accepted: 28 November 2020; Published: 1 December 2020

Abstract: Photosensitive compounds found in herbs have been reported in recent years as having a variety of interesting medicinal and biological activities. In this review, we focus on photosensitizers such as hypericin and its model compounds emodin, quinizarin, and danthron, which have antiviral, antifungal, antineoplastic, and antitumor effects. They can be utilized as potential agents in photodynamic therapy, especially in photodynamic therapy (PDT) for cancer. We aimed to give a comprehensive summary of the physical and chemical properties of these interesting molecules, emphasizing their mechanism of action in relation to their different interactions with biomacromolecules, specifically with DNA.

Keywords: natural photosensitive compounds; anticancer activity; hypericin; emodin; quinizarin; danthron; interaction; DNA

---

1. Photodynamic Therapy

Photodynamic therapy (PDT) is part of photochemotherapy and requires the presence of a photosensitive substance (drug, PS), oxygen, and a powerful light source in the area of absorption of the PS used. The main requirements for activating the properties of a PS are its selective accumulation in tumor tissue, high intensity of absorption in the visible and near-infrared region of the spectrum, low level of dark toxicity, and absence of side-effects [1,2]. Selective accumulation and retention of PS in tumor tissues rather than in the surrounding healthy tissue lead to selective destruction of the tumor in PDT, while the surrounding healthy tissue remains intact. Such selectivity is one of the biggest advantages of this method, which may be substituted in some cases for chemotherapy, radiotherapy, or surgery in the treatment of cancer. Due to drug excretion [3] and redistribution, the effective therapeutic dose entering tumor cells is only a fraction of the administered PS. Administration of increased amounts of therapeutics is not possible because they have cytotoxic effects, which could cause significant toxicity in healthy cells. It is very important therefore to find alternative approaches, which increase the efficacy of the drug dose in the tumor and decrease the dose in healthy tissue [4]. Higher selectivity of PSs for tumor cells can be achieved by combining them with transport agents, which preferentially interact with tumor cells, ensure the selective accumulation of the drug within the diseased tissue, and deliver the desired therapeutic drug concentration to a targeted site in the patient's body. Transport systems commonly used for photosensitizers are polymers, liposomes, oil emulsions, certain metals, some proteins, and carbon-based nanoparticles [5–10]. Stable and biocompatible transport systems with a long half-life in the blood are ideal. Selective drug delivery to tumor tissue, transport of nanoparticles containing a PS, and a tumor cell with a receptor is the objectives for achieving high selectivity and low drug concentration [11]. Several research groups have confirmed the hypothesis that one possible approach to achieving these goals is to prepare low-density lipoprotein (LDL)-based particles [12–17].

1.1. PDT in Cancer Therapy

PDT has been a promising, non-invasive method for the treatment of certain types of cancer for more than 25 years [18–23]; in addition, PDT has also been studied in various non-oncological applications; leishmaniasis [24], psoriasis [25,26], age-related macular degeneration [27,28], hyperplasia [27], restenosis [25], and in cardiology [29], urology [30], immunology [31], ophthalmology [28,32], dental medicine [33,34], and dermatology [35,36]. The combination of light and PS also shows promising results in the treatment of bacterial, fungal, parasitic, and viral infections [37–40].

The basic process in tumor PDT is shown in the diagram in Figure 1. The photosensitizer is usually administered intravenously (systematically) or topically, and the administration of PS has also been studied orally, mostly in patients with extensive damage (not always caused by cancer) [1,41].

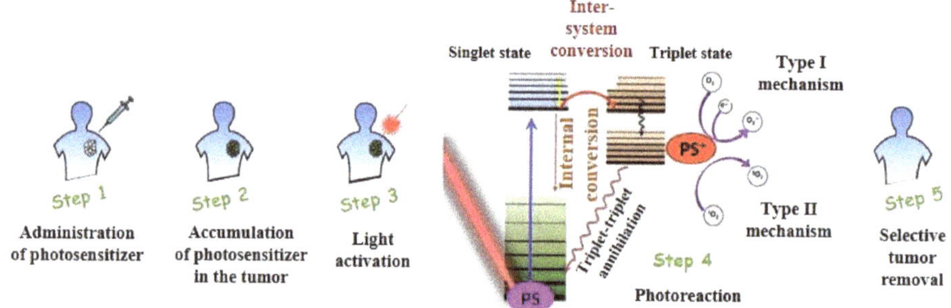

Figure 1. Scheme of photodynamic therapy (PDT; edited by [42,43]).

PS accumulates in the tumor tissue, which is subsequently irradiated with a light source of suitable wavelength, leading, in the presence of oxygen, to the formation of reactive oxygen species (ROS) and destruction of tissue and whole cells [1,41], while deactivation of important enzymes occurs and changes in the properties of biomacromolecules [44]. After activation of the photoactive substance by light and its interaction with molecular oxygen, singlet oxygen is formed. The singlet oxygen released in this photochemical reaction is highly toxic and can directly cause the death of tumor cells, through apoptosis, necrosis, or autophagy mechanism [45]. The release of singlet oxygen evokes oxidative stress, increases cytotoxicity, and DNA damage in cancer cells, modifies cellular metabolism, alters cancer cell death signaling pathways [46]. It also damages the blood vessels in tumor cells, resulting in indirect cell death through hypoxia (deficiency of oxygen) or aging (cell starvation). Diffuse distance of singlet oxygen obtained from photobleaching experiments has been appointed as 10–20 nm [47]. More recent time-resolved phosphorescence measurements show higher diffuse distance corresponding to 100 nm [48]. An explanation of this discrepancy was given by Hatz et al. [48] considering a fact that singlet oxygen behaves as a selective rather than reactive intermediate upon encountering other molecules in the cell.

Tissue damage depends on the depth of light penetration used to activate the PS. Membranes are damaged by cell death, triggering a number of inflammatory and immune processes, which cascade and cause the death of other tumor cells [49]. However, it was found out that singlet oxygen might cause membrane destruction directly. The cell membrane can be the main target of the singlet oxygen reaction for the bacterial membrane [50] and eukaryotic plasma membrane [51].

After surface irradiation PDT, the deeper connective tissues are only slightly damaged, so the patient's body can begin to restore the structure and flexibility of the damaged cells, and subsequently the tissues [52,53].

1.2. Physicochemical Mechanism of PDT

The photodynamic effect can be induced by two mechanisms called Type I and Type II (Figure 1—Step 4). After photon absorption, the PS molecule goes from the ground state (S_0) to the singlet excited state (S_1). From this excited state, the PS can be returned to the ground state by energy emission through non-radiative and/or radiant processes (fluorescence). In its excited state, the PS can also spontaneously move from the singlet state S_1 to the excited triplet state (T_1) by means of the intersystem conversion process. In this state, the transition to the ground state through a phosphorescence process can occur [54].

The type I mechanism involves electron transfer reactions between the PS molecule in the excited states of S_1 and T_1 and the substrate. This process results in the formation of ionic radicals, which tend to react immediately with oxygen to form a mixture of highly reactive oxygen radicals, such as superoxide radical ($\cdot O_2$), hydrogen peroxide (H_2O_2), and hydroxyl radical ($\cdot OH$), which oxidize a wide range of biomacromolecules [2,55].

The type II mechanism is characterized by energy transfer reactions between PS in the excited triplet state T_1 and molecular oxygen, which is also in the triplet ground state (T_0). These reactions cause the formation of singlet oxygen ($^1O_2$), which is able to rapidly oxidize cellular structures such as proteins, lipids, nucleic acids [56,57], and organelles leading to tumor cell death [58,59]. This also means that PDT may be a useful alternative treatment for cancer cells resistant to chemotherapy [60,61].

The reaction mechanism depends on the following conditions. First, the location of the PS is crucial because most of the ROS are highly reactive and cannot move far from the point of origin before disappearing. Second, the relative number of target biomolecules is important [43]. Davies (2003) calculated the percentage of $^1O_2$ responses in leukocytes: protein 68.5%, ascorbate 16.5%, RNA 6.9%, DNA 5.5%, beta-carotene 0.9%, NADH/NADPH 0.69%, tocopherols 0.5%, reduced glutathione 0.4%, lipids 0.2%, and cholesterol 0.1% [62]. This means that the distribution of $^1O_2$ may vary in different cell targets.

Both mechanisms can occur simultaneously. Their proportional representation is significantly influenced by the PS, the substrate, the oxygen concentration, and the binding of PS to the substrate. In addition, the type II mechanism appears to be more efficient as it has a higher rate constant than electron transfer reactions (type I mechanism). As a result, energy transfer to other compounds that can compete with oxygen is less important, so type II is more often dominant [58,63].

1.3. PDT Applications

PDT currently occupies an increasingly important place in clinical medicine. Its most promising application is the treatment of small and superficial tumors, e.g., some types of skin, neck, stomach [64], lung [65–67], bronchial [68] or oral cancer [69], esophagus [70,71], bladder [71,72], brain [73], prostate and pleura [74] tumors, breast [75] but also skin disease [76], atherosclerosis, and viral diseases therapy [77], including AIDS [78].

The future of PDT lies in efforts to find or synthesize new photosensitizers with properties allowing their greater selectivity for tumor tissues, and to find new approaches for the specific localization of already-known PS in tumor cells. This should contribute to the wider and more effective application of PDT in clinical practice. Targeted PDT has several advantages they are important in participation in the treatment of oncological diseases. Some of them are listed below: high selectivity, negligible side effects, low level of complications during and after treatment, high quality of life for patients, diagnostics and therapy in one step, and lower financial costs [79]. It can be a convenient complementary therapeutic method to classical surgery, chemotherapy, radiotherapy, and independent limited diagnosis.

2. Plant-Derived Photosensitive Substances

Photoactive compounds occurring in medicinal plants with potential utilization in PDT have been found to be less toxic than synthetic agents. The reduction of side effects using natural PSs in cancer treatment is another advantage of this therapeutic approach. However, their clinical applications have been limited by several imperfections such as accumulation in tissues, a lack of chemical purity, or low penetration [80]. Muniyandi et al. [81] have published a comprehensive review article about the role of photoactive phytocompounds in PDT. Phototoxic effects, potential applications in the PDT of cancer of the main natural PSs groups (furanocoumarins, thiophenes, alkaloids, curcumins, polyacetylenes, and anthraquinones) have been described [81]. Our paper focuses on the four anthraquinones, of which hypericin is the most promising PS in the PDT of cancer. With regard to the hydrophobicity of studied anthraquinones, which is important for their penetration through membranes, the following types of PSs have been distinguished.

Types of Photosensitive Substances by Hydrophobicity

PDT either uses chemotherapeutics commonly applied in chemotherapy, which must also be photosensitive, or new PSs are proposed. All photosensitizers (except uroporphyrin and photophrin), which have been proposed as drugs for use in PDT, interact more or less with serum proteins after intravenous administration. From the point of view of PDT, however, the interaction of the drugs with DNA is also quite important, as it is necessary to disrupt and stop the division of tumor cells, and this can be achieved only through their interaction with DNA [62,82]. In some cases, this interaction is not very significant (as these drugs have a greater affinity for proteins), so it is necessary to find a transporter that will help deliver the drug to the cell nucleus, thereby mediating the drug–DNA interaction [83,84]. Success in the treatment of cancer requires sufficiently hydrophobic drugs to cross the lipid membrane. For this reason, the hydrophobicity of drugs plays an important role in their distribution, metabolism, and excretion from the patient´s body. Due to these facts, we distinguish four groups of drugs with the ability to localize and accumulate in the tumor. Moreover, there are no rigid boundaries between these groups of drugs, and there is some overlapping between them, in some cases a continuous transition.

1. Hydrophobic PSs—compounds requiring the presence of transporters, such as liposomes or Cremophor EL, or Tween 80. They have the ability to localize in the inner lipid part of lipoproteins, mainly in LDL and high-density lipoproteins (HDLs), but also in very-low-density lipoproteins (VLDLs). This group includes phthalocyanines (ZnPC, C1A1PC), naphthalocyanines (isoBOSINC), tin-etiopurpurine (SnET2) [84], and hypericin [85].
2. Amphiphilic PSs—asymmetric compounds, which can be incorporated into the outer phospholipid and apoprotein layer of lipoprotein particles, e.g., disulfonates (TPPS$_{2a}$, C1A1PCS$_{2a}$), lutetium teraphyrin (LuTex), and monoaspartyl chlorine (MACE), which forms a barrier between albumin and HDL [84]. Emodin can be included in this group [86].
3. Hydrophilic PSs—drugs that predominantly bind to albumins and globulins, e.g., tetra-sulfone derivates of tetraphenylporfin (TPPS$_3$ and TPPS$_4$) and chloroaluminum phthalocyanine (C1A1PCS$_3$ and C1A1PCS$_4$) [84].
4. Intercalators—drugs that are used mainly in chemotherapy, which intercalate into DNA and are also photoactive, e.g., doxorubicin [87], daunorubicin [88], adriamycin [89], quinizarin [90], and danthron [91].

In this work, we focused in more detail on a very prospective PS in PDT of cancer hypericin. During a study of hypericin molecule incorporation into biomacromolecules (we focused on DNA) model compounds are using for simplification of the problem. Anthraquinones emodin, quinizarin, and danthron represent a significantly smaller part of the larger hypericin molecule with the same chemical groups. This fact facilitates the creation of a proper model for interaction between hypericin and biomacromolecules. Moreover, chosen hypericin derivatives themselves originate from medicinal plants, as the PS can be utilized in PDT and their anticancer effects are known. With respect to the above-mentioned sorting of PSs into groups, they can be representatives of highly hydrophobic (hypericin), mildly hydrophobic (emodin), and intercalating molecules (quinizarin and danthron).

Scientists and physicians are currently working on how to increase the effectiveness of cancer treatment. One option that has been shown to be very effective is a combination of therapies (PDT and chemotherapy), which involve the direct or mediated interaction of anticancer drugs with DNA and other bioactive macromolecules (serum albumins, lipoproteins) [92–94]. Hypericin, emodin, quinizarin, and danthron are examples of anticancer drugs which are chemotherapeutics synthesized by medicinal plants and PSs, and which can be used, in PDT. The discovery of new natural drugs is very important because they have many benefits for the patients. Drugs derived from medicinal plants are less toxic to the body, their use poses less risk of adverse side effects, does not depends on them, they are suitable for all age groups of patients, and can be easily combined with conventional drugs, i.e., do not show contraindications.

3. Hypericin

Hypericin, a naturally occurring pigment, is found in certain plant species of the genus *Hypericum*. The most important representative is Saint John´s Wort (*Hypericum perforatum*) from the family *Hypericaceae*, a plant with golden-yellow flowers that grows to a height of 30–90 cm [95]. Hypericin has also been reported as occurring in some species of Australian insects [96]. Hypericin, 7,14 dione-1,3,4,6,8,12-hexahydroxy 10,11 dimethyl-phenanthrol [1,10,9,8-opgra] perylene (Figure 2) is an aromatic polycyclic dione, which exhibits a wide range of biological activities.

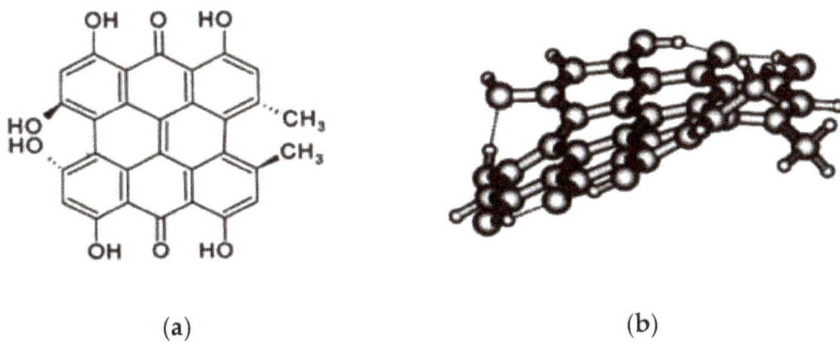

(a)   (b)

Figure 2. Two-dimensional (**a**) and three-dimensional (**b**) structure of hypericin [97].

It has long been used to treat depression caused by a disorder of the brain neurotransmitters responsible for human moods [98,99]. Hypericin has antiviral [100,101] and antitumor activity [102–105] and is used to heal wounds, neuralgic sites, and hypertrophic scar [106–108]. Hypericin has been demonstrated to be very effective on bacteria [109–114]. Studies of hypericin´s antiviral activity confirm that it is known to effectively deactivate enveloped viruses, but it is ineffective against non-enveloped viruses. The antiviral activity of hypericin has been demonstrated against several types of viruses: herpes simplex virus, murine cytomegalovirus, sindbis virus [115,116], hepatitis B [117], anemia, some types of leukemic viruses [118], and HIV [119–121], so it has also been used in the clinical treatment of AIDS patients [122]. Hypericin activity depends to a large extent on the presence of light and oxygen. These two factors determine the extent of its antiviral activity and both antiviral and antitumor activities increase significantly after visible light irradiation [123].

Hypericin, together with two other hydroxyquinones, hypocrellin A and calphostin, is characterized by negligible dark toxicity, significant photocytotoxicity to tumor cells, intense absorption at higher wavelengths (>550 nm), higher selectivity for tumor tissues, and faster leaching than hematoporphyrins [124]. Hypericin is soluble in most solvents. In organic solvents (DMSO, ethanol) its solution is red; in basic media, it is green; and in aqueous solutions, it forms purple dispersed particles [125]. The absorption spectrum of hypericin in neutral organic solvents has two main transitions in the visible region, $S_0 \rightarrow S_1$ (500–600 nm) and $S_0 \rightarrow S_2$ (425–485 nm), and two intense absorption bands. Hypericin forms non-fluorescent aggregates in aqueous solutions. The absorption and fluorescence spectra of hypericin in hexane, chloroform, and toluene are similar to those in water, so it can be assumed that hypericin molecules also form aggregates in these solvents. An indicator of the formation of aggregates in solution is the absence of fluorescence [126]. Several physical and chemical properties of hypericin are shown in Table 1.

Table 1. Physical and chemical properties of selected photosensitizers (hypericin, emodin, quinizarin, and danthron).

Photosensitizer		Hypericin	Emodin	Quinizarin	Danthron
Hydrophobicity		high	amphiphilic with mild hydrophobicity	low	Low
AlogP		5.040	2.568	2.324	2.324
Absorption maximum in DMSO λ_{max} (nm)		560 600	440	475	430
Fluorescence maximum $\lambda_{exc.}$ (nm)		603 560	520 440	540 475	575 430
Dissociated form (nm)	PS^0	560	440	475	430
	PS^{1-}	600 [127]	480	560	475
	PS^{2-}	650 [127]	525	595	500
	PS^{4-}	630 [127]			
	PS^{6-}	640 [127]			
Dissociated constant	pKa_1	2 [127]	7.2	11.3	10.5
	pKa_2	7.8 [127]	10.6	12.7	12.9
	pKa_3	11.5 [127]			
	pKa_4	13 [127]			

The photophysics of hypericin is very complicated, and some effort to interpret its photophysical and photochemical properties is necessary to elucidate its pharmacological and biological activities. Phenols are known to be good electron donors, while quinones act as electron acceptors. Hypericin contains both phenolic and quinone moieties. This structural characteristic places it in the group of amphi-electron compounds. The results of experiments suggest that hypericin has the potential to be a good oxidizing and reducing agent [128]. At present, the mechanisms leading to the selective accumulation of hypericin in tumor tissue compared to normal tissue are not known, nor the mechanism of its antiviral and antitumor action has been clearly determined. Thomas et al. proposed a mechanism of hypericin activity that depends on the presence of molecular oxygen, consisting of transferring energy from the excited triplet state of hypericin to the ground state of molecular oxygen, thereby generating oxygen in the singlet state, which is a highly-reactive molecule [129]. Singlet oxygen production has been theoretically confirmed [118] and experimentally demonstrated in organic solvents and lipid media [130–132]. Other studies suggest that although oxygen may play an important role in some cases, it is not always necessary for the antiviral activity of hypericin [133]. According to the work of the J.W. Petrich group, the origin of the photoinduced antiviral activity of hypericin may lie in its ability to produce a local pH drop in cells after irradiation [134]. The close relationship between the mechanism of hypericin antitumor activity and photoactivated local acidification in cells has been confirmed in experiments on tumor cell membrane experiments [102].

Hypericin is currently considered as a potential antitumor drug in PDT. It belongs in the group of highly hydrophobic PSs (AlogP = 5.04) [135]. Its cytotoxic and antitumor activity has been demonstrated, and its inhibitory effects on epidermal growth factor-R and a wide range of protein kinases (protein tyrosine kinase, MAP-kinase, and protein kinase C) [136]. The significant photosensitivity properties of hypericin together with its selective uptake in tumor tissues, especially in bladder cancer cells and its minimal cytotoxicity in the dark, are major positive indications for the clinical use of hypericin in the photodynamic cancer treatment [137]. Hypericin is a photosensitizer found in the endoplasmic reticulum, which, when activated by light, mediates the rapid emptying of calcium Ca^{2+} from the endoplasmic reticulum, which is linked to programmed cell death [138]. Although hypericin has been shown to be a potent inducer of cell death in PDT, some in vivo studies suggest that cells treated with PDT may activate rescue-signaling pathways, which may ultimately lead to tumor survival and recurrence [18,139].

Hypericin is a very good fluorescent probe for super-resolution microscopy [140]. Knowledge of the distribution of this molecule in the individual components of a living system is essential in detecting the modes of action of hypericin in biological organisms. The accumulation of hypericin in cells is determined by the diffuse and soluble properties of the molecule. It is known that after the initial accumulation in the cell membranes (20 min), an increased concentration of hypericin in the cell nucleus is observed with increasing time [141]. This indicates the possible interaction of hypericin with nucleic acids, the most essential component of the cell nucleus. Ultraviolet resonance Raman spectroscopy with 257 nm excitation was used to study nucleic acid-hypericin complexes. The excitation used amplifies the vibrational modes of nucleic acid bases. Resonance Raman spectra for complexes of polynucleotides with hypericin showed that hypericin preferentially interacts with purines at the N7 position, and this interaction is stronger in guanine than in adenine [142]. In addition, the spectra of the complex of hypericin with polyrG and calf thymus DNA, which were studied by means of surface-enhanced Raman spectroscopy, are dominated by bands corresponding to hypericin vibrations. Based on the spectra obtained, the importance of the guanine base in the interaction of hypericin with nucleic acid was confirmed. The spectrum of the polyrG complex contains changes similar to those in the spectrum of the DNA complex. Based on these results, it can be assumed that DNA induces changes in the stretching vibrations of the hypericin skeleton associated with the vibrations of the hydroxyl groups, suggesting that the hydroxyl groups of hypericin participate in the interaction with nucleic acid [143].

The mechanism of hypericin distribution in tissues, cells, and cellular organelles is also thought to be a relatively complex process, which is affected by hypericin concentration, incubation time, and properties of biological units. The bio distribution of hypericin in human organisms also depends on the mode of transport in the bloodstream. Hypericin binds to serum proteins, where lipoprotein particles appear to be a more efficient transporter than albumin. The interaction of hypericin with albumin is specific [144], while its interaction with lipoproteins is non-specific [145]. After binding with macromolecules (lipid structures, serum proteins, and sugars), hypericin forms biologically active monomers with an emission band around 600 nm, thus overcoming problems with its solubility in physiological solutions, which is very important in terms of its activity [97,137].

4. Emodin

Emodin, (1,3,8-trihydroxy-6-methylanthraquinone, Figure 3) is an anthraquinone derivate, a naturally-occurring pigment isolated from the underground part of the traditional Chinese medicinal plant *Rheum palmatum*, which is used mainly for its antitumor, immunosuppressive, antiinflammatory [146], antibacterial, diuretic, laxative [147], antiulcerogenic (antiulcer) [148], and vasorelaxant effects [149]. It is also found in other plant species, such as *Cassia*, *Aloe*, and *Rhamnus* [150–154]. In practice, it is available in the form of an orange crystalline substance. It is insoluble in aqueous media, and it forms aggregates (mostly in the acidic pH range). Emodin shows poor solubility in chloroform, ether, and benzene, but is highly soluble in DMSO and ethanol.

The antiviral properties of emodin have also been confirmed, mainly on enveloped viruses. This property of emodin is related to its high affinity for the phospholipid membrane, in which it inhibits hydrophobic interactions between individual hydrocarbon chains [155]. Emodin has been proposed as a major precursor of the microbial metabolic pathway in the isolation of hypericin [156,157]. It is a redox catalyst and plays an important role in many processes, such as electron transport, photosynthesis, or cellular respiration [158]. Its effect on cell death has been widely studied both in the dark and after light activation [155,159]. The stimulatory effects and the prokinetic effect of emodin on gastrointestinal smooth muscle have been described in several studies [147,160]. In recent years, its effect on the contractility of smooth muscle cells has been studied [161,162].

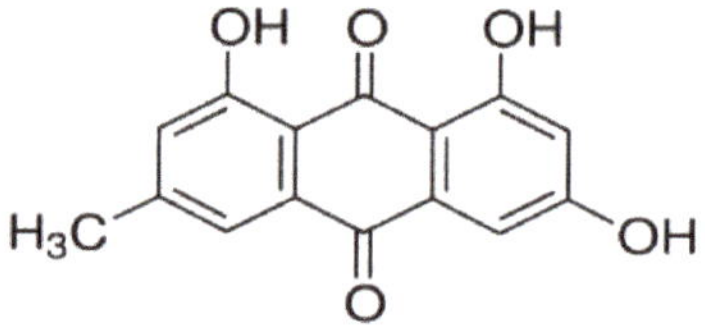

Figure 3. Structural formula of emodin.

In addition to the pharmacological properties mentioned above, emodin also has toxicological effects. Under aerobic conditions, it is photolabile in visible light and phototoxic in vitro and it can cause testicular toxicity in mice leading to hypospermatogenesis [163,164]. It inhibits protein kinase casein kinase II (CK2). Genetic disorder of the catalytic subunits of CK2 protein kinase leads to changes in sperm shape in mice during spermatogenesis and is also responsible for male infertility [165,166]. Several authors have studied the interaction of emodin with DNA [167–169]; for example, it causes the formation of DNA double-strand breaks by stabilizing topoisomerase II-DNA cleavage complexes and inhibiting ATP hydrolysis by topoisomerase II [170]. Emodin has a high binding affinity for serum proteins and is known to bind non-covalently to DNA in intact cells in the presence of serum proteins [159]. The interaction of emodin with proteins has been studied using several techniques surface-enhanced Raman spectroscopy, NMR, fluorescence spectroscopy, circular dichroism, and the stopped-flow method [171–174]. The formation of emodin–protein complexes is very important with regard to understanding the mode of drug delivery and transport in tumor cells [174]. Emodin belongs in the drug of amphiphilic PSs with mild hydrophobic properties (Alog P = 2.568) [135]. It has an anticancer effect on some types of human liver and lung tumors. However, the molecular mechanisms of emodin-mediated tumor regression have not been precisely defined [150]. Emodin has specific antineuroectodermal tumor activity in vitro and in vivo [175]. It suppresses tyrosine kinase activity on HER-2 breast cancer cells, inhibits the transformation of the phenotype of these cells, and can affect androgen receptors directly by inhibiting cell growth in prostate cancer cells [146]. The antitumor activity of emodin on the human chronic myeloid leukemia cell line K562 has also been demonstrated in vitro and in vivo [176]. Emodin is toxic for glioma cells by inhibiting their proliferation and inducing apoptosis of C6 cells [177]. It is known to increase the sensitivity of tumor cells to chemotherapeutics, but the mechanism of the emodin-mediated chemotherapeutic effect on cancer cells has been the subject of intense study for many years [150,178]. Important physical properties of emodin are listed in Table 1.

5. Quinizarin

Quinizarin (1,4-dihydroxyanthraquinone, Figure 4) is a polycyclic aromatic hydrocarbon containing two opposite carboxyl groups (C=O) at positions 9,10. It takes the form of a yellow crystalline substance and occurs in nature in plants (*Aloe*, Cascara Sagrada, Senna, and rhubarb), fungi, some lichens, and insects [179]. Quinizarin is soluble in basic solutions, acetone, chloroform, and DMSO. It is almost insoluble in aqueous solutions, but its solubility increases with increasing temperature. It is used as a pesticide as a fungicide and as an additive in lubricants [180]. In addition, it is used as a colorant in the food, textile, dyeing, and photographic industries [181]. Quinizarin inhibits HIV proteinase [182] and acts as a mutagenic agent on some mammalian bacteria [183]. Immunological analysis studies have shown that it stimulates the P450 enzyme in rat epithelial and liver cells [184].

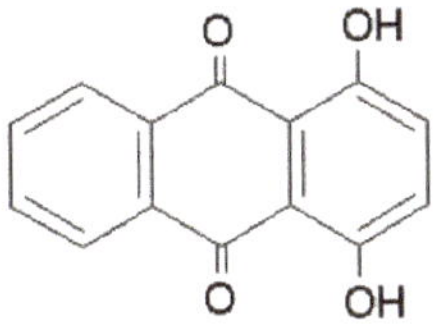

Figure 4. Structural formula of quinizarin.

The structure of quinizarin has been studied using several spectroscopic methods, such as fluorescence measurements [185], resonance Raman and infrared spectroscopy [186], and X-ray structural analysis [187]. It has planar C_{2v} molecular symmetry and crystallizes in a monoclonal system. Quinizarin is a highly fluorescent substance. The fact that it is almost insoluble in water significantly complicates the application of Raman spectroscopy in its study. However, infrared FT Raman and surface-enhanced Raman spectroscopy can also be used to characterize it in aqueous media [188].

Quinizarin is the simplest model molecule of a chromophore typical of some biologically and pharmaceutically significant compounds, including the antitumor anthracycline antibiotics doxorubicin, daunorubicin, and adriamycin, which are used in antineoplastic therapy. The quinizarin-like quinoid moiety is probably responsible for the cytotoxicity and cardiotoxicity of anthracycline drugs [180]. It is thought to be able to intercalate into DNA [189,190]. We can classify it among the photosensitive intercalators with low hydrophobicity (AlogP = 2.324), which could be applied in PDT [135]. The physical properties of quinizarin are shown in Table 1.

6. Danthron

Danthron (1,8-dihydroxyanthraquinone or chrysazine, Figure 5) is a hydroxyanthraquinone that occurs wild in nature found in many plants but also in some insect species (e.g., *Pyrrhalta luteola* larvae). It is very often isolated from dry leaves and stems of the plant *Xyris semifuscata* growing in Madagascar [191]. It is commercially available in the form of an orange, red, or red-yellow crystalline powder. Like quinizarin, it is soluble in basic solutions, chloroform, and DMSO. It is practically insoluble in water. Danthron and quinizarin are derivatives of emodin and hypericin. Danthron is part of the anticancer drug aclacinomycin. It is thought to intercalate into DNA and therefore can be used as an example of a low hydrophobicity intercalator (AlogP = 2.324) in PDT. However, it can also be a carcinogen, causing the development of adenoma or adenocarcinoma of the colon and increased incidence of liver cancer cells [191,192]. Danthron is one of the hydroxyanthraquinone derivatives causing topoisomerase II inhibition [193], which is involved in various cellular processes, including chromosome segregation [194], and it is essential for maintaining genome stability [195].

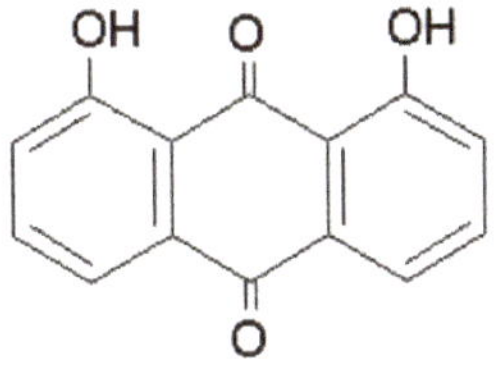

Figure 5. Structural formula of danthron.

Danthron is used as a laxative [159,196]. It is currently applied as an antioxidant, a fungicide to control powdery mildew, and has an irreplaceable role in the research into anticancer agents. Together with emodin, it is the basic structure of aglycones, which are naturally occurring laxative glycosides [191]. Danthron shows mutagenic activity [197] and indicates mutations in cells of the lymphatic system [192].

Its structure has been studied using fluorescence, resonance Raman and infrared spectroscopy. The danthron crystal is tetragonal and its planar molecule has an asymmetric structure with two different intramolecular O … ..O distances. The optical properties of danthron can be discussed in terms of C_{2v} pseudosymmetry [188]. Danthron and quinizarin form two intramolecular hydrogen bonds, which cause small changes in the structure of anthraquinone molecules. The presence of two hydrogen bonds in quinizarin allows for double proton transfer, which can take place in two steps or a one-step process. Conversely, the geometry of the danthron molecule allows only single proton transmission. The structure of danthron and quinizarin has been intensively studied and experimentally determined. For quinizarin, X-ray structural analysis identified the values of the distance O … ..O and O … ..H bounds, which are 2.57 and 1.77–1.79 Å respectively. This technique helped to define the O … ..O distance for danthron and setting its value at 2.49 Å, but it did not provide any information about proton binding [198]. Table 1 shows characteristic physical properties.

7. Hypericin and Its Derivatives Interaction with DNA

In recent years the interaction of hypericin, emodin, quinizarin, and danthron with biomacromolecules has been studied, especially with DNA, LDL, and human serum albumin (HSA) has been investigated [199–202]. The results of our studies in this area supplement the scientific knowledge on this subject. The measurements indicate that PSs incorporate into DNA, LDL, and HSA, but interact most easily with LDL particles. Distribution of molecules has been studied at two different conditions: (i) DNA–ligand complexes were exposed to the presence of free LDL particles or HSA molecules, respectively and (ii) LDL- or HSA-ligand, respectively, complexes were in the presence of free DNA molecules. From measurements at the first conditions, it is clear that while the fluorescence intensity for the DNA-hypericin complex increases with the addition of LDL to the solution, a less pronounced effect can be observed for the other PSs, DNA–emodin, DNA–quinizarin, and DNA–danthron. Similarly, HSA molecules added to the solution slightly increase the fluorescence intensity in the DNA–hypericin complex. In a competitive environment, they redistribute and rebind from DNA macromolecule to LDL (hypericin) or HSA (emodin, quinizarin, and danthron). In contrast, the fluorescence intensity did not change for the DNA–emodin complex, and a slight decrease in fluorescence has been observed in the DNA–quinizarin and DNA–dantron complexes. The presence of free DNA leads to a decrease in the fluorescence intensity of the LDL–emodin, LDL–quinizarin, and LDL–danthron/HSA–quinizarin and HSA–danthron complexes. For LDL–hypericin and HSA–hypericin complexes, an increase in fluorescence with the addition of DNA was recorded. The fluorescence intensity of the HSA–emodin complex remains unchanged [135]. It is noteworthy that the quenching of fluorescence in the presence of DNA has been observed for intercalating molecules [190]. These results allowed us to state that the molecules of the studied PSs can be extracted from DNA by HSA molecules rather than LDL particles, with the exception of hypericin, where there is a more significant binding from DNA in the presence of LDL particles. Experiments in fluorescence analysis of native and denatured DNA were designed to determine if and how the interaction of hypericin, emodin, quinizarin, and danthron with native and denatured DNA differed. Significant changes were detected for quinizarin and danthron, where fluorescence increased significantly in the presence of denatured DNA. It can be assumed that DNA denaturation increased the distances between the individual quinizarin and danthron molecules (there is a decrease in fluorescence quenching) and helped to better visualize the molecules inside the DNA. This applies in particular to molecules that intercalate into the DNA structure, i.e., quinizarin and danthron. For hypericin and emodin, it has been noted only a slight change in the intensity of fluorescence in the presence of denatured DNA, respectively from which it can again predict a different interaction mode of these PSs with DNA [135].

As a representative of highly hydrophobic PSs, hypericin binds to DNA by means of a large groove [203,204]. A binding constant, whose value was also determined, is 4.0×10^4 L/mol [204]. Emodin, with mild hydrophobic properties, interacts with DNA by binding into a small groove. This interaction is characterized by binding constant 8.1×10^4 L/mol. These ligands are incorporated into the DNA groove with hydrophobic or hydrogen bounds [204]. An intercalating mode of interaction with DNA has been confirmed for quinizarin and danthron, which is very clearly supported by atomic force microscopy measurements. Quinizarin and dathron cause DNA to unwind due to intercalation, increasing the contour length of linear DNA by 10%. By the action of natural photosensitizers studied by us, more rigid molecules of linear DNA have been detected. It is known that this is not only the type of reaction they are involved in. Quinizarin and danthron can also bind to the surface of DNA molecules by means of electrostatic forces. The size of the binding constant of this interaction type for both PSs is 1.1×10^4 L/mol [190].

Cell experiments have been performed with the aim of better understanding the mechanisms of the interaction of hypericin, emodin, quinizarin, and danthron with DNA. Success in the treatment of cancer requires sufficient hydrophobic drugs to cross the lipid membrane. The hydrophobicity of drugs therefore plays an important role in their distribution, metabolism, and excretion from the patient´s body. The main barrier preventing the interaction of nuclear DNA with anticancer drugs is the nuclear envelope. On the other hand, the detoxification process is accompanied by a decrease in drug concentration during the active outflow of these drugs. P-glycoprotein, a member of the transport protein family, is involved in this process [205]. The function of these proteins is to remove toxins from cells [206], so it is possible to regulate drug resistance by manipulating P-glycoprotein activity [207,208]. Another approach to preventing cancer cell resistance to drugs could be based on photochemical activation at a well-defined site/organelle [209]. With this in mind, the intracellular interaction of the drugs discussed here with DNA directly in the nucleus was studied using selective photoactivation and regulation of P-glycoprotein function. Selective photoactivation facilitates entry of emodin, quinizarin, and danthron molecules into the nucleus of tumor cells by means of passive diffusion (this phenomenon has not been observed in the case of hypericin). Inhibition of P-glycoprotein may increase the intracellular concentration of the molecules studied and aid their interaction with DNA. Inhibition of P-glycoprotein stimulates the passive transport of these molecules to the nucleus of tumor cells. In addition, emodin is known to significantly promote the entry of other drugs into the nucleus of tumor cells, with the exception of strongly hydrophobic ones (hypericin-like), which are characterized by a high binding affinity for P-glycoprotein [135].

The treatment of tumor cells with natural photosensitive drugs significantly depends on the degree of hydrophobicity, lipophilicity, and DNA intercalation properties of the drugs used.

8. Conclusions

Cancer is a group of diseases characterized by abnormal and uncontrolled cell growth, which can affect a particular tissue or organ in a very aggressive manner and are able to spread to other parts of the body, they can metastasize. Cancer can be treated by classical therapeutic procedures (surgery, chemotherapy, and radiotherapy) and modern, very promising therapeutic methods. These include PDT. These studies show that the most effective way to treat cancer is to combine PDT with traditional therapies, respectively in some cases, PDT can even replace conventional treatment. It is very likely that PDT will be used in the future as the therapeutic approach in the treatment of specific cancers and non-cancerous diseases. This method minimizes the side effects of traditional therapeutic methods, is less invasive, more effective, and affordable. The condition for the successful treatment of cancer with PDT is the disruption of DNA macromolecule, which stops the division of tumor cells. Therefore, it is necessary to find suitable drugs that interact with DNA directly, respectively mediated by transporters that transport them to the nucleus of the tumor cell. A great advantage of newly discovered drugs is that they are natural, easier to prepare, and do not burden the patient´s body as much as synthetically produced drugs. The success of cancer treatment is highly dependent on the application of hydrophobic drugs, which can more easily cross the lipid membrane of tumor cells. Therefore, the interaction of drugs with different hydrophobicity with DNA macromolecule is the subject of many studies.

Author Contributions: Conceptualization, J.S. and V.V.; methodology, J.B.; software, J.B.; validation, J.S., V.V. and J.B.; formal analysis, V.V.; investigation, V.V.; resources, V.V.; data curation, J.S.; writing—original draft preparation, V.V.; writing—review and editing, J.S.; visualization, J.B.; supervision, J.S.; project administration, J.S.; funding acquisition, J.S. All authors have read and agreed to the published version of the manuscript.

Funding: This study was supported by Slovak Research Grant Agency through KEGA project No. 012 UVLF-4/2018.

Conflicts of Interest: The authors declare no conflict of interest.

References

1. Castano, A.P.; Demidova, T.N.; Hamblin, M.R. Mechanisms in photodynamic therapy: Part one—Photosensitizers, photochemistry and cellular localization. *Photodiagn. Photodyn. Ther.* **2004**, *1*, 279–293. [CrossRef]
2. Deda, D.K.; Araki, K. Nanotechnology, light and chemical action: An effective combination to kill cancer cells. *J. Braz. Chem. Soc.* **2015**, *26*, 2448–2470. [CrossRef]
3. Allen, T.M.; Hansen, C.B.; Demenzes, D.E.L. Pharmacokinetics of long circulating liposomes. *Adv. Drug Deliv. Rev.* **1995**, *16*, 267–284. [CrossRef]
4. Loomis, K.; McNeeley, K.; Bellamkonda, R.V. Nanoparticles with targeting triggered release and imaging functionality for cancer application. *Soft Matter* **2011**, *7*, 839–856. [CrossRef]
5. Reddi, E. Role of delivery vehicles for photosensitizers in the photodynamic therapy of tumors. *J. Photochem. Photobiol. B* **1997**, *37*, 189–195. [CrossRef]
6. Huntošová, V.; Buzová, D.; Petrovajová, D.; Kasak, P.; Naďová, Z.; Jancura, D.; Sureau, F.; Miškovský, P. Development of a new LDL-based transport system for hydrophobic/amphiphilic drug delivery to cancer cells. *Int. J. Pharm.* **2012**, *436*, 463–471. [CrossRef]
7. Hally, C.; Delcanale, P.; Nonell, S.; Viappiani, C.; Abbruzzetti, S. Photosensitizing proteins for antibacterial photodynamic inactivation. *Transl. Biophotonics* **2020**, e201900031. [CrossRef]
8. Ghorbani, J.; Rahban, D.; Aghamiri, S.; Teymouri, A.; Bahador, A. Photosensitizers in antibacterial photodynamic therapy: An overview. *Laser Ther.* **2018**, *27*, 293–302. [CrossRef] [PubMed]
9. Buriankova, L.; Buzova, D.; Chorvat, D.; Sureau, F.; Brault, D.; Miskovsky, P.; Jancura, D. Kinetics of hypericin association with low-density lipoproteins. *Photochem. Photobiol.* **2011**, *87*, 56–63. [CrossRef]
10. Lenkavska, L.; Blascakova, L.; Jurasekova, Z.; Macajova, M.; Bilcik, B.; Cavarga, I.; Miskovsky, P.; Huntosova, V. Benefits of hypericin transport and delivery by low- and high-density lipoproteins to cancer cells: From in vitro to ex ovo. *Photodiagn. Photodyn. Ther.* **2019**, *25*, 214–224. [CrossRef]
11. Konan, Y.N.; Gurny, R.; Allemann, E. State of the art in the delivery of photosensitizers for photodynamic therapy. *J. Photochem. Photobiol. B Biol.* **2002**, *66*, 89–106. [CrossRef]
12. Firestone, R.A. Low-density lipoprotein as a vehicle for targeting antitumor compounds to cancer cells. *Bioconjugate Chem.* **1994**, *5*, 105–113. [CrossRef] [PubMed]
13. Versluis, A.J.; van Geel, P.J.; Oppellar, H.; van Berkel, T.J.; Bijsterbosch, M.K. Receptor-mediated uptake of low-density lipoprotein by B16 melanoma cells in vitro and in vivo in mice. *Br. J. Cancer* **1996**, *4*, 525–532. [CrossRef] [PubMed]
14. Rensen, P.C.; de Vrueh, R.L.; Kuiper, J.; Bijsterbosch, M.K.; Biessen, E.A.; van Berkel, T.J. Recombinant lipoproteins: Lipoprotein-like lipid particles for drug targeting. *Adv. Drug Deliv. Rev.* **2001**, *47*, 251–276. [CrossRef]
15. Kader, A.; Pater., A. Loading anticancer drugs into HDL as well as LDL has little affect on properties of complexes and enhances cytotoxicity to human carcinoma cells. *J. Control. Release* **2002**, *80*, 29–44. [CrossRef]
16. Zheng, G.; Chen, J.; Li, H.; Glickson, J.D. Rerouting lipoprotein nanoparticles to selected alternate receptors for the targeted delivery of cancer diagnostic and therapeutic agents. *Proc. Natl. Acad. Sci. USA* **2005**, *102*, 17757–17762. [CrossRef]
17. Song, L.; Li, H.; Sunar, U.; Chen, J.; Corbin, I.; Yodh, A.G.; Zheng, G. Naphthalocyanine-reconstituted LDL nanoparticles for in vivo cancer imaging and treatment. *Int. J. Nanomed.* **2007**, *2*, 767–774.
18. Dougherty, T.J.; Gomer, C.J.; Henderson, B.W.; Jori, G.; Kessel, D.; Korbelik, M.; Moan, J.; Peng, Q. Photodynamic therapy. *J. Natl. Cancer Inst.* **1998**, *90*, 889–905. [CrossRef]
19. Wilson, B.C.; Patterson, M.S. The physics, biophysics and technology of photodynamic therapy. *Phys. Med. Biol.* **2008**, *53*, R61–R109. [CrossRef]
20. Yano, S.; Hirohara, S.; Obata, M.; Hagiya, Y.; Ogura, S.; Ikeda, A.; Kataoka, H.; Tanaka, M.; Joh, T. Current states and future views in photodynamic therapy. *J. Photochem. Photobiol. C Photochem. Rev.* **2011**, *12*, 46–67. [CrossRef]

21. Lim, C.K.; Heo, J.; Shin, S.; Jeong, K.; Seo, Y.H.; Jang, W.D.; Park, C.R.; Park, S.Y.; Kim, S.; Kwon, I.C. Nanophotosensitizers toward advanced photodynamic therapy of cancer. *Cancer Lett.* **2013**, *334*, 176–187. [CrossRef] [PubMed]
22. Bechet, D.; Mordon, S.R.; Guillemin, F.; Barberi-Heyob, M.A. Photodynamic therapy of malignant brain tumours: A complementary approach to conventional therapies. *Cancer Treat. Rev.* **2014**, *40*, 229–241. [CrossRef] [PubMed]
23. Benov, L. Photodynamic therapy: Current status and future direction. *Med. Princ. Pract.* **2015**, *24*, 14–28. [CrossRef] [PubMed]
24. Miguel-Gomez, L.; Vano-Galvan, S.; Perez-Garcia, B.; Carrillo-Gijon, R.; Jaen-Olasolo, P. Treatment of folliculitis decalvans with photodynamic therapy: Results in 10 patients. *J. Am. Acad. Dermatol.* **2015**, *72*, 1085–1087. [CrossRef]
25. Sharman, W.M.; Allen, C.M.; van Lier, J.E. Photodynamic therapeutics: Basic principles and clinical applications. *Drug Discov. Today* **1999**, *4*, 507–517. [CrossRef]
26. Jin, Y.; Zhang, X.; Kang, H.; Du, L.; Li, M. Nanostructures of an amphiphilic zinc phthalocyanine polymer conjugate for photodynamic therapy of psoriasis. *Colloids Surf. B* **2015**, *128*, 405–409. [CrossRef]
27. Sharman, W.M.; van Lier, J.E.; Allen, C.M. Targeted photodynamic theory via receptor mediated delivery systems. *Adv. Drug Deliv. Rev.* **2004**, *56*, 53–76. [CrossRef]
28. Cruess, A.F.; Zlateva, G.; Pleil, A.M.; Wirostko, B. Photodynamic therapy with verteporfin in age-related macular degeneration: A systematic review of efficacy, safety, treatment modifications and pharmacoeconomic properties. *Acta Ophthalmol.* **2009**, *87*, 118–132. [CrossRef]
29. Rockson, S.G.; Lorenz, D.P.; Cheong, W.F.; Woodburn, K.W. Photoangioplasty: An emerging clinical cardiovascular role for photodynamic therapy. *Circulation* **2000**, *102*, 591–596. [CrossRef]
30. Bozzini, G.; Colin, P.; Betrouni, N.; Nevoux, P.; Ouzzane, A.; Puech, P.; Willers, A.; Mordon, S. Photodynamic therapy in urology: What can we do now and where are we heading. *Photodiagn. Photodyn. Ther.* **2012**, *9*, 261–273. [CrossRef]
31. Panzarini, E.; Inguscio, V.; Dini, L. Immunogenic cell death: Can it be exploited in photodynamic therapy for cancer. *Biomed. Res. Int.* **2013**, *2013*, 482160. [CrossRef] [PubMed]
32. Lang, G.E.; Mennel, S.; Spital, G.; Wachtlin, J.; Jurklies, B.; Heimann, H.; Damato, B.; Meyer, C.H. Different indications of photodynamic therapy in ophthalmology. *Klin. Monbl. Augenheilkd.* **2009**, *226*, 725–739. [CrossRef] [PubMed]
33. Trindade, A.C.; De Figueiredo, J.A.P.; Steier, L.; Weber, J.B.B. Photodynamic therapy in endodontics: A literature review. *Photomed. Laser Surg.* **2015**, *33*, 175–182. [CrossRef] [PubMed]
34. Vohra, F.; Al-Kheraif, A.A.; Qadri, T.; Hassan, M.I.A.; Ahmedef, A.; Warnakulasuriya, S.; Javed, F. Efficacy of photodynamic therapy in the management of oral premalignant lesions. A systematic review. *Photodiagn. Photodyn. Ther.* **2015**, *12*, 150–159. [CrossRef]
35. Karrer, S.; Kohl, E.; Feise, K.; Hiepe-Wegener, D.; Lischner, S.; Philipp-Dormston, W.; Podda, M.; Prager, W.; Walker, T.; Szeimies, R.M. Photodynamic therapy for skin rejuvenation: Review and summary of the literature-results of a consensus conference of an expert group for aesthetic photodynamic therapy. *J. Dtsch. Dermatol. Ges.* **2013**, *11*, 137–148. [CrossRef]
36. Fuhrmann, G.; Serio, A.; Mazo, M.; Nair, R.; Stevens, M.M. Active loading into extracellular vesicles significantly improves the cellular uptake and photodynamic effect of porphyrins. *J. Control. Release* **2015**, *205*, 35–44. [CrossRef]
37. Baptista, M.S.; Wainwright, M. Photodynamic antimicrobial chemotherapy (PACT) for the treatment of malaria, leishmaniasis and trypanosomiasis. *Braz. J. Med. Res.* **2011**, *44*, 1–10. [CrossRef]
38. Goslinski, T.; Piskorz, J. Fluorinated porphyrinoids and their biomedical applications. *J. Photochem. Photobiol. C Photochem. Rev.* **2011**, *12*, 304–321. [CrossRef]
39. Hanakova, A.; Bogdanova, K.; Tomankova, K.; Pizova, K.; Malohlava, J.; Binder, S.; Bajgar, R.; Langova, K.; Kolar, M.; Mosinger, J.; et al. The application of antimicrobial photodynamic therapy on *S. aureus* and *E. coli* using porphyrin photosensitizers bound to cyclodextrin. *Microbiol. Res.* **2014**, *169*, 163–170. [CrossRef]
40. Baltazar, L.M.; Ray, A.; Santos, D.A.; Cisalpino, P.S.; Friedman, A.J.; Nosanchuk, J.D. Antimicrobial photodynamic therapy: An effective alternative approach to control fungal injections. *Front. Microbiol.* **2015**, *6*, 202. [CrossRef]
41. Bonnett, R. Progress with heterocyclic photosensitizers for the photodynamic therapy (PDT) of tomours. *J. Heterocyclic Chem.* **2002**, *39*, 455. [CrossRef]

42. Photolitec, LLC. Tumor Specific Imaging Therapy. *Photodynamic Therapy.* Available online: http://photolitec.org/Tech_PDT.html (accessed on 10 August 2020).

43. Vatansever, F.; de Melo, W.C.M.A.; Avci, P.; Vecchio, D.; Sadasivam, M.; Gupta, A.; Chandran, R.; Karimi, M.; Parizotto, N.A.; Yin, R.; et al. Antimicrobial strategies centered around reactive oxygen species—bactericidal antibiotics, photodynamic therapy, and beyond. *Microbiol. Rev.* **2013**, *37*, 955–989. [CrossRef] [PubMed]

44. Henderson, B.W.; Dougherty, T.J. How does photodynamic therapy work. *Photochem. Photobiol.* **1992**, *55*, 145–157. [CrossRef]

45. Agostinis, P.; Berg, K.; Cengel, K.A.; Foster, T.H.; Girotti, A.W.; Gollnick, S.O.; Hahn, S.M.; Hamblin, M.R.; Juzeniene, A.; Kessel, D.; et al. Photodynamic therapy of cancer: An update. *CA Cancer J. Clin.* **2011**, *61*, 250–281. [CrossRef] [PubMed]

46. Senapathy, G.J.; George, B.P.; Abrahamse, H. Exloring the role of phytochemicals as potent natural photosensitizers in photodynamic therapy. *Anticancer Agents Med. Chem.* **2020**, *20*, 1831–1844. [CrossRef] [PubMed]

47. Moan, J.; Berg, K. The photodegradation of porphyrins in cells can be used to estimate the lifetime of singlet oxygen. *Photochem. Photobiol.* **1991**, *53*, 549–553. [CrossRef]

48. Hatz, S.; Poulsen, L.; Ogilby, P.R. Time-resolved singlet oxygen phosphorescence measurements from photosensitized experiments in single cells: Effects of oxygen diffusion and oxygen concentration. *Photochem. Photobiol.* **2008**, *84*, 1284–1290. [CrossRef]

49. Nyst, H.J.; Tan, I.B.; Stewart, F.A.; Balm, A.J.M. Is photodynamic therapy a good alternative to surgery and radiotherapy in the treatment of head and neck cancer. *Photodiagn. Photodyn. Ther.* **2009**, *6*, 3–11. [CrossRef]

50. Schafer, M. High sensitivity of Deinococcus radiodurans to photodynamically-produced singlet oxygen. *Int. J. Radiat. Biol.* **1998**, *74*, 249–253. [CrossRef]

51. Kochevar, I.E.; Lambert, C.R.; Lynch, M.C.; Tedesco, A.C. Comparison of photosensitized plasma membrane damage caused by singlet oxygen and free radicals. *Biochim. Biophys. Acta* **1996**, *1280*, 223–230. [CrossRef]

52. Barr, H.; Tralau, C.J.; Boulos, P.B.; MacRobert, A.J.; Tilly, R.; Bown, S.G. The contrasting mechanisms of colonic collagen damage between photodynamic therapy and thermal injury. *Photochem. Photobiol.* **1987**, *46*, 795–800. [CrossRef] [PubMed]

53. Hopper, C. Photodynamic therapy. A clinical reality in the treatment of cancer. *Lancet. Oncol.* **2000**, *1*, 212–219. [CrossRef]

54. Allison, R.R.; Bagnato, V.S.; Sibata, C.H. Future of oncologic photodynamic therapy. *Future Oncol.* **2010**, *6*, 929–940. [CrossRef] [PubMed]

55. Foote, C. Mechanisms of photo-oxygenation. In *Porphyrin Localization and Treatment of Tumors*, 1st ed.; Doiron, D.R., Gomer, C.J., Eds.; Alan R. Liss: New York, NY, USA, 1984; pp. 3–18.

56. Boegheim, J.P.; Scholte, H.; Dubbehman, T.M.; Beems, E.; Raap, A.K.; van Steveninck, J. Photodynamic effects of hematoporphyrin-derivative on enzyme activities of murine L929 fibroblasts. *J. Photochem. Photobiol. B* **1987**, *1*, 61–73. [CrossRef]

57. Gibson, S.L.; Hilf, R. Interdependence of fluence, drug, dose and oxygen on hematoporphyrin derivate induced photosensitization of tumor mitochondria. *Photochem. Photobiol.* **1985**, *42*, 367–373. [CrossRef]

58. Dolmans, D.E.; Fukumura, D.; Jain, R.K. Photodynamic therapy for cancer. *Nat. Rev. Cancer* **2003**, *3*, 380–387. [CrossRef]

59. Ochsner, M. Photophysical and photobiological processes in the photodynamic therapy of tumors. *J. Photochem. Photobiol. B* **1997**, *39*, 1–18. [CrossRef]

60. Canti, G.; Lattuada, D.; Morelli, S.; Nicolin, A.; Cubeddu, R.; Taroni, P.; Valentini, G. Efficacy of photodynamic therapy against doxorubicin-resistant murine tumors. *Cancer Lett.* **1995**, *93*, 255–259. [CrossRef]

61. Lofgren, L.A.; Hallgren, S.; Nilsson, E.; Westerborn, A.; Nilsson, C.; Reizenstein, J. photodynamic therapy for recurrent nasopharyngeal cancer. *Arch. Otolaryngol. Head Neck Surg.* **1995**, *121*, 997–1002. [CrossRef]

62. Davies, M.J. Singlet oxygen-mediated damage to proteins and its consequences. *Biochem. Biophys. Res. Commun.* **2003**, *305*, 761–770. [CrossRef]

63. Bicalho, L.S.; Longo, J.P.F.; Pereira, L.O.; Santos, M.F.M.A.; Azevedo, R.B. Photodynamic therapy, a new approach in the treatment of oral cancer. *Rev. Univ. Ind. Santander. Salud.* **2010**, *42*, 167–174.

64. Yoo, J.O.; Lim, Y.C.; Kim, Y.M.; Ha, K.S. Differential cytotoxic responses to low-and high-dose photodynamic therapy in human gastric and bladder cancer cells. *J. Cell Biochem.* **2011**, *112*, 3061–3071. [CrossRef]

65. Moghissi, K.; Dixon, K.; Parson, R.J. A controlled trial of Nd-YAG laser vs photodynamic therapy for odvanced malignant bronchial obstruction. *Laser Med. Sci.* **1993**, *8*, 269–273. [CrossRef]

66. Kato, H.; Okunaka, T.; Shimatani, H. Photodynamic therapy for early state bronchogenic carcinoma. *J. Clin. Laser Med. Surg.* **1996**, *14*, 235–238. [CrossRef] [PubMed]
67. Manoto, S.L.; Abrahamse, H. Effect of a newly synthesized Zn sulfophthalocyanine derivative on cell morphology, viability, proliferation, and cytotoxicity in a human lung cancer cell line (A549). *Lasers Med. Sci.* **2011**, *26*, 523–530. [CrossRef]
68. Jheon, S.; Lee, J.M.; Kim, J.K.; Kim, K.H.; Seo, S.J. Photodynamic therapy for tracheobronchial cancer. *Photodiagn. Photodyn. Ther.* **2011**, *6*, 177. [CrossRef]
69. Grant, W.E.; MacRobert, A.J.; Bown, S.G.; Hopper, C.; Speight, P.M. Photodynamic therapy of oral cancer: Photosensitization with systemic aminolaevulinic acid. *Lancet* **1993**, *342*, 147–148. [CrossRef]
70. Barr, H.; Shepherd, N.A.; Dix, A.; Roberts, D.J.H.; Tan, W.C.; Krasner, N. Eradication of high-grade dysplasia in columnar-lined (Barrett's) oesophagus by photodynamic therapy with endogenously generated protoporphyrin IX. *Lancet* **1996**, *348*, 584–585. [CrossRef]
71. Miller, J.D.; Baron, E.D.; Scull, H.; Hsia, A.; Berlin, J.C.; McCormick, T.; Colussi, V.; Kenney, M.E.; Cooper, K.D.; Oleinick, N.L. Photodynamic therapy with the phtalocyanin photosensitizer Pc 4: The case experience with preclinical mechanistic and early clinical-tranlational studies. *Toxicol. Appl. Pharmacol.* **2007**, *224*, 290–299. [CrossRef]
72. Nseyo, U.O.; DeHaven, J.; Dougherty, T.J.; Potter, W.R.; Merrill, D.L.; Lundahl, S.L.; Lamm, D.L. Photodynamic therapy (PDT) in the treatment of patients with resistant superficial bladder cancer: A long-term experience. *J. Clin. Laser Med. Surg.* **1998**, *16*, 61–68. [CrossRef]
73. Leon, S.P.; Folkerth, R.D.; Black, P. Microvessel density is a prognostic indicator for patients with astroglial brain tumors. *Cancer* **1996**, *77*, 362–372. [CrossRef]
74. Zhu, T.C.; Finlay, J.C. The role of photodynamic therapy (PDT) physics. *Med. Phys.* **2008**, *35*, 3127–3136. [CrossRef] [PubMed]
75. George, B.P.A.; Abrahamse, H. A review on novel breast cancer therapies: Photodynamic therapy and plant derived agent induced cell death mechanisms. *Anticancer Agents Med. Chem.* **2016**, *16*, 793–801. [CrossRef] [PubMed]
76. Kostović, K.; Pastar, Z.; Ceović, R.; Mokos, Z.B.; Buzina, D.S.; Stanimirović, A. Photodynamic therapy in dermatology: Current treatments and implications. *Coll. Antropol.* **2012**, *36*, 1477–1481. [PubMed]
77. Kessel, D. Photosensitization of viral particles. *J. Lab. Clin. Med.* **1990**, *116*, 428. [PubMed]
78. Tardivo, J.P.; Del Giglio, A.; Paschoal, L.H.; Baptista, M.S. New photodynamic therapy protocol to treat AIDS-related Kaposi's sarcoma. *Photomed. Laser Surg.* **2006**, *24*, 528–531. [CrossRef] [PubMed]
79. Mĺkvy, P. Position and possibilities of photodynamic therapy in oncology. *Oncology* **2007**, *5*, 299–301.
80. Castano, A.P.; Demidova, T.N.; Hamblin, M.R. Mechanisms in photodynamic therapy: Part three—Photosensitizer pharmacokinetics, biodistribution, tumor localization and modes of tumor destruction. *Photodiagn. Photodyn. Ther.* **2005**, *2*, 91–106. [CrossRef]
81. Muniyandi, K.; George, B.; Parimelazhagan, T.; Abrahamse, H. Role of photoactive phytocompounds in photodynamic therapy of cancer. *Molecules* **2020**, *25*, 4102. [CrossRef]
82. Doherty, R.E.; Sazanovich, I.V.; McKenzie, L.K.; Stasheuski, A.S.; Coyle, R.; Baggaley, E.; Bottomley, S.; Weinstein, J.A.; Bryant, H.E. Photodynamic killing of cancer cells by a platinum(II) complex with cyclometallating ligand. *Sci. Rep.* **2016**, *6*, 22668. [CrossRef]
83. Pouton, C.W.; Wagstaff, K.M.; Roth, D.M.; Moseley, G.W.; Jans, D.A. Targeted delivery to the nucleus. *Adv. Drug Deliv. Rev.* **2007**, *59*, 698–717. [CrossRef] [PubMed]
84. Sobolev, A.S. Novel modular transporters delivering anticancer drugs and foreign DNA to the nuclei of target cancer cells. *J. Buon.* **2009**, *14* (Suppl. 1), S33–S42.
85. de Melo, W.C.M.A.; Lee, A.N.; Perussi, J.R.; Hamblin, M.R. Electroporation enhances antimicrobial photodynamic therapy mediated by the hydrophobic photosensitizer, hypericin. *Photodiagn. Photodyn. Ther.* **2013**, *10*, 647–650. [CrossRef] [PubMed]
86. Alves, D.S.; Pérez-Fons, L.; Estepa, A.; Micol, V. Membrane-related effects underlying the biological activity of the anthraquinones emodin and barbaloin. *Biochem. Pharmacol.* **2004**, *68*, 549–561. [CrossRef]
87. Du, K.; Xia, Q.; Heng, H.; Feng, F. Temozolomide-doxorubicin conjugate as a double intercalating agent and delivery by apoferritin for glioblastoma chemotherapy. *ACS Mater. Interfaces* **2020**, *12*, 34599–34609. [CrossRef] [PubMed]

88. Mandelli, F.; Vignetti, M.; Suciu, S.; Stasi, R.; Petti, M.C.; Meloni, G.; Muus, P.; Marmont, F.; Marie, J.P.; Labar, B.; et al. Daunorubicin versus mitoxantrone versus idarubicin as induction and consolidation chemotherapy for adults with acute myeloid leukemia: The EORTC and GIMEMA groups study AML-10. *J. Clin. Oncol.* **2009**, *27*, 5397–5403. [CrossRef] [PubMed]

89. Wójcik, K.; Zarebski, M.; Cossarizza, A.; Dobrucki, J.W. Daunomycin, an antitumor DNA intercalator, influences histone-DNA interactions. *Cancer Biol. Ther.* **2013**, *14*, 823–832. [CrossRef] [PubMed]

90. Hu, X.; Cao, Y.; Yin, X.; Zhu, L.; Chen, Y.; Wang, W.; Hu, J. Design and synthesis of various quinizarin derivatives as potential anticancer agents in acute T lymphoblastic leukemia. *Bioorganic Med. Chem.* **2019**, *27*, 1362–1369. [CrossRef] [PubMed]

91. Chen, H.; Zhao, C.; He, R.; Zhou, M.; Liu, Y.; Guo, X.; Wang, M.; Zhu, F.; Qin, R.; Li, X. Danthron suppresses autophagy and sensitizers pancreatic cancer cells to doxorubicin. *Toxicol. In Vitro* **2019**, *54*, 345–353. [CrossRef]

92. Wang, Y.; Yang, M.; Qian, J.; Xu, W.; Wang, J.; Hou, G.; Ji, L.; Suo, A. Sequentially self-assembled polysaccharide-based nanocomplexes for combined chemotherapy and photodynamic therapy of breast cancer. *Carbohydr. Polym.* **2019**, *203*, 203–213. [CrossRef]

93. Lee, H.; Han, J.; Shin, H.; Han, H.; Na, K.; Kim, H. Combination of chemotherapy and photodynamic therapy for cancer treatment with sonoporation effects. *J. Control. Release* **2018**, *283*, 190–199. [CrossRef] [PubMed]

94. He, C.; Liu, D.; Lin, W. self-assembled core-shell nanoparticles for combined chemotherapy and photodynamic therapy of resistant head and neck cancers. *ACS Nano* **2015**, *9*, 991–1003. [CrossRef] [PubMed]

95. Gleason, H.A.; Cronquist, A. *Manual of Vascular Plants of Northeastern United States and Adjacent Canada*, 2nd ed.; New York Botanical Garden: Bronx, New York, NY, USA, 1991; p. 910.

96. Rahman, A. Bioactive natural products (Part C). In *Studies in Natural Products Chemistry*, 1st ed.; Rahman, A., Ed.; Elsevier: Amsterdam, The Netherlands, 2000; Volume 22, p. 647.

97. Miškovský, P. Hypericin—A new antiviral and antitumor photosensitizer: Mechanism of action and interaction with biological macromolecules. *Curr. Drug Targets* **2002**, *3*, 55–84. [CrossRef] [PubMed]

98. Thornett, A. Use of hypericin as antidepressant. Valid measure of antidepressant efficacy in primary care is needed. *Brit. Med. J.* **2000**, *320*, 1141.

99. Butterweck, V.; Winterhoff, H.; Herkenham, M. St John´s wort, hypericin, and imipramine: A comparative analysis of mRNA levels in brain areas involved in HPA axis control following short-term and long-term administration in normal and stressed rats. *Mol. Psychiatry* **2001**, *6*, 547–564. [CrossRef]

100. Eberman, R.; Alth, G.; Kreitner, M.; Kubin, A. Natural products derived from plants as potential drugs for the photodynamic destruction of tumor cells. *J. Photochem. Photobiol. B* **1996**, *36*, 95–97. [CrossRef]

101. Pengelly, A. *The Constituents of Medical Plants: An Introduction to the Chemistry and Therapeutics of Herbal Medicine*, 2nd ed.; CABI Publishing: Cambridge, UK, 2004; pp. 20–50.

102. Miroššay, A.; Mirossay, L.; Tóthová, J.; Miškovský, P.; Onderková, H.; Mojžiš, J. Potentiation of hypericin and hypocrellin-induced phototoxicity by omeprazole. *Phytomedicine* **1999**, *6*, 311–317. [CrossRef]

103. Cavarga, I.; Brezani, P.; Fedorocko, P.; Miskovsky, P.; Bobrov, N.; Longauer, F.; Rybarova, S.; Mirossay, L.; Stubna, J. Photoinduced antitumor effect of hypericin can be enhanced by fractionated dosing. *Photomedicine* **2005**, *12*, 680–683. [CrossRef]

104. Plenagl, N.; Duse, L.; Seitz, B.S.; Goergen, N.; Pinnapireddy, S.R.; Jedelska, J.; Brűßler, J.; Bakowsky, U. Photodynamic therapy—hypericin tetraether liposome conjugates and their antitumor and antiangiogenic activity. *J. Drug Deliv.* **2019**, *26*, 23–33. [CrossRef]

105. Bianchini, P.; Cozzolino, M.; Oneto, M.; Pesce, L.; Pennacchietti, F.; Tognolini, M.; Giorgio, C.; Nonell, S.; Cavanna, L.; Delcanale, P.; et al. Hypericin-apomyoglobin an enhanced photosensitizer complex for the treatment of tumor cells. *Biomacromolecules* **2019**, *20*, 2024–2033. [CrossRef]

106. Fiebich, B.L.; Lieb, A.H. Inhibition of substance P induced cytokine synthesis by St John's Wort extracts. *Pharmacopsychiatry* **2001**, *34*, 526–528. [CrossRef] [PubMed]

107. Samadi, S.; Khadivzadeh, T.; Emami, A.; Moosavi, N.S.; Tafaghodi, M.; Behnam, H.R. The effect of Hypericum perforatum on the wound healing and scar of cesarean. *J. Altern. Complement. Med.* **2010**, *16*, 113–117. [CrossRef]

108. Hajhashemi, M.; Ghanbari, Z.; Movahedi, M.; Rafieian, M.; Keivani, A.; Haghollahi, F. The effect of Achillea millefolium and Hypericum perforatum ointments on episiotomy wound healing in primiparous women. *J. Matern. Fetal Neonatal Med.* **2018**, *31*, 63–69. [CrossRef] [PubMed]

109. García, I.; Ballesta, S.; Gilaberte, Y.; Rezusta, A.; Pascual, Á. Antimicrobial photodynamic activity of hypericin against methicillin-susceptible and resistant Staphylococcus aureus biofilms. *Future Microbiol.* **2015**, *10*, 347–356. [CrossRef] [PubMed]

110. Engelhardt, V.; Krammer, B.; Plaetzer, K. Antibacterial photodynamic therapy using water-soluble formulations of hypericin or mTHPC is effective in inactivation of Staphylococcus aureus. *Photochem. Photobiol. Sci.* **2010**, *9*, 365–369. [CrossRef] [PubMed]

111. Yow, C.; Tang, H.M.; Chu, E.S.; Huang, Z. Hypericin-mediated photodynamic antimicrobial effect on clinically isolated pathogens. *Photochem. Photobiol.* **2012**, *88*, 626–632. [CrossRef]

112. Rodríguez-Amigo, B.; Delcanale, P.; Rotger, G.; Juárez-Jiménez, J.; Abbruzzetti, S.; Summer, A.; Agut, M.; Luque, F.J.; Nonell, S.; Viappiani, C. The complex of hypericin with β-lactoglobulin has antimicrobial activity with perspective applications in dairy industry. *J. Dairy Sci.* **2015**, *98*, 89–94. [CrossRef]

113. Delcanale, P.; Rodríguez-Amigo, B.; Juárez-Jiménez, J.; Luque, F.J.; Abbruzzetti, S.; Agut, M.; Nonell, S.; Viappiani, C. Tuning the local solvent composition at a drug carrier surface: Effect of dimethyl sulfoxide/water mixture on the photofunctional properties of hyperici-β-lactoglobulin. *Mat. Chem. B* **2017**, *5*, 1633–1641. [CrossRef]

114. Pezzuoli, D.; Cozzolino, M.; Montali, C.; Brancaleon, L.; Bianchini, P.; Zantedeschi, M.; Bonardi, S.; Viappiani, C.; Abbruzzetti, S. Serum albumins are efficient delivery systems for the photosensitizer hypericin in photosensitization-based treatments against Staphylococcus aureus. *Food Control.* **2018**, *94*, 254–262. [CrossRef]

115. Lopez-Bazzocchi, I.; Hudson, J.B.; Towers, G.H. Antiviral activity of photoactive plant pigment hypericin. *Photochem. Photobiol.* **1991**, *54*, 95–98. [CrossRef]

116. Hudson, J.B.; Lopez-Bazzocchi, I.; Towers, G.H. Antiviral activities of hypericin. *Antivir. Res.* **1991**, *15*, 101–112. [CrossRef]

117. Moraleda, G.; Wu, T.T.; Jilbert, A.R.; Aldrich, C.E.; Condreay, L.D.; Larsen, S.H.; Tang, J.C.; Colacino, J.M.; Mason, W.S. Inhibition of duck hepatitis B virus replication by hypericin. *Antivir. Res.* **1993**, *20*, 235–247. [CrossRef]

118. Guedes, R.C.; Eriksson, L.A. Effects of halogen substitution on the photochemical properties of hypericin. *J. Photochem. Photobiol. A Chem.* **2006**, *178*, 41–49. [CrossRef]

119. Meruelo, D.; Lavie, G.; Lavie, D. Therapeutic agents with dramatic antiretroviral activity and little toxicity at effective doses: Aromatic polycyclic diones hypericin and pseudohypericin. *Proc. Natl. Acad. Sci. USA* **1988**, *85*, 5230–5234. [CrossRef]

120. Hudson, J.B.; Imperial, V.; Haugland, R.P.; Diwu, Z. Antiviral activities of photoactive perylenequinones. *Photochem. Photobiol.* **1997**, *65*, 352–354. [CrossRef]

121. Xu, Y.; Lu, C. Raman spectroscopy study on structure of human immunodeficiency virus (HIV) and hypericin-induced photosensitive damage of HIV. *Sci. China Ser. C* **2005**, *48*, 117–132.

122. Gulick, R.M.; McAuliffe, V.; Holden-Wiltse, J.; Crumpacker, C.; Liebes, L.; Stein, D.S.; Meehan, P.; Hussey, S.; Forcht, J.; Valentine, F.T. Phase I studies of hypericin, the active compound in St. John's Wort, as an antiretroviral agent in HIV-infected adults: AIDS clinical trials group protocols 150 and 258. *Ann. Intern. Med.* **1999**, *130*, 510–514. [CrossRef]

123. Kerb, R.; Reum, T.; Brockmöller, J.; Bauer, S.; Roots, I. No clinically relevant photosensitization after single-dose and steady state in treatment with hypericum extract in man. *Eur. J. Clin. Pharmacol.* **1995**, *49*, A156.

124. Diwu, Z. Novel therapeutic and diagnostic applications of hypocrellins and hypericins. *Photochem. Photobiol.* **1995**, *61*, 529–539. [CrossRef]

125. Wynn, J.L.; Cotton, T.M. Spectroscopic properties of hypericin in solution and at surfaces. *J. Phys. Chem.* **1995**, *99*, 4317–4323. [CrossRef]

126. Yamazaki, T.; Ohta, N.; Yamazaki, I.; Song, P.S. Excited-state properties of hypericin: Electronic spectra and fluorescence decay kinetics. *J. Phys. Chem.* **1993**, *97*, 7870–7875. [CrossRef]

127. Keša, P.; Antalík, M. Determination of pKa constants of hypericin in aqueous solution of the anti-allergic hydrotropic drug Cromolyn disodium salt. *Chem. Phys. Lett.* **2017**, *676*, 112–117. [CrossRef]

128. Redepenning, J.; Tao, N. Measurement of formal potentials for hypericin in dimethylsulfoxide. *Photochem. Photobiol.* **1993**, *58*, 532–535. [CrossRef] [PubMed]

129. Thomas, C.; Pardini, R.S. Oxygen dependence of hypericin-induced phototoxicity to EMT6 mouse mammary carcinoma cells. *Photochem. Photobiol.* **1992**, *55*, 831–837. [CrossRef]

130. Ehrenberg, B.; Anderson, J.L.; Foote, C.S. Kinetics and yield of singlet oxygen photosensitized by hypericin inorganic and biological media. *Photochem. Photobiol.* **1998**, *68*, 135–140. [CrossRef]

131. Roslaniec, M.; Weitman, H.; Freeman, D.; Mazur, Y.; Ehrenberg, B. Liposome binding constant and singlet oxygen quantum yields of hypericin, tetrahydroxyhelianthrone and their derivatives: Studies in organic solutions and in liposomes. *J. Photochem. Photobiol. B Biol.* **2000**, *57*, 149–158. [CrossRef]

132. Gbur, P.; Dedič, R.; Jancura, D.; Miškovský, P.; Hala, J. Time-resolved luminescence and singlet oxygen formation under illumination of hypericin in acetone. *J. Lumin.* **2008**, *128*, 765–767. [CrossRef]

133. Fehr, M.J.; Carpenter, S.L.; Petrich, J.W. The role of oxygen in the photoinduced antiviral activity of hypericin. *Bioorg. Med. Chem. Lett.* **1994**, *4*, 1339–1344. [CrossRef]

134. Carpenter, S.; Fehr, J.M.; Kraus, G.A.; Petrich, J.W. Chemiluminescent activation of the antiviral activity of hypericin: A molecular flashlight. *Proc. Natl. Acad. Sci. USA* **1994**, *91*, 12273–12277. [CrossRef] [PubMed]

135. Verebová, V.; Belej, D.; Joniová, J.; Jurašeková, Z.; Miškovský, P.; Kožár, T.; Horváth, D.; Staničová, J.; Huntošová, V. Deeper insights into the drug defense of glioma cells against hydrophobic molecules. *Int. J. Pharm.* **2016**, *503*, 56–67. [CrossRef]

136. de Witte, P.; Agostinis, P.; Van Lint, J.; Merlevede, W.; Vandenheede, J.R. Inhibition of epidermal growth factor receptor tyrosine kinase activity by hypericin. *Biochem. Pharmacol.* **1993**, *46*, 1929–1936. [CrossRef]

137. Agostinis, P.; Vantieghem, A.; Merlevede, W.; de Witte, P. Hypericin in cancer treatment: More light on the way. *Int. J. Biochem. Cell Biol.* **2002**, *34*, 221–241. [CrossRef]

138. Buytaert, E.; Callewaert, G.; Hendrickx, N.; Scorrano, L.; Hartmann, D.; Missiaen, L.; Vandenheede, J.R.; Heirman, I.; Grooten, J.; Agostinis, P. Role of endoplasmic reticulum depletion and multidomain proapoptic BAX and BAK proteins in shaping cell death after hypericin-mediated photodynamic therapy. *Faseb. J.* **2006**, *20*, 756–758. [CrossRef]

139. Ferrario, A.; von Tiehl, K.; Wong, S.; Luna, M.; Gomer, C.J. Cyclooxygenase-2 inhibitor treatment enhances photodynamic therapy-mediated tumor response. *Cancer Res.* **2002**, *62*, 3956–3961.

140. Delcanale, P.; Pennacchietti, F.; Maestrini, G.; Rodríguez-Amigo, B.; Bianchini, P.; Diaspro, A.; Iagatti, A.; Patrizi, B.; Foggi, P.; Agut, M.; et al. Subdiffraction localization of a nanostructured photosensitizer in bacterial cells. *Sci. Rep.* **2015**, *5*, 15564. [CrossRef] [PubMed]

141. Miškovský, P.; Sureau, F.; Chinsky, L.; Turpin, P.Y. Subcellular distribution of hypericin in human cancer cells. *Photochem. Photobiol.* **1995**, *62*, 546–549. [CrossRef]

142. Miškovský, P.; Chinsky, L.; Wheeler, G.V.; Turpin, P.Y. Hypericin site specific interactions within polynucleotides used as DNA model compounds. *J. Biomol. Struct. Dyn.* **1995**, *13*, 547–552. [CrossRef]

143. Sánchez-Cortés, S.; Miškovský, P.; Jancura, D.; Bertoluzza, A. Specific interactions of antiretroviraly active drug hypericin with DNA as studied by surface-enhanced resonance Raman spectroscopy. *J. Phys. Chem.* **1996**, *100*, 1938–1944. [CrossRef]

144. Das, K.; Smirnov, A.V.; Wen, J.; Miškovský, P.; Petrich, J.W. Photophysics of hypericin and hypocrellin A in complex with subcellular components: Interactions with human serum albumin. *Photochem. Photobiol.* **1999**, *69*, 633–645. [CrossRef]

145. Senthil, V.; Jones, L.R.; Senthil, K.; Grossweiner, L.J. Hypericin photosensitization in aqueous model system. *Photochem. Photobiol.* **1994**, *59*, 40–47. [CrossRef]

146. Mijatovic, S.; Maksimovic-Ivanic, D.; Radovic, J.; Milijkovic, D.; Kaludjerovic, G.N.; Sabo, T.J.; Trajkovic, V. Aloe emodin decreases the ERK-dependent anticancer activity of cisplatin. *Cell Mol. Life Sci.* **2005**, *62*, 1275–1282. [CrossRef] [PubMed]

147. Ma, T.; Qi, Q.H.; Yang, W.X.; Xu, J.; Dong, Z.L. Contractile effects and antracellular Ca^{2+} signaling induced by emodin in circular smooth muscle cells of rat colon. *World J. Gastroenterol.* **2003**, *9*, 1804–1807. [CrossRef] [PubMed]

148. Janeczko, M.; Masčyk, M.; Kubiński, K.; Golczyk, H. Emodin, a natural inhibitor of protein kinase CK2, suppresses growth, hyphal development, and biofilm formation of Candida Albicans. *Yeast* **2017**, *34*, 253–265. [CrossRef]

149. Huang, H.C.; Lee, C.R.; Lee Chao, P.D.; Chen, C.C.; Chu, S.H. Vasorelaxant effects of emodin, an antraquinone from a Chinese herb. *Eur. J. Pharmacol.* **1991**, *205*, 289–294.

150. Su, Y.T.; Chang, H.L.; Shyue, S.K.; Hsu, S.L. Emodin induces apoptosis in human lung adenocarcinoma cells through a reactive oxygen species-dependent mitochondrial signaling pathway. *Biochem. Pharmacol.* **2005**, *70*, 229–241. [CrossRef] [PubMed]

151. Chukwujelwu, J.C.; Coombes, P.H.; Mulholland, D.A.; van Staden, J. Emodin, an antibacterial anthraquinone from the roots of Cassia occidentalis. *S. Afr. J. Bot.* **2006**, *72*, 295–297. [CrossRef]

152. Srinivas, G.; Babykutty, S.; Sathiadevan, P.P.; Srinivas, P. Molecular mechanism of emodin action: Transition from laxative ingredient to an antitumor agent. *Med. Res. Rev.* **2007**, *27*, 591–608. [CrossRef] [PubMed]

153. Fernand, V.E.; Dinh, D.T.; Washington, S.J.; Fakayode, S.O.; Loss, J.N.; van Ravenswaay, R.O.; Warner, I.M. Determination of pharmacologically active compounds in root extracts of Cassia Alata L. by use of high performance liquid chromatography. *Talanta* **2008**, *74*, 896–902. [CrossRef]

154. Hennebelle, T.; Weniger, B.; Joseph, H.; Sahpaz, S.; Bailleul, F. Senna alata. *Fitoterapia* **2009**, *80*, 385–393. [CrossRef]

155. Andersen, D.O.; Weber, N.D.; Wood, S.G.; Hughes, B.G.; Murray, B.K.; North, J.A. In vitro virucidal activity of selected anthraquinones and anthraquinone derivatives. *Antiviral Res.* **1991**, *16*, 185–196. [CrossRef]

156. Kusari, S.; Lamshöft, M.; Zühlke, S.; Spiteller, M. An endophytic fungus from hypericum perforatum that produces hypericin. *J. Nat. Prod.* **2008**, *71*, 159–162. [CrossRef] [PubMed]

157. Karioti, A.; Bilia, A.R. Hypericins as potential leads for new therapeutics. *Int. J. Mol. Sci.* **2010**, *11*, 562–594. [CrossRef] [PubMed]

158. Bogdanska, A.; Chmurzyński, L.; Ossowski, T.; Liwo, A.; Jeziorek, D. Protolytic equilibria of dihydroxyanthraquinones in non-aqueous solutions. *Anal. Chim. Acta* **1999**, *402*, 339–343. [CrossRef]

159. Mueller, S.O.; Lutz, W.K.; Stopper, H. Factors affecting the genotoxic potency ranking of natural anthraquinones in mammalian cell culture systems. *Mutat. Res.* **1998**, *414*, 125–129. [CrossRef]

160. Li, J.; Yang, W.; Hu, W.; Wang, J.; Jin, Z.; Wang, X.; Xu, W. Effects of emodin on the activity of K channel in guinea pig taenia coli smooth muscle cells. *Acta Pharm. Sin.* **1998**, *33*, 321–325.

161. Ma, T.; Qi, Q.H.; Xu, J.; Dong, Z.L.; Yang, W.X. Signal pathways involved in emodin-induced contraction of smooth muscle cells from rat colon. *World J. Gastroenterol.* **2004**, *10*, 1476–1479. [CrossRef] [PubMed]

162. Chen, D.; Xiong, Y.; Wang, L.; Lv, B.; Lin, Y. Characteristics of emodin on modulating the contractility of jejunal smooth muscle. *Can. J. Physiol. Pharmacol.* **2012**, *90*, 455–462. [CrossRef]

163. Vargas, F.; Fraile, G.; Velásquez, M.; Correia, H.; Fonseca, G.; Marin, M.; Marcano, E.; Sánchez, Y. Studies on the photostability and phototoxicity of aloe-emodin, emodin and rhein. *Pharmazie* **2002**, *57*, 399–404.

164. Oshida, K.; Hirakata, M.; Maeda, A.; Miyoshi, T.; Miyamoto, Y. Toxicological effect of emodin in mouse testicular gene expression profile. *J. Appl. Toxicol.* **2011**, *31*, 790–800. [CrossRef]

165. Xu, X.; Toselli, A.; Russell, L.D.; Seldin, D.C. Globozospermia in mice lacking the casein kinase II α′catalytic subunit. *Nat. Genet.* **1999**, *23*, 118–121. [CrossRef] [PubMed]

166. Escalier, D.; Silvius, D.; Xu, X. Spermatogenesis of mice lacking CK2alpha′failure of germ cell survival and characteristic modifications of the spermatid nucleus. *Mol. Reprod. Dev.* **2003**, *66*, 190–201. [CrossRef] [PubMed]

167. Wang, L.; Lin, L.; Ye, B. Electrochemical studies of the interaction of the anticancer herbal drug emodin with DNA. *J. Pharm. Biomed. Anal.* **2006**, *42*, 625–629. [CrossRef] [PubMed]

168. Bi, S.; Zhang, H.; Qiao, C.; Sun, Y.; Liu, C. Studies of interaction of emodin and DNA in the presence of ethidium bromide by spectroscopic method. *Spectrochim. Acta A Mol. Biomol. Spectrosc.* **2008**, *69*, 123–129. [CrossRef] [PubMed]

169. Saito, S.T.; Silva, G.; Pungartnik, C.; Brendel, M. Study of DNA-emodin interaction by FTIR and UV-Vis spectroscopy. *J. Photochem. Photobiol. B: Biol.* **2012**, *111*, 59–63. [CrossRef]

170. Li, Y.; Luan, Y.; Qi, X.; Li, M.; Gong, L.; Xue, X.; Wu, X.; Wu, Y.; Chen, M.; Xing, G.; et al. Emodin triggers DNA double-strand breaks by stabilizing topoisomerase II-DNA cleavage complexes and by inhibiting ATP hydrolysis of topoisomerase II. *Toxicol. Sci.* **2010**, *118*, 435–443. [CrossRef]

171. Fabriciova, G.; Sanchez-Cortes, S.; Garcia-Ramos, J.V.; Miskovsky, P. Surface-enhanced Raman spectroscopy study of the interaction of the antitumoral drug emodin with serum albumin. *Biopolymers* **2004**, *74*, 125–130. [CrossRef]

172. Bi, S.Y.; Song, D.O.; Kan, Y.H.; Xu, D.; Tian, Y.; Zhou, X.; Zhang, H.Q. Spectroscopic characterization of effective components anthraquinones in Chinese medicinal herbs binding with serum albumins. *Spectrochim. Acta A Biomol. Spectrosc.* **2005**, *62*, 203–212. [CrossRef]

173. Koistinen, K.M.; Soininen, P.; Venalainen, T.A.; Hayrinen, J.; Laatikainen, R.; Perakyla, M.; Tervahauta, A.I.; Karenlampi, S.O. Birch PR-10c interacts with several biologically important ligands. *Phytochemistry* **2005**, *66*, 2524–2533. [CrossRef]

174. Sevilla, P.; Rivas, J.M.; García-Blanco, F.; García-Ramos, J.V.; Sánchez-Cortés, S. Identification of the antitumoral drug emodin binding sites in bovine serum albumin by spectroscopic methods. *BBA-Proteins Proteom.* **2007**, *1774*, 1359–1369. [CrossRef]

175. Pecere, T.; Gazzola, M.V.; Mucignat, C.; Parolin, C.; Dalla, V.F.; Cavaggioni, A.; Basso, G.; Diaspro, A.; Salvato, B.; Carli, M.; et al. Aloe-emodin is a new type of anticancer agent with selective activity against neuroectodermal tumors. *Cancer Res.* **2000**, *60*, 2800–2804.

176. Chun-Guan, W.; Jun-Qing, Y.; Bei-Zhong, L.; Dan-Ting, J.; Chong, W.; Liang, Z.; Dan, Z.; Yan, W. Anti-tumor activity of emodin against human chronic myelocytic leukemia K562 cell lines in vitro and in vivo. *Eur. J. Pharmacol.* **2010**, *627*, 33–41. [CrossRef] [PubMed]

177. Kuo, T.C.; Yang, J.S.; Lin, M.W.; Hsu, S.C.; Lin, J.J.; Lin, H.J.; Hsia, T.C.; Liao, L.; Yang, M.D.; Fan, M.J.; et al. Emodin has cytotoxic and pretoctive effect in rat c6 glioma cells: Roles of Mdr1a and nuclear factor κB in cell survival. *J. Pharmacol. Exp. Ther.* **2009**, *330*, 736–744. [CrossRef] [PubMed]

178. Ko, J.C.; Su, Y.J.; Lin, S.T.; Jhan, J.Y.; Ciou, S.C.; Cheng, C.M.; Lin, Y.W. Suppression of ERCC1 and Rad51 expression through ER1/2 inactivation is essential in emodin-mediated cytotoxicity in human non-small cell lung cancer cells. *Biochem. Pharmacol.* **2010**, *79*, 655–664. [CrossRef] [PubMed]

179. Chemicalland21. Quinizarin. Available online: http://www.chemicalland21.com/specialtychem/finechem/QUINIZARIN.htm (accessed on 7 September 2020).

180. Quinti, L.; Allen, N.S.; Edge, M.; Murphy, B.P.; Perotti, A. A study of the strongly fluorescent species formed by the interaction of the dye 1,4-dihydroxyanthraquinone (quinizarin) with A(III). *J. Photochem. Photobiol. A Chem.* **2003**, *155*, 79–91. [CrossRef]

181. Yohida, M. Chemistry and hair dyes application of dihydroxy derivates of anthraquinone and naphtoquinone. *Prog. Org. Coat.* **1997**, *31*, 63–72. [CrossRef]

182. Brinkworth, R.I.; Fairle, D.P. Hydroxyquinones are competitive nonpeptide inhibitors of HIV-1 proteinase. *Biochim. Biophys. Acta* **1995**, *1253*, 5–8. [CrossRef]

183. Brown, J.P. A review of the genetic effects of naturally occurring flavonoids, anthraquinones and related compounds. *Mutat. Res.* **1980**, *75*, 243–277. [CrossRef]

184. Longo, V.; Amato, G.; Salvetti, A.; Gervasi, P.G. Heterogenous effect of anthraquinones on drug-metabolizing enzymes in the liver and small intestine of rat. *Chem. Biol. Interact.* **2000**, *126*, 63–77. [CrossRef]

185. Carter, T.P.; Gillispie, G.D.; Connoll, M.A. Intramolecular hydrogen bonding in substituted anthraquinones by laser-induced fluorescence. 1. 1,4-dihydroxyanthraquinone (quinizarin). *J. Phys. Chem.* **1982**, *86*, 192–196. [CrossRef]

186. Smulevich, G.; Angeloni, L.; Giovannardi, S.; Marzocchi, M.P. Resonance Raman and polarized light infrared spectra of 1,4-dihydroxyanthraquinone, vibrational studies of the ground and excited electronic states. *Chem. Phys.* **1982**, *65*, 313–322. [CrossRef]

187. Nigam, G.; Deppisch, B. Redetermination of the structure of 1,4-dihydroxyanthraquinone ($C_{14}H_8O_4$). *Z. Kristallogr.* **1980**, *151*, 185–191. [CrossRef]

188. Fabriciová, G.; Garcia-Ramos, J.V.; Miškovský, P.; Sanchez-Cortes, S. Absorption and acidic behavior of anthraquinone drugs quinizarin and danthron on Ag nanoparticles studied by Raman spectroscopy. *Vib. Spectrosc.* **2004**, *34*, 273–281. [CrossRef]

189. Bondy, G.S.; Armstrong, C.L.; Dawson, B.A.; Héroux-Metcalf, C.; Neville, G.A.; Rogers, C.G. Toxicity of structurally related anthraquinones and anthrones to mammalian-cell in vitro. *Toxicol. In Vitro* **1994**, *8*, 329–335. [CrossRef]

190. Verebová, V.; Adamčík, J.; Danko, P.; Podhradský, D.; Miškovský, P.; Staničová, J. Anthraquinones quinizarine and danthron unwind negatively supercoiled DNA and lengthen linear DNA. *Biochem. Biophys. Res. Commun.* **2014**, *444*, 50–55. [CrossRef] [PubMed]

191. IARC. Dantron (chrysazin; 1,8-dihydroxyanthraquinone). In *Pharmaceutical Drugs*; International Agency for Research on Cancer: Lyon, France, 1990; pp. 265–275.

192. Mueller, S.O.; Stopper, H.; Dekant, W. biotransformation of the anthraquinones emodin and chrysophanol by cytochrome P450 enzymes. Bioactivation to genotoxic metabolites. *Drug Metab. Dispos.* **1998**, *26*, 540–546.

193. Müller, S.O.; Eckert, I.; Lutz, W.K.; Stopper, H. Genotoxicity of the laxative drug components emodin, aloe-emodin and danthron in mammalian cells: Topoisomerase II mediated. *Mutat. Res. Genet. Toxicol.* **1996**, *371*, 165–173. [CrossRef]

194. Downes, C.S.; Mullinger, A.M.; Johnson, R.T. Inhibitors of DNA topoisomerase II prevent chromatid separation in mammalians cells but do not prevent exit from mitosis. *Proc. Natl. Acad. Sci. USA* **1991**, *88*, 8895–8899. [CrossRef]

195. Wang, J.R.; Caron, P.R.; Kim, R.A. The role of DNA topoisomerases in recombination and genomic stability: A double-edged sword. *Cell* **1990**, *62*, 403–406. [CrossRef]

196. Neimeikaité-Čénienė, A.; Sergedienė, E.; Nivinskas, H.; Čénas, N. Cytotoxicity of natural hydroxyanthraquinones: Role of oxidative stress. *Z. Naturforsch. C* **2002**, *57*, 822–827. [CrossRef]

197. Lieberman, D.F.; Fink, R.C.; Schaefer, F.L.; Mulcahy, R.J.; Stark, A. Mutagenicity of anthraquinone and hydroxylated anthraquinones in the Ames/Salmonella microsome system. *Appl. Environ. Microbiol.* **1982**, *43*, 1354–1359. [CrossRef]

198. Ferreiro, M.L.; Rodriguez-Otero, J. Ab inition study of the intramolecular proton transfer in dihydroxyanthraquinones. *J. Mol. Struct.* **2001**, *542*, 63–77. [CrossRef]

199. Ansari, S.S.; Khan, R.H.; Naqvi, S. Probing the intermolecular interactions into serum albumin and anthraquinone systems: A spectroscopic and docking approach. *J. Biomol. Struct. Dyn.* **2018**, *36*, 3362–3375. [CrossRef] [PubMed]

200. Jutkova, A.; Chorvat, D.; Miskovsky, P.; Jancura, D.; Datta, S. Encapsulation of anticancer drug curcumin and co-loading with photosensitizer hypericin into lipoproteins investigated by fluorescence resonance energy transfer. *Int. J. Pharm.* **2019**, *564*, 369–378. [CrossRef] [PubMed]

201. Kozsup, M.; Dömötör, O.; Nagy, S.; Farkas, E.; Enyedy, E.A.; Buglyó, P. Synthesis, characterization and albumin binding capabilities of quinizarin containing ternary cobalt(III) complexes. *J. Inorg. Biochem.* **2020**, *204*, 110963. [CrossRef]

202. Crlikova, H.; Kostrhunova, H.; Pracharova, J.; Kozsup, M.; Nagy, S.; Buglyó, P.; Brabec, V.; Kasparkova, J. Antiproliferative, DNA binding, and cleavage properties of dinuclear Co(III) complexes containing the bioactive quinizarin ligand. *J. Biol. Inorg. Chem.* **2020**, *25*, 339–350. [CrossRef]

203. Kočišová, E.; Chinsky, L.; Miškovský, P. Sequence specific interaction of the photoactive drug hypericin depends on the structural arrangement and the stability of the structure containing its specific 5′AG3′target: A resonance Raman spectroscopy study. *J. Biomol. Struct. Dyn.* **1999**, *17*, 51–59. [CrossRef]

204. Staničová, J.; Verebová, V.; Strejčková, A. Potential anticancer agent hypericin and its model compound emodin: Interaction with DNA. *Čes. Slov. Farm.* **2016**, *65*, 28–31.

205. Gottesman, M.M.; Fojo, T.; Bates, S.E. Multidrug resistance in cancer: Role of ATP-dependent transporters. *Nat. Rev. Cancer* **2002**, *2*, 48–58. [CrossRef]

206. Aller, S.G.; Yu, J.; Ward, A.; Weng, Y.; Chittaboina, S.; Zhou, R.P.; Harrell, P.M.; Trinh, Y.T.; Zhang, Q.H.; Urbatsch, I.L.; et al. Structure of P-glycoprotein reveals a molecular basis for poly-specific drug binding. *Science* **2009**, *323*, 1718–1722. [CrossRef]

207. Regina, A.; Demeule, M.; Laplante, A.; Jodoin, J.; Dagenais, C.; Berthelet, F.; Moghrabi, A.; Beliveau, R. Multidrug resistance in brain tumors: Roles of the blood-brain barrier. *Cancer Metastasis Rev.* **2001**, *20*, 13–25. [CrossRef]

208. Szaflarski, W.; Sujka-Kordowska, P.; Januchowski, R.; Wojtowicz, K.; Andrzejewska, M.; Nowicki, M.; Zabel, M. Nuclear localization of P-glycoprotein is responsible for protection of the nucleus from doxorubicin in the resistant LoVo cell line. *Biomed. Pharmacother.* **2013**, *67*, 497–502. [CrossRef] [PubMed]

209. Weyergang, A.; Berstad, M.E.; Bull-Hansen, B.; Olsen, C.E.; Selbo, P.K.; Berg, K. Photochemical activation of drug for the treatment of therapy-resistant cancers. *Photochem. Photobiol. Sci.* **2015**, *14*, 1465–1475. [CrossRef] [PubMed]

Publisher's Note: MDPI stays neutral with regard to jurisdictional claims in published maps and institutional affiliations.

MDPI

St. Alban-Anlage 66

4052 Basel

Switzerland

Tel. +41 61 683 77 34

Fax +41 61 302 89 18

www.mdpi.com

Molecules Editorial Office

E-mail: molecules@mdpi.com

www.mdpi.com/journal/molecules